Nanotechnology: Concepts and Applied Principles

Nanotechnology: Concepts and Applied Principles

Editor: Rich Falcon

NY RESEARCH
P R E S S

New York

Published by NY Research Press
118-35 Queens Blvd., Suite 400,
Forest Hills, NY 11375, USA
www.nyresearchpress.com

Nanotechnology: Concepts and Applied Principles
Edited by Rich Falcon

© 2017 NY Research Press

International Standard Book Number: 978-1-63238-535-2 (Hardback)

Cataloging-in-Publication Data

Nanotechnology : concepts and applied principles / edited by Rich Falcon.
    p. cm.
Includes bibliographical references and index.
ISBN 978-1-63238-535-2
1. Nanotechnology. 2. Nanostructured materials. I. Falcon, Rich.
T174.7 .N36 2017
620.5--dc23

Printed in the United States of America.

Contents

Permissions

List of Contributors

Index

Preface

The main aim of this book is to educate learners and enhance their research focus by presenting diverse topics covering this vast field. This is an advanced book which compiles significant studies by distinguished experts. This book addresses successive solutions to the challenges arising in the area of application, along with it; the book provides scope for future developments.

Nanotechnology is the design and engineering of materials that are sized in the dimension of 1 to 100 nanometers. This book on nanotechnology deals with the principles and practices that are followed in nanotechnology. Different approaches, evaluations, methodologies and advanced studies on nanotechnology have been included in this book. The contents presented herein discuss the various approaches and techniques that are followed in nanoengineering and the technological innovations related to it. This book consists of contributions made by international experts. As this field is emerging at a rapid pace, the contents of this book will help the readers understand the modern concepts and applications of the subject.

It was a great honour to edit this book, though there were challenges, as it involved a lot of communication and networking between me and the editorial team. However, the end result was this all-inclusive book covering diverse themes in the field.

Finally, it is important to acknowledge the efforts of the contributors for their excellent chapters, through which a wide variety of issues have been addressed. I would also like to thank my colleagues for their valuable feedback during the making of this book.

Editor

Bio-synthesis of magnetite nanoparticles by bacteria

Mohamed Abdul-Aziz Elblbesy[1], Adel Kamel Madbouly[2], Thamer Abed-Alhaleem Hamdan[1]

[1]Department of Medical Laboratory Technology, Faculty of Applied Medical Science, University of Tabuk, Saudi Arabia, Tabuk, Saudi Arabia

[2]Department of Biology, Faculty of Science, University of Tabuk, Tabuk, Saudi Arabia

Email address:

melblbesy@ut.edu.sa (M. Abdul-Aziz E.)

Abstract: A promising avenue of research in materials science is to follow the strategies used by Mother Nature to fabricate ornate hierarchical structures as exemplified by organisms such as diatoms, sponges and magnetotactic bacteria. Some of the strategies used in the biological world to create functional inorganic materials may well have practical implications in the world of nanomaterials. The aim of our work is to examine the synthetic of magnetite nanoparticles under different conditions to show the influence in magnetic properties of magnetite nanoparticles. *Magnetospirillum* strain AMB-1 was used in this study in order to produce magnetite nanoparticles. Magnetite nanoparticles of average size~47 nm were obtained. The magnetic properties of magnetite nanoparticles under different incubation temperature were examined and a small influence in magnetic properties of magnetite nanoparticles was indicated.

Keywords: *Magnetospirillum,* Magnetite Nanoparticles, Temperatures, Magnetic Properties

1. Introduction

In recent years, nanotechnology research is emerging as cutting edge technology interdisciplinary with physics, chemistry, biology, material science and medicine. The prefix nano is derived from Greek word nanos meaning "dwarf" in Greek that refers to things of one billionth (10^{-9} m) in size. The primary concept of nanotechnology was presented by Richard Feynman in a lecture entitled "There's plenty of room at the bottom" at the American Institute of Technology in 1959. Nanoparticles are usually 0.1 to 1000 nm in each spatial dimension and are commonly synthesized using two strategies: top-down and bottom-up [1].

Microbes produce inorganic materials either intra- or extracellular often in nanoscale dimensions with exquisite morphology. Microbial resistance to most toxic heavy metals is due to their chemical detoxification as well as due to energy-dependent ion efflux from the cell by membrane proteins that function either as ATPase or as chemiosmotic cation or proton anti-transporters. Alteration in solubility also plays a role in microbial resistance [2, 3]. Therefore, microbial systems can detoxify the metal ions by either reduction and/or precipitation of soluble toxic inorganic ions to insoluble non-toxic metal nanoclusters. Microbial detoxification can be made either by extracellular biomineralization, precipitation or intracellular bioaccumulation. Extracellular production of metal nanoparticles has more commercial applications in various fields. Since the polydispersity is the major concern, it is important to optimize the conditions for monodispersity in a biological process [4].

Magnetite, $Fe^{3+}(Fe^{2+}, Fe^{3+})O_4$, is an "inverse" spinel and the unique electronic and magnetic properties of magnetite are directly associated with the extremely rapid exchange of electrons among the octahedrally-coordinated iron ions. Other divalent and trivalent metal ions readily substitute for the iron atoms in both site types. Magnetite formed naturally inevitably contains impurity cations, the most frequent ones being Ti, Al, Mg, and Mn. The effect of metal substitution in magnetite produces systematic variation in magnetic and physical properties: saturation magnetization, curie temperature change; coercivity; magneto crystalline anisotrophy, cell parameter, and electrical resistivity changes. There are many approaches to the synthesis of magnetic nanoparticles such as size reduction through ball milling, chemical precipitation, and microbial synthesis[5,6,7].

Magnetic nanoparticles are promising as therapeutic or diagnostic tools in medicine. In terms of diagnosis they can be used both for *in vitro* and *in vivo* applications for example: in immobilization and detection of biomolecules [8,9,10], cell separation [11], purification [12] and gene transfer [14], and serve as a contrast agents in magnetic resonance imagining [13]. They can also be applied for drug delivery system in target therapy [9] and for hyperthermia treatment, due to the heat they produce in an alternating magnetic field [l4].

The aim of this study is evaluate the physical conditions at which magnetic bacteria can produce magnetite nanoparticles with the best characterizations.

2. Material and Methods

For the isolation of magnetosomes; approximately 100 ml cell culture of *Magnetospirillum* strain AMB-1 was suspended in 100 ml of 20 mM HEPES-4 mM EDTA, pH 7.4, and then split up (disrupted) by sonication. The unbroken cells and the cell debris were removed from the sample by centrifugation (30 min, 9000 rpm), then the cell extract was placed on magnet (NdFeB-magnets, 1h). The black magnetosomes sediment at the bottom of the tube, whereas the residual contaminating cellular material was retained in upper part tube and then decanted. To eliminate the electrostatically bound contamination, the magnetic particles were rinsed first with 50 ml of 10 mM HEPES-200 mMNaCl, pH 7.4, and subsequently with 100 ml of 10 mM HEPES, pH 7.4. The magnetosome suspension (black sediment) was centrifuged (18000 rpm, 30 min). After centrifugation, the cell extract was placed on the magnet for 30 minutes. The magnetic particles were sediment at the bottom of the tube, whereas residual contaminating cellular material was retained in upper part tube. The last step was repeated ten-times to obtain well purified magnetosomes. The previous procedure had been done under different incubation temperatures (30, 40,50,60,70 °C)

Transmission electron microscopy images were taken for magnetosomes and the magnetite nanoparticles. The size of the magnetite nanoparticles was analyzed by Beckman Coulter Particle Size Analyzer. The degree of magnetism of the nanoparticles was evaluated using vibrating sample magnetometer (VSM-9600-IDSM-LDG-USA) and the saturation magnetism (B_r), Retentivity (B_r), and Coercivity (H_c)

3. Statistical Analysis

All results were represented as mean ± SD. In order to study the statistical significance of the results significance regarding to Pearson's coefficient and sample size had been performed and the $p \leq 0.05$ had been taken as the significance limit.

4. Results

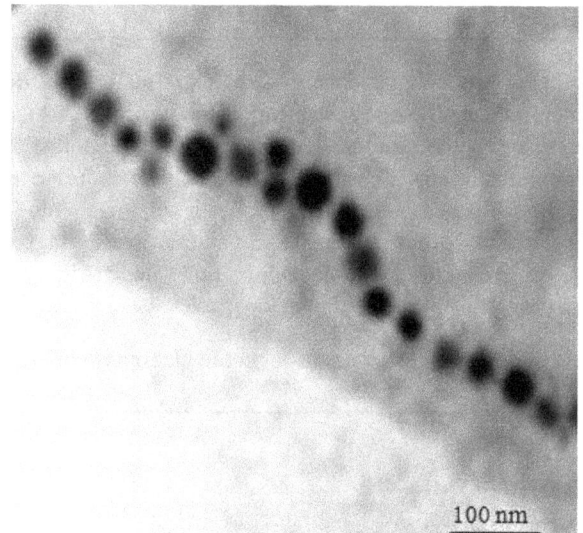

Fig. 1. *Nanoparticles TEM image.*

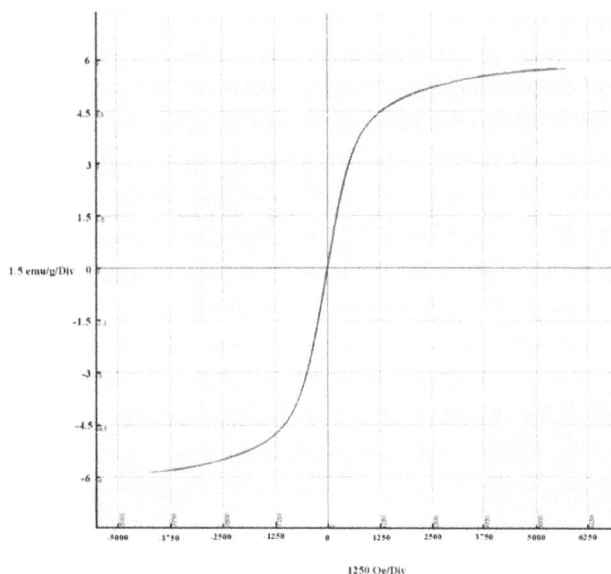

(a)

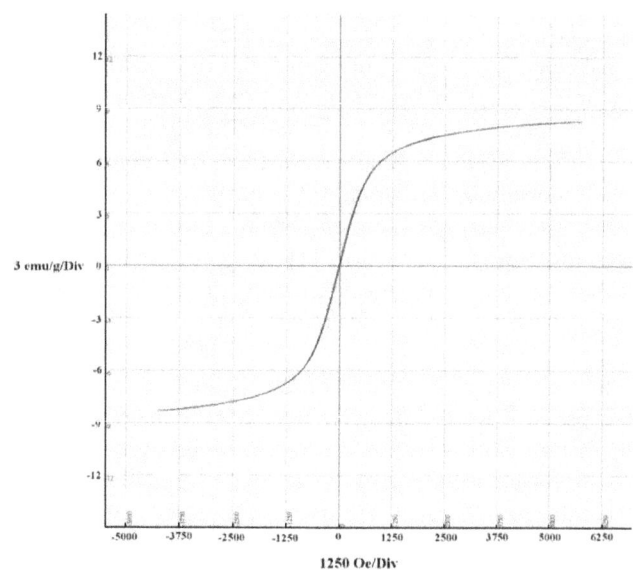

(b)

(c)

(d)

(e)

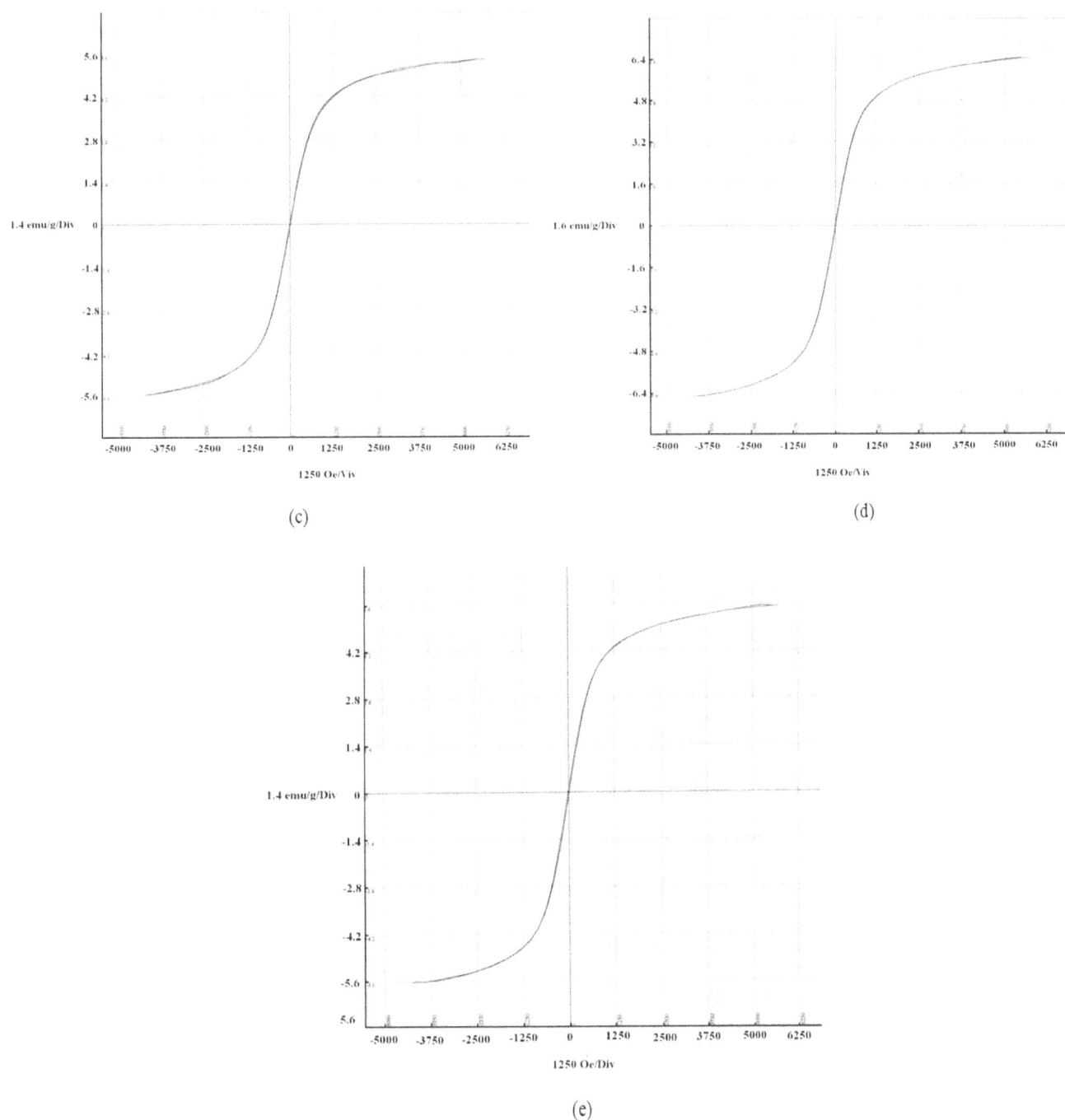

Fig. 2. *The hysteresis loop magnetite nanoparticles prepared at a) 30°C, b) 40 °C, c) 50 °C, d) 60 °C and e) 70 °C.*

Typical electron micrograph of magnetosomes on surface obtained by TEM technique for prepared samples and is shown in Fig.1. For evaluation of different preparation conditions the size distributions of magnetosomes (from 100 particles) according to TEM photographs were prepared. The mean diameter of magnetosome prepared estimated from the size distribution of magnetosomes obtained by cultivation at different incubation temperatures was estimated as to be 45±2.5 nm, 49±2.5 nm, 46±1.5 nm, 48±4.5, nm 47±1.99 nm, respectively. They were the same size of magnetite nanoparticles obtained after separation from the bacteria. It

was observed increased number of magnetosomes in part of higher and lower size of magnetosomes this causing distinct changed of size distribution and size of magnetosomes is more uniform. A small particles size was obtained at 30 °C , but the maximum particles size was obtained at 40 °C.

The magnetic properties including hysteresis loop, saturation magnetization and coercivity of magnetite nanoparticles were measured in this research. Fig.2 shows the hysteresis loops of paramagnetic magnetite nanoparticles. In which the internal area of hysteresis loop represents the capability of magnetic energy storage of magnetic materials, which is an important parameter in electromagnetic absorption field. The hysteresis loop with great area brings on a large loss. The internal areas of hysteresis loops are great, which can be used as electromagnetic absorption materials. The figures represent the hystersis loops of magnetite nanoparticles prepared at different temperature range from 30 to 70 °C. In the first Fig.2(a) which represent magnetite nanoparticles prepared at 30 °C, in which the saturation magnetization was 5.89 emu/g, coercivity was 14.37 Oe and retentivity was 0.7615 emu/g. The second Fig.2(b) which represents magnetite nanoparticles prepared at 40 °C, in which the saturation magnetization was 5.404 emu/g, coercivity was 15.2 Oe and retentivity was 1.01 emu/g. The third Fig.2(c) which represents magnetite nanoparticles prepared at 50 °C, in which the saturation magnetization was 5.6 emu/g, coercivity was 13.91 Oe and retentivity was 0.6726 emu/g. The fourth Fig.2(d) which represents the magnetite nanoparticles prepared at 60 °C, in which the saturation magnetization was 5.626 emu/g, coercivity was 15.85 Oe and retentivity was 0.7133 emu/g. The fifth Fig.2(e) which represents the magnetite nanoparticles prepared at 70 °C, in which the saturation magnetization was 5.65 emu/g, coercivity were 16.49 Oe and retentivity were 0.71 emu/g.

The greatest size of the magnetite nanoparticles was obtained at 40 °C with average value of 48 nm Fig.3. There was no noticeable effect of variation of temperature on B_s values as indicated in Fig.4. A small variation due to the change in temperature was observed on the values of B_r and H_c as shown in Fig.5. and Fig.6.

The comparing means t-test for the obtaining data showed that the relation between temperatures and B_s, B_r, and H_C were significant $p<0.05$. In contrary the relations between temperature and particle size was insignificant $p>0.05$.

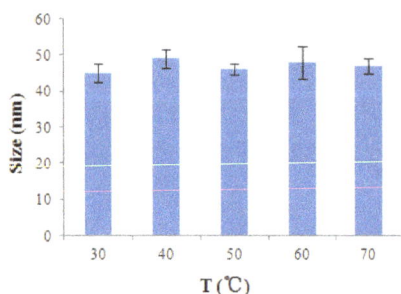

Fig. 3. The variation in magnetite nanoparticles size with temperature.

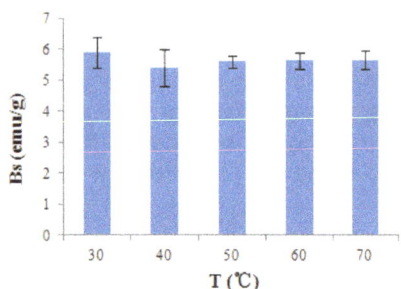

Fig. 4. The variation in measured B_s of magnetite nanoparticles with temperature.

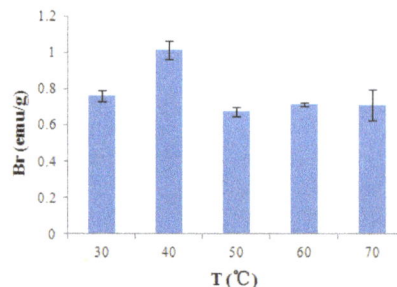

Fig. 5. The variation in measured B_C of magnetite nanoparticles with temperature.

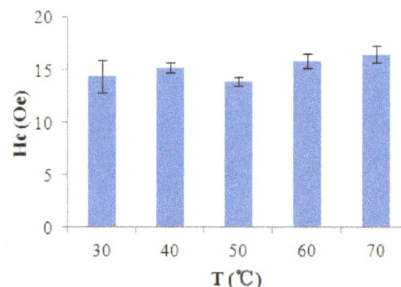

Fig. 6. The variation in measured H_C of magnetite nanoparticles with temperature.

5. Discussion

Biomagnetite production by magnetotactic bacteria and Fe(III)-reducing bacteria has been extensively studied[15]. In contrast, whether magnetite can be formed by Fe(II)-oxidizing bacteria remained still unclear. Here, we experimentally evidence that the nitrate reducing Fe(II)-oxidizing strain BoFeN1 can promote the formation of stable single domain magnetite. This strain can form a diversity of Fe-bearing minerals depending on culture conditions: lepidocrocite is obtained at neutral pH [16-17].

The possibility of using bacteria for the synthesis of oxide nanoparticles has also been explored. Most of the work in this direction has centered towards synthesis of magnetite nanoparticles, by taking inspiration from magnetotactic bacteria found in nature. For instance, laboratory-based studies on magnetite growth have focused mainly on the use of magnetotactic bacteria [18–19] and iron reducing bacteria, such as Geobacter metallireducens (a distant cousin of magnetotactic bacteria) [20]. In these studies, biosynthesis of magnetite was found to be extremely slow (often requiring 1 week) under strictly anaerobic conditions. It was however interesting to observe the ability of bacterium Actinobacter sp to synthesize magnetite (Fe_3O_4) and maghemite (γ-Fe_2O_3) on incubation with suitable aqueous iron precursors under fully aerobic conditions [21]. These nanoparticles were formed extracellularly and showed excellent magnetic properties. The over expression of two inducible proteins was observed in Actinobacter-mediated synthesis of magnetite nanoparticles. When other aerobic bacteria (e.g. Bacillus sp., Aerobacter aero genes, and Micrococcus luteus) were investigated for magnetite synthesis under similar conditions, they did not result in synthesis of magnetite even after one week of reaction. Kumar and co-workers also reported the extracellular synthesis of spinal-structured ferromagnetic Co_3O_4 nanoparticles using a marine cobalt-resistant bacteria strain, obtained from Arabian sea [22]. In agreement of the previous studies we were able to produce magnetite nanoparticle using *Magnetospirillum* strain AMB-1 and obtained average nanoparticles size of 47 nm.

The magnetic properties of these nanoparticles were determined by vibrating sample magnetometry. The hysteresis curve was obtained, the coercitivity was 1.54 Oersted. The low coercivity indicates that the particles are in super paramagnetic state due to their small particle sizes. The resulting saturation magnetization for small particles can be caused by the presence of super paramagnetic relaxation and/or non colinearity of the magnetic moments at the surface of the nanoparticles [23]. The magnetization saturation does not attain saturation at the highest magnetic field of 7 KOe. The fact that Mr/Ms values were below 0.5; where Mr is the remanent moment and Ms is the saturation moment; was explained from the effect of competition between interparticles interaction and intraparticles anisotropy on the spin relaxation process, which produces frustration [24,25]. The magnetic properties of Ni (Ni55 and Ni147) and Fe147 DENs were studied using SQUID magnetometry at temperatures ranging from 5 to 300 K with magnetic field (H) strength of 500 Oe. The effect of thermal energy on the magnetic properties of the DENs becomes apparent at 200 K for both sizes of Ni particles and at 6 K for Fe147. That is, at T>Tb no remanent magnetisation is observed, however, at T<Tb Ni55, Ni147 and Fe147DENs show hysteresis with magnetic saturation values (Ms) of 3.40 emu/gNi for Ni55, 3.95emu/gNi for Ni147 and 70.0 emu/gFe for Fe147. These values are significantly smaller than the bulk value of 55 emu/gNi and 220 emu/gFe at 300 K[26]. The M–H loop for the Fe55 DEN shows a hysteresis-free magnetism and complete saturation of the material was not observed over the range of magnetic fields studied. The absence of hysteresis and a blocking temperature indicate that the Fe55 DENs are super paramagnetic down to the lowest temperature used due their small particle size[27]. Our results showed that there are a small variation in magnetic properties of magnetite nanoparticles prepared using *Magnetospirillum* under different incubation temperatures. This indicted that and with agreement with the previous studies that it may be a slit effect of temperature on the magnetic properties of magnetite nanoparticles.

6. Conclusion

We concluded that synthesis of magnetite naoparticles using *Magnetospirillum* strain AMB-1 is a promising method in order to obtain nanoparticles with ideal size and magnetic properties suitable for biomedical applications. It is clear that the physical conditions under which bio-synthesis of magnetite naoparticles had been done, could have a small effect on their characterizations. Further study on the biocompatibility and toxicity of the bio-synthetic magnetite nanoparticles should be done to evaluate their suitability to be used in both medical and biological applications.

Acknowledgment

The authors would like to acknowledge financial support of this work from the Deanship of Scientific research (DSR), University of Tabuk (Tabuk, Saudi Arabia, under grant no S-0068-1435)

References

[1] Fendler JH. (1998) Nanoparticles and nanostructured films: preparation, characterization and applications. JohnWiley & Son

[2] Bruins RM, Kapil S, Oehme SW. (2000) Microbial resistance to metals in the environment. Ecotoxicol Environ Saf, 45(3),198-207.

[3] Beveridge, T. J., Hughes, M. N., Lee, H., Leung, K. T., Poole, R. K., Savvaidis, I., Silver, S. & Trevors, J. T. (1997) Metal microbe interactions: contemporary approaches. Adv Microb Physiol, 38 ,177-243.

[4] Bao, C., Jin, M., Lu, R., Zhang, T. and Zhao, Y. Y. (2003) Preparation of Au nanoparticles in the presence of low generational poly(am idoamine) dendrimer with surface hydroxyl groups. Mat. Chem. Phys, 81, 160–165.

[5] Y.Roh, R.J.Lauf, A.D.McMillan, C.Zhang, C.J.Ra wn, J.Bai, wn, and T.J.Phelps. (2001) Microbial synthesis and the characterization of metal-substituted magnetites. Solid State Commun. 118, 529-534.

[6] T.Hyeon. (2003) Chemical synthesis of magnetic nanoparticles. Chem. Commun. 8, 927-934.

[7] E. Petrovsky, M.D.Alcala, J.M.Criado, T.Grygar , A.Kapicka, and J.Subrt,(200) Magnetic properties of magnetite prepared by ball-milling of hematite with iron. J. Magnet. Magnet. Mat. 210, 257 -273.

[8] Gu H, Xu K, Xu C, Xu B. (2006) Biofunctional magnetic nanoparticles for protein separation and pathogen detection. Chem Commun (Camb), 7,(9),941-949.

[9] Tamer U, Gundogdu Y, Boyaci IH. Pekmez K. (2010) Synthesis of magnetic core-shell Fe3O4-Au, nanoparticles for biomolecule immobilization and detection. J Nanopart Res, 12, 1187-1196.

[10] Chang JH, Kang KH., Choi J., Jeong YK. (2008) High efficiency protein separation with organosilane assembled silica coated magnetic nanoparticles. Superlattices and Microstructures. 44(4-5), 442-448.

[11] Huang YF, Wang YF, Yan XP. (2010) Amine functionalized magnetic nanoparticles for rapid capture and removal of bacterial pathogens. Environ Sci Technol, 15,44(20),7908-7913.

[12] Dong H, Huang J, Koepsel RR, Ye P, Russell AJ, Matyjaszewski K. (2011) Recyclable antibacterial magnetic nanoparticles grafted with quaternized poly(2-(dimethylamino)ethyl methacrylate) brushes. Biomacromolecules, 11,12(4),1305-11.

[13] Meng X, Seton HC, Lu le T, Prior IA, Thanh NT, Song B. (2011) Magnetic CoPt nanoparticles as MRI contrast agent for transplanted neural stem cells detection, Nanoscale. Mar, 3(3), 977-984.

[14] Laurent S, Dutz S, Häfeli UO, Mahmoudi M. (2011) Magnetic fluid hyperthermia: focus on superparamagnetic iron oxide nanoparticles. Adv Colloid Interface Sci, 10,166(1-2),8-23.

[15] Li J. H., Benzerara K., Bernard S. and Beyssac O. (2013) The link between biomineralization and fossilization of bacteria: insights from field and experimental studies. Chem. Geol, 359, 49–69.

[16] Larese-Casanova P., Haderlein S. B. and Kappler A. (2010) Biomineralization of lepidocrocite and goethite by nitratereducing Fe(II)-oxidizing bacteria: effect of pH, bicarbonate, phosphate, and humic acids. Geochim. Cosmochim. Acta, 74, 3721–3734.

[17] Miot J., Recham N., Larcher D., Guyot F., Brest J. and Tarascon J. M. (2014) Biomineralized a-Fe2O3: texture and electrochemical reaction with Li. Energy Environ. Sci. 7, 451–460.

[18] Bazylinski DA, Frankel RB, Jannasch HW. (1988). Anaerobic Magnetite Production by a Marine, Magnetotactic Bacterium. Nature, 334, 518-519

[19] Sakaguchi T, Burgess JG, Matsunaga T. (1993) Magnetite formation by a sulphate-reducing bacterium. Nature, 365:47-49.

[20] Vali H, Weiss B, Li Y-L, Sears SK, Kim SS, Kirschvink JL. (2004) Formation of tabular single domain magnetite induced by Geobacter metallireducens GS-15. Proc. Nati. Acad. Sci. 101, 16121-16126.

[21] Bharde A, Wani A, Shouche Y, Joy PA, Prasad BLV, Sastry M. (2005) Prasad and M. Sastry, Bacterial Aerobic Synthesis of Nanocrystalline Magnetite. J Am Chem Soc, 127,9326-9327.

[22] Umesh Kumar, Ashvini Shete, Arti Harle, Oksana Kasyutich, W. Schwarzacher, Archana Pundle And Pankaj Poddar. (2008) Extracellular Bacterial Synthesis of Protein Functionalized Ferromagnetic Co3O4 Nanocrystals and Imaging of Self-Organization of Bacterial Cells under Stress after Exposure to Metal Ions. Chemistry of Materials,20,1484 –1491.

[23] A. H. Morr and K. Haneda. (1988) Magnetic structure of small NiFe2O4 particles. J Appl Phys, 52, 4258-4296.

[24] E1-Hilo, M., K. O'Grady, and R.W. Chantrell (1992)Susceptibility phenomena in a fine particle system, I., Concentration dependence of the peak. J. Magn. Magn. Mater., 114, 295-306.

[25] G. A. Held, G. Grinstein, H. Doyle, Shouheng Sun, and C. B. Murray.(2001) Competing interactions in dispersions of superparamagnetic nanoparticles. Phys Rev ,64:012408.

[26] Bozorth RM. Ferromagnetism. New York, NY: D. Van Nostrand Company, Inc.; 1951

[27] Lesli-Pelecky DL, Rieke RD. (1996) Magnetic properties of nanostructured materials. Chem Mater, 8,1770–1783.

Effect of Heat Treatment on Nanoparticle Size and Oxygen Reduction Reaction Activity for Carbon-Supported Pd–Fe Alloy Electrocatalysts

Essam Fadl Abo Zeid[1, 2, 3, *], **Yong Tae Kim**[2]

[1]Physics Department, Faculty of Science, Assiut University, Assiut, Egypt
[2]School of Mechanical Engineering, Pusan National University, Pusan, Korea
[3]Physics Department, Faculty of Science &Arts El Mandaq, Al-Baha University, Al Baha, KSA

Email address:
esabozaid@yahoo.com (E. F. A. Zeid), cabozaid@aun.edu.eg (E. F. A. Zeid)

Abstract: The synthesized carbon-supported Pd-Fe alloy electrocatalysts were characterized for the purpose of the fuel cell cathode oxygen reduction reaction (ORR). The synthesized catalysts were characterized in terms of structural morphology and catalytic activity by XRD and electrochemical measurements. Surface cyclic voltammetry was used to confirm the formation of the Pd–Fe alloy. The catalysts were heat-treated at temperatures ranging from 300 °C to 700 °C for different aging times, in order to improve activity and stability. The average particle size of 10.16 nm, and the highest ORR catalytic activity were obtained at the optimal heat-treatment temperature 300 °C for 3h.

Keywords: Alloys, Chemical Synthesis, Powder Diffraction, Aging

1. Introduction

The high cost arising from the use of expensive noble metal catalysts in the current PEMFC technique is one of the main problems. Electrocatalysts based on Platinum are exclusively used for catalyzing oxygen reduction reaction (ORR) and methanol oxidation reaction (MOR) in Direct Methanol Fuel Cells (DMFCs) [1-4]. The widespread commercialization of DMFCs hindering by several limitations such as the high cost, the low availability of platinum and the irreversible inactivation of the catalysts by CO-like poisoning species [5, 6]. It is therefore desirable to develop low cost catalysts with comparable activity towards methanol oxidation reaction and better CO-tolerance for DMFCs. The important and urgent task for fuel cell investigators is the research on non-Pt catalyst. Unfortunately, in principle, only noble metals can be stable in the acidic environment of PEMFC which limited the choice of catalysts for PEMFC. Pd could be a good candidate, among alternative noble metals, not only because Pd is one of the Pt-group metals and has been applied in many heterogeneous catalysis processes, but it is also less expensive and relatively abundant in comparison with Pt. However, Pd was found to be inferior to Pt towards most fuel cell relevant reactions, except for the electro oxidation of formic acid [7]. In recent years, much effort has been devoted to the development of Pd alloys as alternative catalysts for PEMFC, especially for the reduction of oxygen at the cathode [8–10]. Some of these alloys have exhibited activities comparable to that of Pt. Though encouraging results have been achieved, there is still a lack of systematic understanding for the rational design of Pd-based catalysts. Usually, in order to promote alloy formation the Pt-based [11, 12] and Pd-based [13] alloy catalysts are prepared and/or post-treated at high temperatures in inert or reducing atmospheres. However, a decrease in the surface area and catalytic activity as results of the thermal treatment at high temperatures leads to an undesired particle growth. Therefore, catalyst preparation methods that can offer high degrees of alloy homogeneity with small particle size and high surface area at moderate temperatures are needed. Theoretical calculations and experimental data demonstrated that, upon annealing at elevated temperatures, Pd-M alloys undergo phase segregation, in which the noble metal Pd migrates to the surface forming a pure Pd over-layer on the bulk alloys [14-18]. The electronic structures of the metal over-layers can alter significantly upon bonding with the substrate metal, and, in turn, their catalytic properties can change [19-21]. Nørskov et al., correlated the electronic structure of the surface metal (represented as the energy centre of the valence d-band density of states) and its catalytic activity; there model has been applied to explain the catalytic activity and electrochemical behaviour of some strained surfaces and metal over-layers [22, 23]. Savadogo et al., [9] reported that, the catalytic activity of Pd_3Fe/C electrocatalyst prepared by thermal treatment surpassed that of the state-of-the-art Pt/C catalyst and that the enhanced catalytic activity is due to the more favourable Pd–Pd interatomic distance. The particle size of Pd-based catalysts, as reported in literatures [24–29], is large, thus there proves to be significant room for improvement in ORR mass activity. Challenges to be met for the preparation of improved Pd alloy catalysts include the need for synthesis procedures resulting in catalysts with desirable composition, small particle size and a narrow size distribution. Abo Zeid et al., [30] investigated the effect of the heat-treatment temperature on catalytic activity of Pd-Co alloy in the range of temperature 300-700 °C and they concluded that the optimal heat-treatment temperature was 300 °C. The heat-treated catalyst at 300 °C exhibited an enhanced ORR activity due to the smaller average mean particle size of ca, 12 nm, compared to those treated at other temperatures. In a continuing effort to improve the catalytic activity of Pd–Fe alloys, this paper focuses on the combined effect of the ethylene glycol (EG) and sodium borohydride ($NaBH_4$) as a synthetic reducing agents with the presence of polycation (PDDA) on catalyst morphology and on the corresponding ORR catalytic activity. The effects of heat treatment during the temperature range from 300 to 700 °C on the catalyst morphology are characterized by X-ray diffraction (XRD), high resolution transmission electron microscopy (HRTEM), energy dispersive spectroscopy (EDS), Cyclic Voltammetry, and electrochemical polarization measurements in rotating disk electrodes (RDE) and single-cell PEMFC for ORR.

2. Experimental

2.1. Catalyst Synthesis

Carbon-supported Pd_{70}-Fe_{30} catalysts with 20 wt% metal in carbon were synthesized by a modified polyol reduction process. Required amounts of $(NH_4)_2PdCl_4$ and $FeCl_2\cdot6H_2O$ to obtain 100 mg of Pd_{70}-Fe_{30}/C (20 wt%) were dissolved in deionized water. 30 ml from ionic polycation (PDDA) was added to 30 mL ethylene glycol and sonicated for 15 min. 40 ml of ethylene glycol refluxed at 130 °C under stirring, PDDA was added drop wise to the ethylene glycol under stirring in 10 times with the appropriate amounts of $Pd_{70}Fe_{30}$ to give an atomic ratio of PDDA: $Pd_{70}Fe_{30} = 7:1$. The mixture was kept under stirring for 2h at 130 °C. A fresh solution containing 200 mg of $NaBH_4$ in 40 ml of deionized water was added. The colour of the solution was observed to change from yellow to black, indicating the processing of the reduction reactions. The mixture was kept under stirring and

refluxed at 130 °C for 1h, cooled to room temperature, an appropriate amount of carbon (Vulcan XC 72R) was added, stirred overnight, and the slurry was filtered, washed with water and ethanol, and dried overnight in vacuum oven at 70 °C. These, as prepared samples are denoted as $Pd_{70}Fe_{30}/C$ - ASP. The synthesized samples were heated at 300, 500, and 700 °C in a flowing mixture of 10% H_2-90% Ar for 3 hours, followed by cooling to room temperature in order to study the effect of heat-treatment temperature on the catalytic activity. In order to study the effect of aging time on the catalytic activity, the samples were aged for 1 h, 2 h, 3 h, 4 h and 5 h at 300 °C in a flowing mixture of 10% H_2-90% Ar followed by cooling to room temperature.

2.2. Material Characterization

XRD measurements of $Pd_{70}Fe_{30}/C$ catalysts were carried out on a Philips Pan analytical X-ray diffractometer with (Cu K_α and $\lambda = 0.154$ nm) in The Korea Basic Science Institute. The detailed description of the XRD measurements was indicated in [30]. In order to estimate the particle size from XRD, Scherrer's equation was used [31]. For this purpose, the (111) peak of the Pd face-centered-cubic (fcc)/fct structure around $2\theta = 40°$ was selected. Morphological and particle distribution studies were carried out with a JEOL 2010F high-resolution transmission electron microscope (HRTEM) operated at 200 keV [30].

2.3. Electrochemical Measurements

Cyclic Voltammetric (CV) characterizations were carried out with a standard single-compartment three-electrode cell having a Pt mesh counter electrode, a glassy carbon (5 mm dia.) working electrode and a leak-free (Ag/AgCl, 3.5 M KCl) with a double-junction chamber (Cypress) reference electrode, employing a biologic VSP potentiostat (France) [30]. The CV plots were conducted in N_2 purged 0.1 M $HClO_4$ at a scan rate of 50 mV/s between -0.2 and 0.8 V (vs. Ag/AgCl) at ambient conditions. Before recording the voltammograms, the catalyst surface was cleaned by cycling 50 times between -0.2 and 0.8 V (vs. Ag/AgCl). Rotating disk electrode (RDE) experiments were conducted with a glassy carbon disk electrode (5 mm dia.) mounted onto an interchangeable RDE holder (Pine Instruments, France) in O_2 saturated 0.1 M $HClO_4$.

3. Results and Discussion

3.1. Physical Characterization of Pd-Fe/C Bimetallic Catalysts

The XRD patterns of the carbon-supported Pd–Fe, (a) aged at different temperatures for 3h and (b) aged at 300 °C for different aging times are shown in Fig. 1. Five main characteristic peaks of the fcc crystalline Pd (JCPDS Card 00-005-0681) [32], namely the planes (111), (200), (220), (311), and (222) was observed in Fig. 1 (a). The five diffraction peaks in the Pd–Fe (70:30 atom %) alloy catalysts are shifted to higher 2θ values compared to those of Pd–Fe upon heat-treatment, suggesting incorporation of Fe into the Pd lattice. A shift of diffraction peaks to higher angles with increasing aging temperature indicating the contraction of the lattice and an increase in the degree of alloying of Fe with Pd [30]. The reflections correspond to only a single fcc phase suggestive of the formation of a binary Pd–Fe alloy phase.

The alloy constituents were thoroughly mixed in the crystal system which indicated by the absence of peaks for Fe or its oxides. The diffraction peak at around $2\theta = 25°$ corresponds to the (0 0 2) plane diffraction of the hexagonal structure of the carbon Vulcan.

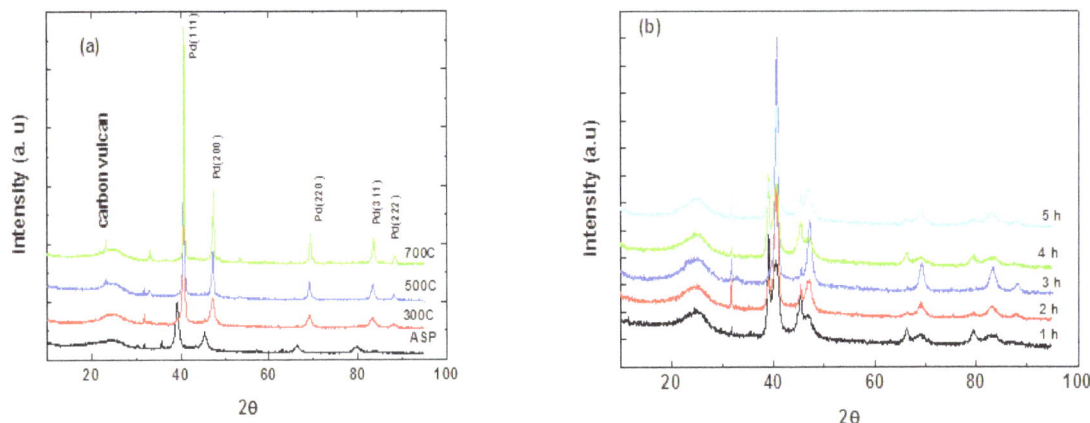

Fig. 1. XRD patterns of the carbon-supported Pd–Fe heated, (a) at different aging temperatures for 3h, (b) at 300 ∘C for different aging times.

In Fig. 1b, there are also five main diffraction peaks for fcc PdFe/C catalysts aged at 300 °C during different times. With increasing the aging time, these diffraction peaks shift to a higher angle. Such angle shifts reveal alloy formation between Pd and Fe with increase in aging times during the aging temperature, and indicate lattice contractions, which are caused by the incorporation of Fe into the Pd fcc structure. However; reflections corresponding to only a single fcc phase are found on aging at 700 °C, indicating the formation

of a binary Pd–Fe alloy phase at higher temperatures.

From Fig. 1(a) it was observed that, the optimal temperature should be thus around 300 °C for studied catalysts. Higher temperatures, such as 500 °C and 700 °C, resulted in an increase in particle size and a decrease in active surface area (Fig.2). The mean particle sizes calculated from XRD patterns (PSXRD) for the catalyst alloy are shown in Table 1.

Table 1. Characteristics of prepared Pd–Fe/C (Pd:Fe = 70:30 atom%) alloy catalyst by XRD.

Heat treatment	Aging temperatures				Aging times at 300 ºC				
Property	ASP	300 ºC for 3h	500 ºC for 3h	700 ºC for 3h	1h	2h	3h	4h	5h
PSXRD (nm)	10.64	16.15	25.3	34.07	12.07	14.87	16.15	16.89	19.71
SXRD(m^2 g^{-1})	50.21	33.08	21.11	15.68	44.26	35.93	33.08	31.63	27.10

It's observed that, the mean particle size increases with increase the aging temperature and time for the as-prepared catalysts. The particle active surface area SXRD in m^2 g^{-1} was calculated using the equation $S = 6000/d\rho$ for spherical particles [33], where d is the crystallite size (diameter) in nm obtained from the (111) diffraction line, XRD data (Fig. 1) using the Scherrer equation $d(A°) = \kappa\lambda / \beta Cos\theta$, [31], and ρ is the density of the Pd-Fe alloy (~11.23 g cm−3). The XRD-determined active surface areas (SXRD) are also

provided in table 1. As indicated, SXRD decreases as the aging temperature for the as-prepared catalysts is increased. This decrease in the SXRD attributed to the agglomeration of the particles which results in the increase of particle size Fig. 2. From the results of the mean particle size and active surface area in the table 1 we can conclude that the optimal aging time ranges from 3h to 4h at the optimal aging temperature 300∘C.

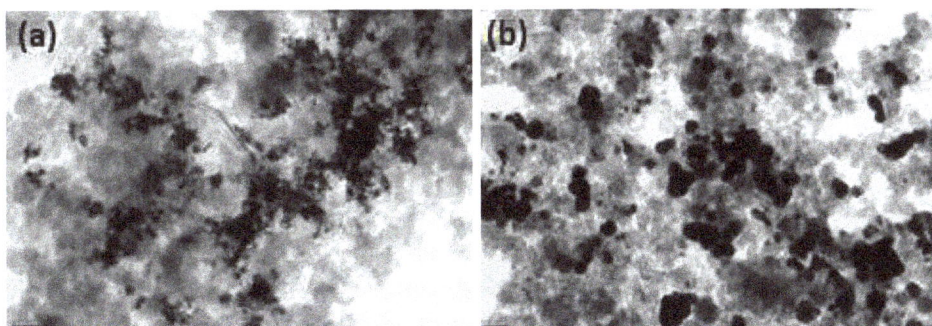

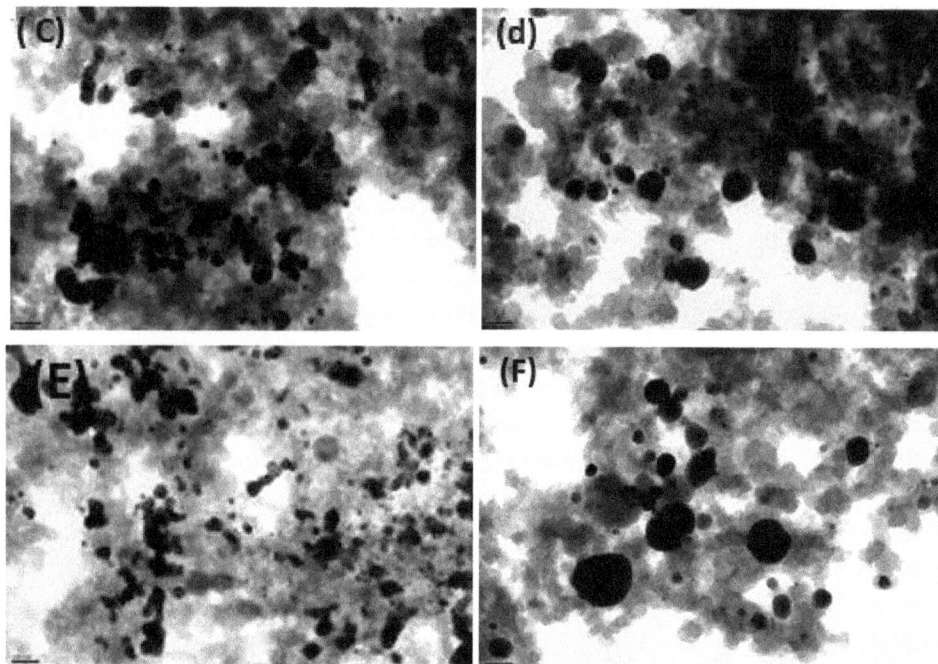

Fig. 2. TEM micrographs of the carbon-supported Pd–Fe (Pd:Fe = 70:30 atom%) (a) ASP sample and followed by aging for 3 h (b) at 300 ℃; (c) at 500 ℃ and (d) at 700 ℃, in (E) and (F) the studied sample aged at 300 ℃ for 1h and 5h respectively, all of these images with 100 nm magnification.

As a comparison, a uniform distribution of catalyst particles with a predominant and regular spherical shape can be observed in all samples after aging at various temperatures. An increase in particle size with increasing aging temperature (Fig. 2b–d) may suggest agglomeration during heat treatment. These images indicate that, all the catalysts have a good dispersion on the carbon surface with a narrow particle size distribution. It is also observed that the average particle size is slightly higher than the untreated one (Fig.2a) and the heat-treatment appears to favour agglomeration as reported earlier [12, 31]. Particles with a few large sized (≤ 35 nm) are also observed, which are formed as a result of the particles aggregation at higher temperatures. This observation was consistent with those calculated from XRD data. The obtained mean particle size is smaller than those reported by others for Pd-based catalysts [34, 35], which may be beneficial for increasing in the ORR mass activity. Hence, the procedure for the catalyst preparation via a modified polyol reduction route may be a method for obtaining nano sized alloy catalysts with a good dispersion and a narrow particle distribution on a support. The aging for a long time increases the agglomeration process which results in an increase in particle size (Fig. 2 E, F).

3.2. Surface Cyclic Voltammograms of Pd–Fe/C Alloy Catalysts

CVs of the Pd–Fe/C synthesized catalysts are shown in (Fig. 3-a, b). These CVs were recorded in a 0.1M $HClO_4$ solution under N_2 atmosphere at 27 ℃ after aging the samples for 3h in the lower temperature range from 300 ℃ to 700 ℃. The CV of Pd–Fe/C aged at 300 ℃ for 3h (Fig. 3-a, b) shows large peaks in the potential range of -0.193 to –0.131V and -0.196 to-0.135 respectively, versus Ag/AgCl sat KCl 3.5

M, which correspond to the hydrogen adsorption/desorption processes. However, the other aged samples all exhibit smaller hydrogen peaks compared to those of the 300 ℃ aged sample. It was observed that, the dissolution of hydrogen into bulk Pd–Fe/C might be restrained by the existence of the iron in the Pd lattice [29]. In the case of the aged sample (Pd–Fe/C) at 300 ℃, the degree of alloying for iron in Pd lattice to form the core shell is less than the other two aged samples (500 ℃ and 700 ℃). Therefore, the larger peaks of the 300 ℃ aged sample than the 500 ℃ and 700 ℃ samples may be ascribed to the lower degree of alloying for iron in this sample than the others. Normally, the areas under the hydrogen adsorption/desorption peaks in CVs can be used to estimate the electrochemically active surface areas (ESA) of a pure Pd catalyst. But in the case of alloy catalysts, such quantitative estimation may not be feasible. As a qualitative estimation, it can be seen that the Pd–Fe/C aged at 300 ℃ sample shows the largest ESA compared to the other catalyst samples, which may be ascribed to the lower degree of aggregation and smaller particle size of this catalyst. Therefore, the aging at temperatures higher than 300 ℃ for this kind of Pd-Fe/C alloys could negatively affect the morphology and electrocatalytic activity of the synthesized catalysts. In the case of an aged sample (Pd-Fe/C) at 300 ℃ for 3 h, the degree of alloying for iron in Pd lattice to form the core shell is less than the other aged samples at higher aging times (4 h and 5 h). Therefore, the decrease of the peaks of the samples aged for 4 h and 5 h at 300 ℃ lower than that sample aged for 3 h may be attributed to the higher degree of alloying for iron in the samples aged for 4 h and 5 h. While aging for long times might be changes both the degree of alloying and crystallite size which influence the activity. Therefore, aging times longer than 4 h for this kind of Pd-Fe/C alloy catalyst, could negatively affect the morphology and electrocatalytic activity.

This result agrees with that reported by L. Zhang et al., [32] that, the optimal aging temperature was found to be as low as 300 °C. There are no iron peaks apparent in the voltammetry, which might indicate that iron was fully incorporated into palladium to form an alloy, and a Pd-rich skin was formed on the alloy's surface.

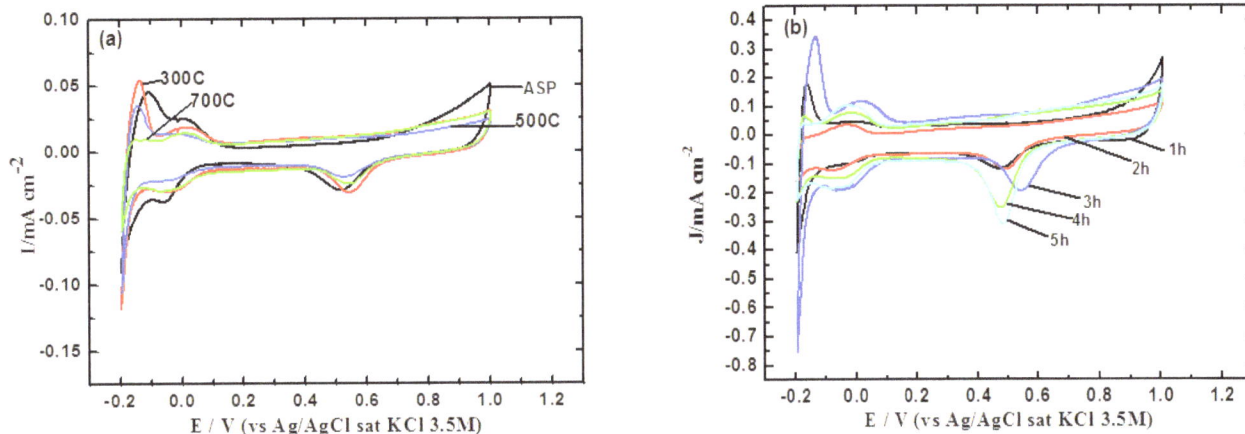

Fig. 3. *Cyclic voltammograms of Pd–Fe/C catalysts (a) aged at different temperatures ASP (as prepared sample), 300 °C, 500 °C and 700 °C and (b) aged at 300 °C for different aging times 1h, 2h, 3h, 4h and 5h.*

3.3. Catalyst Activity Towards ORR as a Function of Aging Temperatures and Time's

Fig. 4a and b, shows the single scan voltammograms for the Pd–Fe/C alloys coated glassy carbon disc electrode at different aging temperatures and times, in an oxygen-saturated 0.1M $HClO_4$ solution, and at ambient conditions. For comparison, an ORR curve for the as prepared Pd–Fe/C (Pd:Fe = 70:30 atom%) catalyst was also plotted in Fig. 4a and b. The ORR activity order was found from Fig. 4a, as follows: Pd–Fe/C (at 300 °C) > (at 700 °C) > (at 500 °C) > (without heat treatment ASP). The Pd–Fe/C alloy electrocatalyst, which was aged at 300 °C, shows the highest ORR activity. This behaviour ascribed to that, the catalyst which aged at higher temperatures has a larger particle size (smaller surface area) compared to those aged at lower temperatures. It's observed that, the order of ORR performance is consistent with the particle size distribution order.

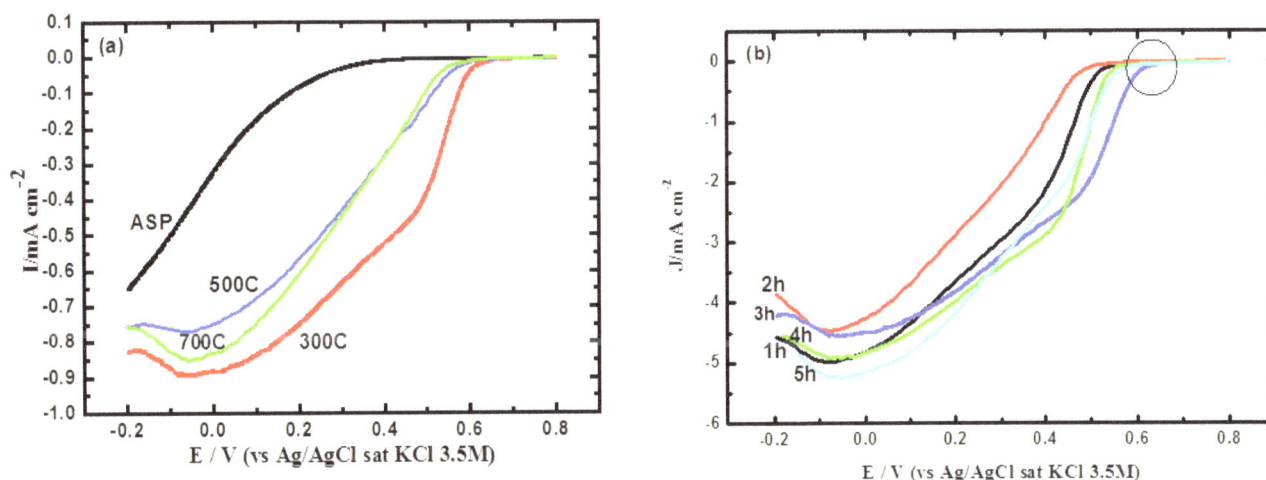

Fig. 4. *Single scan voltammograms for Pd–Fe/C catalyst (a) aged at different temperatures ASP (as prepared sample), 300 °C, 500 °C and 700 °C and (b) aged at 300 °C for different aging times 1h, 2h, 3h, 4h and 5h.*

Increasing the aging temperature from 300 °C to 700 °C, causes an increase in particle size, leads to a decrease in ORR activity of the Pd–Fe/C alloy catalysts. After aging at a high-temperature, the palladium atoms tend to migrate to the surface of the alloy nanoparticles because palladium and iron exhibit a strong trend toward segregation due to the large segregation energy difference between them [36]. Thus, a Pd-rich "skin" should be formed on the Pd–Fe/C nanoparticles.

Also, the ORR activity order is found from Fig. 4(b) as follows: Pd-Fe/C aged at (3 h) > 4 h>5 h >1 h>2 h. The electrocatalyst alloy which is aged at 300 °C for 3 h, showed the highest ORR activity. This behaviour is attributed to that, the catalyst with high aging time has a larger particle size (smaller active surface area) compared to those aged for small periods at the same temperature. Increasing the aging time from 3 h to 5 h caused an increase in particle size, thereby leading to a decrease in ORR activity of the PdFe/C alloy catalysts.

The data in Fig. 4 clearly demonstrates a decrease in catalytic activity for ORR with increasing aging temperature due to a decreasing in active surface area. Additionally, differences in surface characteristics (e.g., crystallographic plane) and particle size distribution depending upon the synthesis method and heating temperature may influence the electrochemical activity. Thus, synthetic approaches that can give a high degree of alloying and homogeneity at lower temperature while keeping the particle size small with optimal surface characteristics have the possibility of improving the catalytic activity further.

4. Conclusions

The effect of aging temperature and time on the catalyst morphology and catalytic ORR activity are studied in more details for the Pd-Fe/C alloy.

It's found that, heat-treatments at appropriate temperatures 300 °C, 500 °C and 700 °C for different aging times improve the activity and stability of the catalysts. The optimal aging temperature and time are found to be 300 °C for 3 h in the studied alloy. Before heat-treatment, a Pd–Fe/C alloy at room temperature showed a weak ORR activity. The Pd-Fe/C alloys contained unreduced Pd ions rather than Pd metal that were revealed by XRD and TEM measurements. It was found that, the increase in crystallite size and the degree of alloying significantly improves the catalyst durability. The kinetic study of ORR revealed that, this reaction catalysed by the Pd-Fe/C alloy electrocatalyst synthesized by this combined reducing agents.

Acknowledgments

This work was supported by the Korea Research Foundation Grant funded by the Korean Government (MOEHRD) (KRF-2008-331-D00094) and a grant (M2009010025) from the Fundamental R&D Program for Core Technology of Materials funded by the Ministry of Knowledge Economy, Republic of Korea.

References

[1]   N. d. L. Heras, E. P. L. Roberts, R. Langton and D. R. Hodgson, "A review of metal separator plate materials suitable for automotive PEM fuel cells", Energy Environ. Sci., Issue, 2, pp. 206-214, 2009

[2]   D.-S. Kim, E. F. Abo Zeid, and Y.-T. Kim, "Additive treatment effect of TiO2 as supports for Pt-based electrocatalysts on oxygen reduction reaction activity", Electrochim. Acta, vol. 55, pp. 3628-3633, 2010

[3]   S. Meenakshi, P. Sridhar and S. Pitchumani, "Carbon supported Pt–Sn/SnO2 anode catalyst for direct ethanol fuel cells", RSC Adv., vol. 4, pp. 44386-44393, 2014

[4]   C. Xu, Y. Liu, Q. Hao and H. Duan, "Nanoporous PdNi alloys

[5]   T. Huang, J. Liu, R. Li, W. Cai and A. Yu," A novel route for preparation of PtRuMe (Me = Fe, Co, Ni) and their catalytic performance for methanol electrooxidation", Electrochem. Commun., vol. 11, Issue 3, pp. 643-646, 2009

[6]   H. A. Gasteiger, S. S.Kocha, B. Sompalli, and F. T. Wagner, "Activity benchmarks and requirements for Pt, Pt-alloy, and non-Pt oxygen reduction catalysts for PEMFCs", Appl. Catal., B, vol. 56, Issues 1-2, pp. 9-35, 2005

[7]   W. Wang, R. Wang, S. Ji, H. Feng, H. Wang and Z. Lei, "Pt overgrowth on carbon supported PdFe seeds in the preparation of core–shell electrocatalysts for the oxygen reduction reaction", J. Power Sources, vol. 195, Issue 11, pp. 3498-3503, 2010

[8]   V. R. Stamenkovic, B. S. Mun, K. J. J. Mayrhofer, P. N. Ross, and N. M. Markovic, "Effect of Surface Composition on Electronic Structure, Stability, and Electrocatalytic Properties of Pt-Transition Metal Alloys: Pt-Skin versus Pt-Skeleton Surfaces", J. Am. Chem. Soc., vol. 128, Issue 28, pp. 8813-8819, 2006

[9]   O. Savadogo, K. Lee, K. Oishi, S. Mitsushimas, N. Kamiya and K. Ota, "New palladium alloys catalyst for the oxygen reduction reaction in an acid medium", Electrochem. Commun., vol. 6, Issue 2, pp. 105-109, 2004

[10]   J. L. Fernandez, V. Raghuveer, A. Manthiram and A. J. Bard, "Pd−Ti and Pd−Co−Au Electrocatalysts as a Replacement for Platinum for Oxygen Reduction in Proton Exchange Membrane Fuel Cells" J. Am. Chem. Soc., vol. 127, Issue 38, pp. 13100-13101, 2005

[11]   V. Raghuveer, A. Manthiram and A. J. Bard, "Pd−Co−Mo Electrocatalyst for the Oxygen Reduction Reaction in Proton Exchange Membrane Fuel Cells", J. Phys. Chem. B., vol. 109, Issue 48, pp. 22909-22912, 2005.

[12]   V. Raghuveer, P. J. Ferreira, A. Manthiram, "Comparison of Pd–Co–Au electrocatalysts prepared by conventional borohydride and microemulsion methods for oxygen reduction in fuel cells", Electrochem. Commun., vol. 8, Issue 5, pp. 807-814, 2006

[13]   M. Shao, K. Sasaki and R. Adzic, "Pd−Fe Nanoparticles as Electrocatalysts for Oxygen Reduction", J. Am. Chem. Soc., vol. 128, Issue 11, pp. 3526-3527, 2006

[14]   M. Shao, P. Liu, J. Zhang and R. Adzic,"Origin of Enhanced Activity in Palladium Alloy Electrocatalysts for Oxygen Reduction Reaction", J. Phys. Chem. B, vol. 111, Issue 24, pp. 6772-6775, 2007

[15]   H. Wang, R. Wang, H. Li, Q. Wang, J. Kang and Z. Le, "Facile synthesis of carbon-supported pseudo-coreshell PdCu/Pt nanoparticles for direct methanol fuel cells ", Int. J. H. Ener., vol. 36, Issue 1, pp. 839–848, 2011

[16]   L. Xiong and A. Manthiram, Influence of atomic ordering on the electrocatalytic activity of Pt–Co alloys in alkaline electrolyte and proton exchange membrane fuel cells", J. Mater. Chem., vol. 14, pp. 1454-1460, 2004

[17]   H. Wang, S. Ji, W. Wang, V. Linkov, S. Pasupathi, and R. Wang, "Pt decorated PdFe/C: Extremely High Electrocatalytic Activity for Methanol Oxidation", Int. J. Electrochem. Sci., vol. 7, pp. 3390-3398, 2012

[18] H. Liu and A. Manthiram, "Controlled synthesis and characterization of carbon-supported Pd4Co nanoalloy electrocatalysts for oxygen reduction reaction in fuel cells", Energy Environ. Sci., vol. 2, Issue 1, pp. 124-132, 2009

[19] B. Hammer and J. K. Norskov, " Theoretical Surface Science and Catalysis—Calculations and Concepts", Adv. Catal., vol. 45, pp.71-129, 2000

[20] J. R. Kitchin, J. K. Norskov, M. A. Barteau and J. G. Chen," Modification of the surface electronic and chemical properties of Pt (111) by subsurface 3d transition metals", J. Chem. Phys. vol. 120, pp. 10240-10246, 2004

[21] V. Stamenkovic, B. S. Mun, K. J. J. Mayrhofer, P. N. Ross, N. M. Markovic, J. Rossmeisl, J. Greeley and J. K. Norskov," Changing the activity of electrocatalysts for oxygen reduction by tuning the surface electronic structure ", Angewandte Chemie., vol. 118, Issue 18, pp. 2963-2967, 2006

[22] K. Shimizu, I. F. Cheng and C. M. Wai, " Aqueous treatment of single-walled carbon nanotubes for preparation of Pt–Fe core–shell alloy using galvanic exchange reaction: Selective catalytic activity towards oxygen reduction over methanol oxidation", Chem. Commun., vol. 11, Issue 3, pp. 691-694, 2009

[23] G. Bozzolo, R. D. Noebe, J. Khalil and J. Morse, "Atomistic Analysis of Surface Segregation in Ni-Pd Alloys", A. Surf. Sci., vol. 219, pp. 149-157, 2003

[24] D.-S. Kim, T.-J. Kim, J.-H. Kim, E. F. Abo Zeid and Y.-T. Kim,"Fine Structure Effect of PdCo electrocatalyst for Oxygen Reduction Reaction Activity: Based on X-ray Absorption Spectroscopy Studies with Synchrotron Beam", J. Electrochemi. Sci. Techn., vol. 1, pp. 31-38, 2010

[25] J. L. Zhang, M. B. Vukmirovic, Y. Xu, M. Mavrikakis and R. R. Adzic, " Lattice-strain control of the activity in dealloyed core–shell fuel cell catalysts", Angew. Chem., Int. Ed., vol. 44, pp. 2132-2135, 2005

[26] E. F. Abo Zeid and Y.-T. kim, " kinetics and mechanism of morphology and oxygen reduction reaction at PdCo electrocatalysts synthesized on XC72", Int. J. Nanotech. Appl., (IJNA), vol. 3, Issue 4, pp. 31-38, 2013

[27] J-Y. Lee, D-H. Kwak, Y-W. Lee, S. Lee and K-W. Park, " Synthesis of cubic PtPd alloy nanoparticles as anode electrocatalysts for methanol and formic acid oxidation reactions", Phys. Chem. Chem. Phys., vol. 17, pp. 8642-8648, 2015

[28] K. Lee, O. Savadogo, A. Ishihara, S. Mitsushima, N. Kamiya and K. Ota, "Methanol-Tolerant Oxygen Reduction Electrocatalysts Based on Pd - 3d Transition Metal Alloys for Direct Methanol Fuel Cell", J. Electrochem. Soc., vol. 153, pp. A20-A24, 2006

[29] D. Wang, S. Lu, P. J. Kulesza, C. M. Li, R. D. Marco and S. P. Jiang, " enhanced oxygen reduction at Pd catalytic nanoparticles dispersed onto heteropolytungstate-assembled pol (diallyldimethylammonium)–functionalized carbon nanotubes", Phys. Chem. Chem. Phys., vol. 13, pp. 4400–4410, 2011

[30] E. F. Abo Zeid, D.-S. Kim, H. S. Lee and Y.-T. Kim,"Temperature dependence of morphology and oxygen reduction reaction activity for carbon-supported Pd–Co electrocatalysts", J. Appl. Electrochemi., vol. 40, pp. 1917-1923, 2010

[31] N. Tian, Z. Zhou, S. Sun, Y. Ding and Z. Wang,"Synthesis of tetrahexahedral platinum nanocrystals with high-index facets and high electro-oxidation activity", Science, vol. 316, pp. 732-735, 2007 and V. R. Stamenkovic, B. Fowler, B. S. Mun, G. Wang, P. N. Ross, C. A. Lucas and N. M. Markovic, " Improved Oxygen Reduction Activity on Pt3Ni(111) via Increased Surface Site Availability", Science, vol. 315, pp. 493-497, 2007

[32] L. Zhang, K. lee and J. Zhang, "The effect of heat treatment on nanoparticle size and ORR activity for carbon-supported Pd-Co alloy electrocatalysts", Electrochim. Acta, vol. 52, Issue 9, pp. 3088-3094, 2007

[33] C. Zhang , W. Sandorf , and Z. Peng,"Octahedral Pt2CuNi Uniform Alloy Nanoparticle Catalyst with High Activity and Promising Stability for Oxygen Reduction Reaction", ACS Catal., vol. 5, Issue 4, pp 2296–2300, 2015

[34] M. Neergat, G. Varadarajan and R. Ramesh,"Carbon-supported Pd–Fe electrocatalysts for oxygen reduction reaction (ORR) and their methanol tolerance" J electrochem., Vol. 658, pp. 25-32, 2011.

[35] Y. Pan, F. Zhang, K. Wu, Z. Lu, Y. Chen, Y. Zhou, Y. Tang and T. Lu "Carbon supported Palladium–Iron nanoparticles with uniform alloy structure as methanol-tolerant electrocatalyst for oxygen reduction reaction". Int. J. Hydrogen Energy, vol. 37, pp. 2993-3000, 2012.

[36] V. Chellasamy and R. Manoharan, "The role of Nanostructured Active support Materials in Electrocatalysis of Direct Fuel cell Reactions", Mater. Sci. Forum, Vol. 710, pp. 709-714, 2011.

3

The Effect of Aggressive Biological Materials on a Painted Automotive Body Surface Roughness

Mohammad Shukri Alsoufi[1, *], Tahani Mohammad Bawazeer[2]

[1]Mechanical Engineering Department, Collage of Engineering and Islamic Architecture, Umm Al-Qura University, Makkah, Saudi Arabia
[2]Chemistry Department, Collage of Science, Umm Al-Qura University, Makkah, Saudi Arabia

Email address:

mssoufi@uqu.edu.sa (M. S. Alsoufi), tmbawazeer@uqu.edu.sa (T. M. Bawazeer)

Abstract: There are different aggressive biological materials which may potentially deposit on a painted automotive body surface during its service life, causing possible local damage, loss of appearance and loss of protective aspects of the system. In this study, the effect of two types of aggressive biological materials on a painted automotive body surface, i.e., natural bird droppings and raw eggs were studied and subsequently explained in more detail. Furthermore, two different testing conditions approaches including in-door and out-door were utilized in order to investigate the surface roughness, R_a, and also to study the behavior of biologically degraded automotive body surface at nano-level scale. The effects of these biological materials on a painted automotive body surface and its appearance were investigated by Atomic Force Microscopy (AFM) and a stylus-based inductive gauge (Taly-surf®, from Taylor Hobson, Inc.), having electromagnetic control of the contact force. Engaged vertically on the top of the specimens, the force could be set much lower than the weight. Results showed that natural bird droppings and raw eggs have a dramatic effect on the appearance and surface roughness of a painted automotive surface body. It was also found that the degradation which occurred due to the natural bird droppings was more severe than that of the samples exposed to raw eggs.

Keywords: Biological, Automotive, Bird Droppings, Raw Eggs, Roughness

1. Introduction

Automotive paints are more complex and precisely engineered than is often appreciated. They often consist of many different layers or 'parts', each chosen to give a precise function in order to provide the desired balance of properties [1]. Appearance of a painted automotive body surface is certainly highly significant for the consumer. This not only seems to be a detrimental property, its retention during service life is of great importance [2]. Primarily, the basic objectives of the painting process are to protect and decorate functions of the motor vehicle body, in which its purpose is to enhance or change an object's appearance, to change its colour or level of gloss or to draw attention to a particular region on the object [1, 3]. However, just as important, there is a need for protection from any environmental conditions (e.g., mechanical, weathering, biological) in terms of degradation of automotive coatings or to lengthen their lifetime [4]. Since the introduction of automotive coatings systems to luxury car lines in the early 1980s, their use has increased dramatically to the point where they are used on all

cars nowadays, as protective prescriptions for items as diverse as airplanes, automotives, bridges, houses and machinery [5]. An automotive coatings system contains different layers or 'parts', each of which plays a key role in its overall performance. These layers or 'parts' typically consist of phosphating conversion pre-treatment, followed by cathodic electro-deposited coating to provide anti-corrosive properties (together with the pre-treatment layer of the car body). These are then coated by a primer surface to enhance mechanical properties. After that, the base-coat layer intended to provide colour and effect is also of high importance in supporting the shape of the car body, and finally, the clear-coat layer for appearance and protection [6]. Fig. 1 shows a typical automotive coating system with approximated dry film thickness and an image of a typical cross-section through a painted panel. Considering all variables, it is inevitable that the clear-coat chemical, the mechanics and appearance can be significantly influenced [7]. However, in different parts of the world, there is often variation in product technology in practical terms as regards topcoat (base-coat and clear-coat), and this can have a significant influence on paint performance, specification and particular details of the process [1].

Moreover, there are different parameters which may affect the clear-coat surface chemistry and can directly influence the anti-corrosive performance of this layer during their exposure to outdoor factors [8]. These mainly include weathering environments (e.g., UV radiation from sunlight, water and humidity, acid rain, hot-cold shocks) [9-11] and biological materials (e.g., bird droppings, raw eggs, tree gum, insect bodies) [2, 4, 12-17]. In response to this, a variety of exposure methods and test protocols have been developed to anticipate the long-term weathering behavior of coating failure and predict performance [10-12, 15, 18-23].

In the literature, published data relating to the impact of external factors on the degradation process of top coatings have been major topics during recent years. Perhaps surprisingly, when seeking ways to precisely investigate the impact of surface roughness, R_a, at nano-level scale on a painted automotive body surface during exposure to aggressive biological materials (e.g., bird droppings and raw eggs), there appears to be no authoritative public data available. Therefore, the present work aims to study two aggressive biological materials, i.e., natural bird droppings and raw eggs, under two different testing conditions approaches including in-door and out-door conditions with the intention of investigating the surface roughness, R_a, on a painted automotive body surface on a small scale.

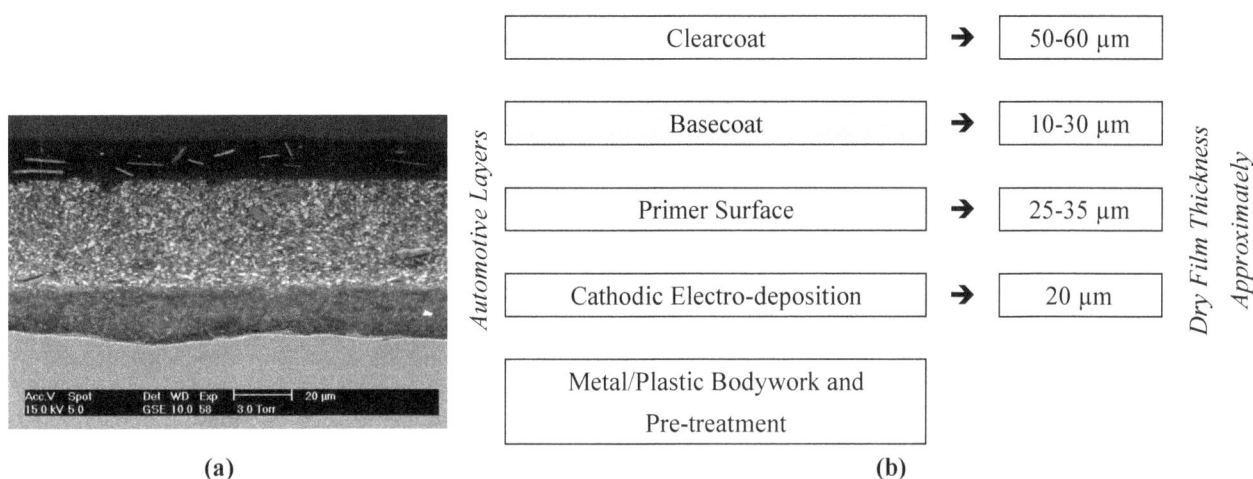

Figure 1. *(a) a typical cross-section through a painted panel and (b) paint layers on vehicles with approximated dry film thickness, adapted from [4]*

1.1. Biological Materials

Throughout history, various types of biological materials have been responsible for degradation of a painted automotive body surface. The most important of these are bird droppings, raw eggs, tree gum and insect bodies [7, 12, 23, 24]. Biological materials have various chemical compositions that can affect coatings' degradation differently [4, 6, 16]. Among the biological materials, the influence of natural bird droppings and raw eggs will be explained in more details, as they are the most important factors which affect the chemical, mechanical and visual performance of automotive coatings. It must be emphasized that no attempt is made to be comprehensive and just a few specific areas in natural bird droppings and raw eggs are discussed.

1.2. Natural Bird Droppings

In general, birds have no bladder; they do not store liquid waste separately from solid waste. All of their waste is mixed together in an organ called the 'cloaca', (the cloaca is the terminal chamber of the gastrointestinal and urogenital systems, opening at the vent). Solid waste from the intestines is mixed with concentrated uric acid ($C_5H_4N_4O_3$) from the kidneys and everything is eliminated together. The uric acid levels in natural bird droppings is relatively high, reaching pH of somewhere between 3.0 and 4.5, which is quite acidic. So, in terms of a painted automotive body surface, these extra Hydrogen (H^+) ions found in the uric acid of bird droppings will react with the hydrocarbons of the body surface and slowly break down the clear-coat of the body surface in the order of a few nano-meter levels. Thus, natural bird droppings contain uric acid, a chemical that is corrosive enough to quickly eat through a coating of wax or paint sealant and begin to etch.

It is worth mentioning here that it also contains different types of amylase, lipase and protease enzymes, which can hydrolyze 'ester' and 'ether' linkages of the clear-coat resin which leads to the formation of soluble products and subsequently to release from the coating, resulting in the formation of cracks and holes in the film surface [12].

- *Amylase enzyme:* (pH ≈ 7, neutral) breaks down starch/carbohydrates into simple sugars.
- *Lipase enzyme:* (pH ≈ 8, slightly alkaline) breaks down lipids (fat) into glycerol and fatty acids.
- *Protease enzyme:* (pH ≈ 2, highly acidic) breaks down protein into amino acids or into smaller protein molecules by breaking the peptide bond joining them.

Enzymes are protein molecules that speed up some chemical reactions and begin others. Without enzyme catalysts, the biochemical actions required for each body function would not occur fast enough and the body would cease to live. Most enzymes are much larger than the substrate they act upon, and only a small portion of the enzyme (around 3.0 to 4.5 amino acids) is directly involved in the hydrolytic catalysis [7, 12, 13, 15, 16]. It was also illustrated that an enzymes activity is affected by temperature, chemical environment and the substrate composition [25-27]. It was found that natural bird droppings decrease the appearance parameters of the clear-coat, i.e., gloss, distinctness of image and colour value hence negatively affecting the aesthetic properties of the coating system [23]. Indeed, the mechanism of this type of degradation seems quite controversial. Whatever the mechanism of this failure, chemical, and/or physical, one can expect that the biological resistance of automotive coating depends on different parameters such as the chemical composition of the biological substances, chemistry of the curd coating, as well as on the physical and/or chemical properties of the coating system itself, and more importantly, on the synergistic effect of environmental factors combined with these aggressive chemicals [12].

Fig. 2 shows the image of pigeon and natural bird droppings on a painted automotive body surface (colour: red car).

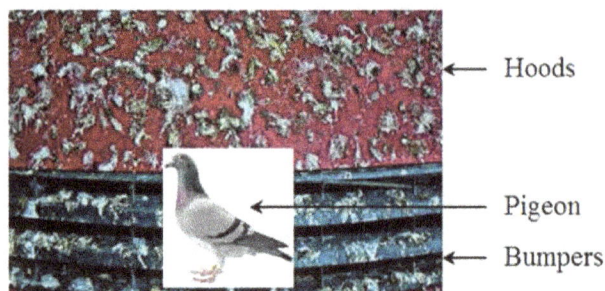

Figure 2. Image of pigeon and natural bird dropping on a painted automotive body surface

1.3. Raw Eggs

In addition to natural bird droppings, raw eggs (as biological material) also play a key role in rendering the coating to gradually diminish its appearance leading to catastrophic failure.

Weight and composition of hen eggs vary, and is dependent on species, feed, age, habitat, and several other factors. Among all the factors, poultry feed significantly influences the chemical and physical composition of hen eggs. A hen egg is composed of three major parts: white, yolk, and shell. According to the published data (% of total weight), the whole egg consists of (9 - 11% shell), (60 - 63% white) and (28 - 29% yolk) and the average weight is 57g [28, 29].

Additionally, eggs are good source of several important nutrients including protein, total fat, monounsaturated fatty acids, polyunsaturated fatty acids, cholesterol, choline, folate, iron, calcium, phosphorus, selenium, zinc and vitamins A, B2, B6, B12, D, E and K. They are also a good source of the antioxidant carotenoids, lutein and zeaxanthin [30].

Stevani and his colleagues in [22], studied the influence of dragon-fly eggs on an acrylic melamine automotive clear-coat. They found that the hydrogen peroxide released during hardening of eggs, oxidizes the cysteine and cysteine residues present in the egg protein, leading to the formation of sulfinic and sulfonic acids. The acids produced then catalyze the hydrolytic degradation.

It seems that dragon-fly eggs show the same damaging mechanism of biological materials as natural bird droppings and pancreatin (simulated natural bird droppings) on clear-coat. Both dragon-fly eggs and natural bird droppings may catalyze the hydrolysis reaction of clear-coat in hot-humid conditions but with different mechanisms. Dragon-fly eggs produce an acidic pH whereas natural bird droppings catalyze the hydrolytic reaction of clear-coat using their enzymatic structure [12, 22, 31]. Fig. 3 shows the image of raw eggs on a painted automotive body surface and the general structure of the egg.

 **(a)**

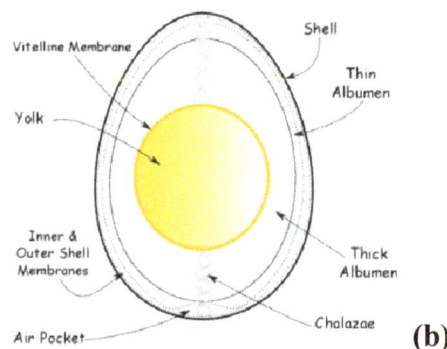 **(b)**

Figure 3. Image of (a) raw eggs on a painted automotive body surface and (b) general structure of egg

2. Experimental

2.1. Biological Materials Selection

Various types of biological materials are responsible for degradation of a painted automotive body surface. However,

the two biological substances utilized were natural bird droppings and raw eggs. The natural bird droppings were collected from a pigeon and the raw egg dropping were purchased from the farm. All biological materials involved in this experiment were collected from a single city namely Makkah (as the feed significantly influences the chemical and physical composition of both pigeon and hen egg). Table 1 shows the specification of biological materials used in this study. Ramezanzadeh and his co-workers in [12] used pancreatin (which is a mixture of several digestive enzymes produced by the exocrine cells of the pancreas) as the synthetic equivalent of natural bird droppings. On the other hand, although the natural bird droppings contain different impurities and effects, all the chemical composition of the natural bird droppings was used in order to stimulate the real live data and investigate the impact of the surface roughness on a painted automotive body surface on a small scale.

Table 1. *Specification of biological materials used in this study (approximate pH number)*

Biological Materials	Specifications			
	pH	Bird Name	State	Colour
Natural Bird Droppings	4.0	Pigeon	Liquid	Gray
Raw Eggs	8.0	Hen	Liquid	White/Yellow

2.2. Sample Preparations

More than 300 samples (a white painted automotive body surface, from the Toyota Company, KSA, model 2013) were used in this study with identical dimensions of length of 30 × 30 × 0.85 mm. A Waterjet machine (from TecnoCut waterjet cutting systems) was used to cut the front of the painted automotive body surface into identical dimensions. Each sample was mounted onto an aluminum sample holder (of a standard commercial and inexpensive nature) using the commercial adhesive. This was accomplished by dispensing an equal amount from two tubes onto a clean disposable surface and mixing this thoroughly for 45 seconds and then using it for 4 minutes. To eliminate any misalignment between the samples and the surface of the aluminum sample holder, the adhesive was added at the surrounding end of the sample holder. This is vital in order to avoid error in the measurement being introduced by misalignment of the measurement systems. Fig. 4 shows the image of the sample to be tested and the aluminum sample holder.

(a)   (b)

Figure 4. *Image of (a) sample to be tested and (b) aluminum sample holder*

It is of the utmost significance before starting the experiments to clean the specimens of any surface contaminants, such as dust, grease, or any other soluble organic particles so that there will be no adverse effect on the results. To achieve this, all specimens were ultrasonically cleaned in two five minute steps: (1) water with detergent to remove dust and oil, and (2) distilled water to remove detergent, and this was followed by warm air drying. After cleaning, the specimens were stored for 24 hours in the same environment (in-door and out-door) that would be used for the testing to allow the sample surface condition to equilibrate with the environment. The procedure described above was judged to be adequate at this stage.

2.3. Testing Procedures

In this work, the process of degradation on a painted automotive body surface takes place under two different conditions during the first 24 hours of the impact as follows:

- *In-door conditions:* testing took place in an artificial atmosphere, with an ambient temperature of $20\pm1^{\circ}C$ and a relative humidity of greater than $40\pm5\%$ RH.
- *Out-door conditions:* testing took place in weathering environments (e.g., UV radiation of sunlight, humidity, hot-cold out-door shocks, acid rain) which impose different kinds of degradations on a painted automotive body surface during the experiment.

According to these testing conditions, all samples were covered by the natural bird droppings and raw egg. The surface profile of a painted automotive body surface was quantitatively analyzed in order to determine the statistical standard parameter of average roughness, R_a, by using Taly-surf® (from Taylor Hobson Precision, Inc) which delivers 0.8 nm resolution over 12.5 mm seamless measuring range and includes 0.125 μm horizontal data spacing. A nominal 2 μm stylus was used with a normal load of 0.7 mN and selectable traverse speed down to 0.5 mm s^{-1} and which conforms to British Standards see Fig. 5. Surface roughness errors were calculated from the standard deviation of the absolute values of height deviation (absolute values). The traces were auto-leveled to a linear least-squares straight line and then filtered with a standard 0.8 mm cut-off. The surface parameters were selected according to the recommendations in the literature and also with respect to the data processing facilities available [32-36].

Figure 5. *Image of Taly-surf® and x-y stages for mounting the sample holder and specimen.*

Every test condition was repeated at least three times at different "*new*" locations on a painted automotive body surface in order to ensure reproducibility of the results. The new location was ±100 μm from the previous one. This approach should have avoided any alteration of the counterbody surface, e.g., due to wear, which might occur during the test and affect the measurements in the following tests. All experiments were performed with a typical "*ball-on-flat*" arrangement applying a linear sliding contact at constant velocity over a specific distance. Tests were performed by using single scan mode (forwards motion). The profiler had a scan length of 10 mm, which is close to the size of a human fingertip.

2.4. Calibration Procedure

Standard calibration ball radius D = 22.0161 mm, 112/1844, Serial No. 639-506-B (from Taylor Hobson Precision, Ltd.) was used to calibrate the test-rig. For convenience, ten calibration trials have been carried out. This is adequate as these trials are predominantly about relative behavior; design interpretation to other systems is always vulnerable to variations in terms of materials and dimensions. Calibration showed the cantilever was a linear spring ($R^2 > 0.99$), under operating and environmental conditions typical for this type of device, with absolute uncertainties of <1% of reading and realizable measurement resolution down to at worst 50 nm. Fig. 6 shows the set-up of the standard calibration ball and the systematic diagram of the ball and a nominal 2 μm stylus with a normal load of 0.7 mN and selectable traverse speed down to 0.5 mm s^{-1}.

This method of calibration ensures that the gauge travels through (and therefore, is calibrated over) most of its range.

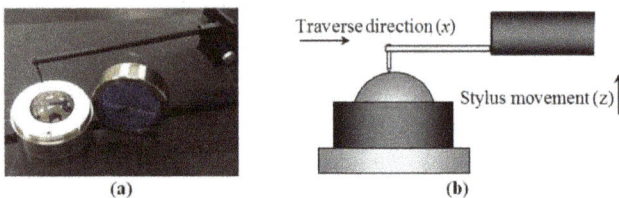

Figure 6. (a) image of standard calibration ball radius D = 22.0161 mm, 112/1844, Serial No. 639-506-B (b) ball with nominal 2 μm stylus

For further investigation of the degradation process AFM microscope Park NX10 were utilized to investigate the effect of biological materials on the surface morphology of the automotive body surface.

3. Results and Discussions

As has been previously mentioned, the present work aims to study the effect of two types of aggressive biological materials on a painted automotive body surface, i.e., natural bird droppings and raw eggs under two different conditions (out-door and in-door) with time. In all, over 300 separate experiments (not all successful) and samples were tested. Surface roughness, R_a, has been extensively used to determine the total average roughness and the degraded net (before and after the attack) during 24 hrs. Fig. 7 illustrates an example of a painted automotive body surface roughness

before and after the attack of biological materials (i.e., natural bird droppings and raw eggs), which also indicated that even the smoothest surfaces are rough on the atomic scale and that contact only occurs at the tips of asperity peaks. It is worth mentioning here that mathematically, R_a, is the arithmetic average value of the profile departure from the mean line of the surface, within a sampling length.

Besides the conventional roughness parameters, R_a, the statistical parameters R_{sk} (skewness) and R_{ku} (kurtosis) were determined. R_{ku} is a measure of the randomness of profile heights. A perfectly random surface has a value of 3. The further the value is from 3, the more repetitive the surface is. Spiked surfaces exhibit high value, whereas bumpy surfaces possess lower value. From the R_{sk} value, a conclusion can be drawn about the symmetry of the profile about the mean line. Negative R_{sk} values point to a predominance of valleys, while positive values correspond to a peaked surface.

The surface analysis revealed that a varying predominance of valley surface features indicated by the negative R_{sk} values was measured on these samples. Additionally, it was observed that the structures of these samples have more repetitive features ($R_{ku} > 3$).

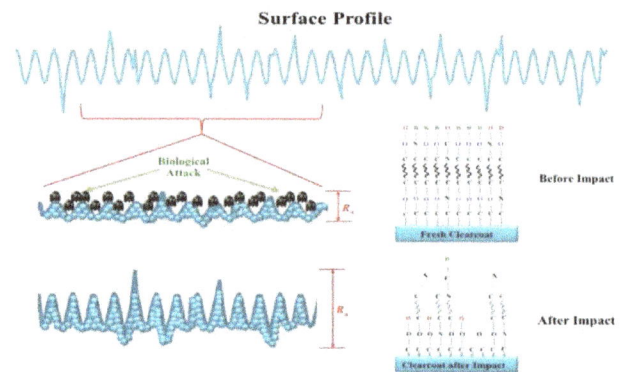

Figure 7. A typical example of a painted automotive body surface roughness before and after the attack of biological materials

3.1. Effect of Natural Bird Droppings on a Painted Automotive Body Surface

A painted automotive body surface was exposed to natural bird droppings for both (in-door and out-door) conditions. The effect of the natural bird droppings on a painted automotive body surface before and after exposure is shown in Fig. 8(a) and 9(a). In real conditions, almost always, a clearcoat layer of body car surface experiences both biological and natural environmental processes (sunlight and humidity) conditions before and after the attack of the natural bird droppings. These figures show that the surface of the clearcoat before any exposure is smooth. However, both environmental and biological attacks roughened the surface. Also, it is evident that this roughening phenomenon becomes more severe when both environmental and biological conditions are introduced simultaneously. The author in [37] mentioned that the UV portion of sunlight causes degradation, because it has energy content sufficient to break chemical bounds.

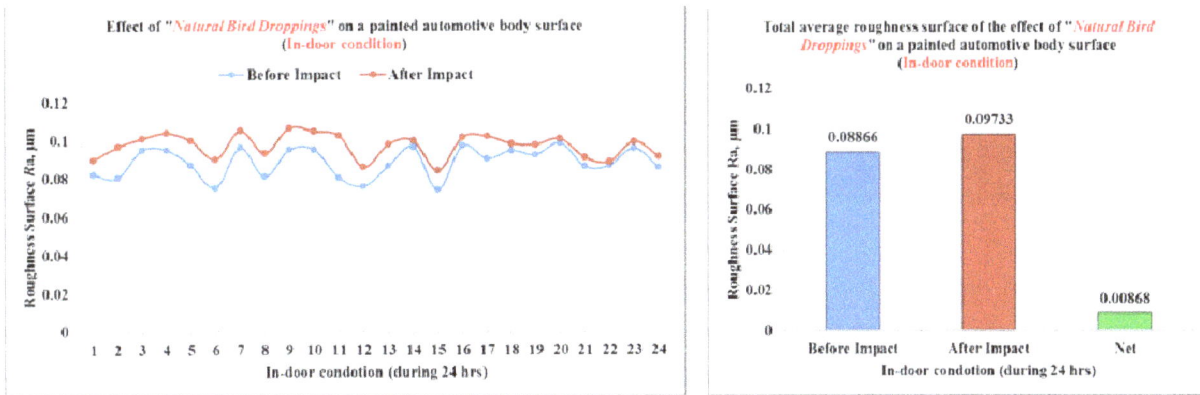

Figure 8. (a) effect of natural bird droppings on a painted automotive body surface during 24 hrs (b) total average roughness with degraded net (In-door condition)

As can be seen from Fig. 8(b) and 9(b), the degraded net of the total average roughness surface of the effect of natural bird droppings on a painted automotive body surface in (out-door condition), $R_a \approx 12.7$ nm, were higher than those obtained in identical testing (in-door condition), $R_a \approx 8.6$ nm. Back to Fig. 1(b), which represents the typical cross-section through a painted panel, the dry film thickness of the first automotive layer of the clearcoat is approximately $50 - 60$ μm. So, it is

clearly during the first 24 hrs of the impact of the biological material (i.e., natural bird droppings) for both conditions (in-door and out-door), the degradation was only in the clearcoat layer of the automotive body car surface to a depth of 12.7 nm and 8.6 nm, respectively. As a result, the appearance and the protection layer of the body car surface degraded by a few nano-levels of a total micro-level.

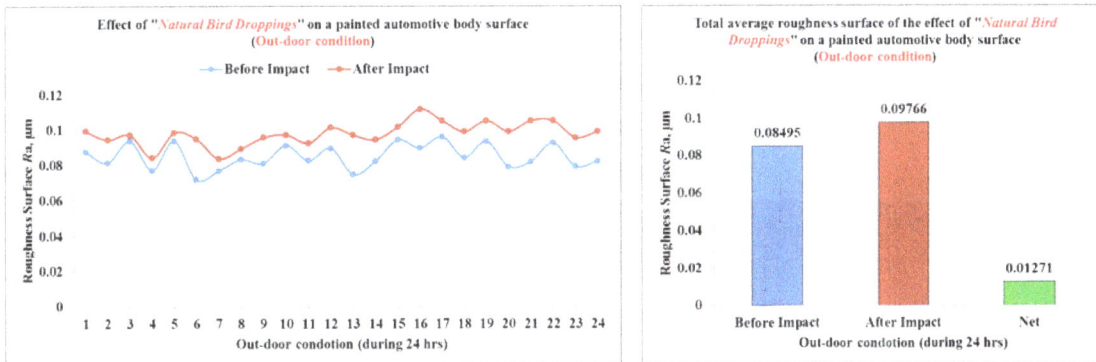

Figure 9. (a) effect of natural bird droppings on a painted automotive body surface during 24 hrs (b) total average roughness with degraded net (Out-door condition)

3.2. Effect of Raw Eggs on a Painted Automotive Body Surface

It is clear from Fig. 10(a) and 11(a), that the total average roughness curves, R_a, were affected in the presence of raw eggs on the samples at different times and conditions. Each sample represents a specific time, this is why the average roughness curve, R_a, fluctuated during 24 hrs before and after impact. However, the total average roughness of the sample at each hour during 24 hrs increased after the impact of the biological material (i.e., raw eggs) on all samples depending on the atomic structure of the surface and leading to the degradation on a painted automotive body surface.

As can be seen from Fig. 10(b) and 11(b), the degraded net of the total average roughness surface of the effect of raw eggs on a painted automotive body surface in (out-door condition), $R_a \approx 8.5$ nm, were approximately twice as high as those obtained in identical testing (in-door condition), $R_a \approx 5.3$ nm.

It was felt that this might be due to the effect of the environmental parameters such as UV radiation of sunlight, humidity and hot-cold out-door shocks, which may accelerate the process of the degradation on a painted automotive body car surface during the same period of experiment (24 hrs).

Refer to Fig. 1(b), which represents the typical cross-section through a painted panel, the dry film thickness of the first automotive layer of the clearcoat is approximately 50 – 60 μm. So, it is clearly during the first 24 hrs of the impact of the biological material (i.e., raw eggs) for both conditions (in-door and out-door), the degradation was only in the clearcoat layer of the automotive body car surface by 5.3 nm and 8.5 nm, respectively. As a result, the appearance and the protection layer of the car body surface degraded by a few nano-levels of a total micro-level. This relationship between the dry film thickness and average total profile roughness can indeed be used to monitor the total average degradation of the effect of the aggressive biological materials on a painted   automotive body surface.

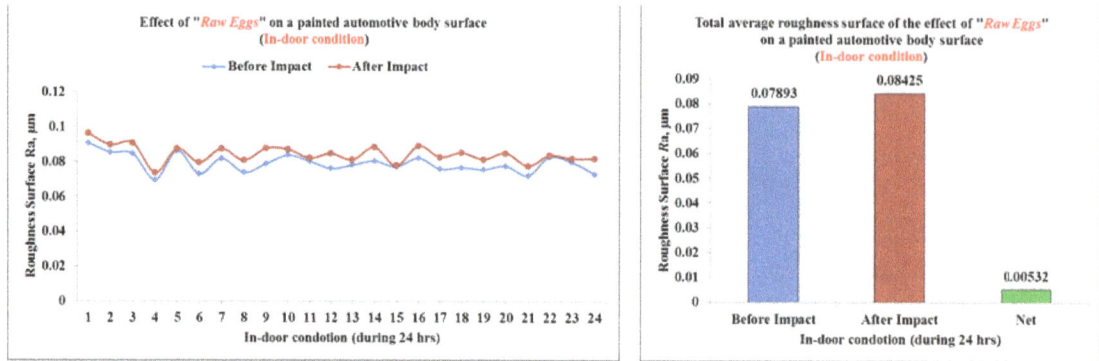

Figure 10. *(a) effect of raw eggs on a painted automotive body surface during 24 hrs (b) total average roughness with degraded net (In-door condition)*

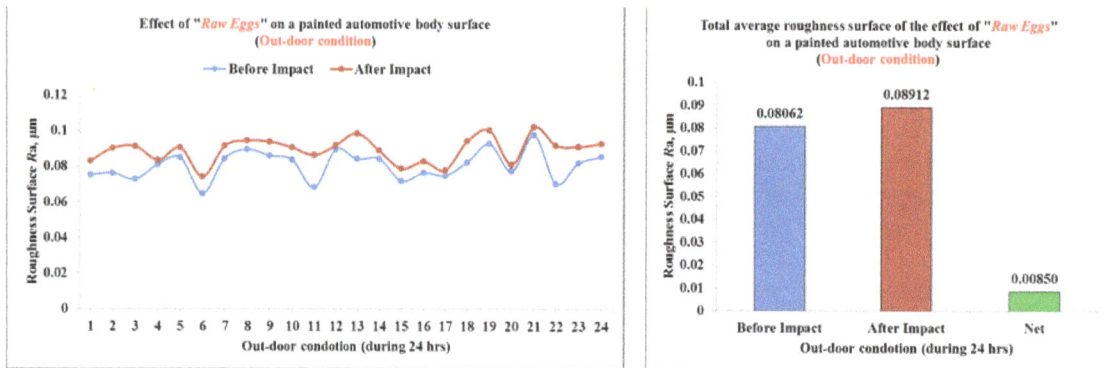

Figure 11. *(a) effect of raw eggs on a painted automotive body surface during 24 hrs (b) total average roughness with degraded net (Out-door condition)*

3.3. *Visual Evaluation*

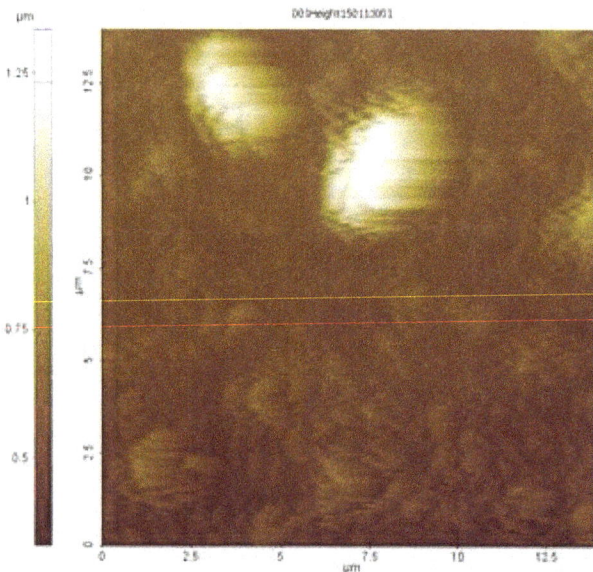

Figure 12. *typical 2-D AFM micrographics painted automotive body surface before impact*

Visual performance of an automotive body surface can be regarded as the main objective to follow its behavior up on exposure to outdoor conditions. To study the effect of biological material (i.e., raw eggs) on a painted automotive body surface, Atomic Force Microscopic (AFM) technique

was utilized. Fig. 12 shows the typical 2-D AFM micrographics of sample before impact. Also, Fig. 13 shows a typical 2-D AFM micrographics on painted automotive body surface after 1 hrs impact of raw eggs (out-door condition). Fig. 14 shows typical 2-D AFM micrographics on painted automotive body surface after 24 hrs impact of raw eggs (out-door condition).

The comparison between the AFM micrographics painted automotive body surface before impact Fig. 12 with the other samples Fig. 13 and Fig.14 shows that the biologically exposed samples have been roughened and this may possibly be attributed to biological degradation. More severe degradations were discerned for a sample exposed to 24 hrs impacts of raw eggs (out-door condition). However, no noticeable surface defects can be observed for the blank (painted automotive body surface before impact).

As can be seen from Fig. 13 and 14, although the surfaces have been exposed to varying short/long times to biological effect, this substance has etched the surfaces of both samples (after 1 hrs impact of raw eggs (out-door condition) and after 24 hrs impact of raw eggs (out-door condition)) catastrophically. It can be observed that the scales at which the effect of biological materials can be clearly distinguished are quite different in Figs. 13 and Figs 14. The observations made in these two figures, which depict the presence of cracks and roughened areas, as well as the formation of blisters and white stains on samples may explain the morphology changes observed in samples as a result of etching made by biological materials. These images may confirm that chemical reactions have occurred on the surface, leading to dissolved and etched areas. Gerbig and his colleagues in [38], originate the same results.

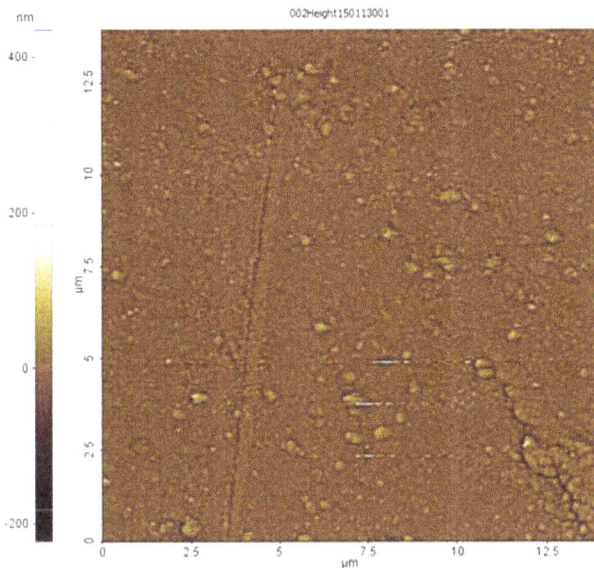

Figure 13. typical 2-D AFM micrographics on painted automotive body surface after 1 hrs impact of raw eggs (out-door condition)

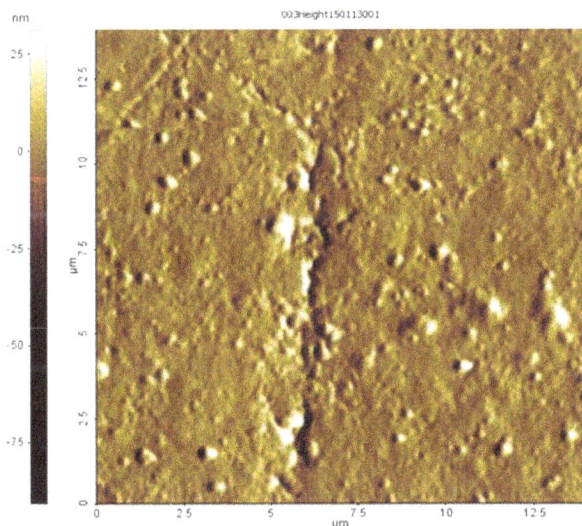

Figure 14. typical 2-D AFM micrographics on painted automotive body surface after 24 hrs impact of raw eggs (out-door condition)

4. Concluding Remarks

The effect of two types of aggressive biological materials on a painted automotive body surface, i.e., natural bird droppings and raw eggs were studied and further explained in more detail, which has a significant influence on economics and safety, especially for metals. Taking into consideration the weathering environments (e.g., UV radiation of sunlight, humidity, hot-cold out-door shocks, acid rain), results showed that mechanical, chemical and physical properties might be alert as a result of damage to the appearance and the surface roughness of a painted automotive body surface. According to the results of the present study, both mechanical and chemical

properties of the clear-coat are important parameters influencing the level of clear-coat resistance to natural bird droppings and raw eggs.

The general conclusions obtained are shown below:

- In all cases, $R_{a \text{ (bird droppings, out-door)}}$, represents the rougher structured samples and indeed the more degraded samples by $R_a \approx 12.7$ nm.
- There was no significant variation of degradation in any of the tests between natural bird droppings (in-door condition) and raw eggs (out-door condition), where, $R_{a \text{ (bird droppings, in-door)}} = 8.6$ nm $\approx R_{a \text{ (raw eggs, out-door)}} = 8.5$ nm.
- $R_{a \text{ (raw eggs, in-door)}}$, represents the smoothest structured samples and indeed the less degraded samples by $R_a \approx 5.3$ nm.
- As expected, R_a, fluctuated in nano-level during the first 24 hrs before and after the attack of specifically natural bird droppings and raw eggs on a painted automotive body surface due to surface structure, distribution of asperities, manufacturing process or environmental factors.
- As a function of time, it was found that the material structure defect formation before and after impact, results in conventional surface roughness, R_a, effects: $R_{a \text{ (bird droppings, out-door)}} > R_{a \text{ (bird droppings, in-door)}} > R_{a \text{ (raw eggs, out-door)}} > R_{a \text{ (raw eggs, in-door)}}$.

In conclusion, we believe that the new methodology that was used in this paper to measure the degradation on a painted automotive body surface has enormous potential in the field of micromechanics and nanotechnology.

References

[1] Ansdell, D.A., 10 - Automotive paints, in Paint and Surface Coatings (Second Edition), R. Lambourne and T.A. Strivens, Editors. 1999, Woodhead Publishing. p. 411-491.

[2] Yari, H., et al., Investigating the degradation resistance of silicone-acrylate containing automotive clearcoats exposed to bird droppings. Progress in Organic Coatings, 2012. 75(3): p. 170-177.

[3] Nichols, M.E., 20 - Paint Weathering Tests, in Handbook of Environmental Degradation of Materials (Second Edition), M. Kutz, Editor. 2012, William Andrew Publishing: Oxford. p. 597-619.

[4] Mohseni, M., B. Ramezanzadeh, and H. Yari, Effects of Environmental Conditions on Degradation of Automotive Coatings, New Trends and Developments in Automotive Industry. 1st ed, ed. P.M. Chiaberge. 2011, Croatia: In Tech Publisher. 394.

[5] Tahmassebi, N., et al., Effect of addition of hydrophobic nano silica on viscoelastic properties and scratch resistance of an acrylic/melamine automotive clearcoat. Tribology International, 2010. 43(3): p. 685-693.

[6] Ramezanzadeh, B., M. Mohseni, and H. Yari, On the electrochemical and structural behavior of biologically degraded automotive coatings; Part 1: Effect of natural and simulated bird droppings. Progress in Organic Coatings, 2011. 71(1): p. 19-31.

[7]   Ramezanzadeh, B., M. Mohseni, and H. Yari, The role of basecoat pigmentation on the biological resistance of an automotive clearcoat. Journal of Coatings Technology and Research, 2010. 7(6): p. 677-689.

[8]   Rezvani Moghaddam, A., et al., Studying the rheology, optical clarity and surface tension of an acrylic/melamine automotive clearcoat loaded with different additives. Progress in Organic Coatings, 2014. 77(1): p. 101-109.

[9]   Yari, H., M. Mohseni, and B. Ramezanzadeh, Comparisons of weathering performance of two automotive refinish coatings: A case study. Journal of Applied Polymer Science, 2009. 111(6): p. 2946-2956.

[10] Nguyen, T., et al., Relating laboratory and outdoor exposure of coatings: II. Journal of Coatings Technology, 2002. 74(932): p. 65-80.

[11] Schulz, U., et al., The effects of acid rain on the appearance of automotive paint systems studied outdoors and in a new artificial weathering test. Progress in Organic Coatings, 2000. 40(1–4): p. 151-165.

[12] Ramezanzadeh, B., et al., An evaluation of an automotive clear coat performance exposed to bird droppings under different testing approaches. Progress in Organic Coatings, 2009. 66(2): p. 149-160.

[13] Ramezanzadeh, B., et al., A study of thermal–mechanical properties of an automotive coating exposed to natural and simulated bird droppings. Journal of Thermal Analysis and Calorimetry, 2010. 102(1): p. 13-21.

[14] Ramezanzadeh, B., M. Mohseni, and N. Naseh, Effects of different silicon-based surface active additives on degradability of clearcoats exposed to bird droppings. Journal of Coatings Technology and Research, 2014. 11(4): p. 533-543.

[15] Ramezanzadeh, B., M. Mohseni, and H. Yari, The Effect of Natural Tree Gum and Environmental Condition on the Degradation of a Typical Automotive Clear Coat. Journal of Polymers and the Environment, 2010. 18(4): p. 545-557.

[16] Yari, H., et al., Use of analytical techniques to reveal the influence of chemical structure of clearcoat on its biological degradation caused by bird-droppings. Progress in Organic Coatings, 2009. 66(3): p. 281-290.

[17] Rabea, A.M., M. Mohseni, and S.M. Mirabedini, Investigating the antigraffiti properties of a polyurethane clearcoat containing a silicone polyacrylate additive. Journal of Coatings Technology and Research, 2011. 8(4): p. 497-503.

[18] Mori, K., et al., Mechanism of acid rain etching of acrylic–melamine coatings. Progress in Organic Coatings, 1999. 36(1–2): p. 34-38.

[19] Nguyen, T., et al., Relating laboratory and outdoor exposure of coatingsIII. Effect of relative humidity on moisture-enhanced photolysis of acrylic-melamine coatings. Polymer Degradation and Stability, 2002. 77(1): p. 1-16.

[20] Nguyen, T., J. Martin, and E. Byrd, Relating laboratory and outdoor exposure of coatings: IV. Mode and mechanism for hydrolytic degradation of acrylic-melamine coatings exposed to water vapor in the absence of UV light. Journal of Coatings Technology, 2003. 75(941): p. 37-50.

[21] Palm, M. and B. Carlsson, New accelerated weathering tests including acid rain. Journal of Coatings Technology, 2002.

74(924): p. 69-74.

[22] Stevani, C.V., et al., Mechanism of automotive clearcoat damage by dragonfly eggs investigated by surface enhanced Raman scattering. Polymer Degradation and Stability, 2000. 68(1): p. 61-66.

[23] Yari, H., M. Mohseni, and B. Ramezanzadeh, A mechanistic study of degradation of a typical automotive clearcoat caused by bird droppings. Journal of Coatings Technology and Research, 2011. 8(1): p. 83-95.

[24] Naseh, N., M. Mohseni, and B. Ramezanzadeh, Preparation of surface energy controlled automotive clearcoats loaded with functional silicon additives: Studying the resistance against tree gum attack. Journal of Industrial and Engineering Chemistry, 2014. 20(4): p. 1402-1410.

[25] Nagata, M., et al., Synthesis, characterization, and enzymatic degradation of network aliphatic copolyesters. Journal of Polymer Science Part A: Polymer Chemistry, 1999. 37(13): p. 2005-2011.

[26] Matheson, L.A., J.P. Santerre, and R.S. Labow, Changes in macrophage function and morphology due to biomedical polyurethane surfaces undergoing biodegradation. Journal of Cellular Physiology, 2004. 199(1): p. 8-19.

[27] Chapanian, R., et al., The role of oxidation and enzymatic hydrolysis on the in vivo degradation of trimethylene carbonate based photocrosslinkable elastomers. Biomaterials, 2009. 30(3): p. 295-306.

[28] Majumder, K. and J. Wu, Angiotensin I Converting Enzyme Inhibitory Peptides from Simulated in Vitro Gastrointestinal Digestion of Cooked Eggs. Journal of Agricultural and Food Chemistry, 2009. 57(2): p. 471-477.

[29] Mine, Y. and M.K. Roy, 4.46 - Egg Components for Heart Health: Promise and Progress for Cardiovascular Protective Functional Food Ingredient, in Comprehensive Biotechnology (Second Edition), M. Moo-Young, Editor. 2011, Academic Press: Burlington. p. 553-565.

[30] Kerver, J.M., Y. Park, and W.O. Song, The Role of Eggs in American Diets: Health Implications and Benefits, in Eggs and Health Promotion. 2008, Iowa State Press. p. 9-18.

[31] Stevani, C.V., et al., Automotive clearcoat damage due to oviposition of dragonflies. Journal of Applied Polymer Science, 2000. 75(13): p. 1632-1639.

[32] Dong, W.P., P.J. Sullivan, and K.J. Stout, Comprehensive study of parameters for characterizing three-dimensional surface topography I: Some inherent properties of parameter variation. Wear, 1992. 159(2): p. 161-171.

[33] Dong, W.P., P.J. Sullivan, and K.J. Stout, Comprehensive study of parameters for characterizing three-dimensional surface topography II: Statistical properties of parameter variation. Wear, 1993. 167(1): p. 9-21.

[34] Dong, W.P., P.J. Sullivan, and K.J. Stout, Comprehensive study of parameters for characterising three- dimensional surface topography: III: Parameters for characterising amplitude and some functional properties. Wear, 1994. 178(1–2): p. 29-43.

[35] Dong, W.P., P.J. Sullivan, and K.J. Stout, Comprehensive study of parameters for characterising three-dimensional surface topography: IV: Parameters for characterising spatial and hybrid properties. Wear, 1994. 178(1–2): p. 45-60.

[36] Thomas, T.R., Characterization of surface roughness. Precision Engineering, 1981. 3(2): p. 97-104.

[37] Braun, J.H., Titanium dioxide's contribution to the durability of paint films. Progress in Organic Coatings, 1987. 15(3): p. 249-260.

[38] Y. B. Gerbig, A.R. Phani, H. Haefke, Influence of nanoscale topography on the hydrophobicity of fluoro-based polymer thin films, Applied Surface Science 242 (2005) 251–255.

4

How the Drilling Fluids Can be Made More Efficient by Using Nanomaterials

Mortatha Saadoon Al-Yasiri[1, *], Waleed Tareq Al-Sallami[2]

[1]Chemical Engineering Department, College of Engineering, University of Baghdad, Baghdad, Iraq
[2]Department of Air conditioning& Refrigeration, Technical College, Mosul, Iraq

Email address:

mortathasaaadoon@gmail.com (M. S. Al-Yasiri), waleed_salammi90@yahoo.com (W. T. Al-Sallami)

Abstract: Drilling fluids serve many objectives in a drilling process, including the elimination of cuttings, lubricating and cooling the drill bits, supporting the stability of the hole and preventing the inflow-outflow of fluids between borehole and the formation. However, with increasing production from non-conventional reservoirs, the stability and effectiveness of traditional drilling fluids under high temperature and high pressure (HTHP) environment have become big concerns. Both water and oil based drilling fluids are likely to experience a number of deteriorations such as gelation, degradation of weighting materials and breakdown of polymeric additives under HTHP conditions. Recently, nanotechnology has shown a lot of promise in the oil and gas sectors, including nanoparticle-based drilling fluids. This paper aims to explore and assess the influence of various nanoparticles on the performance of drilling fluids to make the drilling operation smooth, cost effective and efficient. In order to achieve this aim, the article will begin by explaining the important role that drilling fluid plays during the drilling process with a historical review of drilling fluid industry development. Then, definitions, uses and types of drilling fluid will be demonstrated as well as, the additives that are appended in order to enhance drilling fluid performance. Moreover, the maturation of the oil production industry from unconventional wells will be discussed after which the limitations and degradation of the traditional drilling fluid will be cleared up. Finally, this essay will discuss the great potential of nanotechnology in solving drilling problems in addition to the technical and the economic benefits of using nanomaterials in drilling fluids before offering a brief conclusion.

Keywords: Drilling Fluids, Reservoir, Nanotechnology, HTHP Conditions, Nanoparticles, Nano-Fluid

1. Introduction

Every rotary drilling operation has three systems that work at the same time in boring hole: a rotating system which rotates the drill bit, a lifting system that raises and lowers the drill string into the hole, and a circulating system which performs the function of moving a fluid around from the drill stem, out of the drill bit and up again to the hole at the surface, this fluid is called drilling fluid (Van Dyke & Baker 1998). Drilling fluids are necessary for drilling success as they increase oil recovery and minimize the amount of time needed to achieve first oil (Nasser et al., 2013).

The drilling fluids in the drilling process can be considered the same as the blood in the human physical structure. The mud pump is the heart; the cuttings that are transferred from the borehole by drilling fluid represent the unwanted materials that are removed from the body by blood and the mud cleaning system works as the kidney and lungs.

Recent investigations have demonstrated that nano-fluids have engaging features for applications where heat transfer, gel formation, drag reduction, binding ability for sand consolidation, wettability alteration, and corrosive control is of interest.

Nano-fluids can be produced by adding nano-sized particles in low volumetric fractions to a fluid. The nanoparticles promote the fluid's rheological, mechanical, optical, and thermal characteristics. Fluids with nano-sized particles may provide the following supports:

(1) Nano-sized particles can have enhanced stability against sedimentation since surface forces easily balance the gravity force.

(2) Thermal, rheological, optical, electrical, mechanical and magnetic properties of nanoparticles, which depend significantly on size and shape, can be designed during manufacture and are often superior to the base material.

2. Historical Review of Drilling Fluid Development

It is useful to recognize the chronological succession of issues that contribute to the various developments in a drilling fluid industry before moving deeper into details. In the past, early drilling operations used water to remove the cuttings from the hole. This was reported in 1846 when Fauvelle drilled a well in France by using flushed water. Nevertheless, the role of water alone as a drilling fluid was only partly successful in removing cuttings and achieved limited drilling depth. Hence; the inability of water in this application promoted the researchers and oil companies to produce new fluids.

The real basis of drilling fluid science was started by the Chapman idea of using water and plastic material as drilling fluid which successfully formed a strong wall around the reservoir (Chapman, 1890). Moreover, the first development was produced in 1901, when a well was drilled by using a mixture of clay and water (Offshore technology report). As a result, in 1935, Harth (1935) introduced bentonite clay and this innovation takes the basis of current drilling fluids.

3. Drilling Fluid Definitions, Functions and Types

There is a wide range of drilling fluid definitions, depending on function, composition and complexity. The American Petroleum Institute (API) defines the drilling fluid as a circulating fluid employed to save any or all of the various responsibilities involved in drilling operations in rotary drilling (Fink 2011). The drilling fluid can also be defined as all compositions that used to remove the cuttings from a borehole (Apaleke et al., 2012). In addition, it can be defined as a complex fluid that consists of a multitude of additives (Shah et al., 2010). The form and the quantity of additives are based on the drilling technique and the formations of a reservoir (Ibid).

In drilling operations, drilling fluids are utilized to remove cuttings from a borehole and transport it to the surface, to stabilize and support wellbore, and to cool and lubricate the drill bit. In addition, drilling fluids play a role in suspending cuttings when not circulating and controlling formation pressure. Furthermore; drilling fluids protect the environment by preventing inflow-outflow of fluids between a borehole and the reservoir formation (Nabhani & Emami, 2012; Abdo & Haneef, 2012).

The challenges during drilling operations in the petroleum industry have contributed to the formulation of different types of drilling fluids. All the same, the primary ingredients of these fluids are water, oil and gas in addition to chemical additives (Apaleke et al., 2012). More often than not, the drilling fluids are classified as water-based mud, oil-based mud and gas-based mud. According to Shah et al. (2010) the most popular drilling mud used in drilling is water-based mud (WBM) while, oil-based mud (OBM) is usually used in

swelling shale formation. In WBM, particles are suspended in water or brine (Caenn et al., 2011). OBM is used in swelling shale formation because with WBM, the shale will absorb the water, as a consequence of this it expands and this expansion may cause stuck pipe (Shah et al., 2010). However, there are drilling special conditions under which a liquid drilling fluid is not a suitable circulating medium. Therefore; foam, air and gases may be employed in drilling some wells when these conditions exist (Ibid). In this situation, the drilling fluid type is called gas based mud (GBM).

4. Chemical Additives

Chemical additives are added to drilling fluids in order to enhance its performance by changing the properties and composition, particularly when circumstances need mud with special capabilities to optimize the oil production process. Several mud additives exist some performing more than one function (Awele, 2014).

The most common additives are: pH control to control the acidity and alkalinity of the fluids, bactericides to reduce the bacterial count and corrosion inhibitors to prevent corrosion and the formation of scale in drilling fluids. In addition, defoamers are used to reduce foaming action, emulsifier to make a mixture of two liquids and a filtrate loss to reduce water loss to the formation. Also, flocculants are used to settle out the solids, lubricants to reduce the friction coefficient, and lost circulation materials to plug the zone in the formation (Skalle 2010; Hawker 2001).

5. Limitations of Drilling Fluids and Challenges

There are still limitations during use these traditional drilling fluids in spite of the chemicals added to improve the drilling fluid performance. The main limitations of water based drilling fluids are: the ability of WBM to dissolve salts which may result in an unwanted jump in density. Moreover, the WBM is capable of interfering with the flow of gas and oil through porous media. Other limitations are the ability of WBM to promote the disintegration and dispersion of clays and the inability of WBM to drill through water sensitive shale. As well as the ability of WBM to corrode iron such as drill pipes, drill collars and drill bits (Mellot, 2008).

Just like water based mud, oil-based drilling fluids have limitations such as the fluids are very expensive from several aspects, as the constituents of this type of mud are very expensive and the high cost of treatment cuttings and disposal of it (Oakley et al., 1991). On the other hand, this type is not favourable to the surroundings because their disposal may result in the pollution of water bearing aquifers, pollution of lands, and the decimation of the coral reefs. Furthermore; this type of fluids is unsuitable for use in dry gas reservoirs (Apaleke et al., 2012).

Not only do WBM and OBM have limitations, but GBM is also likely to experience a number of limitations. The most common one is a high risk of explosion due a high pressure that may be generated because the phase of SBM is gas or foam. Besides, this type causes drilling string corrosion. As well the SBM cannot be used through water bearing formations because the cuttings will aggregate together in these formations, therefore it is impossible to carry out by air or gas (Apaleke et al., 2012).

Oil well drilling technology has evolved from vertical, horizontal to sub-sea and deep-sea wells. These specific drilling techniques require specialized drilling fluids to fulfil the objectives (Shah et al. 2010). The traditional drilling fluids are suitable for low and medium temperature and pressure conditions. Although oil based drilling fluids were used in high temperature and pressure because of their stability, but these fluids are likely to experience a number of deteriorations such as gelation, degradation of weighting materials and breakdown of polymeric additives under HTHP conditions (Oakley et al., 2000).

This debasement of the fluids in these conditions decreases drilling performance by deceleration of the rates of penetration and this creates intractable problems that lead to leaving behind most of the oil unrecovered (Nasser et al., 2013).

6. Drilling Fluids for HTHP Conditions

Drilling fluid should have appropriate high temperature transfer and flow properties. Besides, it must be friendly to the environment in order to perform the functions in an effective responsible (Gupta & Walker, 2007). Recently, these specifications have been achieved, with some limitations as mentioned before, by water-based and oil-based muds. Both have bentonite clay and some of the chemical additives (Shah et al., 2010).

These additives may improve density, decrease corrosion rate, change viscosity, and stop bacterial growth (Hawker, 2001). However, for deep-well drilling the temperatures and pressures can be very high and the heat transfer requirements on the drilling fluid impossible to meet (Oakley et al., 2000). In this situation, to design a drilling fluid has capability to work successfully, it is required to significantly enhance the fluid's thermal properties.

Nanotechnology offers light, strong and corrosion-resistant materials which is what the drilling fluid industry needs (Ragab & Noah, 2014). The application of nanoparticles in drilling fluids will enable the drilling engineers to adjust the drilling fluid rheology by modifying the composition, type, or size distribution of nanoparticles in drilling fluid to accommodate any special situation (Abdo & Haneef, 2012). Materials manufactured from nanoparticles are not like those prepared using their larger equivalents. Nanomaterials are stronger and more reactive than other materials. They also conduct heat efficiently (Singh & Ahmed, 2010). The reason behind that is the increased surface interaction. As, for given quantities of material, there are a higher number of particles as a result of their size reduction as well as there is more surface area to bear the heat (Shah et al., 2010).

7. Nanomaterials Based Drilling Fluid

The beginning of nanotechnology has revolutionized the science and engineering sectors due to its vast range of applications. The oil production industry like every other industry can take out massive benefits from nanotechnology (Abdo & Haneef, 2012). One of the most encouraging prospects is the use of nanoparticles in drilling muds so as to have a clear operational performance, stability, and suitability. These features make drilling fluids adopt well with a wide range of operating conditions by minor changes in composition and sizes (Ibid).

Amanullah and Al-Tahini (2009) define nanomaterials based drilling fluids as mud containing additives with particle sizes between 1 to 100 nanometres; also they classified the nano-fluids into simple and advanced nano-fluids based on the concentration of the nanoparticles in drilling fluids. Nanoparticles in drilling fluids can play a major role in fixing the most common problems during drilling like wellbore instability, lost circulation, pipe sticking, toxic gases, high torque and drag.

8. Prospective Performances

1. Wellbore Instability

It is known, each year, millions of dollars are spent due to wellbore instability problems which are happening from exposure of shale to drilling fluid (Nabhani & Emami, 2012). The drilling fluids that contain nanoparticles have the power to depreciate wellbore instability (Singh & Ahmed, 2010). The nanoparticles size is less than the pore throat sizes of rocks that lead to plug the pore throats (Ibid). According to Suri and Sharma (2004), the particle size should not be higher than one-third of the pore throat to build a bridge and plug the pores.

2. Lost Circulation

One of the most popular drilling problems is loss circulation (Nabhani & Emami, 2012). It is a partial or complete loss of the drilling fluid to the formation. This situation occurs due to naturally fractured, crevices and channels (Abdo & Haneef, 2012). The loss of circulation leads to increase the cost and time required for drilling to reach the target depth (Nabhani & Emami, 2012). The loss of circulation also causes loss of pressure control and increasing safety concerns (Ibid).

Therefore; a lot of time and effort has been spent to control the loss circulation through produced additives materials or muds. The use of micro and macro particles have shown limited success (Mostafavi et al., 2011). The utilization of nanoparticles led to reduce loss circulation by raising carrying capacity sufficiently to carry the cuttings efficiently and to maintain drilling fluid density and pressure over a wide range of operational conditions (Bicerano, 2009).

3. Pipe sticking

Sometimes the drill pipe stuck to the wall of the borehole due to cutting accumulation when drilling fluid circulation stops or because the filtrate loss in the wall of the wellbore

(Palaman &Bander, 2008). Pipe sticking has a significant impact on the drilling performance and well costs. Many parameters are affected on the pipe sticking which are dependent on drilling fluid rheology (Nabhani &Emami, 2012). Any change in any rheology property can cause pipe sticking (Ibid).

The nano-fluids can play a role in recovering the stuck pipe. Nanomaterials based drilling mud have the potential to decrease the sticking tendency of mud cakes by making a thin film covering the drill pipe that lead to cutting down the pipe sticking problem (Amanullah & Ashraf, 2009). Also, nano-fluids have excellent carrying capacity, thus reducing the pipe sticking by cleaning the wellbore from cuttings (Nabhani &Emami, 2012).

4. Reduction Torque and Drag

There is a noteworthy boost in torque and drag difficulties due to the clash between the drill string and the borehole. Micro and macro materials based drilling muds have limited power to overcome torque and drag problems (Wasan & Nikolove, 2003). On the other hand, the application of nanoparticles leads to a significant reduction of the friction between the pipe and the borehole (Donald & Frank, 2007). Nano-fluids have the potential to form slightly lubricating film in the wall pipe interface (Ibid).

5. Toxic Gases

Nanoparticles can be employed in drilling fluids to rid of toxic and corrosive gases, like hydrogen sulphide. This gas should take away from the drilling fluids in order to cut environmental contamination as well as to protect the health of drilling staff and to prevent corrosion of drilling equipment's (Singh & Ahmed, 2010). Sayyadnejad et al. (2008) have found that the addition of 14 to 25nm zinc oxide particles into drilling muds removes hydrogen sulphide completely while bulk zinc oxide remove only 2.5% and take more time.

9. Technical and Economic Benefits from Using Nanomaterials in Drilling Fluid

In addition to the enhance drilling fluid performance, the most engaging features of nanomaterial are their low cost (Abdo & Haneef, 2012). This is imputable to the amount of nanomaterials required for any utilization is, much less which results from the fact that nanoparticles have a very huge surface area to mass ratio that enhances the reactivity of nanomaterials (Shah et al., 2010).

Also, the using of nanoparticles in drilling fluids has technical and economic benefits (Abdo & Haneef, 2012). Technically, nanofluids are suitable to use in new oil production techniques and to overcome severe drilling operations (Nasser et al., 2013). Economically, the use of nanoparticles can affect through three aspects. The first one is the use of nanoparticles instead of expensive additives reduces the drilling fluids cost (Nabhani &Emami, 2012). Also, the use of nanofluids as drilling fluids was enhanced oil recovery by reaching deep challenge formations (Abdo &

Haneef, 2013). As well as, the non-productive time was shortened due to an elimination of troubles, thus saving huge costs (Abdo & Haneef, 2012).

10. Conclusion

To sum up, drilling fluids have been used to serve many purposes in the drilling process and it as blood in human bodies. However, there are problems with wellbore instability, lost circulation, pipe sticking, toxic gases and high torque; with continued using of these fluids with unconventional reservoirs. During the last decades, scientists and researchers discovered nanotechnology and nowadays there are attempts to apply this technology in the drilling process.

This paper has explained the drilling fluid functions, types and the purpose of adding additives. In addition, it has clarified the degradation of drilling fluids during high-temperature and high-pressure conditions.

From this article, it can be inferred that nanoparticles can enhance drilling fluids due to their stability of the rheological properties at high pressure and high temperature conditions. The nano drilling fluid can cause a revolution in oil and gas drilling industry because it can fulfil the specific needs of new drilling technologies and it can hit the target depth in less time.

One of the main limitations in this paper is that has ignored a mention of the sizes and the concentrations of nanoparticles that used in drilling fluid because the article's purpose is review of important role that nanoparticles did in drilling fluids rather than research. Likewise, each circumstance or problem would need to use nanomaterials with specific sizes and concentrations.

Future work could be borne out in the field of property measurements to establish a better comparison study. The cost feasibility of using nanoparticles in drilling fluids can also be research.

References

[1] Abdo, J. and Danish, M. (2010). Nanoparticles: Promising solution to overcome stern drilling problems. In Nanotech Conference and Exhibition, Anaheim, California.

[2] Abdo, J. and Haneef, M. (2012). Nano-Enhanced Drilling Fluids: Pioneering Approach to Overcome Uncompromising Drilling Problems. J. Energy Resour. Technol., 134(1), p.014501.

[3] Abdo, J. and Haneef, M. D. (2013). Clay nanoparticles modified drilling fluids for drilling of deep hydrocarbon wells. Applied Clay Science, 86, pp. 76-82.

[4] Amanullah, M., and Al-Tahini, A. M. (2009). Nano-technology-its significance in smart fluid development for oil and gas field application. In SPE Saudia Arabia Section Technical Symposium. Society of Petroleum Engineers.

[5] Apaleke, A. S., Al-Majed, A. A., and Hossain, M. E. (2012). Drilling Fluid: State of The Art and Future Trend. In North Africa Technical Conference and Exhibition. Society of Petroleum Engineers.

[6]   ASME Shale Shaker Committee. (2011). Drilling fluids processing handbook. Elsevier.

[7]   Awele, J. (2014). Investigation of additives on drilling mud performance with "tønder geothermal drilling" as a case study. Master. Aalborg University Esbjerg.

[8]   Bicerano, J. (2008). Drilling fluid, drill-in fluid, competition fluid, and work over fluid additive compositions containing thermoset nancomposite particles; and applications for fluid loss control and wellbore strengthening. U.S. Patent Application 12/178,785.

[9]   Caenn, R., Darley, H. C., and Gray, G. R. (2011). Composition and properties of drilling and completion fluids. Gulf Professional Publishing.

[10]   Chapman, M.T., 1890. U.S. Patent Records, U.S. Patent No. 443,069 (Dec. 16).

[11]   Darley, H. C., and Gray, G. R. (1988). Composition and properties of drilling and completion fluids. Gulf Professional Publishing.

[12]   Fauvelle, M. (1846). On a new method of boring for artesian springs. Journal of the Franklin Institute, 42(6), pp. 369-372.

[13]   Fink, J. (2011). Petroleum engineer's guide to oil field chemicals and fluids. Gulf Professional Publishing.

[14]   Gupta, D. V. S., and Pierce, R. G. (1998). A New Concept for On-the-Fly Hydration of Water-based Fracturing Fluids. In SPE Gas Technology Symposium. Society of Petroleum Engineers.

[15]   Harth, P.E. (1935). Application of mud-laden fluids to oil or gas wells. U.S. Patent No. 1,991,637 (Feb.19).

[16]   Health and Safety Laboratory (2000) Drilling Fluids Composition and use with the UK Offshore Drilling Industry, United Kingdom: Health and Safety Executive.

[17]   Jimenez, M. A., Genolet, L. C., Chavez, J. C., and Espin, D. (2003). U.S. Patent No. 6,579,832. Washington, DC: U.S. Patent and Trademark Office.

[18]   Mellot, J. (2008). Technical Improvements in Wells Drilled with a Pneumatic Fluid. In SPE paper 99162, presented at the SPE/IDAC drilling Conference, Miami, Florida, USA, February (pp. 21-23).

[19]   Mostafavi, V. Ferdous, M.Z. Hareland, G.Husein, M. Design and Application of Novel Nano Drilling Fluids to Mitigate Circulation Loss Problems During Oil Well Drilling Operations, J clean Technology, ISBN: 978-1-4398-8189-7, 2011.

[20]   Nabhani, N., & Emami, M. THE POTENIAL IMPACT OF NANOMATERIALS IN OIL DRILLING INDUSTRY.

[21]   Nasser, J., Jesil, A., Mohiuddin, T., Al Ruqeshi, M., Devi, G., & Mohataram, S. (2013). Experimental Investigation of Drilling Fluid Performance as Nanoparticles. World Journal of Nano Science and Engineering, 2013.

[22]   Oakley, D. J., James, S. G., and Cliffe, S. (1991). The Influence of Oil-Based Drilling Fluid Chemistry and Physical Properties on Oil Retained on Cuttings.Offshore Europe.

[23]   Oakley, D. J., Morton, K., Eunson, A., Gilmour, A., Pritchard, D., and Valentine, A. (2000). Innovative Drilling Fluid Design and Rigorous Pre-Well Planning Enable Success in an Extreme HTHP Well. In IADC/SPE Asia Pacific Drilling Technology. Society of Petroleum Engineers.

[24]   Palaman, A, N and Bander, D.A.A. (2008). Using Nanoparlides to Decrease Differential Pipe Stricking and Its Feasibility in Iranian Oil Fields, J.Oil and Gas Business.

[25]   Ragab A. M. Salem, and Noah A. (2014). Reduction of Formation Damage and Fluid Loss using Nano-sized Silica Drilling Fluids. Petroleum Technology Development Journal (2), PP. 75-88

[26]   Sayyadnejad, M. A., Ghaffarian, H. R., and Saeidi, M. (2008). Removal of hydrogen sulfide by zinc oxide nanoparticles in drilling fluid. International Journal of Environmental Science & Technology, 5(4), pp. 565-569.

[27]   Shah, S. N., Shanker, N. H. and Ogugbue, C. C. (2010). Future challenges of drilling fluids and their rheological measurements. In AADE fluids conference and exhibition, Houston, Texas.

[28]   Singh, S., Ahmed, R. and Growcock, F. (2010). Vital Role of Nanopolymers in Drilling and Stimulations Fluid Applications. In Paper SPE 130413 presented at the SPE Annual Technical Conference and Exhibition, Florence, Italy, pp. 19-22

[29]   Skalle, P. (2010). Drilling fluid engineering. Book boon.

[30]   Suri, A. and Sharma, M. M. (2004). Strategies for sizing particles in drilling and completion fluid. SPE Journal, 9(01), pp.13-23.

[31]   Van Dyke, K. (1998). Drilling Fluids, Mud Pumps, and Conditioning Equipment. University of Texas at Austin Petroleum.

[32]   Wasan, D. T., and Nikolov, A. D. (2003). Spreading of nanofluids on solids. Nature, 423(6936), pp. 156-159.

[33]   Wawrzos, Frank A., and Donald J. Weintritt. (2007). Drilling fluid lubricant and method of use." U.S. Patent Application 11/957,634.

5

A Review on Role of Nanofluids for Solar Energy Applications

Suresh Sagadevan

Department of Physics, Sree Sastha institute of Engineering and Technology, Chennai, India

Email address:

sureshsagadevan@gmail.com

Abstract: The sun is a nature source of renewable energy. Solar energy consumption is very important in the backdrop of global warming and decrease of carbon dioxide secretion. Solar energy has been explored through solar thermal exploitation, photovoltaic power invention, and so on. Solar thermal consumption is the most accepted utilization surrounded by them. In conservative solar thermal collectors, plates or tubes coated with a layer of selectively absorbing material are used to take up solar energy, and then energy is carried away by working fluids in the form of warm. This type of collector exhibits several shortcomings, such as restrictions on incident flux density and relatively high heat losses. The shortage of fossil fuels and environmental considerations motivated the researchers to use alternative energy source such as solar energy. Therefore, it is essential to improve the effectiveness and recital of the solar thermal systems. In addition, some reported works on the applications of nanofluids in thermal energy storage, solar cells, and solar stills are reviewed. Dispersing outline amounts of nanoparticles into common base-fluids has a significant impact on the optical as well as thermo-physical properties of the base-fluid. Enhancement of the solar irradiance assimilation capacity leads to a higher heat convey rate resulting in more capable heat transmit. Nanofluids are suspension of nanoparticles in base fluids, a new challenge for thermal sciences provided by nanotechnology. Nanofluids have unique features different from conventional solid-liquid mixtures in which mm or μm sized particles of metals and non-metals are dispersed. Due to their excellent characteristics, nanofluids find wide applications in enhancing heat transfer. The aim of this appraisal manuscript is the study of the nanofluids in solar Energy applications. In order to overcome these drawbacks, direct solar absorption collector has been used for solar thermal exploitation.

Keywords: Nanofluid, Nanoparticle, Solar Water Heater, Thermoelectric Cells, and Solar Thermal Energy

1. Introduction

Nanofluids are the novel splitting up of fluids engineered by dispersing nanometer-sized materials (Nanoparticles, nanofibers, nanotubes, nanowires, nanorods, nanosheet, or droplets) in base fluids. In other words, nanofluids are nothing but the nanoscale colloidal suspensions which contains the diluted nanomaterials. There are two-phase systems in which one corresponds to solid phase and the other is related to liquid phase. Nanofluids have been initiated to acquire the improved thermo physical properties such as thermal conductivity, thermal diffusivity, viscosity, and convective heat transfer coefficients which were compared to those of base fluids like oil or water. It has been established with great impending applications in many fields.

The idea behind development of nanofluids is to use them as thermo fluids in heat exchangers for enhancement of heat transfer coefficient and thus to minimize the size of heat transfer equipments. The important parameters which influence the heat transfer characteristics of nanofluids are its properties which include thermal conductivity, viscosity, specific heat and density. The thermo physical properties of nanofluids also depend on operating temperature of nanofluids. Hence, the accurate measurement of temperature dependent properties of nanofluids is essential. Common fluids such as water, ethylene glycol, and heat transfer oil plays a momentous position in many manufacturing processes such as power generation, heating or cooling processes, chemical processes, and microelectronics. Though, these fluids have comparatively stumpy thermal conductivity and thus cannot reach elevated heat substitute rates in thermal engineering devices. An approach in the direction of rise above over this impenetrability in using ultra-fine solid particles balanced in frequent fluids to advance their thermal conductivity. The suspension of nano-sized particles (1–100 nm) in a conventional base fluid is called a nanofluid. Choi first used the term "nanofluid" in 1995 [1]. Nanofluids, compared to suspensions with particles of millimeter-or-micrometer size, show better stability, rheological properties, and considerably higher thermal conductivities. Researchers have in addition applied a selection of research methods, characteristics, and dissimilar models used for the computation of thermo physical properties of nanofluids (i.e., thermal conductivity, viscosity, density, specific heat capacity) [2–9]. Various investigators have also summarized the possessions of nanofluids on stream and heat transfer in usual and compulsory convection in different systems [10–13]. The improved thermal performance of nanofluids which can supply a starting point massive modernization of heat transfer intensification, which is of foremost significance to a measure of industrial sectors together with transportation, power generation, micro manufacturing, thermal therapy for cancer treatment, chemical and metallurgical sectors, as well as heating, cooling, ventilation and air-conditioning. Nanofluids are also important for the production of nanostructured materials for the manufacture of complex fluids as well as for clean-up oil from surfaces due to their exceptional wetting and spreading conduct [14]. An additional significance of the nanofluid stream is in the release of nano-drugs as suggested by Kleinstreuer et al. [15]. In recent years, the utilization of solar energy have amazing border. The apparent deficiency of fossil fuels as well as ecological considerations will confine the use of fossil fuels in the future. Therefore, researchers are motivated to find substitute sources of energy. This has been a turn out to be still an extra trendy as the cost of fossil fuels continues to ascend. Most solar energy applications are economically feasible whereas tiny system for entity use just a few kilowatts of power [16, 17]. It is appropriate to use solar energy to an extensive variety of applications and make available solutions through the alteration of the energy fraction, improving energy firmness, growing energy sustainability, and enhancing system effectiveness [18]. This paper presents a review of the application of nanofluids in solar energy applications. Therefore, this appraisal mainly investigates the special effects of nanofluids on the capability development of solar collectors as well as on profitable and environmental considerations concerning the handling of these systems. An additional application of nanofluids in thermal energy storage, solar cells, and solar stills are also reviewed.

2. Nanoparticles and Nanofluids Properties

Nanoparticles are sized between 1 and 100 nanometers. Nanoparticles may or may not exhibit size related properties that differ significantly from those observed in fine particles or bulk materials. Nanoclusters have at least one dimension between 1 and 10 nanometers and a narrow size distribution. Nanopowders are agglomerates of ultra-fine particles, nanoparticles, or Nanoclusters. Nanometer-sized single crystals, or single-domain ultra-fine particles, are often referred to as nanocrystals. Nanoparticles are of great scientific

interest as they effectively form a bridge between bulk materials and atomic or molecular structures. A bulk material should have constant physical property regardless of its size, but at the nano-scale size- dependent properties are often observed. Thus, the properties of materials change as their size approaches the nanoscale and as the percentage of atoms at the surface of a material becomes significant. Suspensions of nanoparticles are possible since the interaction of the particles surface with the solvent is strong enough to overcome density differences, which otherwise usually result in a material either sinking or floating in a liquid. Nanoparticles also often possess unexpected optical properties as they are small enough to confine their electrons and produce quantum effects. For example, gold nanoparticles appear deep red to black in solution. Nanoparticles have a very high surface area to volume ratio, which provides a tremendous driving force for diffusion, especially at elevated temperatures. Sintering can take place at lower temperature, over shorter time scales than for larger particles. The fluids with nanosized solid particles suspended in them are called "nanofluids." The suspended metallic or nonmetallic nanoparticles change the transport properties and heat transfer characteristics of the base fluid. Heat transfer enhancement in solar devices is one of the key issues of energy saving and compact designs. Solar energy is widely used in applications such as electricity generation, chemical processing, and thermal heating due to its renewable and nonpolluting nature. Most solar water heating systems have two main parts: a solar collector and a storage tank. The most common collector is called flat-plate collector but this suffer from relatively low efficiency. There are so many methods introduced to increase the efficiency of the solar water heater. But the novel approach is to introduce the nanofluids in solar collector instead of conventional heat transfer fluids (like water).

3. Preparation Methods for Nanofluids

3.1. Two-Step Method

Two-step technique is the mass extensively used technique for preparing nanofluids. Nanoparticles, nanofibers, nanotubes, or other nanomaterials used in this method. Initial method produced were dry powders by chemical or physical methods. Then, the nanosized powder will be detached into a fluid in the second processing step with the assist of exhaustive magnetic force agitation, ultrasonic agitation, high-shear mixing, homogenizing, and ball milling. Two-step method is the majority of economic technique to fabricate nanofluids in outsized scale, because nanopowder synthesis techniques have been already scaled up to industrial fabrication levels. Due to the high surface region and surface motion, Nanoparticles have the affinity to aggregate. The important technique to enhance the stability of Nanoparticles in fluids is in the use of surfactants. However, the functionality of the surfactants under high temperature is also a big concern, especially for high-temperature applications. Due to the difficulty in preparing stable nanofluids by two-step method, several advanced techniques are developed to produce nanofluids, including one-step method.

3.2. One-Step Method

To reduce the agglomeration of nanoparticles developed a one-step physical vapor condensation method to prepare Cu/ethylene glycol nanofluids [19]. The one-step process consists of simultaneously making and dispersing the particles in the fluid. In this method, the processes of drying, storage, transportation, and dispersion of Nanoparticles are avoided, so the agglomeration of Nanoparticles is minimized, and the stability of fluid is increased [20]. The one-step processes can be prepared uniformly dispersed Nanoparticles, and the particles can be stably suspended in the base fluid. The vacuum-SANSS (submerged arc nanoparticle synthesis system) is the capable technique to prepare nanofluids by means of dissimilar dielectric liquids [21, 22]. The different morphologies are mostly subjective and indomitable by a range of thermal conductivity properties of the dielectric liquids. The Nanoparticles prepared by this method exhibits needle-like, polygonal, square, and circular morphological shapes. This method avoids the undesired particle aggregation comparatively well. One-step physical method cannot manufacture nanofluids in large scale and the cost is also high, so the one-step chemical method is mounted rapidly. Zhu et al. offered a novel one-step chemical technique for preparing copper nanofluids by dropping $CuSO_4 \cdot 5H_2O$ with $NaH_2PO_2 \cdot H_2O$ in ethylene glycol under microwave irradiation [23]. Well-dispersed and firmly balanced copper nanofluids were obtained. Mineral oil-based nanofluids containing silver Nanoparticles with a narrow-size distribution were also equipped by this method [24]. The particles might be stabilized by Korantin, which is synchronized to the silver particle surfaces via two oxygen atoms forming a dense layer around the particles. The silver nanoparticle suspensions were stable for about 1 month. Stable ethanol-based nanofluids containing silver Nanoparticles can be prepared by microwave-assisted one-step method [25]. In this method, polyvinylpyrrolidone (PVP) was employed as the stabilizer of colloidal silver and dipping agent for silver in solution. The cationic surfactant octadecylamine (ODA) is in addition of capable phase-transfer agent to produce silver colloids [26]. The phase shift of the silver nanoparticles arises owing to pairing of the silver nanoparticles with the ODA molecules in presence in organic phase through each coordination bond configuration or weak covalent interaction. Phase transfer method has been developed for preparing homogeneous and stable graphene oxide colloids. Graphene oxide nanosheets (GONs) were effectively transferred from water to n-octane following the alteration by oleylamine, and the schematic illustration of the phase transmit process is shown in Fig. 1 [27]. On the other hand, there are several disadvantages for one-step method. The most significant role is that the residual reactants are left in the nanofluids due to unfinished reaction or stabilization. It is complex to make clear the nanoparticle effect without eliminating this contamination effect.

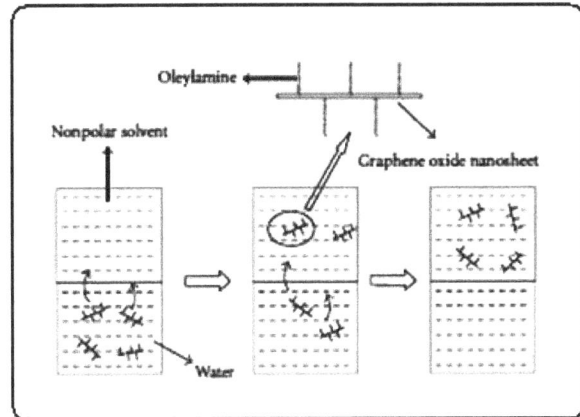

Fig 1. Schematic illustration of the phase transfer process[27]

3.3. Other Novel Methods

Wei et al. developed a continuous flow micro fluidic micro reactor to synthesize copper nanofluids. By this method, copper nanofluids can be continuously synthesized, and their microstructure and properties can be varied by adjusting parameters such as reactant concentration, flow rate, and additive. CuO nanofluids with high solid volume fraction (up to 10 vol %) can be synthesized through a novel precursor transformation method with the help of ultrasonic and microwave irradiation [28]. The precursor Cu $(OH)_2$ is completely transformed to CuO nanoparticle in water under microwave irradiation. The ammonium citrate prevents the growth and aggregation of nanoparticles, resulting in a stable CuO aqueous nanofluid with higher thermal conductivity than those prepared by other dispersing methods. Phase-transfer method is also a facile way to obtain monodisperse noble metal colloids [29]. In a water cyclohexane two-phase system, aqueous formaldehyde is transferred to cyclohexane phase via reaction with dodecylamine to form reductive intermediates in cyclohexane. The intermediates are capable of reducing silver or gold ions in aqueous solution to form dodecylamine-protected silver and gold nanoparticles in cyclohexane solution at room temperature. Feng et al. used the aqueous organic phase transfer method for preparing gold, silver, and platinum nanoparticles on the basis of the decrease of the PVP's solubility in water with the temperature increase [30]. Phase transfer method is also applied for preparing stable kerosene based Fe_3O_4 nanofluids. Oleic acid is successfully grafted onto the surface of Fe_3O_4 nanoparticles by chemisorbed mode, which lets Fe_3O_4 nanoparticles have good compatibility with kerosene [31]. The Fe_3O_4 nanofluids prepared by phase-transfer method do not show the previously reported "time dependence of the thermal conductivity characteristic". The preparation of nanofluids with controllable microstructure is one of the key issues. It is well known that the properties of nanofluids strongly depend on the structure and shape of nanomaterials. The recent research shows that nanofluids synthesized by chemical solution method have both higher conductivity enhancement and better stability than those produced by the other methods [32]. This method is distinguished from the others by its controllability. The nanofluid microstructure can be varied and manipulated by adjusting synthesis parameters such as temperature, acidity, ultrasonic and microwave irradiation, types and concentrations of reactants and additives, and the order in which the additives are added to the solution.

4. Nanofluids in Solar Energy Applications

Initially, the application of nanofluids in collectors and water heaters are investigated for the efficiency, economic, and environmental aspects. Some studies conducted on thermal conductivity and optical properties of nanofluids are also briefly reviewed, because these parameters can determine the capability of nanofluids to enhance the performance of solar systems.

4.1. Collectors and Solar Water Heaters

A Nanofluid poses the following advantages as compared to conventional fluids which make them suitable for use in solar collectors: Absorption of solar energy will be maximized with change of the size, shape, material and volume fraction of the nanoparticles. The suspended nanoparticles increase the surface area and the heat capacity of the fluid due to the very small particle size. The suspended nanoparticles enhance the thermal conductivity which results in improvement in efficiency of heat transfer systems. Properties of the fluid can be changed by varying concentration of nanoparticles. Extremely small size of nanoparticles ideally allows them to pass through pumps. The fundamental difference between the conventional and nanofluids-based collector lies in the mode of heating of the working fluid. In the former case the sunlight is absorbed by a surface, where as in the later the sunlight is directly absorbed by the working fluid (through radiative transfer). On reaching the receiver the solar radiations transfer energy to the nanofluids via scattering and absorption. The nanofluid based solar water heater and collector are shown in Fig.2.

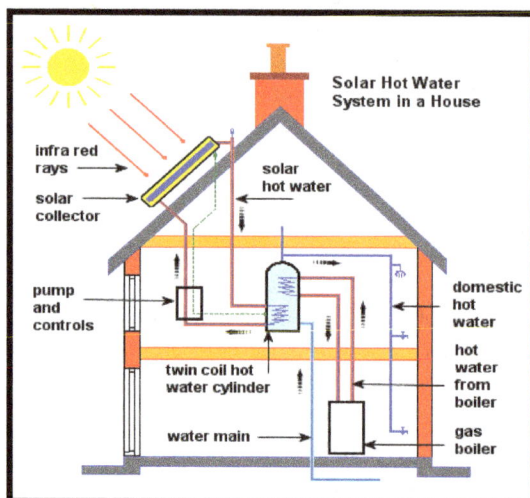

Fig 2. The nanofluid based solar water heater and collector

These devices absorb the incoming solar radiation, convert it into heat, and transfer the heat to a fluid (usually air, water, or oil) flowing through the collector. The energy collected is carried from the working fluid, either directly to the hot water or space conditioning equipment or to a thermal energy storage tank, from which it can be drawn for use at night or on cloudy days [33]. Solar water heaters are the most popular devices in the field of solar energy. Solar collector is used to collect the solar energy and transfers the collected solar energy to a fluid passing in contact with it, so it is always a matter of investigation to know that how efficiently solar collectors are converting solar energy into thermal energy. Solar collectors are classified as: Non-Concentrating or flat plate type solar collectors and Concentrating Solar Collectors. Flat plate collectors are very simple in construction and are mostly used as household equipment as water heater, air heater, and solar cooker whereas; concentrating solar collectors are mostly used for power generation, heating up water with higher mass flow rate. Concentrating Solar Collectors are more efficient then flat plate collectors, on the other hand they require a tracking system and require higher installation cost compared to flat plate collectors. The performance of solar collector depends upon the physical properties of the fluid flowing through it. It has been found that the conventional fluids used in solar collectors suffer from poor thermal properties. A new class of working fluids called "Nanofluid" can be used instead of conventional fluids, which have the improved thermal properties and thereby increase the thermal performance of the solar collector. Nonomaterial have unique mechanical, optical, electrical, magnetic and thermal properties in which average size of the nanoparticles is below 100 nm. A very small amount of nanoparticles when dispersed in any host fluids (e.g. water, oil, ethylene glycol) can improve the thermal properties of fluids dramatically. Commonly used material making nanofluids are as: oxide ceramics (Al_2O_3, CuO), nitride ceramics (AlN, SiN), carbide ceramics (SiC, TiC), metals (Cu, Ag, Au), semiconductors (TiO2, SiC), carbon nanotubes, composite materials ($Al_{70}Cu_{30}$). Nanoparticles can be manufactured by mainly two processes; those are Physical Processes and Chemical Processes. Physical Processes include Inert Gas Condensation (IGC) and mechanical grinding whereas Chemical Processes include Chemical Vapor Deposition (CVD), Chemical precipitation and micro emulsion. For making Nanofluids, nanoparticles are suspended in conventional heat transfer fluids by two methods called single step method and two step method. As described in section.3 above in single step method making and dispersion of nanoparticle happens simultaneously where as in two step method first nanoparticles are fabricated and then nanoparticles are dispersed into the base fluids.

4.2. Solar Stills

A lot of research is carried out on solar stills and different methods are invented to improve their efficiency. In recent times, Gnanadason et al. [34] stated that solar stills efficiency can be increased by using nanofluids. Their results showed that the efficiency is improved by 50% with addition of nanofluids. However, it was not clear that the exact amount of nanofluid added to the water for the solar still. As cost of nanofluid is so high, so the economic capability should be considered. It is also suggested that solar stills efficiency can be improved by adding dyes in fluid [35] have reported that solar stills efficiency is increased by 29% by adding violet dye to water, which is remarkable. It is clear that nanofluids are more expensive than dyes. Hence, it is a challenge to use nanofluids in solar stills. It can be used to minimize the production of greenhouse gas emissions from the production of fresh water. The solar stills illustrated in Fig. 3

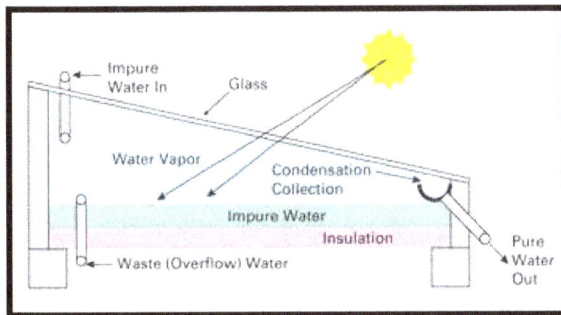

Fig 3. Nanofluids based Solar Stills

4.3. Solar Cells

Efficiency of solar cell can be improved by cooling solar cells (illustrated in Fig.4). Elmir et al. [36] numerically simulated the cooling a solar cell by forced convection in the presence of a nanofluid. The Al_2O_3 and water nanofluid was used for analysis purpose. The thermal conductivity and viscosity of the nanofluid are calculated using the models of Brinkman and Wasp respectively. It has been concluded that the average Nusselt number increase by use of nanofluids which leads to the improving the rate of cooling. But in whole analyses the thermal conductivity and viscosity of nanofluid was not considered.

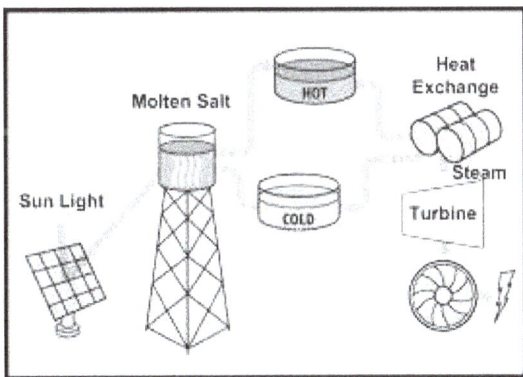

Fig 4. Schematic diagram for solar cells

4.4. Solar PANELS

Nanofluids are a simple product of the emerging world of nanotechnology. Suspensions of nanoparticles (nominally 1 to 100nm in size) dispersed in fluids such as water, oils,

glycols and even air and other gases can rightly be called nanofluids. The first decade of nanofluid research was primarily focused on measuring and modeling fundamental thermo physical properties of nanofluids (thermal conductivity, density, viscosity, heat transfer coefficient). Due to its renewable and non-polluting nature, solar energy is often used in applications such as electricity generation, thermal heating, and chemical processing. Solar power plants with surface receivers have low overall energy conversion efficiencies due to large emissive losses at high temperatures. Nanofluids have recently found relevance in applications requiring quick and effective heat transfer such as industrial applications, cooling of microchips, microscopic fluidic applications, etc. The normal efficiency of the solar panels being used is recently found out to be 44.7% and the use of nanofluids it can be increased by 10-15%, also by Plasmonic Nanofluids by 20%.

4.5. Thermal Energy Storage

Conventional solar thermal energy storage system needs the storage medium to have high thermal conductivity and heat capacity. But, very few materials are available with such properties and can be used in high temperatures. In recent times, Shin and Banerjee [37] stated that the abnormal improvement of specific heat capacity can be possible for high-temperature nanofluids than conventional one. It has been found that the specific heat capacity of the nanofluid increase by 14.5% when Alkali metal chloride salt eutectic is doped with silica nanoparticles at 1% mass concentration . So this can be appropriate for the use in solar thermal energy storage system. Paraffin is the suitable because of its advantageous characteristics, since it has high latent heat capacity with minor super cooling and also low cost. Phase Change Material can also use for solar energy storage. Wua et al. [38] numerically simulate the thermal energy storage behavior of Cu/paraffin nanofluids PCMs. The results showed that with 1 wt.% Cu/paraffin, the melting time can be saved by 13.1%. Hence, it has been concluded that the use of nanoparticles is an efficient method to improve the heat transfer rate in latent heat thermal energy storage system as illustrated in Fig.5.

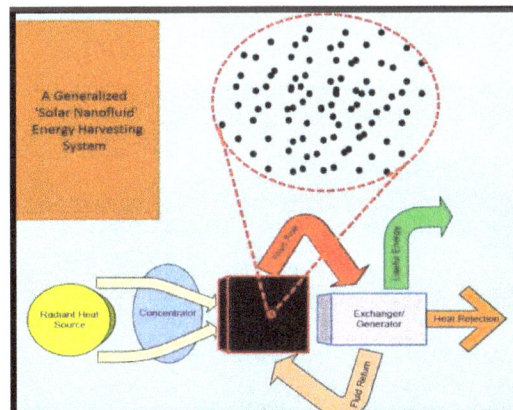

Fig 5. Schematic diagram of Thermal energy storage

4.6. Photovoltaic/Thermal Systems

A photovoltaic/thermal (PV/T) system is a hybrid structure that converts part of the solar radiation to electricity and part to thermal energy [39]. One can investigate experimentally the effects of using different nanofluids on the cooling rate, and, hence, the efficiency of the PV/T systems. The effects of different volume fractions, nanoparticle size on the efficiency of the system can be studied. A review of the literature shows that many researches have been carried out on the potential of nanofluids for cooling of different thermal systems such as electronic devices [40-42], automobile radiator [43], and micro channel heat sinks [44]. Therefore, using nanofluids to cool the PV/T system may be reasonable (illustrated in Fig.6).

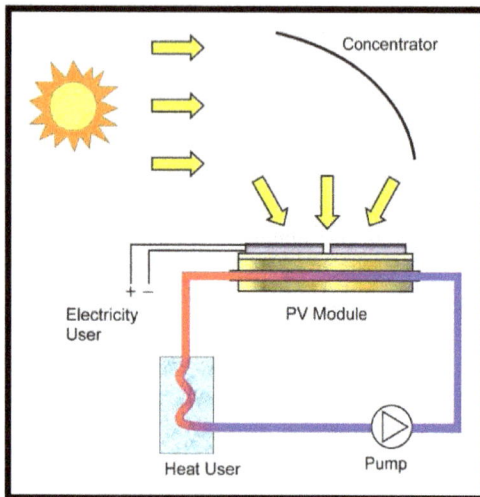

Fig 6. Schematic diagram of Photovoltaic/thermal systems

4.7. Solar Thermoelectric Cells

In recent years, interest in the development of solar thermoelectric systems has considerably increased [45]. The thermoelectric cells can be used to convert the solar energy to electricity due to the temperature difference between two hot and cold surfaces. A greater temperature difference between the hot and cold surfaces of the thermoelectric cell leads to a bigger electricity production. In this way the effects of different nanofluids with various mass flow rates on the efficiency of the solar thermoelectric cell can be studied (illustrated in Fig.7).

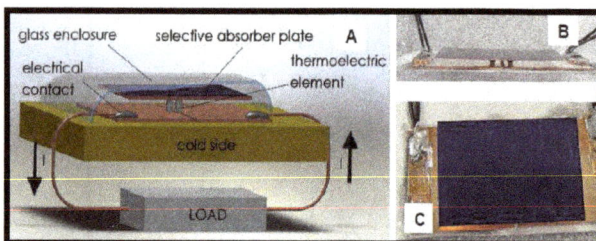

Fig 7. Nanofluids based solar thermoelectric cells

4.8. Thermal Conductivity of Nanofluids

Thermal properties of nanofluids very much depend on

size as well as volume fraction of nanoparticles in the base fluid. The heat transfer characteristics of nanofluids also depend on the sonication time of nanoparticles and the quality of sonication used. So, as stated the higher value of thermal conductivity and thermo physical characteristics e.g. density, viscosity etc. enhance the heat absorption capacity, the specific heat capacity and heat carrying properties of nanofluids. It can be observed that same as the temperature difference collector efficiency of the nanofluid increases at lower volume fraction. Higher the volume concentration, higher will be the density, viscosity as well as heat capacity. It is known that thermal conductivity of solids is greater than liquids. Commonly used fluids in heat transfer applications such as water and ethylene have low conductivity when compared to thermal conductivity of solids especially metals. So the addition of solid particles in a fluid can increase the conductivity of liquids. When nanofluids are used as working fluids of the direct solar absorbers, the thermal properties of nanofluids are critical to the solar utilization. Photo thermal property is very important to the assessment of solar energy absorption of nanofluids because it directly reflects the solar absorption ability of nanofluids. Viscosity and rheological behaviors not only are essential parameters for nanofluid stability and flow behaviors but also affect the heat transfer efficiency of direct solar absorbers. Thermal conductivity is an important parameter for heat transfer fluids. It also affects the collectors' heat transfer efficiency. Great efforts have been made to the rheological behaviors and thermal conductivities of nanofluids, and these studies are helpful to research on nanofluids as solar absorption working fluids.

4.9. Electronic Applications

Due to higher density of chips, design of electronic components with more compact makes heat dissipation more difficult. Advanced electronic devices face thermal management challenges from the high level of heat generation and the reduction of available surface area for heat removal. Therefore, the reliable thermal management system is vital for the smooth operation of the advanced electronic devices. In general, there are two approaches to improve the heat removal for electronic equipment. One is to find an optimum geometry of cooling devices; another is to increase the heat transfer capacity. Nanofluids with higher thermal conductivities are predicted convective heat transfer coefficients compared to those of base fluids. Recent researches illustrated that nanofluids could increase the heat transfer coefficient by increasing the thermal conductivity of a coolant. Jang and Choi designed a new cooler, combined microchannel heat sink with nanofluids [46]. Higher cooling performance was obtained when compared to the device using pure water as working medium. Nanofluids reduced both the thermal resistance and the temperature difference between the heated microchannel wall and the coolant. A combined microchannel heat sink with nanofluids had the potential as the next-generation cooling devices for removing ultrahigh heat flux. Nguyen et al. designed a closed liquid-circuit to investigate the heat transfer enhancement of a liquid cooling system by replacing the base fluid (distilled water) with a nanofluid composed of distilled water and Al_2O_3 nanoparticles at various concentrations [47]. Silicon microchannel heat sink performance using nanofluids containing Cu nanoparticles was analyzed [48]. It was found that nanofluids could enhance the performance as compared to the use of pure water as the coolant. The enhancement was due to the increase in thermal conductivity of coolant and the nanoparticle thermal dispersion effect. The other advantage was that there was no extra pressure drop, since the nanoparticle was small, and particle volume fraction was low.

4.10. Energy Storage

The temporal difference of energy source and energy needs made necessary the development of storage system. The storage of thermal energy in the form of sensible and latent heat has become an important aspect of energy management with the emphasis on efficient use and conservation of the waste heat and solar energy in industry and buildings [49]. Latent heat storage is one of the most efficient ways of storing thermal energy. Wu et al. evaluated the potential of Al_2O_3-H_2O nanofluids as a new phase change material (PCM) for the thermal energy storage of cooling systems. The thermal response test showed the addition of Al_2O_3 nanoparticles remarkably decreased the super cooling degree of water, advanced the beginning freezing time, and reduced the total freezing time. Only adding 0.2 wt% Al_2O_3 nanoparticles, the total freezing time of Al_2O_3-H_2O nanofluids could be reduced by 20.5%. Liu et al prepared a new sort of nanofluid phase change materials (PCMs) by suspending small amount of TiO_2 nanoparticles in saturated $BaCl_2$ aqueous solution [50]. The nanofluids PCMs possessed remarkably high thermal conductivities compared to the base material. The cold storage/supply rate capacity increased greatly than those of $BaCl_2$ aqueous solution without adding nanoparticles. The higher thermal performances of nanofluids PCMs indicate that they have a potential for substituting conventional PCMs in cold storage applications. Copper nanoparticles are efficient additives to improve the heating and cooling rates of PCMs [51]. For composites with 1 wt% copper nanoparticle, the heating and cooling times could be reduced by 30.3 and 28.2%, respectively. The latent heats and phase-change temperatures changed very little after 100 thermal cycles.

4.11. Solar Absorption

Solar energy is one of the best sources of renewable energy with minimal environmental impact. The conventional direct absorption solar collector is a well-established technology, and it has been proposed for a variety of applications such as water heating; however, the efficiency of these collectors is limited by the absorption properties of the working fluid, which is very poor for typical fluids used in solar collectors. Recently, this technology has been combined with the emerging technologies of nanofluids and liquid-nanoparticle suspensions to create a new class of

nanofluid-based solar collectors. Otanicar et al. reported the experimental results on solar collectors based on nanofluids made from a variety of nanoparticles (CNTs, graphite, and silver) [52]. The efficiency improvement was up to 5% in solar thermal collectors by utilizing nanofluids as the absorption media. In addition, the experimental data has been with a numerical model of a solar collector with direct absorption nanofluids. The experimental and numerical results demonstrated an initial rapid increase in efficiency with volume fraction, followed by a leveling off in efficiency as volume fraction continues to increase. Theoretical investigation on the feasibility of using a non-concentrating direct absorption solar collector showed that the presence of nanoparticles increased the absorption of incident radiation by more than nine times over that of pure water [53]. Under the similar operating conditions, the efficiency of an absorption solar collector using nanofluid as the working fluid was found to be up to 10% higher (on an absolute basis) than that of a flat-plate collector. Otanicar and Golden evaluated the overall economic and environmental impacts of the technology in contrast with conventional solar collectors using the life-cycle assessment methodology [54]. Sani et al investigated the optical and thermal properties of nanofluids consisting of aqueous suspensions of single-wall carbon nanohorns [55]. The observed nanoparticle-induced differences in optical properties appeared promising, leading to a considerably higher sunlight absorption. Both these effects, together with the possible chemical functionalization of carbon nanohorns, make this new kind of nanofluids very interesting for increasing the overall efficiency of the sunlight exploiting device.

5. Conclusion

Solar energy is one of the cleaner forms of renewable energy resources. The conventional solar collector is a well established technology which has various applications such as water heating, space heating and cooling. However, the thermal efficiency of these collectors is limited by the absorption properties of the working fluid, which is very poor for typical conventional solar flat plate collector. Recently, usage of nanofluids, which is basically liquid-nanoparticle colloidal dispersion as a working fluid has been found to enhance the thermal efficiency of solar flat plate collector by 30 percent. Nanofluids are advanced fluids containing nano-sized particles that have emerged during the last two decades. Nanofluids are used to improve system performance in many solar energy applications. This paper presents an overview of the recent developments in the study of nanofluids, including the preparation methods. From the economic and environmental point of view, the review showed that using nanofluids in collectors leads to a reduction in CO_2 emissions and annual electricity and fuel savings. Many researchers have reported, works on the applications of nanofluids in solar cells, solar thermal energy storage, and solar stills are also reviewed. Hence it will be a promising effort to develop research projects on the use of nanofluids in different solar systems such as solar absorption, solar thermoelectric cells, and thermal conductivity of nanofluids, electronic applications, energy storage and solar absorption.

Acknowledgement

The author thanks the Management and Principal of Sree Sastha Institute of Engineering and Technology, Chembarambakkam, Chennai-600123 for their encouragements throughout this work.

References

[1] U.S. Choi, *ASMEFED*. 231 (1995) 99–103.

[2] Y. Li, J. Zhou, S. Tung, E. Schneider, S. Xi, *Powder Technol.* 196 (2009) 89–101.

[3] J.H. Lee, S.H. Lee, C.J. Choi, S.P. Jang, S.U.S. Choi, *Int. J. Micro–Nano Scale Transport.* 1 (2010) 269–322.

[4] A. Ghadimi, R. Saidur, H.S.C. Metselaar, *Int. J. Heat Mass Transfer.* 54 (2011) 4051–4068.

[5] G. Ramesh, N.K. Prabhu, *Nanoscale Res. Lett.* 6 (2011) 334.

[6] K. Khanafer, K. Vafai, *Int. J. Heat Mass Transfer.* 54 (2011) 4410–4428.

[7] J. Fan, L. Wang, *J. Heat Transfer.* 133 (2011) 040801.

[8] R.S. Vajjha, D.K. Das, *Int. J. Heat Mass Transfer.* (2012),

[9] V. Trisaksri, S. Wongwises, *Renew. Sustain. Energy Rev.* 11 (2007) 512–523.

[10] W. Daungthongsuk, S. Wongwises, *Renew. Sustain. Energy Rev.* 11 (2007) 797–817.

[11] S. Kakaç, Pramuanjaroenkij, *Int. J. Heat Mass Transfer.* 52 (2009) 3187–3196.

[12] L. Godson, B. Raja, D. Mohan, S.Wongwises// *Renew. Sustain. Energy Rev.* (2009)

[13] J. Sarkar, *Sustain. Energy Rev.* 11 (2011) 3271–3277.

[14] Y. Ding, H. Chen, L. Wang, C.-Y. Yang, Y. He, W. Yang, W.P. Lee, L. Zhang, R. Huo, *Kona, Nr.* 25 (2007) 23–38.

[15] C. Kleinstreuer, J. Li, J. Koo// *Int. J. Heat Mass Transfer.* 51 (2008) 5590–5597.

[16] M. Thirugnanasambandam, S. Iniyan, R. Goic, *Renew. Sustain. Energy Rev.* 14 (2010) 312–322.

[17] A. Sharma, *Renew. Sustain. Energy Rev.* 15 (2011) 1767–1776.

[18] S. Mekhilef, R. Saidur, *Renew. Sustain. Energy Rev.* 15 (2011) 1777–1790.

[19] J. A. Eastman, S. U. S. Choi, S. Li, W. Yu, *Applied Physics Letters.* 78(2001) 718–720

[20] Y. Li, J. Zhou, S. Tung, E. Schneider, and S. Xi, *Powder Technology.* 196 (2009) 89–101

[21] C. H. Lo, T. T. Tsung, and L. C. Chen, *Journal of Crystal Growth.* 277(2005) 636–642

[22] C. H. Lo, T. T. Tsung, L. C. Chen, C. H. Su, and H. M. Lin, *Journal of Nanoparticle Research.* 7(2005) 313–320

[23] H. T. Zhu, Y. S. Lin, and Y. S. Yin, *Journal of Colloid and Interface Science.* 277(2004) 100–103

[24] H. B¨onnemann, S. S. Botha, B. Bladergroen, and V. M. Linkov, *Applied Organometallic Chemistry.* 19 (2005) 768–773

[25] A. K. Singh and V. S. Raykar// *Colloid and Polymer Science.* 286(2008)1667–1673,

[26] A. Kumar, H. Joshi, R. Pasricha, A. B. Mandale, and M. Sastry, *Journal of Colloid and Interface Science.* 264 (2003) 396–401

[27] W. Yu, H. Xie, X. Wang, and X. Wang, *Nanoscale Research Letters.* 6 (2011) 47,

[28] H. T. Zhu, C. Y. Zhang, Y. M. Tang, and J. X. Wang, *Journal of Physical Chemistry C*, 111 (2007)1646– 1650.

[29] Y. Chen and X. Wang, *Materials Letters.* 62 (2008) 2215–2218.

[30] X. Feng, H. Ma, S. Huang, *Journal of Physical Chemistry B.* 110 (2006) 12311–12317

[31] W. Yu, H. Xie, L. Chen, and Y. Li, *Colloids and Surfaces A.* 355 (2010) 109–113

[32] L. Wang and J. Fan, *Nanoscale Research Letters.* 5 (2010) 1241–1252

[33] S.A. Kalogeria, Solar Energy Engineering: Processes and Systems, Elsevier, Oxford, (2009)

[34] M.K. Gnanadason, P.S. Kumar, S. Rajakumar, M.H.S. Yousuf, *I.J.AERS.* 1(2011) 171–177

[35] S. Nijmeh, S. Odeh, B. Akash, *Int. Commun. Heat Mass Transfer.* 32 (2005)565–572

[36] M. Elmir, R. Mehdaoui, A. Mojtabi, *Energy Procedia.* 18 (2012)594–603

[37] D. Shin, D. Banerjee, *Int. J. Heat Mass Transfer.* 54 (2010)1064–1070

[38] S. Wua, H. Wanga, S. Xiaoa, D. Zhub, *Energy Procedia.* 31(2012)240–244

[39] Omid Mahian, Ali Kianifar, Soteris Kalogirou, Ioan Pop, Somchai Wongwises, *Heat Mass Transfer.* 57 (2012)582-594

[40] A. Bouzoukas, Ph.D Thesis, University of Nottingham (2008)

[41] C.T. Nguyen, G. Roy, C. Gauthier, N. Galanis, *Appl. Therm.Eng.* 27 (2007) 1501–1506.

[42] M. Elmir, R. Mehdaoui, A. Mojtabi// *Energy Procedia.* 18 (2012) 724–732.

[43] A. Ijam, R. Saidur, *Appl. Therm. Eng.* 32 (2012) 76–82.

[44] S.M. Peyghambarzadeh, S.H. Hashemabadi, M. Seifi Jamnani, S.M. Hoseini, *Appl. Therm. Eng.* 31 (2011) 1833–1838.

[45] T. Hung, W. Yan, *Int. J. Heat Mass Transfer.* 55 (2012)3225–3238.

[46] H. Fan, R. Singh, A. Akbarzadeh, J. *Electron. Mater.* 40 (2011)1

[47] S. P. Jang and S. U. S. Choi, *Applied Thermal Engineering*. 26 (2006)2457–2463

[48] C. T. Nguyen, G. Roy, N. Galanis, and S. Suiro, In Proceedings of the 4th WSEAS International Conference on Heat Transfer, Thermal Engineering and Environment, Elounda, Greece (2006) 103–108

[49] H. Shokouhmand, M. Ghazvini, and J. Shabanian, In Proceedings of the World Congress on Engineering (WCE '08), London, UK, 3 (2008)

[50] M. F. Demirbas, *Energy Sources Part B*. 1 (2006)85–95, 2006.

[51] S. Wu, D. Zhu, X. Zhang, and J. Huang, *Energy and Fuels*. 24 (2010)1894–1898

[52] Y. D. Liu, Y. G. Zhou, M. W. Tong, and X. S. Zhou, *Microfluidics and Nanofluidics*. 7(2009) 579–584

[53] T. P. Otanicar, P. E. Phelan, R. S. Prasher, G. Rosengarten, and R. A. Taylor,*Journal of Renewable and Sustainable Energy*, 2 (2010) 13

[54] H. Tyagi, P. Phelan, and R. Prasher, *Journal of Solar Energy Engineering*.131 (2009) 0410041–0410047

[55] T. P. Otanicar and J. S. Golden, *Environmental Science and Technology*.43 (2009)6082–6087

[56] E. Sani, S. Barison, C. Pagura, *Optics Express*. 18 (2010) 4613

6

Three Quantum Particles Hardy Entanglement from the Topology of Cantorian-Fractal Spacetime and the Casimir Effect as Dark Energy – A Great Opportunity for Nanotechnology

Mohamed S. El Naschie

Dept. of Physics, University of Alexandria, Alexandria, Egypt

Email address:

Chaossf@aol.com

Abstract: The present work brings together three different fields which depend crucially upon nano hardware under the umbrella of E-infinity theoretical framework. We start by following E-infinity topological methodology by dividing Hardy's entanglement into two parts, a global 'counterfactual' part given by Φ^3 where $\Phi = 2/(1+ \sqrt{5})$ and a 'local' part Φ^n where n is the number of quantum particles. For Hardy's celebrated gedankenexperiment with two quantum particles, which was moreover experimentally confirmed with high accuracy, the quantum probability is found for n = 2 to be P(2) (Hardy) = $\Phi^{3+2}= \Phi^5$ exactly as calculated by Hardy using orthodox quantum mechanics. Applying the same topological E-infinity entanglement theory to three quantum particles give a maximal Φ^6 as well as a three partite much smaller value equal $\Phi^3(1- \Phi^3)/ = 0.018033989$. We conclude by outlining the relevant and extremely timely ideas and remarks on the possible connection, via a state of the art nanotechnology, to the Casimir effect as a conjectured origin of dark energy.

Keywords: Three Quantum Particles Entanglement, Hardy's Topological Entanglement, E-Infinity, Hilbert Space, Cantorian Spacetime, Golden Mean Number System, Casimir Effect Connection to Dark Energy, Nanotechnology

1. Introduction

Hardy's quantum entanglement [1-12] is, without a trace of a doubt, all that Sir Roger Penrose considered it to be and more [6], namely an ingenious gedankenexperiment [4-8] which revealed a striking feature of nature and the building blocks of our spacetime with far reaching physical and cosmological consequences [5]. In fact it is in the meantime an understatement to call Hardy's quantum entanglement a gedankenexperiment or a model since it is generic, was verified with high accuracy in numerous experiments [3-7] and was found to explain various fundamental mysterious phenomena in nature, notably dark energy [5] as well as theoretical findings with a claim to reality. In this context we mention Hawking radiation [13-15], Unruh temperature [13], Rindler's wedge [14] in addition to the Casimir effect which may be the origin of dark energy as sophisticated nanotechnology aided experiments may eventually reveal.

In the present work we show first how this result of three Hardy particles may be found from orthodox quantum mechanics in Hilbert space using Gram-Schmidt orthonormalization procedure [1-3]. In particular we point out that interpreted as a three-partite Hardy entanglement one finds P(Hardy) to be a relative minimum equal to 'tHooft dimensional regularization order parameter $k = \phi^3(1-\phi^3) = 0.18033989$ divided by ten, i.e. $P_{min}(H) = 0.018033989$. On the other hand for 3 actual quantum particle entanglement P(H) corresponds to a maximal quantum probability almost three times larger than $k/10$, namely P(H) = $\phi^{3+3} = \phi^6 = 0.05572809014$. This value $3+(k/2) = 3.0901699$ is interpreted within fuzzy set theory as the fractal weight of the number of the three entangled quantum particles Immirzi parameter ϕ^6 [16] and $k = 2\phi^5$ could be drawn in to explain why we could really have many compactified dimensions [15] in addition to various other aspects of high energy physics and quantum cosmology [15-18]. As is well known, in its original form Hardy's entanglement is concerned with two quantum particles [1-4]. Using Dirac's formalisms and a number of conventional mathematical tools of orthodox quantum mechanics Hardy found that the maximal probability for entanglement is almost nine percent [1-5]. Hardy did not notice at first that this nine percent is actually $\phi^5 = 0.09016994393$ and means the golden mean to the power of five that is until Prof. Mermin [1-5] and much later the present author [1-5] noticed this remarkable and intriguing fact linking the most irrational KAM theoretical number which is ϕ to quantum entanglement [19-21]. Consequently and in turn, quantum entanglement is linked directly with topological entanglement [4]. The chain-like reasoning propagated from there in a natural, predictable manner to mean that quantum entanglement [24-29] could be interpreted as a topological consequence of the zero measure topology [4] of our real spacetime geometry at the microscopic scale [19-21]. In short this became the background against which the quantum entanglement theory of E-infinity theory was developed and the simple quantum entanglement formula of n quantum particles was established [4,11], namely $P_n(H) = \phi^{3+n}$. The last five years or so saw a flurry of papers published on the topological side of Hardy's entanglement and its topological interpretation [10-19].

From the above it could not pass unnoticed that the dark energy density is equal 1 minus half of Hardy's ϕ^5 divided by 2. In addition dark energy is the energy of the quantum wave. Consequently it must be related to the Casimir effect [30-31]and a possible source of near to infinite clean energy. This extremely important technological breakthrough innovation [32-44] will also be discussed at the end of the present paper.

2. The Three Quantum Particles Problem

From E-infinity theory [4,7][20-23] it is a well established result that Hardy's probability of quantum entanglement would be decomposed into two parts as mentioned earlier on. The first is the global part due to the core of spacetime. This part is exactly equal to the inverse of the Hausdorff dimension of the Cantor spacetime core [19-21], i.e. $1/(4+\phi^3) = \phi^3$. The second part is the local topological entanglement of two quantum particles, each with a zero set Hausdorff dimension ϕ. That way the total local probability is ϕ^2 [4,11]. Consequently Hardy's probability is the multiplication of the global ϕ^3 with the local ϕ^2 giving the gross total. This is in this case equal to ϕ^5 as should be [4,11]. However as far as the author is aware none of the previous papers dealt explicitly with the case of three particles which in view of its classical non-integrable analogue and the findings of quantum chaos theory [19-21] could harbour some surprises [6,7]. For these reasons the present paper is devoted to analysing explicitly Hardy's entanglement for three particles.

We may start from the outset that the present analysis confirmed the validity of the E-infinity general formula mentioned earlier [4,11] as a maximal quantum probability [4-11]. However a minor discovery was waiting for us in store, namely a minimal quantum probability relevant to quantum information science problems which require the division of the E-infinity probability by the fractal weight of the number of quantum particles concerned as will be explained in the following sections.

It is needless to reiterate that Hardy obtained the same results reported here for two particles by using conventional quantum mechanics [4] and we will now proceed in the same way for the case of 3 particles. As for the global part, it remains of course as it should be, namely ϕ^3. On the other hand for 3 particles the local part changes to $\phi \; \phi \; \phi = \phi^3$. Consequently the total Hardy entanglement is no more ϕ^5 but $\phi^3 \phi^3 = \phi^6$ which is the same value behind the Immirzi parameter reconciling the result of the blackhole entropy found using superstring theory with that found using loop quantum gravity [16]. It is now a nontrivial question to ask oneself if a smaller value than ϕ^6 could be realized in an actual situation. The answer is somewhat non-standard and is "yes, we can" but only when we introduce a twist to the entire set up by looking at the problem not as three 'geometrical-topological' quantum particles but as three-partite Hardy entanglement. The minimum in this case means simply the share of each particle from the union of all the three 'particles', that is to say a Hardy probability of exactly ϕ^6 divided by $3+(k/2)$ where k is 'tHooft's order parameter of dimensional regularization which is given by $k = \phi^3(1-\phi^3) = 0.18033980$ [22]. In this sense $3+(k/2) = 3.090169945$ is a fractal weight for the number of the three particles and $k/2 = \phi^5$ is the Hardy entanglement for two particles. The final result is thus

$$P(Hardy) = \phi^6/(3+\phi^5) = k/10 = 0.018033989. \qquad (1)$$

3. Solution in Hilbert Space

Now we could have worked everything backwards and started from a three-partite Hardy setting and solved the problem using conventional Hilbert space techniques [4,11]. This leads us to various conditional probability densities which upon close examination could be extremized by letting the first derivative vanish and finding that

$$P(Hardy) = 0.018033989. \qquad (2)$$

Subsequently by examining the second derivative we find that the value is a relative maximum but could be described as a minimum with respect to the first result, namely ϕ^6 where k is 'tHooft's order parameter of dimensional regularization. Applying the same strategy to one particle only one finds $P(Hardy)_{min} = \phi^3$. That means $P(Hardy)_{min}$ becomes equal to the global entanglement of spacetime. Going one step further and asking what is the minimal probability for no particles at all would mean division by zero [9] and leads to the minimal Hardy probability to become infinite. There is a deep meaning for this superficially nonsensical result. This limiting result means that spacetime at the quantum resolution is completely interconnected and nonlocal [23].

4. The Casimir Effect Behind Dark Energy and the Role of Nanotechnology

Topological quantum entanglement seen from E-infinity and noncommutative geometry perspective could be the solution to many hard theoretical problems ranging from wormholes to how we could obtain a near to infinite amount of clean energy. We start by recalling our result about the ordinary energy density of the cosmos

$$E(O) = \left(\phi^5 / 2\right) mc^2 \cong mc^2 / 22. \tag{3}$$

We note that $\gamma(O)$ is simply the probability for quantum entanglement for one particle only of the two particles Hardy set up, i.e.

$$\gamma(O) = P(H) / 2 = \phi^5 / 2. \tag{4}$$

Consequently the dark energy density is give by

$$\gamma(D) = 1 - \gamma(O) = 1 - \left(\phi^5 / 2\right) = \left(5\phi^2\right) / 2 \tag{5}$$

where ϕ^2 is the Hausdorff dimension of the empty set modelling the quantum wave. We also notice that for n interpreted as internal dimensions there is a vital duality between $\gamma(O)$ and $\gamma(D)$. To show this we compare

$$\gamma(O) = \phi^n / 2 \tag{6}$$

with

$$\gamma(D) = n\phi^2 / 2 \tag{7}$$

for very large n. It is clear from equations 6 and 7 that for $n \to \infty$ we have $\gamma(O) \to 0$ and $\gamma(D) \to \infty$. This leads us to make an educated guess to scientifically speculate on the possibility that the Casimir effect and dark energy are intimately connected or may be one and the same thing. It would seem that while our result that 95.5% of the energy of the cosmos is confirmed from the endophysical experiment [45,46] involving the entire universe, i.e. COBE, WMAP and Planck [10-19], the small amount of Casimir energy which we were able to measure is found by contrast from local exophysical experiments [45,46]. For this reason we think it is more than feasible to initiate a large scientific program utilizing nanotechnology [32-44] as well as E-infinity and K-theory to find out all what we can about how to utilize the Casimir-dark energy reservoir of vacuum energy [30,31] for the benefit of humanity and that way solve forever the colossal energy problems threatening our civilization.

5. Nanotechnology as the Key to the Infinite Reservoir of Empty Spacetime Clean Energy

It is curious or even more than curious that gravity within E-infinity theory was conceived from the very beginning as a fluctuation phenomena [13] caused by the "passing" of fractal time [47,48]. The idea was inspired by an old side remark due to R. Feynman and is analogous to Van der Waals forces as noted by the author as well as Professor M. Agop [49]. From this perspective the so called Feynman-El Naschie conjecture [47-49] could be extended to mean that the Casimir effect is nothing but a strong van der Waals effect and in turn dark energy is nothing but again the amplification of Casimir quantum wave forces brought about by the intricate topology and geometry of a high dimensional Banach spaces [47-54] where measure concentration of Dvoretzky's theorem [54] and holography are operative. Even more general than that, and as is obvious from the E-infinity topological conception of Hardy's quantum entanglement ϕ^5, ordinary energy density $\phi^5 / 2$ and dark energy density $5\phi^2 / 2$ one could easily conclude that these are all quantum vacuum fluctuation phenomena and include but are not restricted to Unruh temperature and Hawking's various effects [13,50]. That way we must recognize empty spacetime as the most abundant 'substance' in the universe and the origin of everything starting from energy. The scientific question which borders on science fiction and suggests itself whether we find it realistic or not, is if nanotechnology could bring us near to harnessing this vacuum energy in an efficient and economical way? Our answer to this spacetime question is a definite scientific-engineering yes. To motive our ideas we have to first recall that the Casimir forces are essentially the difference between the "quantum wave" density between the two Casimir plates and the same density outside in the surrounding space with dependence on the plate surface, material and shape. Now the universe in its entirety could be seen as analogous to the tiny space within the Casimir plates while the immensely infinite "empty set space" constituting the outside of our universe is analogous to the outside unbounded space in the Casimir classical experiments. In other words our cosmos is a one sided Casimir set up, i.e. we have Casimir pressure inside and there is no outside with outside pressure to balance it. This is essentially what we call dark energy. It is produced by the fact that the empty boundary of our universe is like a complex one sided so called Mobius strip and an extension of J.A. Wheeler's boundary of a boundary is zero principles. From that we could have an initial hunch on how nanotechnology and material surfaces technologies could simulate a holographic boundary for Casimir-dark energy experiments. We can also imagine performing experiments resembling classical reactor designs to create a gradient of Casimir-dark energy. Vaguely speaking at this very early stage, we guess that we might need to create an inverted Faraday cage so to speak in order to experiment with phenomena of vacuum fluctuation but it is fair to say that at this point, everything is in flux due to the novelty and the excitement of a completely new existence not bounded with traditional ways of thinking about matter and energy. One thing is however certain in this brand new world of post modernistic theoretical physics; nanotechnology is the key and interface between the classical world of man and the said post modernistic physical world of vacuum energy of empty space.

6. Conclusion

Computer information science, cosmological measurement in dark energy and experimenting with the Casimir effect are three different fields which depend crucially upon nanotechnology hardware. In the present work we bring all these fields together under the theoretical framework of E-infinity theory and the technical possibilities offered by nanotechnology. In the present short paper we looked first upon the three-partite quantum Hardy problem as a topological entanglement. We applied two solution strategies with different subtle interpretations. In the first we see the problem as that of three quantum particles and obtain a maximal quantum entanglement probability using E-infinity Cantorian spacetime theory amounting to $P(\text{Hardy})_{max} = \phi^6$. The second strategy is to start from a Hilbert space setting for a three-partite Hardy gedankenexperiment and find a minimal probability of about one third of ϕ^6. The exact result in the three partite case is $P(\text{Hardy})_{min} = k/10$. Motivated and encouraged by these results linking entanglement with dark energy we made a few proposals regarding the connection between the Casimir effect and dark energy as well as the role of nanotechnology in harnessing both the Casimir energy and/or dark energy which may be different sides of the same coin [30,31]. With the new possibilities offered by nanotechnology, the door is open now to utilize dark energy and Casimir energy as an unheard of source of free energy.

References

[1]   L. Hardy, Nonlocality of two particles without inequalities for almost all entangled states. Phys. Rev. Lett. Vol. 71(11), 1993, pp. 1665-1668.

[2]   J.S. Bell, Speakable and Unspeakable in Quantum Mechanics. Cambridge University Press, Cambridge 1991)

[3]   Ji-Huan He et al, Quantum golden mean entanglement test as the signature of the fractality of micro spacetime. Nonlinear Sci. Lett B, Vol. 1(2), 2011, pp. 45-50.

[4]   Mohamed S. El Naschie, Quantum Entanglement as a Consequence of a Cantorian Micro Spacetime Geometry. Journal of Quantum Information Science, Vol. 1(2), 2011, pp. 50-53.

[5]   Mohamed S. El Naschie, Quantum Entanglement: Where Dark Energy and Negative, Accelerated Expansion of the Universe Comes from. Journal of Quantum Information Science, Vol. 3(2), 2013, pp 55-57.

[6]   R. Penrose, The Road to Reality. J. Cape, London, UK. 2004

[7]   M.S. El Naschie, M.A. Helal, L. Marek-Crnjac and Ji-Huan He, Transfinite corrections as a Hardy type quantum entanglement. Fractal Spacetime & Noncommutative Geometry in Quantum & High Energy Physics, Vol. 2(1), 2012, pp. 98-101.

[8]   M.S. El Naschie, Ji-Huan He, S. Nada, L. Marek-Crnjac and M. Helal, Golden mean computer for high energy physics. Fractal Spacetime and Noncommutative Geometry in Quantum and High Energy Physics. Vol. 2(2), 2012, pp. 80-92.

[9]   M.S. El Naschie and S.A. Olsen, When zero is equal one: A set theoretical resolution of quantum paradoxes. Fractal Spacetime & Noncommutative Geometry in Quantum High energy Physics, Vol. 1(1), 2011, pp. 11-24.

[10]   M.S. El Naschie, Electroweak connection and universality of Hardy's quantum entanglement. Fractal Spacetime & Noncommutative Geometry in Quantum High energy Physics, Vol. 1(1), 2011, pp. 25-30.

[11]   L. Marek-Crnjac, Ji-Huan He and M.S. El Naschie, On the universal character of Hardy's quantum entanglement and its geometrical-topological interpretation. Fractal Spacetime & Noncommutative Geometry in Quantum & High Energy Phys. Vol. 2(2), 2012, pp. 118-12.

[12]   M.S. El Naschie, L. Marek-Crnjac and Ji-Huan He, Using Hardy's entanglement, Nash embedding and quantum groups to derive the four dimensionality of spacetime. Fractal Spacetime & Noncommutative Geometry in Quantum & High Energy Phys. Vol. 2(2), 2012, pp. 107-112.

[13]   Mohamed S. El Naschie, Experimentally Based Theoretical Arguments that Unruh's Temperature, Hawking's Vacuum Fluctuation and Rindler's Wedge Are Physically Real. American Journal of Modern Physics, Vol. 2(6), 2013, pp. 357-361.

[14]   Mohamed S. El Naschie, A Rindler-KAM Spacetime Geometry and Scaling the Planck Scale Solves Quantum Relativity and Explains Dark Energy. International Journal of Astronomy and Astrophysics, Vol. 3(4), 2013, pp. 483-493.

[15]   Mohamed S. El Naschie, Topological-Geometrical and Physical Interpretation of the Dark Energy of the Cosmos as a "Halo" Energy of the Schrödinger Quantum Wave Journal of Modern Physics, Vol. 4(5), 2013, pp. 591-596.

[16]   M.S. El Naschie, The quantum gravity Immirzi parameter – A general physical and topological interpretation. Gravity and Cosmology, Vol. 19(3), 2013, pp. 151-153.

[17]   Mohamed S. El Naschie, Compactified dimensions as produced by quantum entanglement, the four dimensionality of Einstein's smooth spacetime and 'tHooft's 4-ε fractal spacetime. American Journal of Astronomy & Astrophysics, Vol. 2(3), 2014, pp. 34-37.

[18]   Mohamed S. El Naschie, Electromagnetic—pure gravity connection via Hardy's quantum entanglement. Journal of Electromagnetic Analysis and Applications, Vol. 6(9), 2014, pp. 233-237.

[19]   L. Marek-Crnjac and Ji-Huan He, An Invitation to El Naschie's theory of Cantorian space-time and dark energy. International Journal of Astronomy and Astrophysics, Vol. 3(4), 2013, pp. 464-471.

[20]   M.S. El Naschie, A review of E-infinity and the mass spectrum of high energy particle physics. Chaos, Solitons & Fractals, Vol. 19(1), 2004, pp. 209-236.

[21]   M.S. El Naschie, Superstrings, knots and noncommutative geometry in E-infinity space. International Journal of Theoretical Physics, Vol. 37(12), 1998, pp. 2935-2951.

[22] Mohamed S. El Naschie, On a new elementary particle from the disintegration of the symplectic 't Hooft-Veltman-Wilson fractal spacetime. World Journal of Nuclear Science and ATechnology, Vol. 4(4), 2014, pp. 216-221.

[23] W. Tan, Y. Li, H.Y. Kong and M.S. El Naschie, From nonlocal elasticity to nonlocal spacetime and nanoscience. Bubbfil Nano Technology, Vol. 1(1), 2014, pp. 3-12.

[24] D. Heiss (Editor):, Fundamentals of Quantum Information. Springer, Berlin 2002.

[25] I. Gengtsson and K. Zyczkowski, Geometry of Quantum States. Cambridge University Press, Cambridge 2006.

[26] M.A. Nielsen and I.L. Chuang, Quantum Computation and Quantum Information. Cambridge University Press, Cambridge, 2010.

[27] A. Furusawa and P. van Loock, Quantum Teleportation and Entanglement. Wiley-VCH, Weinheim, Germany, 2011.

[28] J.K. Pachos, Introduction to Topological Quantum Computation. Cambridge University Press, Cambridge, 2012.

[29] D.G. Marinescu and G.M. Marinescu, Classical and Quantum Information. Elsevier, Amsterdam, 2012.

[30] Mohamed S. El Naschie, Casimir-like energy as a double Eigenvalues of quantumly entangled system leading to the missing dark energy density of the cosmos. International Journal of High Energy Physics, Vol. 1(5), 2014, pp. 55-63.

[31] J. Matsumoto, The Casimir effect as a candidate of dark energy. arXiv: 1303.4067[hep-th]26December 2013.

[32] K. Eric Drexler, Engines of Creation. Fourth Estate Ltd., London, 1990.

[33] P. Day (Editor), Unveiling The Microcosmos. Oxford University Press, Oxford, 1996.

[34] S.E. Lyshevski, Nano and Microelectromechanical Systems. CRC Press, Boca Ratan, 2001.

[35] L.E. Foster, Nanotechnology, Science, Innovation and Opportunity. Prentice Hall, Boston, 2006.

[36] M. Krummenacker and J. Lewis (Editors), Prospects in Nanotechnology. John Wiley, New York, 1995.

[37] M.S. El Naschie, Nanotechnology for the developing world. Chaos, Solitons & Fractals, Vol. 30(4), 2006, pp. 769-773.

[38] M.S. El Naschie, Chaos and fractals in nano and quantum technology. Chaos, Solitons & Fractals, Vol. 9(10), 1998, pp. 1793-1802.

[39] M.S. El Naschie, Nanotechnology and the political economy of the developing world. International, May 2007, pp. 7-12. (Periodical International Economic Magazine by AS&S Publishing Ltd, Camden Town, London, UK, Registration No. 04761267).

[40] M.S. El Naschie, Can nanotechnology slow the aging process by inteferring with the arrow of time. International, June 2007, pp. 10-15. (Periodical International Economic Magazine by AS&S Publishing Ltd, Camden Town, London, UK, Registration No. 04761267).

[41] M. Aboulanan, The making of the future via nanotechnology (in Arabic). International, July 2007, pp. 32-35. (Periodical International Economic Magazine by AS&S Publishing Ltd, Camden Town, London, UK, Registration No. 04761267).

[42] M.S. El Naschie, From relativity to deterministic chaos in science and society. International, August 2007, pp. 11-17. (Periodical International Economic Magazine by AS&S Publishing Ltd, Camden Town, London, UK, Registration No. 04761267).

[43] M.S. El Naschie, The political economy of nanotechnology and the developing world. International Journal of Electrospun Nanofibers and Application. I(I), 2007, pp. 41-50. Published by Research Science Press, India.

[44] D. Brito and J. Rosellon, Energy and Nanotechnology: Prospects for solar energy in the 21st century. The James A. Baker III Institute for Public Policy of Rice University, December 2005, pp. 1-16.

[45] O.E. Rössler and M.S. El Naschie, Interference through causality vacillation. In Symposium on the Foundations of Modern Physics, Helsinki, Finland, June 1994, pp. 13-16.

[46] O.E. Rössler and M.S. El Naschie: Interference is Exophysically Absent. In Endophysics – The World As An Interface. World Scientific, Singapore, 1998, pp. 159-160.

[47] M.S. El Naschie, A note on quantum gravity and Cantorian spacetime. Chaos, Solitons & Fractals, 8(1), p. 131-133 (1997).

[48] M.S. El Naschie, The symplictic vacuum exotic quasi particles and gravitational instantons. Chaos, Solitons & Fractals, 22(1), 2004, pp. 1-11.

[49] M. Agop, E-infinity Cantorian spacetime, polarization gravitational field and van der Waals-type forces. Chaos, Solitons & Fractals, 18(1), 2003, pp. 1-16.

[50] Mohamed S. El Naschie, A Rindler-KAM spacetime geometry and scaling the Planck scale solves quantum relativity and explains dark energy. International Journal of Astronomy and Astrophysics, Vol. 3(4), 2013, pp. 483-493.

[51] J. Cugnon, The Casimir effect and the vacuum energy. Few-body System, 53(1-2), 2012, pp. 181-188.

[52] K.A. Milton, Resource Letter VWCPF-1 van der Waals and Casimir-Polder forces. American Journal of Physics, 79, 2011, pp. 697.

[53] M. Ito, Gravity, higher dimensions, nanotechnology and particles physics. Journal of Physics, Conference Series, 89(1), 2007, p. 1-8.

[54] M.S. El Naschie, Casimir-like energy as a double Eigenvalue of quantumly entangled system leading to the missing dark energy density of the cosmos. International Journal of High Energy Physics, 1(5), 2014, pp. 55-63.

7

Fabrication of Sintered Si Nano-polycrystalline with Reduced Si Nanoparticles and Properties of Photoluminescence in Visible Regime for Sintered Si Nano-polycrystalline by Violet Light Excitation

Taku Saiki, Yukio Iida

Department of Electrical and Electronic Engineering, Faculty of Engineering Science, Kansai University, Osaka, Japan

Email address:

tsaiki@kansai-u.ac.jp (T. Saiki)

Abstract: Si oxide powder is reduced by highly repetitive pulse laser ablation in liquid, and Si nanoparticles are produced efficiently with a low cost in a short time. A Si nanopaste with highly doped Si nanoparticles was sintered by using a hot plate. We succeeded in fabricating a sintered Si nano-polycrystalline for the first time. The structure and components of the fabricated sintered Si nano-polycrystalline were investigated by SEM and EDX analysis. Furthermore, the reduced Si nanoparticles and the sintered Si nano-polycrystalline were excited by violet light and stable photoluminescence (PL), which were observed in the visible regime. The peak wavelengths of the PL were 550 nm and 560 nm. Particularly, the intensity of the observed PL of the sintered Si nano-polycrystalline was five times higher than that of the reduced Si nanoparticles powder. This result is attributed to the PL being amplified inside the sintered Si nano polycrystalline. These experiments show that because the mean diameters of the Si nanocrystals in the reduced Si nanoparticles were below 2 nm, the structure of the Si nanocrystals changed to a direct-transition type; the bandgap energy of the Si nanocrystals changed from 1.1 eV to 2.25 eV, and PL in the visible regime was generated. Moreover, the possibility of Si photonics is discussed. The sintered Si nano-polycrystalline will be applicable to light waveguides, optical switches using a free carrier effect, and light amplifiers.

Keywords: SiO$_2$, Si, Nanoparticles, Polycrystalline, Optical Waveguide, Optical Switch, Free Carrier Effect, Light Amplification, Photoluminescence, Laser Ablation in Liquids

1. Introduction

Electrons in bulk Si make an indirect transition between bandgaps. The bandgap energy of bulk Si is 1.1 eV. The wavelength is close to near-infrared. Thus, it is difficult for bulk Si to generate photoluminescence. However, due to the quantum effect that occurs when electrons and holes are confined in the area below their wavelength, Si nanoparticles have larger bandgap energy than that of bulk Si, and the bandgap structure changes to a direct transition type. A theory stating that Si nanoparticles can emit photoluminescence in the visible regime has already been proposed [1]. Photoluminescences in both the visible and violet regime have been observed in some experiments [2–9]. The observed experimental results are consistent with the theoretical ones [8–9]. Applications using Si nanoparticles are expected to become alternatives for conventional illumination.

Si nanoparticles are produced by many kinds of methods. The production of Si nanoparticles by pulse laser has been extensively researched. Si nanoparticles work as quantum dots, and they generate photoluminescence in the visible regime [1–6].

Currently, calculations for information processing are mainly performed by using semiconductors. Because of advancements in information technology (IT), the number of calculations has become enormous. Thus, more rapid calculations have been required.

Si photonics, in which arithmetic processing is done for all the information gained by optical signals using light controlling devices based on Si, has been proposed. Calculations by optical signals have also been widely proposed. We are the first to have presented a proposal on a Si optical switch [10] using a free carrier effect [11-12].

Following that proposal, NICT Japan has reported that the Si optical switch was fabricated and that the device actually worked [13]. We have researched Si optical switches to develop an optical computer as shown in reference [10].

Until now, the main focus of research has been on the synthesis of Si nanoparticles. As previously reported, we could produce a sufficient number of Si nanoparticles to work in emission devices. However, Si nanoparticles can barely work as an optical waveguide or device for generating laser light [14]. The nanoparticles should be close to each other to efficiently propagate electromagnetic waves (light) as an optical waveguide. For example, when an electromagnetic wave with a wavelength of 0.5 µm propagates in a nano-polycrystalline with a mean diameter of a few nm, the structure should not negatively influence propagation.

In this study, we used Si oxide powder to synthesize Si nanoparticles. Si oxide powder is reduced by highly repetitive pulse laser ablation in liquid, and Si nanoparticles are produced. A Si nanopaste with highly doped Si nanoparticles was sintered by using a hot plate, and we obtained a Si nano-polycrystalline for the first time. The properties of the photoluminescence in the visible regime of the reduced Si nanoparticles, and the sintered Si nano-polycrystalline are shown. The similar results have not been previously reported as far as we know. The Si nano-polycrystalline has a temporally stable photoluminescence when it is excited by light. If the surface conditions of the Si nanoparticles change with time, and we can not obtain stable photoluminescence. The structure and components of the fabricated sintered Si nano-polycrystalline were investigated by SEM and EDX analysis. We also discuss the possibility of Si photonics.

2. Experimental Setup

2.1. Synthesis of Reduced Si Nanoparticles and Fabrication of Sintered Si Nano-polycrystalline

Si oxide powder (purity 99.9%, mean diameter 4 µm, Kojundo Chemical Co., Ltd.) is reduced by highly repetitive pulse laser ablation in liquid. Here, as shown in reference [7], using the same method, highly repetitive Nd: YAG laser pulses (repetitive rate 18 kHz, wavelength 1064 nm) were focused by a lens. The focused laser pulses were directly irradiated into pure water with the Si oxide powder in a small glass bottle. The water was continuously mixed by a magnetic stirrer, and the Si oxide powder was reduced to Si, and Si nanoparticles were produced at the same time. The irradiating time was ten minutes. The weight of the total Si oxide powder was chosen to be 1 g. The reduced Si nanoparticles were dried, and as many reduced Si nanoparticles as possible were mixed with an Ag paste weighing 5 mg (NAG10, Ag 82 wt.% density, Daiken Chemical Co., Ltd., Japan). Finally, the prepared nanopaste was sintered by using a hot plate.

To compare the sintered Si nanopaste, Si powder (purity 99.9%, mean diameter 5 µm, Kojundo Chemical Co., Ltd.) was mixed with the Ag nanopaste, and the prepared Si nanopaste was sintered.

The Si paste was pasted on a glass plate and sintered using an electrical hot plate (CHP-170AN, ASONE) in an atmosphere of air at 250 deg. for one minute and at 300 deg. for four minutes. When using reduced Si nanoparticles, the prepared Si nanopaste was sintered in an atmosphere of air at 200 deg. for five minutes. Gaps between Si nanoparticles are vanished with keeping the size of Si nanocrystal when the Si pastes are sintered. Also, components except for Ag, Si, and impurity metals are removed from the pastes because they become gas.

The structure of the sintered Si nanopastes was observed by using an S-4700 scanning electron microscope (SEM) (Hitachi High Technologies, Japan). An EMAX7000 energy-dispersive X-ray spectrometer (EDX) (Horiba, Japan) was used for analyzing the distributions of silver, mixed oxides, and impurities.

2.2. Observation of Photoluminescence Excited by Violet Light

In this section, we describe the principles of the PL of the Si nanocrystal line. Si nanoparticles have larger bandgap energy than that of bulk Si by the quantum size effect, and the bandgap structure changes to a direct transition type. The Si nano-crystalline can emit PL in the visible regime (as mentioned in the introduction of this paper). From the results described in references [6, 9], we have considered that the PL is due to the quantum size effect of Si nano-crystalline.

Experimental setup for excitation of the sintered Si nano-polycrystalline is shown in Fig. 1.

LD (NDHV220APAE1, peak wavelength 404 nm, Nichia, Japan) as a source for exciting the sintered Si nano-polycrystalline was used. The emitting image was transferred using a single lens, and the output laser light was focused at a distance of 35 mm from the focusing lens as shown in Fig. 1. The output laser power was 1 mW. The output laser light was focused on the edge of the sintered Si nano-polycrystalline. Here, the intensity of the focused laser light was estimated to be 13 W/cm^2. Moreover, the PL of the reduced Si nanoparticles and the sintered Si nano-polycrystalline were observed and measured using a spectrometer with a fiber light guide (USB4000, Ocean Photonics). The PL of the sintered Si nano-polycrystalline was measured at a distance of 5 cm from the opposite side of the pump surface of the sintered Si nano-polycrystalline.

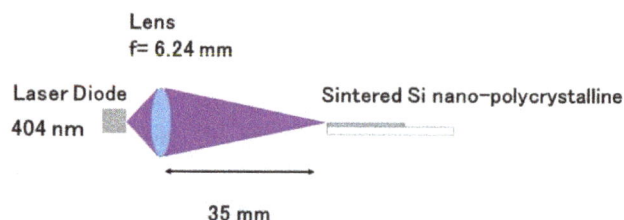

Fig. 1. Experimental setup for excitation of the sintered Si nano-polycrystalline.

3. Results and Discussion

3.1. Sintered Si Nanopastes

The color of Si oxide is white. The color of the reduced Si nanoparticles after irradiation with laser pulses in pure water is gray. The color was close to that of Si powder. However, the surface of the reduced Si nanoparticles is oxidized [4, 6–7]. It has been recognized that the mean diameters of the reduced Si nanoparticles were evaluated to be 10 nm using the Scherrer formula and the bandwidth of the X-ray spectrum in XRD analysis.

We have succeeded in fabricating a Si nano-polycrystalline by sintering prepared Si nanopaste. Sintered polycrystallines by using a hot plate are shown in Fig. 2. The Si polycrystalline obtained by sintering a Si paste mixed with Si micro powder is shown in Fig. 2 (a). Also, the Si nano-polycrystalline obtained by sintering a Si nanopaste mixed with the reduced Si nanoparticles is shown in Fig. 2 (b). The size of the Si polycrystalline obtained by sintering a Si paste mixed with Si micro powder is 5 mm x 5 mm. The thickness is around 50 μm. The sintered Si polycrystalline has a clear luster while the sintered nano-polycrystalline has only a slight luster. The size is 6 mm x 13 mm, and the thickness is around 50 μm.

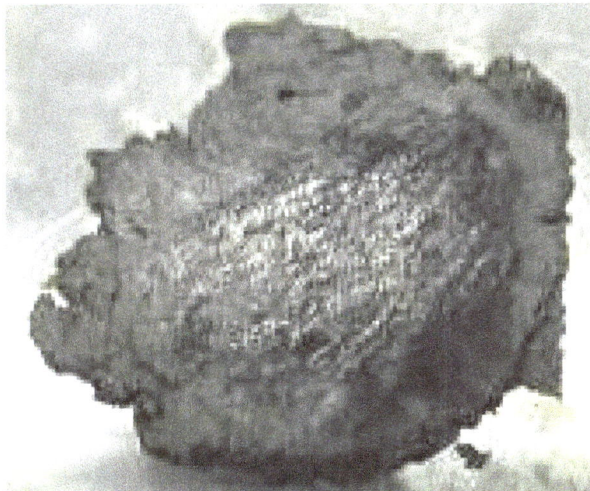

(a)

(b)

Fig. 2. *Sintered Si polycrystalline. (a) Using Si powder, (b) Using reduced Si nanoparticles.*

The resistances of the sintered Si nano-polycrystalline were evaluated by the four-terminal method, and their volume resistivity was estimated. The volume resistivity of the sintered Si nano-polycrystalline was 4.5×10^{-7} (Ω · m). The evaluated resistance of the sintered nano-polycrystalline per 1 cm was 0.01Ω. The value is close to that of metal. The resistance of the Si polycrystalline obtained by sintering a Si paste mixed with Si micro powder is also as low as that of the sintered Si nano-polycrystalline.

3.2. SEM and EDX Analysis

(a)

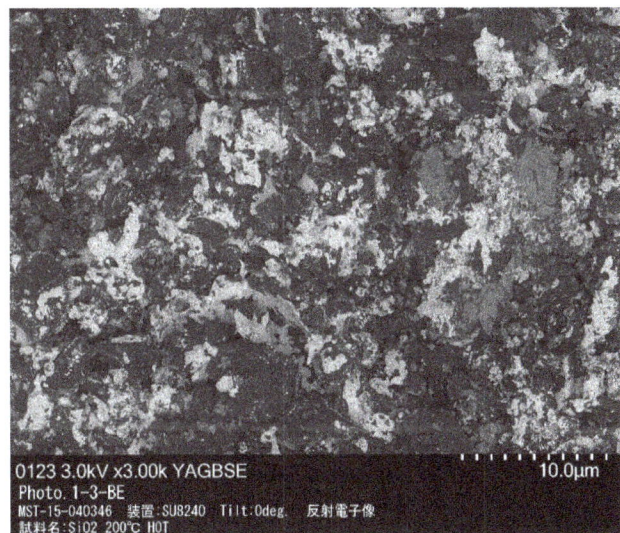

(b)

Fig. 3. *SEM image (Reflection image). (a) Using Si powder. (b) Using reduced Si nanoparticles.*

The photo image (reflection image) obtained by SEM analysis is shown in Fig. 3. A photo image (reflection image) obtained by SEM analysis for sintered Si polycrystalline using Si powder is shown in Fig. 3 (a). A photo image (reflection image) obtained by SEM analysis for sintered Si nano-polycrystalline using reduced Si nanoparticles is shown in Fig. 3 (b). Both photo images are results of observing the surfaces of the sintered Si polycrystalline. It has been proven from the specific X-ray spectrum of EDX analysis that the black parts mainly contain Si. Large Si particles with diameters of 3–5 μm were observed on the surface of the Si polycrystalline as shown in Fig. 3 (a). However, as shown in Fig. 3 (b), fewer large Si particles were observed, and smaller Si particles, whose sizes were below 1 μm, were observed. The sintered Si seemed to be melted, and this melted Si consisted of very small Si nanocrystals with diameters of a few nm.

As for impurities, it has been found that the gray part contains mainly Fe. The gray parts are found as shown in Fig. 3. It has been proven from the specific X-ray spectrum of EDX analysis that the white parts mainly contain Ag. This is because the purity of both the Si powder and the Si oxide powder is 99.9%. The Si and Si oxide are metal grade, and they slightly contain C, Fe, Ni, and Cr.

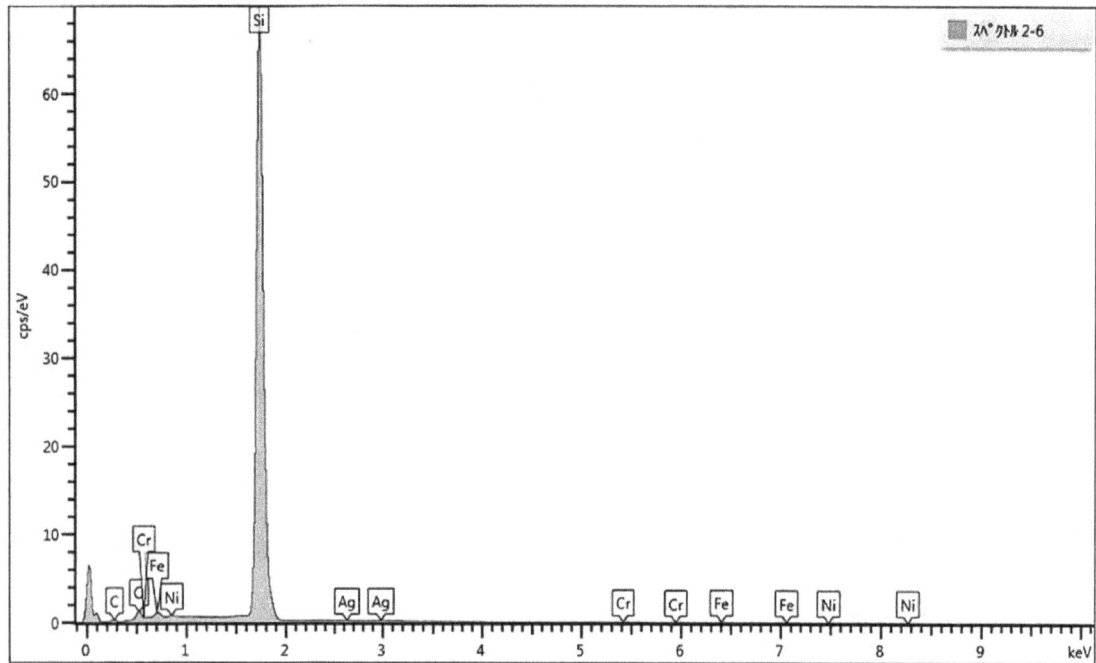

(a)

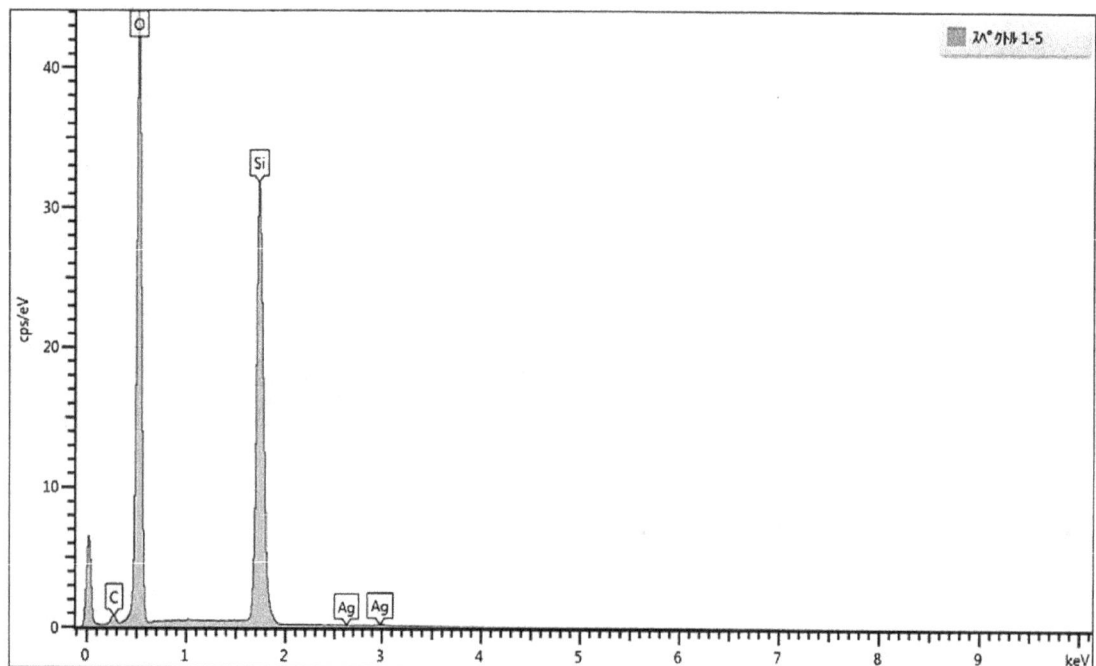

(b)

Fig. 4. *Specific X-ray spectral intensity obtained by EDX analysis. (a) Using Si powder. (b) Using reduced Si nanoparticles.*

Specific X-ray spectral intensity obtained by EDX analysis is shown in Fig. 4. Specific X-ray spectral intensity for the sintered Si polycrystalline using Si powder is shown in Fig. 4 (a). Also, specific X-ray spectral intensity for the sintered Si nano-polycrystalline using reduced Si nanoparticles is shown in Fig. 4 (b). As shown in Fig. 4 (a) and (b), strong specific X-ray spectral intensities of the Si were measured. As shown in Fig. 4 (b), the specific X-ray spectral intensity of the oxygen is stronger than that of the Si. Actually, the number of the oxygen atoms should be one hundredth as small as that of the Si atoms. This is because the oxygen should slightly remain when the Si oxide is reduced to Si. The fact that the resistance of the sintered Si nano-polycrystalline is very low should prove that less oxygen remains in the sintered Si nano-polycrystalline.

3.3. Spectrum of Photoluminescence

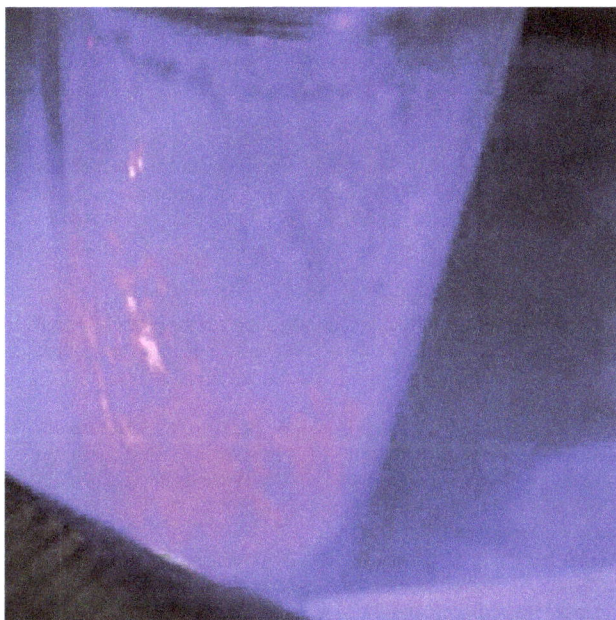

Fig. 5. *Photo of PL from reduced Si nanoparticles.*

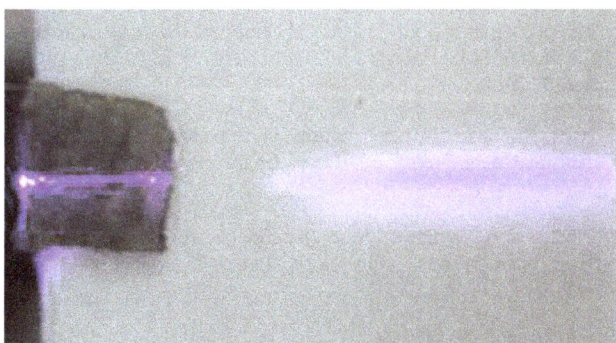

Fig. 6. *Photo of PL from sintered Si nano-polycrystalline. Violet light for excitation was injected from the left side edge of the sintered Si nano-polycrystalline.*

The PL from the reduced Si nanoparticles under the irradiation of weak violet light is shown in Fig. 5. The PL contains a mixture of the colors blue, yellow, and red. When the reduced Si nanoparticles were irradiated with weak violet light, each Si nanoparticle emitted PL uniformly.

Fig. 6 shows a photo of the PL from the sintered Si nano-polycrystalline when the edge of the sintered Si nano-polycrystalline was irradiated with the focused violet light. It has been found that the PL emitted from the surface was below 1 mm. Thus, the absorption coefficient at an excitation wavelength of 404 nm was evaluated to be a few 10 cm^{-1}. Violet light was also focused and irradiated on the sintered Si polycrystalline using Si particles. However, less PL was observed. The observed PL was very weak and could barely be observed. The PL spectrum was broadened from 500 nm to nearly 1000 nm. The intensity was one tenth times lower than that of the sintered Si nano-polycrystalline. The reason is presumed to be because the sintered Si polycrystalline mainly contains Si particles with diameters of around 5 μm and few small Si particles with various diameters below 1 μm. It is prospected that the sintered Si nano-polycrystalline mainly contains small Si nanoparticles with evidently uniform diameters. The diameter of the emitting beam expanded to 2 mm at a distance of 0.8 cm from the end of the sintered Si nano-polycrystalline as shown in Fig. 6. The emission has a certain divergence angle, and the evaluated divergence angle was 0.25 rad.

The refractive index of the sintered Si nano-polycrystalline should be larger than 5.58 [15]; although this is not accurate because the dielectric constant should be changed by the quantum size effect. Thus, the injected excitation light is confined within the sintered Si nano-polycrystalline. Also, the loss due to Fresnel reflection is evaluated to be 48% when the refractive index is 5.58 because the Si nano-polycrystalline has no coating. This value is large.

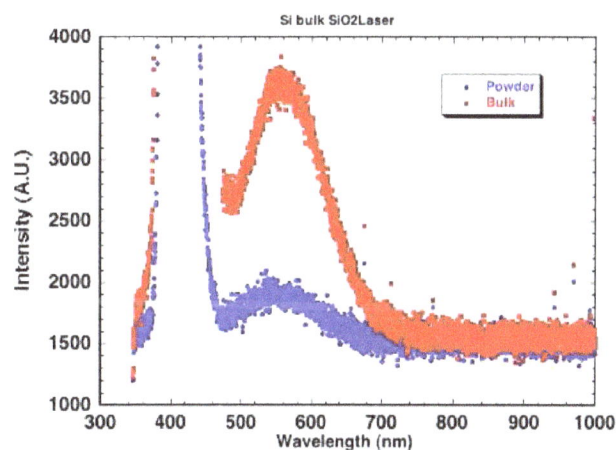

Fig. 7. *Observed spectrum of photoluminescence. Blue dots: Reduced Si nanopowder. Red dots: Sintered Si nano-polycrystalline. PL from the opposite side of the pump surface.*

The measured spectrum of the PL from the reduced Si nanoparticles and the sintered Si nano-polycrystalline by 404 nm violet light excitation is shown in Fig. 7. The peak wavelengths of the PL were 550 nm and 560 nm. The evaluated bandgap energy of the sintered Si nano-polycrystalline from Fig. 7 was 2.25 eV, and the mean diameter of the sintered Si nano-polycrystalline was below 2 nm. This result shows that the reduced Si nanoparticles with mean diameters of 10 nm prepared by pulse laser ablation are secondary particles. Thus, the secondary Si nanoparticles contain small Si nanocrystals with diameters below 2 nm.

The spectral intensity of PL for sintered Si nano-polycrystalline was five times higher than that of the reduced Si nanopowder when it was observed at the same distance and with the same excitation intensity. To compare the spectra of the PL for the sintered Si nano-polycrystalline with that of the reduced Si nanopowder, the bandwidth for the sintered Si nano-polycrystalline was narrower than that of the reduced Si nanopowder. At this point, we cannot provide a definitive conclusion. However, it may be said from these experimental results that the PL for sintered Si nano-polycrystalline is amplified inside it.

Here, we discuss improvement methods for this sintered Si nano-polycrystalline and a method for tuning the peak wavelength of PL. Basic optical parameters, such as emission efficiency, fluorescence lifetime, and stimulated emission cross-sections, should be obtained by conducting experiments. Improving the transparency of the sintered Si nano-polycrystalline is also required.

The mean diameter of the prepared reduced Si nanoparticles by laser ablation in liquid should be selected to narrow the bandwidth of PL and obtain high optical gain. It has been decided that a stable PL should be obtained from the sintered Si nano-polycrystalline rather than from Si nanoparticles because the surface of the sintered Si nano-polycrystalline is chemically stable.

Highly repetitive pulse laser ablation in liquids makes it possible to produce amount of reduced and emission-capable Si nanoparticles from Si oxide powder. We can also fabricate sintered and emission-capable Si nano-polycrystalline with a low cost in a short time.

The sintered Si nano-polycrystalline will be applicable to optical waveguides, optical switches using a free carrier effect, and optical amplifiers. These results should also indicate that all optical circuits, such as laser light generators and optical signal detectors, can be fabricated on Si substrates.

4. Conclusion

We have succeeded in fabricating a sintered Si nano-polycrystalline for the first time. Si oxide powder is reduced by highly repetitive pulse laser ablation in liquid to produce Si nanoparticles. Si nanopaste with highly doped Si nanoparticles was sintered by using a hot plate, and a Si nano-polycrystalline was obtained. The structure and components of the sintered Si nano-polycrystalline were observed by SEM and EDX analysis. Furthermore, the reduced Si nanoparticles and the sintered Si nano-polycrystalline were excited by violet light, and stable PL in the visible regime was observed. The peak wavelengths of the PL were 550 nm and 560 nm. Strong PL of the sintered

Si nano-polycrystalline was observed especially. PL should be amplified in the sintered Si nano-polycrystalline when it is excited by violet light. These experiments show that because the mean diameters of the Si nanocrystals in the reduced Si nanoparticles were below 2 nm, the structure of the Si nanocrystals changed to a direct-transition type; the bandgap energy of the Si nano-crystalline was enlarged to 2.25 eV, and PL in the visible regime was generated. The sintered Si nano-polycrystalline should work as an optical waveguide, an optical switch using a free carrier effect, and an optical amplifier.

References

[1] L. T. Canham, "Silicon Quantum Wire Array Fabrication by Electrochemical and Chemical Dissolution of Wafers", Appl. Phys. Lett., 57, pp.1046-1048 (1990).

[2] V. Švrček, T. Sasaki, Y. Shimizu, and N. Koshizaki, "Blue Luminescent Silicon Nanocrystals Prepared by Ns Pulsed Laser Ablation in Water", Appl. Phys. Lett., 89, 213113-1-3 (2006).

[3] K. Saitow and T. Yamamura, "Effective Cooling Generates Efficient Emission: Blue, Green, and Red Light-Emitting Si Nanocrystals", J. Phys. Chem. C, 113(19), pp.8465-8470 (2009).

[4] N. Suzuki, Y. Yamada, T. Makino, T. Yoshida, "Laser Processing for Fabrication of Silicon Nanoparticles and Quantum Dot Functional Structures", The Review of Laser Engineering, 31(8), pp.548-551 [in Japanese] (2003).

[5] I. Umezu, A. Sugimura, M. Inada, T. Makino, K. Matsumoto, and M. Takata, "Formation of Nanoscale Fine-Structured Silicon by Pulsed Laser Ablation in Hydrogen Background Gas", Phys. Rev. B, 76, 045328-1-10 (2007).

[6] M. Hirasawa, T. Orii, and T. Seto, "Synthesis of Visible-Light Emitting Si Nanoparticles by Laser Nano-prototyping", J. of Aerosol Res., 20, pp.103-107 [in Japanese] (2005).

[7] T. Saiki, T. Okada, K. Nakamura, T. Karita, Y. Nishikawa, and Y. Iida, "Air Cells Using Negative Metal Electrodes Fabricated by Sintering Pastes with Base Metal Nanoparticles for Efficient Utilization of Solar Energy", Int. J. of Energy Science, 2(6), pp. 228-234 (2012).

[8] L. Brus, "Electrinic Wave Functions in Semiconductor Clusters: Experiment and Theory", J. Phys. Chem., 90, pp.2555-2560 (1986).

[9] N. Hill and K. Whaley, "Size Dependence of Excitations in Silicon Nanocrystals", Phy. Rev. Lett., 75, pp.1130-1133 (1995).

[10] H. Kobayashi, Y. Iida, and Y. Omura, "Single-Mode Silicon Optical Switch with T-Shaped SiO2 Optical Waveguide as a Control Gate", Jpn. J. Appl. Phys., 41, pp.2563-2565 (2002).

[11] D. K. Schroder, R. N. Thomas, and J. C. Swartz, "Free Carrier Absorption in Silicon", IEEE J. of Solid-state Circuits, SC-13, pp.180-187 (1978).

[12] R. A. Soref and B. R. Bennet, "Electrooptical Effects in Silicon", IEEE J. Q. E., 23(1), pp.123-129 (1987).

[13] L. R. Nuenes, T. K. Liang, K. S. Abedin, D. V. Thourhout, P. Dumon, R. Baets, H. K.Tsang, T. Miyazaki, and M. Tsuchiya, "Low Energy Ultrafast Switching in Silicon Wire Wavegide", ECOC 2005, 25-29 Sep. Glasgow, Scotland, Proceedings, 6, Paper Th 4.2.3.

[14] R. Terawaki, Y. Takahashi, M. Chihara, Y. Inui, and S. Noda, "Ultrahigh-Q Photonic Crystal Nanocavities in Wide Optical Telecommunication Bands", Opt. Express, 20(20), pp. 22743-22752 (2012).

[15] E. D. Palik, "Handbook of Optical Constants of Solids III", Academic Press, Massachusetts (1998) p529.

8

Self Cleaning PET Fabrics Treated with Nano TiO$_2$ Chemically Cross-Inked with Xanthenes Gum or Cyclodextrin

Amr Atef Elsayed[1], Omaima Gaber Allam[1, *], Sahar Hassan Salah Mohamed[2], Hussain Murad[2]

[1]Textile Research Division, National Research Centre, 33 Bohouth st. Dokki, Giza, Egypt
[2]Dairy Science, Food industry, Nutrition, National Research Centre, 33 Bohouth st. Dokki, Giza, Egypt

Email address:

omaimaalaam@yahoo.com (O. G. Allam)

Abstract: This paper would like to compare the ability of two cellulosic polymers to bind nano titania to polyester fabrics, in order to provide the fabric a self-cleaning property. The fixation of the nano titania on the polyester fabric was explored using Cyclodextrin or Xanthan gum. The photocatalytic activity of TiO$_2$ nanoparticles deposited on the polyester fabric was followed by the degradation of methylene blue as a model of an organic stain on the polyester fabric surface. The XRD patterns and SEM photographs of polyester fabric coated with nano titania were recorded. The different factors affecting the self-cleaning property as well as the fixation of nano titania was investigated.

Keywords: Nano Titania, Xanthan Gum, Cyclodextrin, Polyester Fabric

1. Introduction

Recent developments of nanotechnology directed to applications in textile areas including fibres are considered. Nanotechnology can provide high durability for fabrics, because nanoparticles have a large surface area-to-volume ratio and high surface energy, thus presenting better affinity for fabrics and leading to an increase in durability of the function. Some of the applications of nanoparticles to textiles are considered [1]. TiO$_2$ is one of the most popular and promising materials in photo catalytic application due to its strong oxidizing power. TiO$_2$ is commercially available and easy to prepare in the laboratory [2].

Several recent studies reported the promising potentials of nontoxic and inexpensive TiO$_2$ nanoparticles (TiO$_2$ NPs) for imparting multifunctional properties to different textile materials [3-8]. The compatibility of TiO$_2$ NPs surface with fiber surface chemical functionalities is one of the most important prerequisites for obtaining stable composite system and long-term durability effects. The tailoring of desirable fiber surface from the standpoint of its chemical functionality for improvement of TiO$_2$ attachment to the fabric surface is very important. NPs binding efficiency is recently gained much scientific interest, generally hydroxyl and particularly carboxylic groups are the potential sites for binding of TiO$_2$ NPs [9, 10]. The alginate was applied as a fiber surface modifier for improvement of binding efficiency between colloidal TiO$_2$ NPs and polyester fabric. The results imply good laundering durability of the fabric [11]. Cationization is a novel treatment on cotton to produce fabric with new characteristics. The nano titanin particles were stabilized on the cotton surface using butane tetra carboxylic acid [12].

Xanthan gum is a high molecular weight polysaccharide. It is used as a rheology control agent in aqueous systems and as a stabilizer for emulsions and suspensions. In textile printing, common thickeners such as guar gum, xanthan gum and sodium alginate are used to control the rheology of the dye pastes; Alginates, guar gum, and xanthan gum are often the thickeners of choice because they are pure non-reactive hydrodynamic thickeners unlike starches and other thickeners. In addition to fashion and comfort demands, the garments today must simultaneously provide self-cleaning properties, antimicrobial and UV protection. Utilizing these advantages is the ideal material for a wide range of uses, including filter materials in industrial applications [13]. The use of cyclodextrins and their derivatives (CDs) in the textile domain is a challenge that rose in the early 80's. The grafting of CDs onto cellulose fibers by using epichlorohydrin as crosslinking agent was reported [14], also it was covalently linked to the fabrics by the intermediate of the poly carboxylic acid (PCA) that esterified (or amidified) at the same time with the OH (or NH_2) groups of fibers and those of CDs [15]. Recently, several studies have reported the presence of nanocrystalline TiO_2 layers on textiles that were prepared from sol– gel at relatively low temperatures [16]. Keeping in mind, the polyester fabrics is one of the most consumed fabrics, so the present paper consists to report the results obtained in a study that aimed to apply the above mentioned chemical path onto polyester made fabrics. We describe the study of the parameters involved in this particular textile processing. These parameters include the curing conditions (temperature, time) and also the nature of the reactants. Citric acid (CTA) used as crosslinking agents; sodium dihydrogen hypophosphite (NaH_2PO_2) was used as catalyst. The methylene blue was used as a model for stain on the fabric surface.

2. Materials and Methods

2.1. Materials

Poly (ethyleneterephtalate) (PET) nonwoven fabrics (surface weight = 55 g/m^2) were supplied by Abou El-Ola for Spinning and Weaving, 10th of Ramadan, Egypt. B-Cyclodextrin (CD), Acros organic, USA. Titanium tetraisopropoxide {Ti ($OCH(CH_3)_2)_4$} 97% from Sigma-Aldrich. Citric acid hydrate (CTA), sodium, hypophosphite (NaH_2PO_2) and other chemicals were of a laboratory grade.

2.2. Synthesis of Xanthan Gum

Xanthan gum obtained from milk permeates by Xanthomonas campestris LIS-4 isolated from infected cabbage seeds [17]. The obtained xanthan gum has the following characteristics: Production yield of xanthan gum from milk permeate is 43.9 g/l, pyruvic acid content is 2.3 mg/100g and the viscosity of the aqueous solutions of xanthan gum is 4.8 CP.

2.3. Preparation of Nano Titanium

Single-phase anatase sol was prepared according to previous methods [18, 19]. Ten milliliter titanium tetraisopropoxide was added dropwise to 100ml acidic aqueous solution containing1ml nitric acid (70%) and 10 ml acetic acid (97%) under vigorous stirring. The mixtures were heated at 60°C and maintained at that temperature while stirring for 16h.

2.4. Finishing of Polyester (PET) Fabrics

The polyester fabrics were impregnated in an aqueous bath containing citric acid (5% w/w), the catalyst (NaH_2PO_2) (0.5% w/w) and the CD or gum) (5% w/w). Polyester fabrics were padded at wet pick up 100%. The fabrics were then dried at 90 °C, and cured at 180 °C for 30 min in case of CD [20], while the curing conditions were studied in case of gum. Samples were washed with water in soxhlet in order to remove unreacted products. The percentage grafting (% wt) of the samples were calculated from the following equation.

$$\% wt = \frac{wt_2 - wt_1}{wt_1} X100$$

Where wt_2 and wt_1 are the weight of sample after and before treatment, respectively.

2.5. Treatment of Grafted Polyester Fabrics with TiO2 Sol

Virgin polyester as well as pretreated polyester fabrics were subjected to a dip-pad–dry-cure process. The substrates were dipped in TiO_2 sol (containing 1.6% Ti) for 1 min, and padded at wet pick up 100%. The substrates were oven dried, and finally cured at 120°C for different periods of time. The fabrics were then washed in an ultrasonic bath to remove the unbound nano titania from the fabric surface.

2.6. Characterization of the Prepared Nano Titania

2.6.1. Transmission Electron Microscopy (TEM) Characterization

The surface morphology of the prepared nano titania was investigated using transmission electron microscope JEM - 1230 JEOL Co. Japan, maximum magnification power is 600kX.

2.6.2. X-ray Diffraction (XRD) Analysis

The XRD patterns of the prepared nano titania were recorded by using PAN Analytical X Pert PRO X-ray diffractometer and Cu Ka radiation (λ=1.5406 Å) as X-ray source. The scanning rate were carried out at 8 °/min in 2θ range of 20 °-70 ° (θ being the angle of Braggs diffraction), using a wavelength of (λ=1.5406 Å). The crystal diameter has been determined according to Deby-Scherrer formula:

$$D=K \lambda/ (\beta \cos \theta)$$

Where K is the Scherrer constant (0.9), the radiation wavelength λ=0.15406 nm, β is the peak width of half maximum; and θ is the Braggs diffraction angle and D is the particle size.

2.7. Characterization of the Modified Polyester Fabrics

Virgin polyester as well as modified polyester fabrics were mounted on aluminum stubs, and sputter coated with gold in a 150 Å sputter (coated Edwards), and examined by Jeol (JXA-840A) Electron Probe Microanalysis (Japan), magnification range 35 – 10,000, accelerating voltage 19kV.

The amount of nano titania absorbed by the PET fabrics was determined by boiling 1 g fabrics in 100 ml concentrated sulfuric acid for 5 hours, the eluted titanium was estimated using atomic absorption spectrometer (VARIAN AA220).

2.8. Photocatalytic Activity of the Nano Titania Estimation

The photocatalytic activity of the nano titanium oxide coated PET fabric was investigated by treating the fabric with 2 ml methylene blue (MB) solution (1%) followed by drying at ambient temperature. One half of each methylene blue stain on the fabric was exposed to 400W high pressure mercury lamp for different periods of time. The other half was enclosed with a black paper to avoid its irradiation from the lamp. K/S (absorption to scattering coefficient where, K is the absorption; S, the scattering coefficient.) was assessed to each stain on the fabric (the enclosed and the unenclosed parts) after the exposure to the UV irradiation, using Hunter Lab Universal Software MiniScan TM XE. The efficiency of nano titanium to remove the methylene blue (MB) stain was calculated according to the following equation.

$$\% \ Removal \ of \ (MB) = \frac{(K/S) \ befor \ irradiation - (K/S) after \ irradiation}{(K/S) \ befor \ irradiation} X100$$

2.9. Washing Durability Test

Adopting the AATCC test method 61 – 2003, the washing fastness of the treated fabrics was assessed. Washing was performed using a home laundering machine. The fabrics were washed in 0.15% aqueous solution of sodium lauryl sulphate as detergent and in the presence of 50 steel balls for 45 min. The fabrics were then rinsed with water, and dried at room temperature.

3. Results and Discussions

3.1. Characterization of Nano Titania

XRD data (Fig. 1) show distinct diffraction peaks at 2Θ of 25.28, 27.41, 46.9, and 54.6. The sharp peaks confirmed the crystallinity of the prepared nano titanium dioxide. The data also indicates that the majority of the prepared nano titanium dioxide is in the anatase form, and according to Scherrer's equation the average nano titanium dioxide crystal size is 20-25 nm. Figure 2 shows the TEM photograph of the prepared nano titania, and the average particle size calculation showed that the prepared titania is in the size of 25 nm, which is in consistence with the value obtained from the XRD analysis.

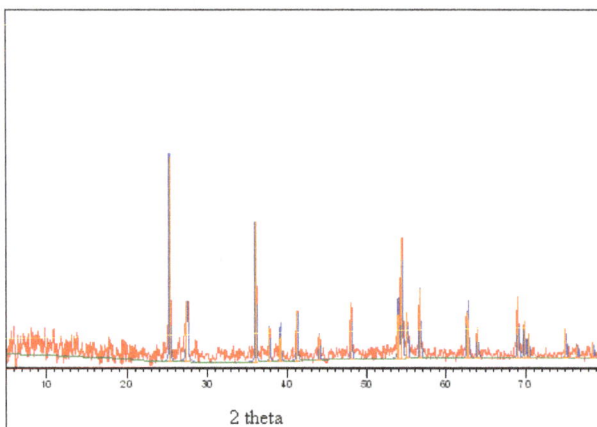

Figure 1. XRD pattern of the prepared nano titania.

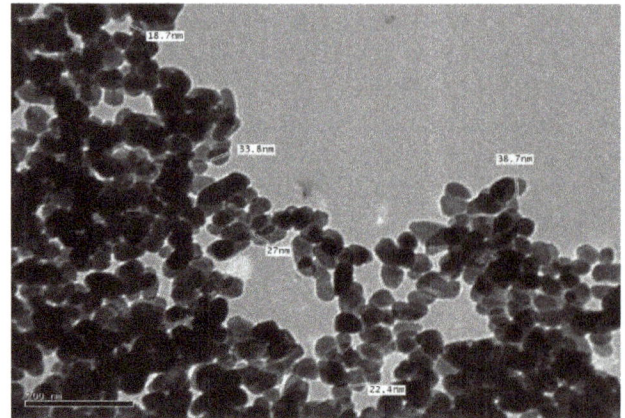

Figure 2. SEM photograph of the prepared nano titania.

3.2. Effect of Curing Conditions

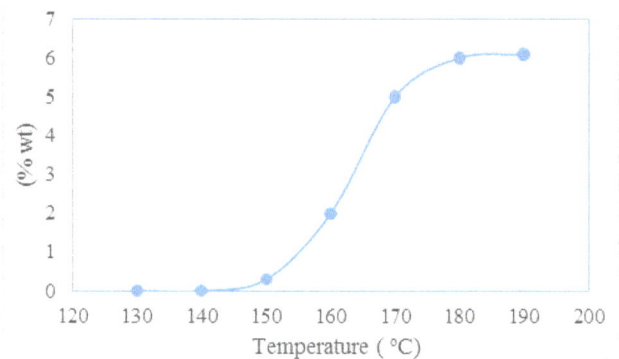

Figure 3. Effect of curing temperature on the % wt of gum added to PET fabrics. (Curing time 10 min).

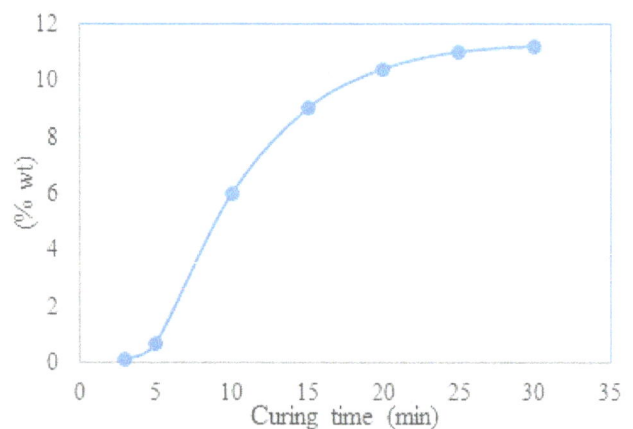

Figure 4. Effect of curing time on the % wt of gum added to PET fabrics. (Curing temperature 180 °C).

The curing conditions between gum, CTA and PET fabrics were studied at variable temperatures and time. It is observed that, for ten minutes curing, as the curing temperature increase, the (%wt) increases to reach its maximum at 180 °C (Figure 3) after which no appreciable increase in the (%wt) was observed. Figure 4 shows that, the maximum (%) grafting (11%wt) could be attained after 25 min curing at 180 °C. It is worth mentioning that, the %wt of CTA and CD added to PET fabrics was 8.1% at curing conditions of 180

°C for 30 min. The PET fabrics treated by a mixture of CTA and CD or gum resulted in a weight increase of the samples. We could confirm that a permanent grafting occurred as this value did not vary upon the successive washings. The direct reaction between CD and PET fibers has been proved to be

impossible. It is concluded that the reaction occurred through the polyesterification between CTA and CD and the formed copolymer is physically adhered or was even entangled into the fibrous network so that grafting was resistant to washings and was thus permanent [15].

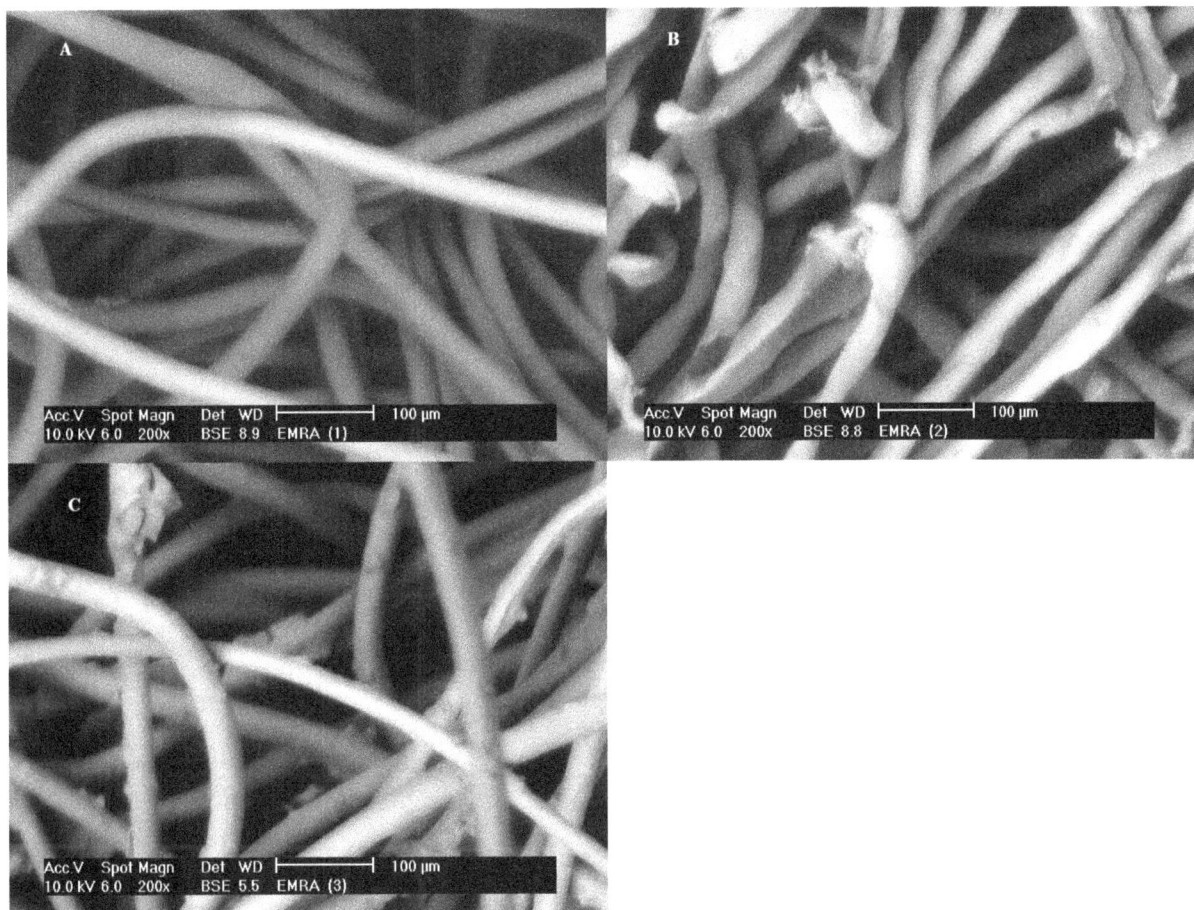

Figure 5. SEM images of PET fibers (A) untreated PET, (B) PET treated with CD & CTA and (C) PET treated with gum &CTA).

3.3. Characterization of PET Fabrics Treated with CD or Gum

The presence of CD or gum on the PET fibers was confirmed by SEM analysis Fig 5. The SEM photographs are also show the formation of homogeneous layer of CD or gum on the BET fibers surface, as well as the presence of small aggregates in case of gum.

3.4. Influence of TiO₂ Concentration and Curing Time on the % MB Removal

The self cleaning property of PET fabrics treated with CTA and CD and after treated with nano titania was studied. Values of % MB dye removal were presented against the nano titania concentration and curing time in Table 1.

Data of Table 1 shows the inability of virgin PET fabric to remove the MB dye. While in both cases of unmodified as well as modified PET fabrics with CTA or CD and after treated with nano titania, it was found that, as the concentration of titanium increased from 0.3% to 0.9%, the % dye removal increased; it is also observed that, higher

titanium concentration leads to a decrease in % dye removal. This observation could be explained as a result of the coagulation of the nano particles at higher concentrations on the fabric surface, and therefore losing the preferential of the nano size. The data also clarify the % dye removal form PET fabric treated with CTA-CD and after treated with nano titania is much higher than that of PET fabrics only treated with nano titania, which could be referred to the ability of the unreacted carboxyl groups of CTA to attract larger amount of nano titania to the fabric surface [13].

Maximum dye removal of (91.7%) was attained when nano titania cured with the modified PET fabrics for 25 min, and that seems to be the required time for the formation of a stable coordination between Ti^{4+} and the created functionalities on PET fabric.

The aforementioned study was repeated for the PET fabrics treated with CTA and gum (Table 2). It should be underlined that PET fabrics treated with CTA-gym and after treated with nano titania shows higher ability to remove MB dye in comparison with those of CD-CTA. Additionally curing for 15 min is quite enough to reach the maximum %

of dye release.

Table 1. Effect of nano TiO_2 and curing time on the% of MB dye removal from PET fabrics pretreated with CTA-CD.

Type of modification of PET fabrics	Conc. Of nano titania (%)	Curing time of nano titania and fabrics (min)	% MB dye removal after UV irradiation for		
			30 min	60 min	120 min
MB dye solution only			2.23	7.38	9.34
Unmodified PET and without nano titania			3.22	8.63	10.29
Unmodified PET	0.3	25	25.46	30.95	56.84
	0.6		28.59	39.19	61.36
	0.9		31.73	59.19	71.91
	1.2		29.38	44.68	62.93
	1.5		25.85	42.72	62.14
PET modified with CTA - CD	0.3	15	47.03	63.51	72.14
	0.6		52.53	65.08	77.63
	0.9		56.06	67.04	87.56
	1.2		44.29	64.29	83.13
	1.5		44.68	63.51	77.63
	0.3	20	52.13	67.04	72.93
	0.6		54.09	75.67	79.99
	0.9		59.19	79.99	90.23
	1.2		56.84	74.89	78.42
	1.5		53.31	72.93	75.67
	0.3	25	60.37	69.79	79.99
	0.6		61.16	73.32	81.56
	0.9		63.12	83.52	91.761
	1.2		50.56	80.77	86.66
	1.5		41.54	77.24	81.56

Table 2. Effect of nano TiO_2 and curing time on the% of MB dye removal from PET fabrics pretreated with CTA-gum.

Type of modification of PET fabrics	Conc. Of nano titania (%)	Curing time of nano titania and fabrics (min)	% MB dye removal after UV irradiation for		
			30 min	60 min	120 min
PET modified with CTA - gum	0.3	15	56.45	64.29	87.44
	0.6		61.16	71.75	89.40
	0.9		61.94	76.06	95.51
	1.2		61.16	73.32	92.5
	1.5		60.37	72.53	91.36
	0.3	20	63.90	74.10	85.48
	0.6		65.47	84.30	90.19
	0.9		67.04	85.48	96.07
	1.2		65.08	84.69	94.50
	1.5		64.29	83.52	94.50
	0.3	25	61.94	73.71	87.83
	0.6		67.83	79.99	92.15
	0.9		69.79	87.83	96.50
	1.2		67.04	86.66	92.54
	1.5		65.47	84.69	92.15

The superior ability of PET fabrics treated with gum and after treated with nano TiO_2 over CD even at lower curing time, could be explained on the bases of the ability of gum to absorb higher amount of nano titania. In order to elucidate that assumption, the titanium existed on PET fabric modified either with gum or CD was leached out; and the obtained amount of titanium was presented in Table 3. It is clear that CTA–gum is able to bind higher amount of titanium than CTA-CD, which could be referred to the availability of unreacted carboxyl groups of CTA [21] in addition to readily existed carboxyl groups in gum macromolecules (pyruvic acid content 2.3 mg/100g), which is known for its affinity toward nano titania. Similar results was observed in previous study about the affinity of PET fabrics treated with alginates to bond to nano titania [11]. Namely, when diameter of nanocrystalline anatase TiO_2 particles becomes around 20 nm or smaller, the surface of Ti atoms adjust their coordination environment from octahedral to more reactive pentacoordinated (square pyramidal) [22,23]. These undercoordinated defect sites are likely the sources of enhanced binding between carboxyl functionalities and Ti atoms. Bearing in mind, CTA-CD only counts on the unreacted carboxyl groups of CTA for the bonding to TiO_2.

Table 3. Concentration of TiO_2 extracted from PET fabrics.

Modification	Initial conc. of TiO_2 sol (%)	Conc. of TiO_2 extracted (g/100g fabric)
PET unmodified (25 min)		0.01
PET+ CTA+CD (15 min)		0.05
PET+ CTA+CD (20 min)		0.09
PET+ CTA+CD (25 min)		0.10
PET+ CTA+CD (30 min)	1.6	0.11
PET+ CTA+ gum (15 min)		0.33
PET+ CTA+ gum (20 min)		0.35
PET+ CTA+ gum (25 min)		0.35
PET+ CTA+ gum (30 min)		0.37

3.5. Washing Durability Evaluation

In order to elucidate the durability of photocatalytic activity of nano TiO_2 deposited on the PET fabrics treated with CTA-gum or CTA-CD, the photo degradation process under the UV irradiation was repeated after different washing cycles as presented in Figure 3. It can be noticed that the xanthan gum is able to maintain the bonded nano titania without noticeable decrease in the self cleaning property of the PET fabrics, even after 10 washing cycles. It is worth mentioning that the PET fabrics treated with CTA-CD and after treated with nano titania can sustain the self cleaning property after the repeated washing cycles but in a manner less than that of gum. On the contrary, unmodified PET fabric and treated with nano titania loses all of its self cleaning ability after 5 washing times. The overall results confirm the formation of strong and stable bond between carboxyl groups of gum with Ti^{4+}.

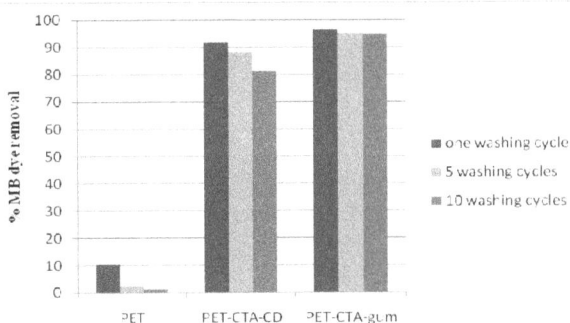

Figure 6. Washing durability test (TiO_2 conc. 0.9% cured at 20min, irradiation time 120 min).

4. Conclusions

The purpose of this study was to investigate the self-cleaning properties of modified PET fabrics with CD or gum and after treated with nano TiO_2. Regarding to the curing conditions, the % wt of CTA and gum added to PET fabrics maximum % grafting (11%wt) could be attained after 25 min curing at 180 °C, while the % wt of CTA and CD added to PET fabrics was 8.1% at curing conditions of 180 °C for 30 min. The XRD data and TEM photograph of the prepared nano titania indicate that the majority of the prepared nano

titanium dioxide was in the anatase form with crystal size is 20-25 nm. Examination of the self-cleaning property of the modified PET fabrics with CTA and CD or gum and after treated with nano titania showed maximum dye removal of using CD was attained when nano titania cured with the modified PET fabrics for 25 min, while curing for 15 min is quite enough to reach the maximum % of dye release in case of gum. The PET fabrics treated with CTA-gym and after treated with nano titania shows higher ability to remove MB dye (95.5%) in comparison with those of CD-CTA (91.7%). The xanthan gum is able to maintain the bonded nano titania without noticeable decrease in the self cleaning property of the PET fabrics, even after 10 washing cycles.

References

[1] Baeyer, H. C. V. The Lotus Effect. Sciences, vol.40, pp12-15, (2000).

[2] Kathirvelu, S., Louis, D., and Bharathi, D. "Nanotechnology applications in textiles". Ind. J. Sci. Tech., vol. 1, pp 1-10, (2008).

[3] Bozzi, A., Yuranova, T., & Kiwi, J. "Self-cleaning of wool-polyamide and polyester textiles by TiO_2-rutile modification under daylight irradiation at ambient temperature. J. Photochem. Photobio. A, vol. 172, pp 27–34, (2005).

[4] Daoud, W. A., & Xin, J. H. "Low temperature sol–gel processed Photocatalytic titanium coating. J. of Sol–Gel Sci Tech" vol. 29, pp25–29, (2004).

[5] Daoud, W. A., & Xin, J. H. "Synthesis of Single-Phase Anatase Nanocrystallites at Near Room Temperatures" Chem. Commun., vol. 16, pp 2110–2112,(2005).

[6] Daoud , b.b., Leung,W. A. Tung, S. K., Xin, W. S. J. H., Cheuk , K., & Qi, K. "Self-cleaning keratins" Chem. Mater., vol. 20, pp1242–1244, (2008)

[7] Dong, Y., Bai, Z., Zhang, L., Liu, R., & Zhu, T. "Finishing of cotton fabrics with aqueous nano-titanium dioxide dispersion and the decomposition of gaseous ammonia by ultraviolet irradiation". J Appl. Poly. Sci., vol. 99, pp 86–291(2006).

[8] Fei, B., Deng, Z., Zhang, Y., & Pang, G. "Room temperature synthesis of Rutile nanorods and their applications on cloth". Nanotechnology, vol. 17, pp 1927–1931, (2006).

[9]   Mihailovic, D., Radetic, M., Ilic, V., Stank Vic, S., Jovancic, P., Potomac, B. "Modification of corona pretreated polyester fabrics with colloidal TiO$_2$ nanoparticles for imparting specific properties". Proceedings of Aachen Dresden International Textile Conference. Dresden, Germany (2008).

[10]   Bozzi, A., Yuranova, T., Guasaquillo, I., Laub, D., & Kiwi, J." Self-cleaning of modified cotton textiles by TiO$_2$ at low temperatures under daylight irradiation" J. Photochem Photobio A, vol. 174, pp 156–164, (2005).

[11]   Darka, M., Zoran , Š., Marija , R.,Tamara , R., Petar , J., Jovan, N., Maja , R. "Functionalization of polyester fabrics with alginates and TiO$_2$ nanoparticles" Carbo. Poly., vol. 79, pp 526–532, (2010).

[12]   Nazaria, A., Montazerb, M., Moghadamc. M.B., Anary-Abbasinejadd, M. "Self- cleaning properties of bleached and cationized cotton using TiO$_2$: A statistical approach". Carb. Poly., vol. 83, pp 1119–1127, (2011).

[13]   Wong, Y. W. H., Yuen, C. W. M., Leung, M. Y. S., Ku, S. K. A., & Lam, L. I. "Selected applications of nanotechnology in textiles" Autex Res. J., vol. 6, pp 1–8, (2006).

[14]   Szejtli, J., Zsadon, B., Fenyvesi,E., Horvarth,O., & Tudos, F.(1982)US Patent 4,357,468 .

[15]   Martel, B., Morcellet, M., Ruffin, D., Ducoroy, L., & weltrowski, M. "Finishing of Polyester Fabrics with Cyclodextrins and Polycarboxylic Acids as Crosslinking Agents". J. Inclu. Phen. Macro. Chem., vol. 44, pp 443–446, (2002).

[16]   Kaihong , Qi., John ,H. X., & Walid, A. D. "Functionalizing Polyester Fiber with a Self-Cleaning Property Using Anatase TiO$_2$ and Low-Temperature Plasma Treatment" Int. J. Appl. Ceram. Technol., vol. 4, pp 554–563, (2007).

[17]   Abd El-Gawad, I.A., Murad, H.A., El-Sayed, E.M., Salah, S.H."Optimum conditions for production of xanthan gum from hydrolyzed UF- Milk permeate by locally isolated Xanthomonas campestris". Egypt J Dairy Sci., vol. 1, pp 29-37, (2001).

[18]   Qi, K. H., Daoud, W. A., Xin, J. H., Mak, C. L., Tang, W. S., & Cheung, W. P. "Self-Cleaning Cotton" J. Mater. Chem., vol. 16, pp 47, 4567–4574, (2006).

[19]   Daoud, W. A., Xin, J. H., & Zhang, Y. H."Surface functionalization of Cellulose fibers with titanium dioxide nanoparticles and their combined bactericidal activities "Surf Sci, vol. 599 .pp 69–75, (2006).

[20]   Ducoroy, L., Martel, B., Bacquet, B., & Morcellet, M. "Ion exchange textiles from the finishing of PET fabrics with cyclodextrins and citric acid for the sorption of metallic cations in water". J. Incl. Phenom. Macrocycl. Chem., vol. 57, pp 271–277, (2007).

[21]   Laurent, D., Maryse, B., Bernard, M., Michel, M. "Removal of heavy metals from aqueous media by cation exchange nonwoven PET coated with b-cyclodextrin-polycarboxylic moieties". Reac. Func. Poly., vol. 68, pp594–600, (2008).

[22]   Chen, L. X., Rajh, T., Jager, W., Nedeljkovic, J., & Thurnauer, N. C." X-ray absorption reveals surface structure of titanium dioxide nanoparticles". J. Synchrotron Rad., vol. 6, pp445–447, (1999).

[23]   Rajh, T., Chen, L. X., Lukas, K., Liu, T., Thurnauer, M. C., & Tiede, D. M. "Surface restructuring of nanoparticles: An efficient route for ligand–metal oxide crosstalk". J. Phys. Chem. B, vol. 106, pp 10543–10552, (2002).

9

Investigation of Surface Structure and Thermostimulated Depolarization Effect of Composite Materials with Aluminum Nano-particles

Eldar Mehrali Gojayev[1], Khadija Ramiz Ahmadova[2], Sevinc Sarkar Osmanova[3], Shujaat Zeynalov Aman[4]

[1]Department of Physics and Research Laboratory of the Department in "Physics and Technology of Nanostructures" Azerbaijan Technical University, Physical and Mathematical Sciences, Honored Scientist of the Republic of Azerbaijan, Baku, Azerbaijan

[2]Senior Laboratory Laboratory, "Thermophysical Properties of Oil and Petroleum Products", Department of Physics, Azerbaijan Technical University, Baku, Azerbaijan

[3]Physics, Physical and Mathematical Sciences, Department of Physics, Azerbaijan Technical University, Baku, Azerbaijan

[4]Physical and Mathematical Sciences of Azerbaijan Technical University, Department of Physics, Baku, Azerbaijan

Email address:

geldar-04@mail.ru (E. M. Gojayev), Yubaba66@hotmail.com (K. R. Ahmadova), daysi68@mail.ru (S. S. Osmanova), sucaetz@mail.ru (S. Z. Aman)

Abstract: To investigate the effect of semiconductor filler - compounds and aluminum nanoparticles: topography and the nature of the spectra of thermally stimulated depolarization (TSD) of composite materials $PE+xvl.\%TlInSe_2$, $PE+xvl.\%TlInSe_2<Al>$. A technology of producing composite materials studied by atomic force microscopy, the surface microrelief. The polarization of the films was carried out at a constant voltage of 6 kV in 5 min, according to the method of corona discharge. To study the spectra of the TSD used conventional techniques. Analysis of histograms AFM showed that the surface uniformity of the samples varies between 25 nm and in the boundary layer of the compositions there are some roughness caused by the fact that the destruction of the binding forces remain on the surface of individual atoms, and their groups - clusters.

Keywords: Aluminum Nano-additive, Fourier Spectra, Composites $PE+xvl.\%TlInSe_2$, $PE+xvl.\%TlInSe_2<Al>$, Thermally Stimulated Depolarization

1. Introduction

In recent years, investigation in the field of creation of materials with special and the almost important electrophysical properties on the basis of polymer structures is of great interest [1-8]. It should be noted that depending on the nature, size, shape, nature of the distribution of the filler the polymer composition may be electroconductive [9], antistatic or dielectric.By these reasons, the findings of various authors differ concerning about nature of occurrence of various electroactive properties of composites when filling poliolefinola different fillings [10-14]. For obtaining new composite materials are used fillers of organic and inorganic nature. In relatively new papers [10, 14, 15], terpolymers semiconductor compounds $TlInSe_2, TlGaSe_2$ and the solid solution $TlIn_{0.98}Ce_{0.02}Se_2$ serve as filler. The investigations show that the compositions $LDPE+TlInSe_2$ where $LDPE$ is a lower density polyethylene are superfine electrets materials with lifetime by $5 \div 13$ times exceeding pure polymers [10, 15]. In [10] microrelief of the surface of compositions $LDPE+xvol.\%TlInSe_2$ and $yvol.\%Al$ was studied by the methods of atomic force microscope. At the same time, the scanning probe microscopy is one of the powerful up-to-date methods for studying morphology and local properties of a solid with higher spatial resolution. For the last 10 years, the method available to limited number of research groups became a widespread and successfully used tool for studying the properties of a surface. Today none of the works in the field of physics of surface and thin-film technologies manage without the methods of scanning probe microscopy. Development of the scanning probe microscopy served also for development of new methods of nanotechnologies, the technologies for creating structures with nanometer scales. In [1, 14] the microrelief of the surface of compositions $LDPE + TlInSe_2$ was studied by the methods of atomic force microscope.

It is known that for detailed study of the mechanism of relaxation processes proceeding in polymers and composite systems, various dielectric methods belonging to relaxation spectrometry, various dielectric methods are used. However, for the frequencies $\leq 10^{-1} Hz$ direct measuring of dielectric losses are connected with great difficulties, therefore by studying molecular mobility in polymers by the dielectric method in frequency range $10^{-5} - 10^{-1}$ the direct current method is used. To this end, it is reasonable to use the data on temperature dependences of currents of thermostimulated depolarization I, depolarization function ψ and other parameters dependent on steady leakage current.

In this connection, the goal of the paper is to study the spectra of thermostimulated depolarization of new composite materials $PE + TlInSe_2$ and $PE + Al$ where PE is polyethylene.

By analyzing the currents of thermostimulated depolarization (TSD) of films on the basis of samples of compositions $PE + TlInSe_2$, $PE + Al$ crystallized at tempering conditions, it was revealed that on TSD curves, the observed depolarization peaks belong to release of charges from a trap connected as separate components $PE, TlInSe_2$ and Al .

2. Techniques of Experiment

Composition materials $PE + TlInSe_2$ with participation of aluminum nano-particle were obtained by the following technology the power of the polymer is mixed with powder $TlInSe_2$. Mixing was realized at the laboratory mill at room temperature, and then by hot pressing under pressure $10^7 Pa$ the samples of dispersity 50 μm were obtained. From the obtained mixture, different thickness films, at melting temperature of polymer matrix and pressure $10 - 15 MPa$ are extruded between aluminum foil. The samples with foil are quickly cooled in water and then the foil is removed [3].

The samples obtained in such a way are suitable for studying the structure and properties. The films $PE + xvol\%TlInSe_2$ with participation of nano-particles were obtained in the same way.

The size of the aluminum nano-particle used in the experiment was 50nm. We investigated the relief of the surface of compositions $LDPE + xvol\%TlInSe_2$ and $yvol\% Al$ by AFM method [14, 15]. The thermostimulated depolarization (TSD) spectra were studied by the scheme indicated in fig 1.

The installation consists of a shielding (8) a heater built therein (2), a camera (1) located between two electrodes (3), thermocouples in the lower hole of the electrode (4). Heating

is carried out by a system consisting of three LATR (5).

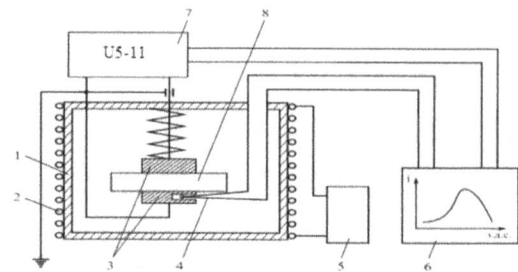

Fig. 1. *Apparatus for measuring the thermally stimulated depolarization currents.*

Thermodepolarization current from the upper electrode located above the sample, through special high-resistance cable flows to the entry of electrometric amplifier (U5-11) (7). From the output of the amplifier, the signal is submitted to the entry of the recording device (6).

According to the spectra of thermally activated current one can define power and concentration parameters of dipoles, electronic or ionic centers of the capture of charge carriers. Activation energy E_a , frequency factor ω and concentration of carriers n belong to these parameters. Most simply these parameters are determined by the curves of the current of TSD. In particular, they may be determined by temperature condition of maximums of the current of TSD by the method of variation of heating velocity, by typical points of the curve of TSD and so on [14]. However, in all of these methods for calculating the activation energy, only separate points of experimental curve are used. This reduces to loss of a part of information and increase of error of measurement of parameters. More precise and although bulky are the methods where all experimental curve is used. Let us in short consider the methods for defining above parameters according to spectra of TSD.

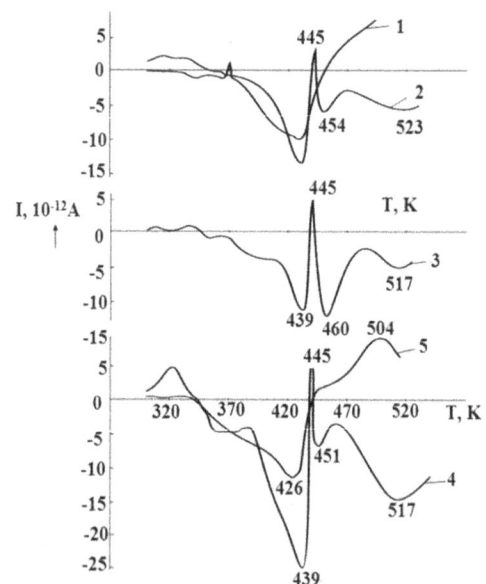

Fig. 2. *The spectra of thermally stimulated depolarization compositions* $PE + xvol.\%TlInSe_2$; *1-x=0; 2-x=1; 3-x=3; 4-x=5; 5-x=10.*

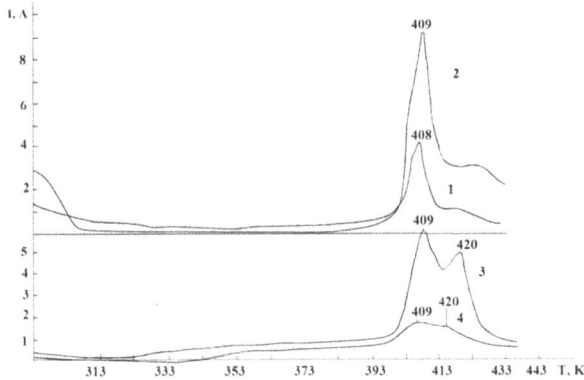

Fig. 3. The spectra of thermally stimulated depolarization compositions PE+xvol.%TlInSe$_2$+yvl.%A; 1– x=7, y=3; 2 – x=5, y=5; 3 – x=3, y=7; 4 – x=10, y=10.

Heating of the sample happened strictly linearly by means of three LATR system with constant speed $2^0C/min$, therefore the two-coordinate self-recording device 6 in which change of current depending on temperature in the coordinates $x - y$ was written.

3. Results and Discussion

As a result we obtained planar images of the surface of composites $LDPE + xvol.\%TlInSe_2$ (where $x = 1;3;5;10$) of size $1,6\cdot10^3 \times 1,6\cdot10^3$.

The analysis of the histogram of the AFM images showed that the homogeneity of the surface varies from 25 nm. However, in the boundary layer compositions remain some roughness is probably due to the fact that at fracture of tractions not separate atoms but their groups, i.e. clusters remain on the surface. It also shows the results of a Fourier analysis of the surface obtained by the AFM. The concentration at the center of the spectrum shows that the surface particles have approximately the same dimensions i.e. they are comparable.

Investigation of the TSD phenomenon was carried out in temperature ranges 293-543K at linear growth of temperature of samples by the method described in the papers [1-7].

The results of investigation of TSD spectra of composite materials with semi-conduction additive are in fig. 2, where the spectra of TSD for PE (curve 1) and for the samples $LDPE + TlInSe_2$ with different content of filler (curves 2, 3, 4, 5). Typical for these spectra is that on the curves of TSD samples of compositions with content of filler 1, 3, 5 vol. % we observe clear inversion narrow peak at temperature 445 K. Furthermore, we observe high-temperature wide peak in the domain of temperature $500 \div 520K$. These structures have three identical maxima. Temperature location of the first maximum corresponds to $432 \div 439K$, of the second one to $450 \div 460K$, of the third one to $517 \div 523K$. The samples of composite containing $10vol\%TlInSe_2$, have two maxima with opposite signs of the accumulated charge. The first maximum of 426K K corresponds in temperature to the first maximum of other composites and the base polymer, the second maximum in temperature to the third maximum of composites.

Analyzing the spectra of TSD we can note that introduction of the filler $1 \div 5$ $vol\%$ into the polymer reduces to appearance of deeper center of capture of carriers. Herewith, the number of injected charges at corona discharge (increase of intensity and area of appropriate maximum) and the depth of their occurrence increases (temperature position shifts in the high temperature area).

The nature of occurrence of the observed inversion peak at temperature 445 K on the background of the large main peak in the $432 \div 455$ K may be explained in the following way. At electrets in the process of action of the corona charge it is formed a space charge, and in the field of space charges (SC), on the boundaries of the particles of $TlInSe_2$ and of a polymer an interfacial polarization (IP) is formed [5]. Direction of this polarization is opposite to the field of space charges. In this case, at depolarization in the spectra of TSD we observe inversion current caused by IP. Explanation of inversion current formation agues in some sense with the Maxwell-Wagner effect where it is considered that accumulation of charges on homogeneous materials (in our case on composites) is due to difference of conductivity in amorphous and crystalline phases. At electrification of such a material, the carrier will gather near the given interfacial boundary, or vice versa will be away from it depending on which of two currents has great conductivity. Difference in local currents of conductivity reduces also to dissipation of charges when removing the currents of TSD, as in this case currents flow already in opposite direction. It should be noted that the given maximum in great degree is related with relaxation of charges on the surfaces of particles of $TlInSe_2$ as with increasing the content of the filler in compositions, the quantity of the peak increases.

Effects of inversion of the sign of the current of TSD were observed also for other electrically active dielectrics, electrets and polymer composites [4]. But interpretation of these phenomena remains discussible. Depending on temperature condition of the inversion current, the state of the surface, nature of a polymeric matrix and filler, nature of polarization and other factors, the observed currents with the inverse sign on the curves of TSD are connected also with reorientation of dipoles existing in a polymer. To our mind, the maxima at temperature $432 \div 439K$ and $450 \div 460K$ have the same nature, i.e. both of these maxima are a part of one and the same maximum related to α-relaxation in PE. The observed inversion in narrow range of temperature is the result of reduction of conductivity of $TlInSe_2$ at polarization of spaces charges in IP in the field of SC. We can suppose that the volume content of $TlInSe_2$ in the composite may influence on temperature position of α-relaxation process.

The third high –temperature maximum for $517 \div 523K$ may be connected with increase of intrinsic conductivity of a filler as the quantity of the peak increases with increasing the filler's content in the composition. For quantitative estimation of the thickness of interfacial layer in heterogeneous polymeric compositions, in a number of cases we use the representation on formation of a double layer (such a point of view is widely held in extrinsic semiconductors, where difference in conductivity reduces to formation of double layer).

In polymeric mixtures and composites, the thickness of interfacial layer may be estimated by the formula of [16].

$$d_M^2 = \frac{2\varepsilon_1\varepsilon_2\varepsilon_0 kT}{n \cdot e^2}$$

where ε_1 and ε_2 are dielectric permeabilities of each phase, ε_0 is a dielectric constant, n is the concentration of carriers of charge (for polymeric dielectrics $n = 10^{21} m^{-3}$ is the charge of the electrode, k is the Boltzmann constant, T is absolute temperature. Calculations of values of d_m by the given formula show that they make up about 0, 4 ÷ 1,2 μm.

The results of investigations of properties of the sample of compositions $LDPE + TlInSe_2$ crystallized at hardening conditions at 273 K are given in fig. 2. The analysis of the obtained results shows that on the curves of TSD we observe a number of depolarization peaks in certain temperature fields. These peaks belong to release of charges from traps connected both separate components (PE and $TlInSe_2$) and shaped interfacial polarization in the field of space charges. At temperature 445 K on the curves of TSD of the samples of compositions we observe inversion peak with semi-length 3 ÷ 5 K. In the spectra of TSD, on electret compositions with $TlInSe_2$ at temperatures 515 ÷ 520 K we find depolarization peak connected with a new center (of traps) of stabilization of electrets charges.

The fillers $TlInSe_2$ with p-conductivity and dispersity 50 ÷ 63 μm in competitions with a polyethylene play the role of structurant observed in the increase of the degree of crystallimity and change of supramolecular structure of a polymer.

The results of investigations of spectra of TSD of compositions with aluminum nanoparticles are given in fig. 3. The composites $3\% Al, 5\% Al, 7\% Al, 10\% Al$ were studied. From fig. 3 it follows that for the composite $3\% Al$ on the spectrum $I(t)$ we observe one strongly marked maximum at temperature 408 K, weak minimum at 413 K. At wide temperature range 298-393 K the current remains constant, then harshly increases to 4,2 A, and in the same way converges to zero. On the spectrum of TSD of the composite $5\% Al$ we also observe a strongly marked maximum at temperature 409 K. At this temperature the current achieves to 9,5 A and with further increase of temperature, it decreases. For 415 K we observe weak minimum and decrease of current for this composite. It was revealed that at temperature range 273-397 K the current about 0,5 A remains constant. On the spectrum TSD $7\% Al$ we observe two bright maxima at temperatures 409 K current 6,2 A and 42K current 4,9 A. Between these maxima at 413 K it was revealed a deep minimum of the current 3,5 A. Note that also for this composite at temperature range 273-400 K the current 0,4 A remains constant. For the composite $10\% Al$ on the curve $I(t)$ we obverse a diffuse maximum at temperature range 405-416 K and for this composite in the wide range 273-403 K the current remains constant.

4. Conclusion

By investigating the three-dimensional image of the microrelief of the surface of composites it was revealed that at boundary layer of compositions the roughness of the surface decreases, smoothens. It was revealed that putting a filler into a polymeric matrix reduces to occurrence of deep centers of capturer of carriers at corona and the occurrence depth. Under the influence of aluminum nanoparticles the current of TSD is stabilized to melting temperature of the composite's matrix.

References

[1]   Smirnov A. V., Fedorov B. A., Temnov D. E. Structural and electret properties of polypropylene with different content of amorphous silica// Nanosystems: Physics, Chemistry, Mathematics, 2012, 3 (2), p. 65-72.

[2]   Dark D. E., Fomichev E. E., Fedorov B. A., Smirnov A. V. Oglomeration filler particles aluminum-based composite films of polypropylene// Proc. XII Intern. conf. "Physics of Dielectrics" 23-26 May 2011, Saint-Petersburg. NIB: WPC, 2011, p. 129-132.

[3]   E. M. Gojayev, S. I. Safarova, G. S. Djafarova, Sh. M. MextiyevaEffect of Aluminum Nano-Particles on Microrelief and Dielectric Properties of PE+TlInSe2 Composite Materials//Open Journal of Inorganic Non-metallic Materials, 2015, 5, p. 11-19.

[4]   Burda V. V., Galikhanov M. F., Gorohovatsky I. Yu., Karulina E. A., Chistyakov O. V. Electret state in composite materials based on high-density polyethylene with fillers nanodispersnymi SiO2.//Tr. XII Intern. conf. "Physics of Dielectrics." 23-26 May 2011. St. Petersburg. NIB: WPC, 2011, p. 165-168.

[5]   Galikhanov M. F., Gorohovatsky I. Yu., Gulyakova A. A., Temnova D. E., Fomichev E. E. Stability study of the electret state in composite polymer films dispersed fillers.//Proceedings RSPU. AI Herzen, 2011, № 138, p. 25-34.

[6]   Temnov D. E., Fomicheva E.E. Electret properties of polyethylene composite films with talc. // Proceedings of the international conference nanomaterials: applications and properties. 2013vol. 2 no 3, 03ncnn48(2pp).

[7]   Fomichev E. E. "Electrical properties of polypropylene with dispersed fillers".// Dissertation for the degree of Cand. Sci. Sciences St. Petersburg, Russian State Pedagogical University. A. N. Gertsenag. 2011, p. 17.

[8]   Ramazanov M. A., Guseinova A. S. Effect of the electrothermo polarization on the electret properties and the charge state of polyethylene nanocomposites with Cr and PbCrO4 additives // Surface Engineering and Applied Electrochemistry, March 2013, Volume 49, pp 97-100.

[9]   Anisimov O. D., Kostikov V. I., Lopatin V. Y., Chebryakova E.V., Shtankin Y. V. // Metallomatrix way to create composites based on aluminum reinforced with nanoparticles. Proceedings of the IV All-Russian Conference on Nanomaterials, "Nano 2011" 1 - March 4, 2011, Moscow: IMET RAS, 2011. Pp. 465-467.

[10] E. M. Gojayev, N. S. Nabiyev, Sh. A. Zeynalov, S. S. Osmanova, E. A. Allahyarov, A. G. Hasanov// Studies of the fluorescence spectra and dielectric properties of composites of HDPE + x. % TlGaSe2. Electronic Materials Processing, № 3, 2013, с. 14-18.

[11] M. Y. Yablokov, A. S. Kechekyan, A. B. Gilman, A. N. Ozerin//Electret properties of nanocomposite materials based on polypropylene. Nanotechnics. 2011. № 2 (26). Pp. 86-88.

[12] I. A. Zhigaeva, V. E. Nikolaev//Study based on fluoropolymer koronoelektretov-32L by thermally stimulated relaxation of the surface potential. Min. Education and Science Minister modern trends in chemistry and technology of polymeric materials International conference abstracts St. Petersburg in 2012, with 38-39.

[13] G. A. Mamedov, E. M. Godzhaev, A. M. Magerramov//Study of surface topography by atomic force and dielectric properties of compositions of high-density polyethylene with an additive TlGaSe2 / / Electronic Materials Processing, 2011.47 (6), 94-98.

[14] E. M. Gojayev, S. S. Safarova, D. M. Kafarova, K. D. Gulmammadov, J. R. Ahmedova// Investigation of surface microrelief and dielectric properties of the compositions of PP + TlIn0, 98Ce0, 02Se2 / / Electronic Materials Processing, 2013, 49 (4), p.267-271.

[15] E. M. Godzhaev, A. M. Magerramov, Sh. A. Zeinalov, S. S. Osmanova, E. A. Allakhyarov Coronoelectrets based on composites of high density polyethylene with a TlGaSe2 semiconductor filler. // Surface Enggineering and applied electrochemistry. December 2011, Volume 46, pp 615-619.

[16] H. A. Sadyhov// Effect of interfacial phenomena on the formation of the piezoelectric effect in the polymer-piezoelectric. Author's abstract on scientific degree of Candidate. nat. mat. -Sciences, Baku, 1992, p. 13.

10

Synthesis of $Zn_{0.5}Co_xMg_{0.5-x}Fe_2O_4$ Nano-Ferrites Using Co-Precipitation Method and Its Structural and Optical Properties

Abdulmajid Abdallah Mirghni[1,2], Mohamed Ahmed Siddig[2,3,*], Mohamed Ibrahim Omer[4], Abdelrahman Ahmed Elbadawi[2], Abdalrawf Ismail Ahmed[2]

[1]Department of Physics, Faculty of Education, Al Fashir University, Al Fashir, Sudan
[2]Department of Physics, Faculty of Science and Technology, Alneelain University, Khartoum, Sudan
[3]Department of Medical Physics, Faculty of Medicine, National University, Khartoum, Sudan
[4]Department of Physics, Faculty of Science and Technology, Nile Valley University, Atbara, Sudan

Email address:

siddig_ma@yahoo.com (M. A. Siddig)

Abstract: In this work, cobalt (Co) substituted magnesium Zinc nanocrystalline spinel ferrites having general formula $Zn_{0.5}Co_xMg_{0.5}-xFe_2O_4$ (with x=0.1, 0.2, 0.3, 0.4, 0.5) were synthesized using chemical co-precipitation method. The Cobalt substituted magnesium was annealed at $450°C$ and characterized using X-ray diffraction (XRD), Fourier transform infrared spectroscopy (FTIR) and UV-visible spectroscopy. XRD analysis confirmed the formation of single phase spinel structure. The crystalline size was calculated using Scherer's formula and wasfound to be in 21.44 – 25.03 nm range. The lattice constant was found to decreases as substitution of Co is further increased. The decrease in lattice constant may attribute to the smallerionic radius of Co as compared to Zinc ion. The FTIR spectra for the samples measured in the range of 4000-400 cm^{-1} exhibited symmetric stretching mode of vibration of tetrahedral and octahedral sites. The energy band gaps of the materials were calculated and were found to be in the range of 4.5 to 4.8eV.

Keywords: FTIR, Nanoferrite, Nanoparticles, Spinel Structure

1. Introduction

Particles in the size range of approximately 1–100 nm can display novel optical, electronic, magnetic, chemical and structural properties because of quantum confinement and surface effects that may find many important technological applications[1,2]. The technology based on these nanoparticles, known as nanotechnology, is possible in theory, but its practical realization requires the solution of quite challenging issues of applied technology, e.g., control the geometry, the particle size, the morphology of nanoparticles and their assembly into structures performing specific functions and delivering specific effects[2,3]. Nanotechnology is well known as a very important key technology in science and industry. In the field of material science and engineering, nano-particles are considered to be a unique unit material made from atoms and molecules, to build ceramics, catalysts, or even nano-machines [4].

Ferrites are well-known magnetic nano materials intensively studied as a recording media due to their superior physical properties [5].These properties make ferrites an ideal candidate for technical applications such as magnetic resonance imaging enhancement, catalysis, sensors and pigments [6]. Spinel nano-ferrites with chemical formula AB_2O_4 are materials of today's research due to their amazing structural, dielectric, electrical and magnetic properties[7].Such properties are dependent on the nature of cations, their charges and their distribution among tetrahedral (A) and octahedral (B) sites [8].

Various physical and chemical methods of preparation have been developed to achieve nano-sized ferrite particles such as sol-gel [9], chemical co-precipitation [10], high-energy ball milling [11], hydrothermal [12], citrate precursor [13], and Chemical combustion route[14]. Among them chemical co-precipitation method seems to be the most convenient method for the synthesis of Zn-Co-Mg ferrites. It is very simple and has better control over crystalline size and other properties of the materials [15]. Several researchers used co-precipitation method to successfully prepare their different samples. Among those, Nikumbh et al. [16] used the method in order to prepare $CoRE_xFe_{2-x}O_4$nano-phase, while P. Kumar et al. [17] used it to prepare $CoFe_{2-x}GdO_4$. The preparation of TiO_2 nanoparticles using a wet chemical technique was carried out by Suresh Sagadevan [18].

In this work, the aim is to to synthesis nano-ferrites using chemical co-precipitation method and to investigate its structural and optical properties. Moreover, the effect of doping of Cobalt (Co) substituted magnesium was investigated. X-ray diffraction (XRD) was used to confirm the formation of single phase spinel structure and to determine the crystalline size. Fourier Transform Infrared Spectroscopy (FTIR), and Ultraviolet visible spectrometer (UV-visible) were used in order to explore the effect of Co substituted zinc magnesium nano-ferrites on the optical properties.

2. Experimental

The $Zn_{0.5}Co_xMg_{0.5-x}Fe_2O_4$ (x = 0.0, 0.1, 0.2, 0.3, 0.4) nano-crystalline ferrites were prepared using chemical co-precipitation method as reported in the literature [19, 20]. Following chemicals were purchase from Sigma Aldrich and used in the preparation of $Zn_{0.5}Co_xMg_{0.5-x}Fe_2O_4$ nano-crystalline ferrites; namely Fe $(NO_3)_29H_2O$ (99%), $Zn(NO_3)_2$(98%), $CONO_3$(99%), and $MgNO_3$(99%). Required volumes of metal salts solutions having concentrations 0.15 M were mixed and stirred on a magnetic hot plate at $80°C$ for 1 hour. Specific amount of oleic acid was used as surfactant and for coating. The pH value was adjusted using 3MNaOH solution and the pH value was maintained 11-13 for all of the samples. The precipitates were washed with deionized water until the pH reduced to 7. Water was, then, evaporated using the oven at 100 °C, and annealing was carried out at $450°C$ for 6 h in a temperature controlled muffle furnace Vulcan A-550 at a heating rate $10°C/min$. The obtained materials were grinded into powder form and were made ready for characterization using various techniques.

The XRD analysis was carried out to confirm the purity of the synthesize materials using Shimadzu 6000 X-ray diffractometer with Cu-Kα radiation of a wavelength of λ= 1.5406 Å source.

FTIR measurements were performed using (Mattson, model 960m0016) spectra, while, the absorption of solution with different concentration was calculated using UV min 1240 spectrometer Shimadzy.

3. Results and Discussion

The crystalline and structure of prepared particles are confirmed by XRD patterns. Figure 1 displays the typical XRD spectra of cobalt substituted magnesium nanoferrites with composition (x= 0). XRD patterns are well indexed using MDI jade 5.0 [21]. Several peaks have been observed and indexed and assigned as (111), (220), (311), (400), (422),

(511), (440) and (533) which are the characteristics planes of single phase cubic spinel structure with space group Fd-3m and most intense (311) reflection [22].

From the XRD data, the crystallite size of $Zn_{0.5}Co_xMg_{0.5-x}Fe_2O_4$ prepared nanoparticles is calculated using Scherer's formula [23].

$$D_c = \frac{0.9\,\lambda}{\beta\,cos\,\theta} \qquad (1)$$

where D_c is crystallite size, β is full width at half maximum (FWHM) of the most intense (311) peak, λ is X-ray wavelength and θ is diffraction angle. The results of X-ray diffraction are listed in Table 1.

Figure 2 (a) shows the intensity versus 2θ for samples with different concentration of cobalt. It can be observed that the crystalline sizes are found to increase from 23.60nm to 25.03nm as the concentration of cobalt increases from x=0 to x=0.4. The lattice constant as a function of cobalt concentrations is shown in figure 2 (b).It is clear, from the figure, that the lattice constant, a, increase from 8.375 to 8.392 Å. The increasing in lattice constant is attributed to the larger ionic radius of Co^{+2} (0.74 Å) in compare with Mg^{+2} (0.72 Å) [11]. Sonal Singhal et al. showed in their results that the lattice parameter, a, was increase with cobalt concentration and attributed to the smaller ionic radius of nickel [24]. Rapolu Sridhar et al., studied Copper Substituted Nickel Nano-Ferrites by Citrate-Gel Technique and they found the lattice parameter was decreasedwhen Cu^{+2} concentration was further increase [25].

The density of composition is estimated using the following relation [26]

$$D_x(ferrite) = \frac{8M}{N \times a^3} \qquad (2)$$

where, M is molecular weight of the ferrite, N is the Avogadro's number and a^3 is the volume of the cubic unit cell. It can be observed from table 1 that, the composition density increases with increasing Co content. The increase in density may be due to the ionic radii of constituent ions causing increase in lattice constant. The density of pure $CoFe_2O_4$ is $(5.29g/cm^3)$ [14], while for pure $MgFe_2O_4$is $(4.52g/cm^3)$ [27].

Figure 3 shows the infrared spectra of synthesized $Zn_{0.5}Co_xMg_{0.5-x}Fe_2O_4$nano ferrite powders where x = 0.0, 0.1, 0.2, 0.3, and 0.4 (as pellets in KBr) in the range of 400 to 4000 cm^{-1}. The spectra of all the ferrites have been used to locate the band positions which are given in the Table 2. In the present study the absorption bands ν1, ν2, ν3, ν4, and ν5 are found to be around 603, 1142, 1405, 1643and 3111 cm^{-1}, respectively for all the compositions. The absorption bands observed within these specific limits revealed the formation of single-phase spinel structure having two sub-lattices tetrahedral (A) site and octahedral (B) site [24]. There is a band, not very clear, appear around 465 cm^{-1}which may be caused by the metal-oxygen vibration in the octahedral side. The (ν1) band usually observed around 600 cm^{-1} is caused by the metal-oxygen vibration in the tetrahedral sides [24]. The difference in the spectral positions may be due to the different of distances of metal ion-O^{-2} for octahedral and tetrahedral sites. The band (v2) around 1142 cm^{-1} is due to formation of Co substituted spinel ferrites. The (v3) around 1405 cm^{-1} is associated with the bending vibration related to the CH_2 groups [11]. (v4, v5) which observed around 1643 and 3111 cm^{-1} are due to the stretching mode of H-O-H bending vibration of free or absorbed water which implies that the hydroxyl groups are retained in ferrites [28].

The absorption as a function of wavelength for $Zn_{0.5}Co_xMg_{0.5-x}Fe_2O_4$ samples is performed. In case of sample 5 $Zn_{0.5}Mg_{0.1}Co_{0.4}Fe_2O_4$ (figure not shown), the maximum absorption is observed at wavelength 235nm. The optical band gap energy of the nanostructured was obtained using the following equation [28]

$$(\alpha h\nu) = A(h\nu - E_g)^m \qquad (3)$$

In Eq. (3) E_g, the optical band gap whereas m represents

the nature of the transition band gap, constant A is an energy-independent constant, $(h\nu)$ is energy of photon. Assuming direct band gap transition for the samples, m was assigned a value of 1/2. To evaluate a precise value for the optical band gap, we plotted $(\alpha h\nu)^2$ versus energy $(h\nu)$ for $Zn_{0.5}Mg_{0.5}Co_{0.1}Fe_2O_4$ (sample 2) as shown in figure 4. The optical band gap was determined by extrapolating the linear portion of the plot to $(\alpha\ h\nu)^2 = 0$ and is found to be 4.8, 4.7, 4.8, 4.8 and 4.5eV for the studied samples, respectively. The results of wave numbers and band gaps of all the compounds are listed in Table 2. The effect of substitution is profound when the Co concentration is further increased to 0.4 as optical band gap energy dropped to 4.5. The reduction in optical band gap may be due to increase in lattice constant with Co concentration.

Table 1. Particle size (Dc), Lattice constant (a), Cell volume (V) and the cell Density of $Zn_{0.5}Co_xMg_{0.5-x}Fe_2O_4$.

$Zn_{0.5}Co_xMg_{0.5-x}Fe_2O_4$	Dc (nm)	Lattice constants (Å)	V (nm³)	Density (g/cm³)
X=0.0	23.60	8.375	0.587	4.52
X=0.1	23.88	8.377	0.588	5.30
X=0.2	21.69	8.387	0.590	4.50
X=0.3	25.03	8.391	0.591	5.18
X=0.4	21.44	8.392	0.591	5.30

Table 2. Wave numbers and band gaps of $Zn_{0.5}Co_xMg_{0.5-x}Fe_2O_4$ nanoferrite.

No	$Zn_{0.5}Co_xMg_{0.5-x}Fe_2O_4$	v_1	v_2	v_3	v_4	v_5	Band gap (eV)
1	X=0.0	603	1142	1405	1646	3132	4.8
2	X=0.1	601	1142	1405	1643	3111	4.7
3	X=0.2	600	1142	1404	1643	3111	4.8
4	X=0.3	603	1142	1405	1643	3101	4.8
5	X=0.4	603	1111	1405	1643	3111	4.5

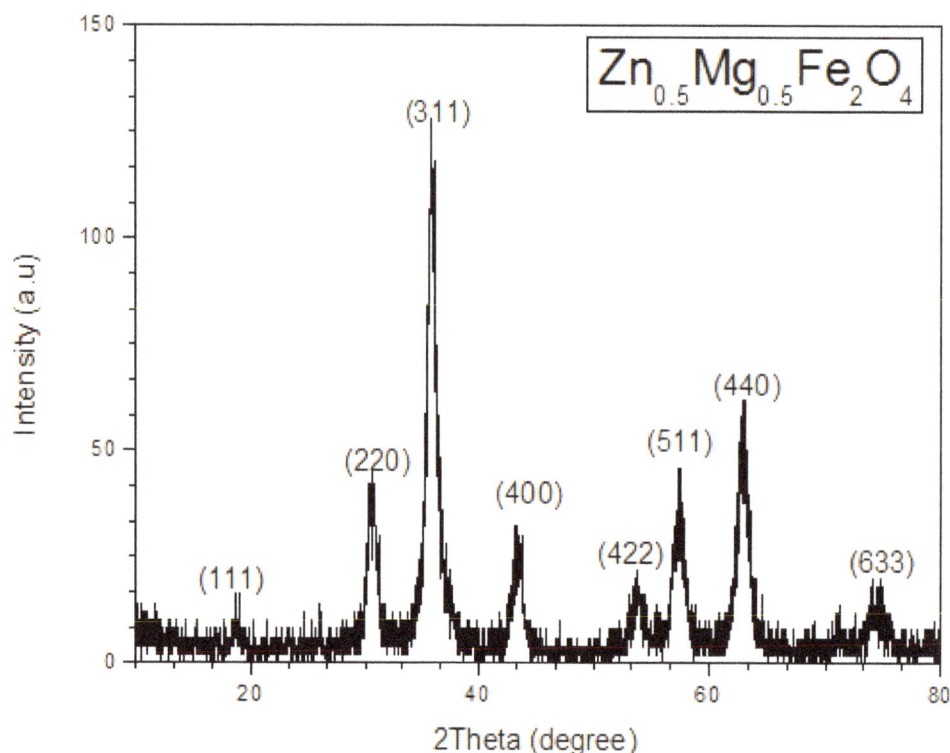

Figure 1. XRD patterns of $Zn_{0.5}Mg_{0.5}Fe_2O_4$ nanoferrite.

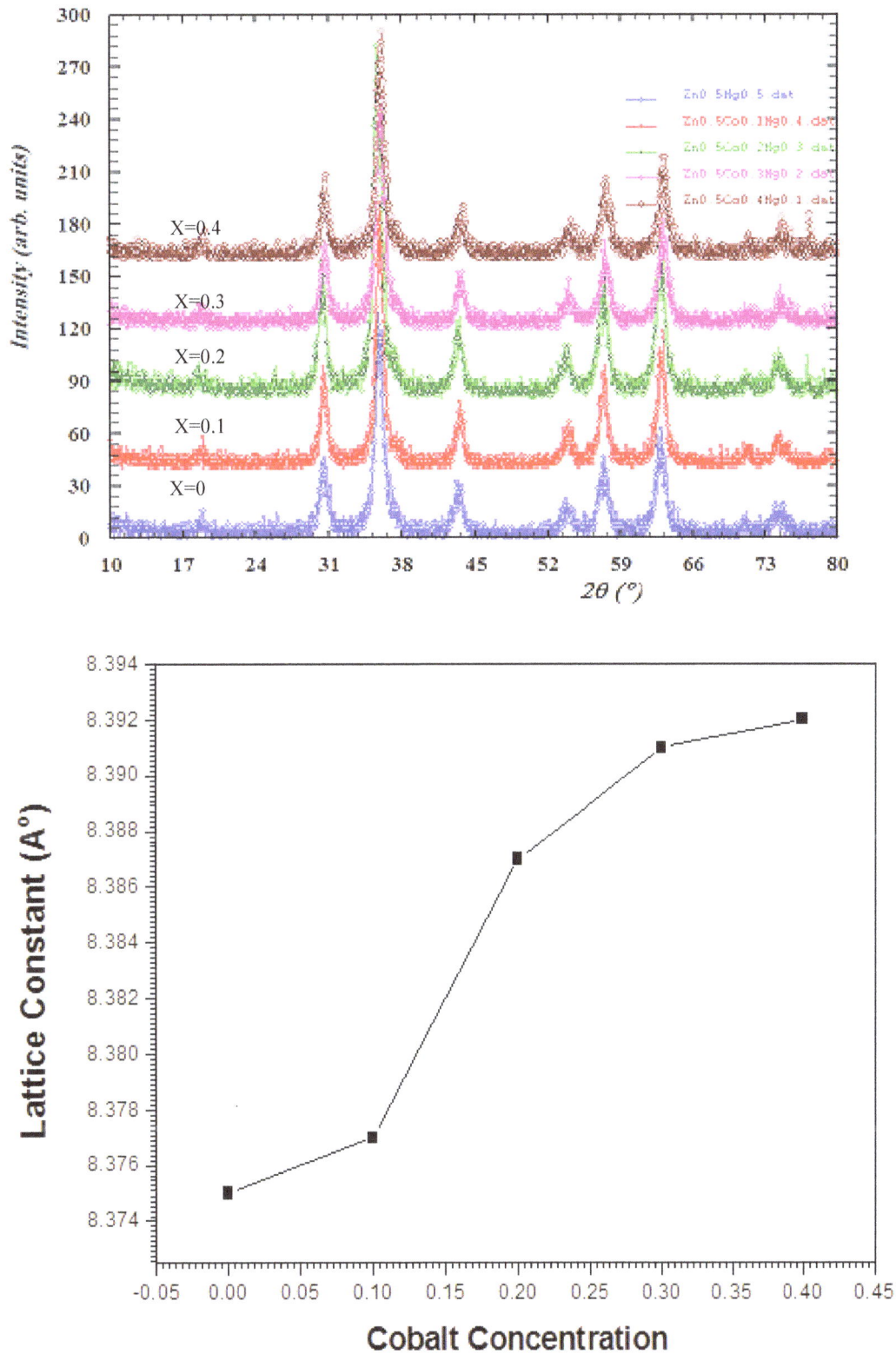

Figure 2. *(a)* *XRD patterns of $Zn_{0.5}Co_xMg_{0.5-x}Fe_2O_4$ nano-ferrites.* *(b)Lattice constant as a function of different concentration of Co of $Zn_{0.5}Co_xMg_{0.5-x}Fe_2O_4$ nano-ferrites.*

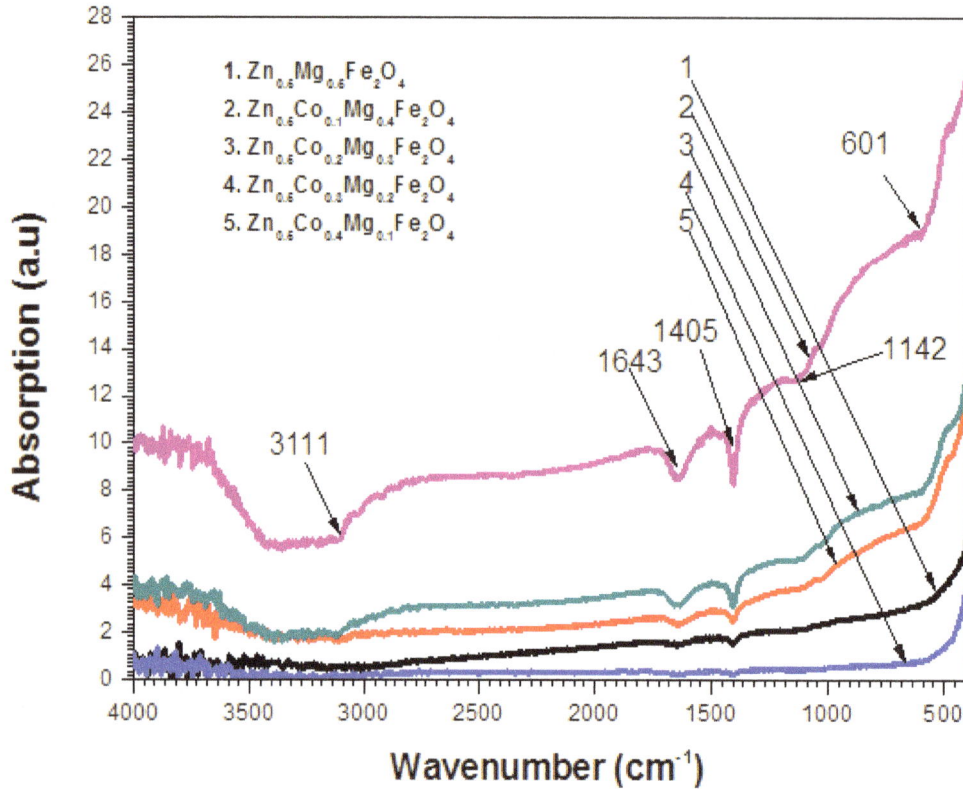

Figure 3. *FTIR spectra of $Zn_{0.5}Co_xMg_{0.5-x}Fe_2O_4$ nanoferrites of different concentration of Co.*

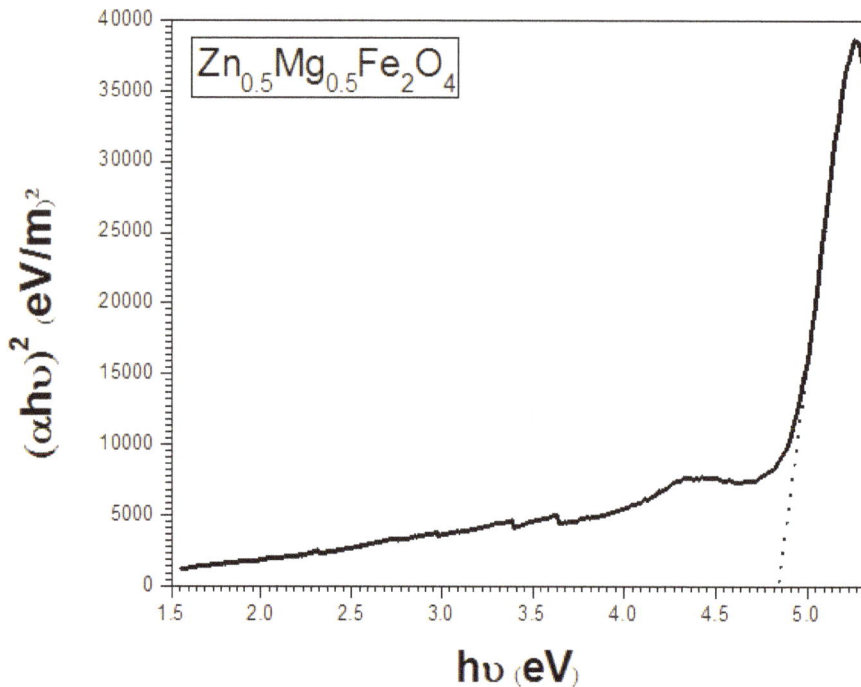

Figure 4. *Plot of $(\alpha h\nu)^2$ versus $h\nu$ for $Zn_{0.5}Mg_{0.5}Fe_2O_4$ nanoferrite*

4. Conclusion

The effect of cobalt substitution on magnesium nanocrystalline ferrites was studied. The formation of single phase crystalline structure with size in the range 23.60 – 25.03 nm was confirmed by X-ray diffraction. FTIR

spectrum exhibited expected main absorption bands, thereby confirming the spinel structure. Optical band gap energy of $Zn_{0.5}Co_xMg_{0.5-x}Fe_2O_4$ nanoferrite was found to be in the range 4.5 to 4.8eV for samples with different cobalt concentration. The substitution was resulted in slight increased in the lattice constant and that sequentially may lead to the slightly decreased in the energy gap. The synthesized nanoferrites are expected to be useful in several technological applications such as soft magnets and magnetic fluids for hyperthermia. The structural and properties of spinel ferrites depend upon the method of preparation, the nature of substitutional element and the concentration of the substitution element. Attempts can be made to prepare the samples by different methods to get desired properties and particle sizes.

Acknowledgment

The authors would like to thank Al Fashir university and the physics department of Alneelain University, particularly material laboratory for supporting this research.

References

[1] Flores-Acosta M., Sotelo-Lerma M., Arizpe-Chavez H., Castillon-Barraza F.F., Ramirez-Bon R., Solid State Commun. 128 (2003) 407

[2] Bognolo G., Adv. Colloid Interface Sci. 106 (2003) 169

[3] Lal Said Jan, Radiman S., Siddig M. A., Muniandy S. V., Hamid M. A., Jamali H. D., Colloids and Surfaces A: Physicochem. Eng. Aspects 251 (2004) 43–52

[4] Pulisova P., Kovac J., Voigtd A., Raschman P., Journal of Magnetism and Magnetic Materials 341 (2013) 93–99

[5] Salunkhe A. B., Khot V. M., Phadatare M. R., Thorat N. D., Joshib R.S., Yadav H. M., Pawar S. H., Journal of Magnetism and Magnetic Materials 352 (2014) 91–98

[6] Ganjkhanlou, Yadolah, Application of Image Analysis in the Characterization of Electrospun Nanofibers, J. Chem. Chem. Eng, Vol. 33, No. 2, 2014

[7] Maria Yousaf Lodhi, Khalid Mahmood, Azhar Mahmood, Huma Malika, Muhammad Farooq Warsi, Imran Shakir, M. Asghar, Muhammad Azhar Khan Current Applied Physics 14 (2014) 716-720

[8] A.K. Nikumbh, R.A. Pawar, D.V. Nighot, G.S. Gugale, M.D. Sangale, M.B. Khanvilkar, A.V. Nagawade, Journal of Magnetism and Magnetic Materials 355 (2014) 201–209

[9] Feng Huixia, Chen Baiyi, Zhang Deyi, Zhang Jianqiang, Tan Lin, Journal of Magnetism and Magnetic Materials 356 (2014) 68–72

[10] Yue Zhang, Zhi Yang, Di Yin, Yong Liu, ChunLong Fei, Rui Xiong, Jing Shi, GaoLin Yan, Journal of Magnetism and Magnetic Materials 322 (2010) 3470–3475

[11] Sukhdeep Singh, Manpreet Singh, N. K. Ralhan, R. K. Kotnala, Kuldeep Chand Verma, Adv. Mat. Lett. 3 (2012) 504-506

[12] Saeed Abedini Khorrami, Qazale Sadr Manuchehri, Journal of Applied Chemical Research, 7 (2013) 15-23

[13] I. V. Kasi Viswanath , Y. L. N. Murthy, Kondala Rao Tata, Rajendra Singh, Int. J. Chem. Sci, 11(2013) 64-72

[14] R.C. Kambale, P.A. Shaikh, N.S. Harale, V.A. Bilur, Y.D. Kolekar, C.H. Bhosale, K.Y. Rajpure, Journal of Alloys and Compounds, 490 (2010) 568–571

[15] S.J. Azhagushanmugam, N. Suriyanarayanan, R. Jayaprakash, Materials Science in Semiconductor Processing 21 (2014) 33–37

[16] A.K. Nikumbh, R.A. Pawar, D.V. Nighot, G.S. Gugale, M.D. Sangale, M.B. Khanvilkar,A.V.Nagawade

[17] P.Kumar,J.Chand,SatishVerma,M.Sing, International journal of theoretical and applied science, 3 (2011) 10-12

[18] Suresh Sagadevan. Synthesis and Electrical Properties of TiO2 Nanoparticles Using a Wet Chemical Technique. American Journal of Nanoscience and Nanotechnology. Vol. 1, No. 1, 2013, pp. 27-30. doi: 10.11648/j.nano.20130101.16

[19] Ihab A. Abdel,Latif, Journal of Physics, 1(2012)50-53

[20] M. Abdullah Dar, Jyoti Shah, W. A. Siddiqui, R. K. Kotnala, Appl Nanosci, 4 (2014) 675–682

[21] Saeed Abedini Khorrami, Qazale Sadr Manuchehri, Applied Chemical Research, 7 (2013) 17-20.

[22] A.B. Salunkhe, V.M. Khot, N.D. Thorat, M.R. Phadatare, S.H. Pawar, C.I. Satish, D.S. Dhawale, Applied Surface Science, 264 (2013) 598–604

[23] Rajjab Ali, Muhammad Azhar Khan, Azhar Mahmood, Adeel Hussain Chughtai, Amber Sultan, Muhammad Shahide, Muhammad Ishaq, Muhammad Farooq Wars, Ceramics International 40 (2014) 3841–3846

[24] Sonal Singhal, Santosh Bhukal, Jagdish Singh, Kailash Chandra, and S. Bansal, Journal of Nanotechnology, 10 (2011) 1-6

[25] Rapolu Sridhar, Dachepalli Ravinder, K. Vijaya Kumar, Advances in Materials Physics and Chemistry, 2 (2012) 192-199

[26] J. Smith, H.P. Wijn, Ferrites, 1959, Wiley, New York

[27] Abd Elkade, N. M. Deraz and Omar H., Int. J. Electrochem. Sci., 8 (2013) 8614-8622

[28] Maria Elena Sanchez-Vergara 1, Juan Carlos Alonso-Huitron, Arturo Rodriguez Gómez and Jerry N. Reider-Burstin, Molecules, 17 (2012) 10000-10013

11

Soluble poly (methyl methacrylate) composites containing covalently associated zirconium dioxide nanocrystals

Natalia Yevlampieva[1, *], Alexander Bugrov[2, 3], Tatiana Anan'eva[3], Mikhail Antipov[1], Evgeny Ryumtsev[1]

[1]Faculty of Physics, Saint Petersburg State University, Saint Petersburg, Russia
[2]Faculty of Chemistry, Saint Petersburg State University, Saint Petersburg, Russia
[3]Institute of Macromolecular Compounds, Russian Academy of Sciences, Saint Petersburg, Russia

Email address:

yevlam@paloma.spbu.ru (N. Yevlampieva), bugrov.an@mail.ru (A. Bugrov)

Abstract: Well soluble composite samples of poly(methyl methacrylate) containing hybrid nanoparticles with covalently associated ZrO_2 nanocrystals of an average size of (20 ± 5) nm have been studied by light scattering, viscometry and absorption spectroscopy methods in diluted solutions. Composites were synthesized by two ways: *in situ* bulk polymerization of methyl methacrylate in a presence of ZrO_2, and by polymerization of methyl methacrylate in toluene solution with the dispersed ZrO_2 nanocrystals. Surface of ZrO_2 was preliminary chemically modified by γ-(trimethoxysilyl)propyl methacrylate in both cases. Weight fraction of ZrO_2 in composite samples was varied in the range 1-3 %. Solution properties of composite polymers revealed that a way of monomer polymerization (in bulk or in solution) affect the type of the produced polymer-inorganic hybrids. Sphere like "core-shell" nanoparticles with a single ZrO_2 nanocrystal as a core are mainly formed when polymerization in solution is carried out. Under the conditions of *in situ* bulk polymerization the organic-inorganic particles of significantly larger size with the irregular number of associated ZrO_2 nanocrystals are produced. The size of hybrid nanoparticles in composite samples was determined. Transmission electron microscopy was applied to visualize the difference of ZrO_2 distribution in thin films of the both type composite samples.

Keywords: Organic-Inorganic Composites, Hybrid Nanoparticles, PMMA, ZrO_2

1. Introduction

Development of new nanocomposite polymer materials and improvement of those available are the important trends in modern materials science. Considerable experience accumulated over the last decade in the field of polymer composites with different type inorganic nanoparticles has led to the conclusion that nanoparticles should be bound covalently by polymer to ensure a homogeneous distribution of inorganic component in its matrix [1-3]. Only in this case it is possible to obtain the material with the controlled properties. A great progress in producing of polymer-inorganic oxide (SiO_2, TiO_2, ZrO_2) composite materials have been achieved when the oxides distribution inhomogeneity problems, such as a prevention of their aggregation and build-up of a sufficiently high adhesion at the polymer-inorganic particle interface, were overcome due to of special chemical modification of particles surface prior to their incorporation into polymer. By

means of the required functional groups formation on the surface of inorganic particles a covalent linkage between the polymeric and inorganic components can be obtained. This approach was successfully realized for a range of polymers, and polymer-inorganic oxides composites with novel physico-chemical properties were developed [3-8].

Poly(methyl methacrylate) (PMMA) is the universal polymer that is often used as a matrix material for the nanocomposites creation [1, 2]. In [9, 10] the methods of the synthesis of PMMA-ZrO_2 composites by means of *in situ* bulk polymerization of methyl methacrylate (MMA) in the presence of nanosized ZrO_2 were described, and data on the properties of PMMA-ZrO_2 in films and in thin blocks were reported. A preliminary treatment of ZrO_2 surface with organosilicon compounds before polymerization MMA had been applied for the formation of PMMA- ZrO_2 networks [9, 10]. Modification of PMMA by relatively small amount of covalently bound zirconium dioxide nanoparticles (less 15 wet. %) results in significant enhancement of its thermomechanical and optical characteristics [3, 9, 10]. The area of potential applications of PMMA-ZrO$_2$ nanocomposites is extremely wide and extends from dentistry to microelectronics [3].

Practical usage of composite materials with the covalently bound nanoparticles is often faced with the necessity to give a preference only soluble well one. This is dictated by a perspective of multiple utilization of the hybrid polymer-inorganic nanoparticles or by possibility of theirs reselection. PMMA-ZrO$_2$ composites have never been considered in this aspect mainly due to they were developed for fabrication of the transparent films with the improved mechanical properties for which a crosslinking of macromolecules by surface-modified inorganic nanoparticles is desirable [9, 10]. However, to find the conditions that lead to formation of PMMA-coated ZrO$_2$ particles soluble well in organic solvents is also possible. This work is dedicated to one of the possible solutions of this problem.

Herein, we report the results of comparative study of dissolvable PMMA-based composites containing covalently bound ZrO$_2$ nanocrystals, 20 nm in diameter, synthesized by two ways: *in situ* bulk polymerization of MMA, and by polymerization of MMA in toluene solution with the dispersed ZrO$_2$ nanocrystals. We'll use the term "nanocrystal" for ZrO$_2$ as individual compound to distinguish between ZrO$_2$ and hybrid polymer-inorganic components of PMMA composites, since both are the nanoparticles. γ-(Trimethoxysilyl)propyl methacrylate (TMSPM) was used as the modificator of the surface of ZrO$_2$.

Our study is mainly aimed at the determination of hydrodynamic properties and structure of the hybrid polymer-inorganic nanoparticles forming in the processes of two different methods of synthesis of PMMA composites. To carry out the above task, we have studied composite samples in diluted solutions by dynamic light scattering, viscometry, and absorption spectroscopy. Transmission electron microscopy (TEM) was utilized for confirmation of the results of study in solution. The unmodified PMMA samples, synthesized at the same conditions as ZrO$_2$ containing analogues, have been studied also.

2. Experimental

2.1. Objects and Materials

Nanosized crystals ZrO$_2$ with the narrow distribution were prepared under the hydrothermal conditions [11, 12]. Nanocrystals which shape was close to sphere with the mean

diameter value of (20±5) nm were used. Chemical structure of TMSPM which was applied for modification of ZrO$_2$ is shown in Fig. 1. The technique of ZrO$_2$ modification by TMSPM was proposed earlier in [7]. TSMPM reacts with OH-groups on ZrO$_2$ surface so that silicon-containing part of TSMPM covalently joints to inorganic surface when it another part may be involved to the MMA polymerization process being fully similar to structure of PMMA monomer (Fig.1). From one to three chemical bonds between TMSPM molecule and ZrO$_2$ surface may be realized according to [10]. 2,2'-*Azo-iso*butyronitrile (AIBN) was utilized as an initiator of MMA polymerization. TMSPM-coated ZrO$_2$ nanocrystals have been placed

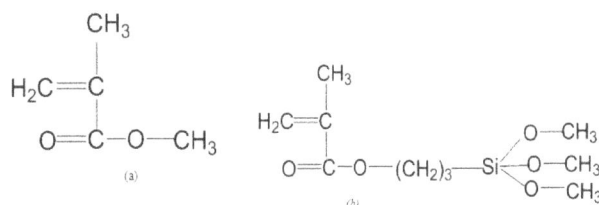

Figure 1. *Chemical structure of MMA (a) and TSMPM (b).*

to the mixture MMA/ AIBN or mixture MMA/ AIBN/ toluene under the argon atmosphere. Reaction mixture was sonicated (at 3.5 KHz) at room temperature during 20 min, and then the sonication was repeated every hour for 10 min at the corresponding temperature of reaction for a maintenance of high uniformity dispersion of ZrO$_2$. The content of the compounds in reaction mixtures, temperatures and duration of radical polymerization processes are shown in Table 1. Synthesized composite polymers were precipitated by methanol, and dried under vacuum to the constant weight. Composite PMMA samples containing ZrO$_2$ presented in Table 1 were completely soluble without any deposition in toluene, benzene, and ethyl acetate.

At the same conditions as the modified samples PMMA-2, 4 (Tabl.1) were received the unmodified samples PMMA-1, 3 have been synthesized in the absence of zirconia component. Composite samples PMMA-5-7 were synthesized in toluene solution under the variable conditions (Tabl.1).

Toluene, ethanol, ethyl acetate (EtAc) ("Vecton", Saint Petersburg, Russia) have been utilized without additional purification. Methanol ("Vecton") was dried and distilled from magnesium methoxide. MMA ("Aldrich") was distilled under normal pressure in argon flow atmosphere. Initiator AIBN was recrystallized from ethanol solution before utilization.

Table 1. *Synthesis conditions and yield of polymers.*

Sample	Initial concentration ZrO$_2$ (wt.%)	Reaction conditions	Polymer yield (mass %)
1	-	Bulk polymerization *C$_{AIBN}$=3,8×10^{-3}; 75°C; 45 min	16
2	1	Bulk polymerization C$_{AIBN}$=3,8×10^{-3}; 75oC; 45 min	14
3	-	MMA : toluene=1:1, C$_{AIBN}$=13×10^{-3}; 75°C; 24 h	92
4	1	MMA : toluene=1:1, C$_{AIBN}$=13×10^{-3}; 75°C; 24 h	92
5	3	MMA : toluene=1:1, C$_{AIBN}$=13×10^{-3}; 75°C; 24 h	93
6	1	MMA : toluene=2:3, C$_{AIBN}$=90×10^{-3}; 60°C; 24 h	54
7	3	MMA : toluene=2:3, C$_{AIBN}$=90×10^{-3}; 60°C; 24 h	68

* C$_{AIBN}$ – concentration of initiator, mol l^{-1}

2.2. Experimental Method

Solution properties of modified and unmodified PMMA samples have been studied in toluene and EtAc at 25 °C. Polymer solutions were prepared at room temperature during 2-3 days.

2.2.1. Optical Properties and Viscosity Determination

UV-V spectra were obtained using spectrophotometer SF-2000 managed automatically by SFScan-program (OKB "Spectr", St.Petersburg, Russia). Quartz cells of 1 cm in length were applied for measurements.

High accuracy microviscometer Lovis-2000 M/ME («Anton Paar», Austria) based on the Heppler method [13] have been used for viscometric measurements. Intrinsic viscosity value [η] of polymers was determined in accordance with Huggins procedure via a graphical extrapolation of η_{sp}/c value to zero concentration of the solute [14]:

$$\eta_{sp}/c = [\eta] + k'\,[\eta]^2 c + \qquad (1)$$

where η_{sp}/c corresponds to the ratio $(\eta - \eta_0)/\eta_0\,c = (t - t_0)/t_0 c$, η and η_0 are the viscosities of a solution and a solvent, respectively; t and t_0 are the time of motion of the ball in a viscometer capillary for solution and solvent, respectively; c is the solute concentration; k' is the Huggins constant corresponding to the slope of the linear dependence $\eta_{sp}/c = f(c)$.

Measurements of solution viscosities have been carried out at the inclination angles of viscometer capillary 45°-65°, where the contribution of gradual dependence to viscosity value was small and comparable with the accuracy of its determination.

Refractometer (Mettler Toledo, Switzerland) with the accuracy ±0.0001 for refractive index n determination has been applied for measuring of refractive index increments dn/dc. There were received dn/dc=0.0109±0.005 for PMMA-toluene system and dn/dc =0.118±0.005 for PMMA-EtAc system. There was not detected any difference in dn/dc values for ZrO_2 containing PMMA samples and for the unmodified PMMA samples in spite of refractive indecies of ZrO_2 and PMMA are significantly different. This fact may be considered as a consequence of small enough weight fraction of ZrO_2 in the modified PMMA samples.

2.2.2. Static and Dynamic Light Scattering

The study of static and dynamic light scattering were performed on PhotoCor Complex (PhotoCor, Russia) installations equipped with a real time correlator (288 channels, 20 ns), a goniometer, and a thermostat with a temperature regulation in accuracy within ±0.05 K. The installations were equipped by the different light sources - lasers with the wavelength 654 nm (power 25mW) and 445 нм (25 mW), correspondingly. The measurements were performed in the range of scattering angles θ = 30°–140°.

Static light scattering method was used only for measuring of the weight-averaged molecular mass M of the unmodified samples PMMA-1, 3 (Table 1). A basic relationship (2) for the

intensity of the excessive light scattering $I(q, c)$ was applied for data analysis [15]:

$$Hc/\,I(q,\ c) = (1/M)\ (1+q^2 R_g^2/3+....) + 2A_2 c + \qquad (2)$$

where $q =(4\pi n_0/\lambda_0)\ sin\ (\theta/2\ )$ is a scattering vector, A_2 is the second virial coefficient, R_g is radius of gyration, $H= 4\pi^2 n_0^2 (dn/dc)^2/\ N_A\lambda_0^4$ is an optical constant, n_0 is a refractive index of the solvent, λ_0 is the wavelength of the light source, N_A is Avogadro's number.

Starting from the relationship (2), a Zimm plot (3) have been used for determination of M value [15].

$$Hc/\ I(0,\ c)= 1/M +2A_2c \qquad (3)$$

Here $I(0, c)$ is a Rayleigh factor determined at $\theta \to 0$ for all solute concentration c at which the measurements of $I(q, c)$ were performed [15].

M values presented in Table 2 were received from the intercept on y-axis of $Hc/\ I(0, c)= f(c)$ according to eq. (3).

The autocorrelation functions of scattered light intensity received in the dynamic light scattering regime were treated with the DynaLS program through the Laplace inverse transform method [16] to obtain the distribution functions of particles in solution over their relaxation times τ.

Translation diffusion coefficients D of the particles in solutions were derived from the slopes of the dependences of the reciprocal relaxation times 1/τ on the squared amplitude of scattering vector q (4) [15].

$$1/\ \tau = Dq^2 \qquad (4)$$

Hydrodynamic radii R of the particles was calculated using experimental D values from the Einstein-Stocks equation (5) [14].

$$R = kT/6\pi\eta_0 D\ , \qquad (5)$$

where k is Boltsmann's constant, T is an absolute temperature.

2.2.3. Transmission Electron Microscopy

TEM samples were prepared by depositing 5µL of PMMA solution in EtAc with the solute concentration 0.01 g cm^{-3} onto a Lacey carbon film-coated copper grid. In comparison with a holey carbon films, the Lacey carbon films offer a greater percentage of an open area due to theirs mesh structure. In drying out process, the polymer sample formed a thin film of approximate thickness of 20 - 40 nm. TEM/STEM imaging was performed on a Zeiss Libra 200 FE transmission electron microscope operated at an acceleration voltage of 200 kV.

3. Results and Discussion

3.1. Solution Properties of Modified /Unmodified PMMA

The vinyl groups of the used organosilicon modifier of ZrO_2 surface are capable of entering into the process of radical copolymerization with MMA (Fig.1), thereby forming a covalent linkage between the inorganic nanocrystals and polymer molecules. During in situ bulk polymerization of

MMA, the nanocrystals with an active surface cannot be individually "suspended" to the polymer chains, but may bring about formation of several bonds/cross-links with one or many polymer chains (particle of the "core-shell" type), or build up the network structures that comprise more than one ZrO_2 nanocrystal cross-linked by a number of different macromolecules [9, 10]. Thus, the formation of covalent bonds between PMMA and ZrO_2 through TMSPM may result in both a branched hybrid polymer-inorganic particles and a cross-linked polymer-ZrO_2 networks.

In this study, PMMA samples modified by ZrO_2 were synthesized at different conditions. Not only a polymerization procedure (in bulk or in solution) was varied, but also the ratio solvent/initiator/monomer, as well as a temperature of MMA polymerizing in toluene solution were varied in order to obtain a high yield of the soluble hybrid material (Table 1).

Performing bulk polymerization of MMA in the presence of surface-modified ZrO_2 for more than 55 min resulted in production of an insoluble polymer system. The polymer yield after 45 min from the onset of polymerization, obtained in both the presence and absence of ZrO_2, was not more than 16 wt. % (Table 1). However, a carrying out of MMA polymerization in solution at the same content of ZrO_2 provided a high yield of soluble composite polymer (Table 1).

The molecular properties of the synthesized polymers were studied only for the completely soluble composite samples, which were specially selected. At the beginning of the study the pairs of the unmodified and modified by ZrO_2 PMMA samples, synthesized at the same conditions, have been compared.

corresponding unmodified PMMA-1 and PMMA-3. In Table 2 are shown full data of the viscometric measurements for the polymers under investigation.

It is well known that an addition of large-sized foreign particles does not affect the [η] value of a polymer if their weight fraction is small or the particles are spherical [14]. Both cases may be the reason of the fact that [η] does not change for the studied pairs of PMMA samples. However, first of all, the results of solution viscosity measurements indicate that the main weight fraction in the modified samples corresponds to PMMA molecules of linear structure, and, the second, that weight fraction of the hybrid organic-inorganic particles in the content of these samples is small enough to influence on [η] value.

It was ascertained from the studied absorption spectra of the PMMA samples that, in contrast to viscometry, the presence of ZrO_2 in the modified polymers is detected in the spectra and, moreover, the spectra are different for the composite samples synthesized in bulk and in solution.

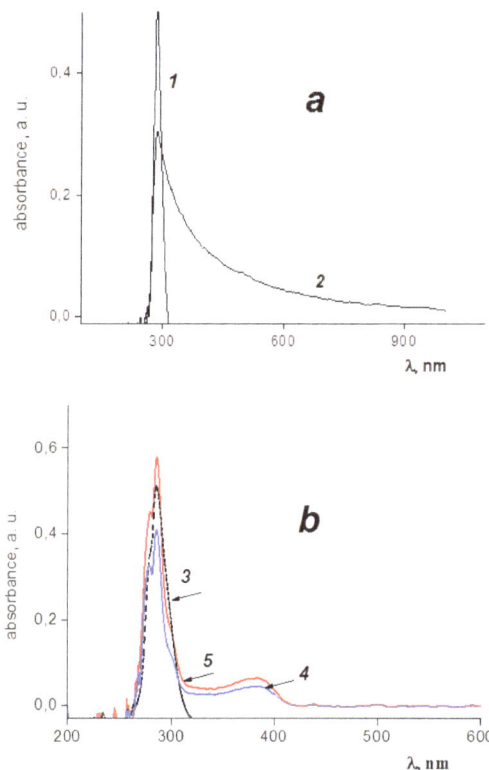

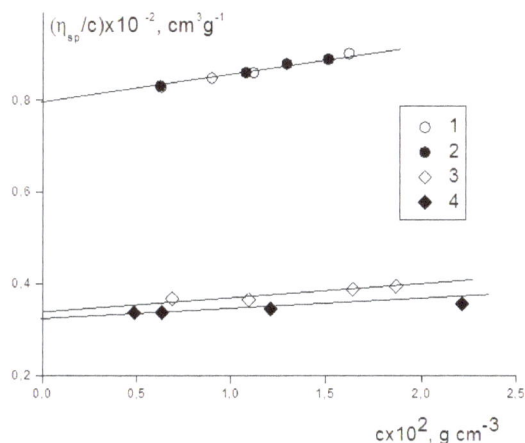

Figure 2. Concentration dependence of reduced viscosity value ($η_{sp}/c$) for pairs unmodified-modified samples PMMA-1, 2 (1, 2) and PMMA-3, 4 (3, 4) in toluene.

Figure 3. Absorption spectra for PMMA-1 and PMMA-2 at closed values of solute concentration (~0.4 x10^{-2} g cm^{-3}) (a) and corresponding spectra for samples PMMA-3, 4, 5 in toluene solutions (b).

Determination of the intrinsic viscosity value [η] in toluene and ethyl acetate (EtAc) revealed that the viscometry is insensitive to presence of ZrO_2 nanoparticles incorporated into the polymer. The concentration dependences of $η_{sp}/c$ value for two pairs of the modified-unmodified PMMA samples in toluene are shown in Figure 2. Both ZrO_2 containing samples PMMA-2(bulk polymerization) and PMMA-4 (polymerization in solution) are characterized, within the accuracy of measurement, the same [η] values as those of the

Figure 3a shows the absorption spectra for a pair of PMMA-1 and PMMA-2 prepared by bulk polymerization. The unmodified PMMA-1 absorbs only in the UV spectral range with a characteristic peak at λ=285,5 nm [17]. Figure 3b shows the absorption spectra of the PMMA-3, 4, 5 synthesized in solution. It can be seen from Figs. 3a and 3b that, for the modified PMMA-4, 5, there is an additional peak in the spectra at λ=383 nm, which is absent in the spectra of the unmodified samples PMMA-1 and PMMA- 3. The appearance of the additional peak in the spectra of the PMMA-4, 5 samples is associated with the absorption by the ZrO$_2$ surface modified by the organosilicon compound. The presence of the additional peak in the spectra of the modified PMMA asserts the covalent bonding of the ZrO$_2$ nanocrystals to the polymer chains.

As compared with the spectrum of the unmodified PMMA-1, in that of modified PMMA-2, a change in the spectrum width and the appearance of the arm in the near-*vis* spectral range are revealed (Fig. 3a). The spectrum of the PMMA-2 does not contain a clearly exhibited additional absorption peak, as in the case of the modified PMMA-4, 5 synthesized by polymerization in solution (Fig. 3b). The noted difference in the absorption spectra may be attributed both to the unequal number of inclusions of ZrO$_2$ nanocrystals in the hybrid polymer-inorganic particles and to a different number of the TMSPM—PMMA bonds in them.

All modified PMMA samples synthesized in solution featured similar absorption spectra. The additional peak at λ=383 nm was observed in the absorption spectra of PMMA-6, 7 solutions as well, which was of higher intensity than that in the spectra of PMMA-4, 5. This could be an evidence of a difference in the structure of the hybrid particles in these polymers. It should be specially noted that, under the identical synthesis conditions, the intensity of the absorption peak at λ=383 nm was higher in the case of the PMMA-5 sample (Fig. 3b), which was prepared at a higher content of ZrO$_2$ than that in the case of PMMA-4 (Table 1). This suggests a potentially possible estimate of the concentration of the hybrid particles in the composition of the modified PMMA samples synthesized in solution from the spectral data.

3.2. A Content of the Modified PMMA Samples

The dynamic light scattering method (DLS) provides determination of the hydrodynamic size and content of various components in solution of heterogeneous polymer [15]. In particular, this can be made for bimodal type distributions in polymer solution. In order to determine weight fractions of the components, the bimodal distributions according to DLS data for spherical particles were theoretically analyzed in [18, 19]. It was also demonstrated that a combination of experimental data from static light scattering (SLS) and DLS could be used for the same task [20, 21].

The algorithm of analysis of the bimodal distributions according to the SLS and DLS data with due account of the peak area ratio was most substantiated and experimentally

validated in [15, 20-22]. It was shown in [20-22] that, when extrapolated to the zero scattering angle q, the S$_1$/S$_2$ peak area ratio in the bimodal distribution of the diffusion coefficients D satisfied the following equation

$$(S_1/S_2)_{q\to 0}=(m_1 M_1)/(m_2 M_2), \qquad (6)$$

where m_i and M_i (i=1, 2) are the mass quantity of the i-th component in solution and its weight averaged molecular mass, respectively.

According to [23], for Gaussian polymer coils Eq. (6) can be transformed into the form convenient for analysis only from the DLS data.

$$(m_1/m_2)=(S_1/S_2)_{q\to 0}\ (R_2/R_1)^{5/3}\ \text{good solvent} \qquad (7a)$$

$$(m_1/m_2)=\ (S_1/S_2)_{q\to 0}\ (R_2/R_1)^2\ \text{θ-solvent} \qquad (7b)$$

Here R_i is the hydrodynamic radius of the i-th component.

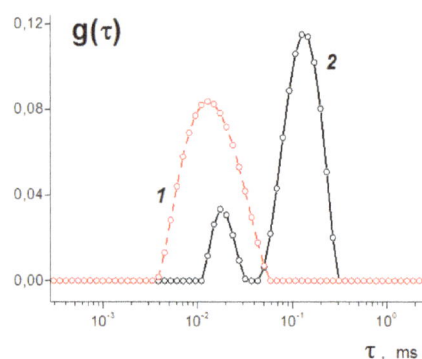

Figure 4. *Relaxation time distribution functions for the unmodified sample PMMA-3 (1) and modified PMMA-5 in EtAc at solute concentration 0.46 x10^{-2} g cm^{-3} (2). Scattering angle value is 110°.*

Figure 4 shows distribution of the relaxation times obtained from the DLS data for the PMMA-3 and PMMA-5 in EtAc. The distribution is unimodal for the unmodified PMMA-3; hence, the peak of this distribution function is attributed to the ordinary linear PMMA molecules. The relaxation time distribution function in solution of PMMA-5 is bimodal. Besides the particles with the relaxation times corresponding to the PMMA-3 molecules, there are particles of sizes substantially larger than those of PMMA-3 that are present in PMMA-5 solution. The slow and fast (molecular) modes of the distribution presented in Fig. 4 for this polymer correspond to the two types of particles in solution of PMMA-5.

Table 2. *Hydrodynamic properties of the studied PMMA samples in toluene and EtAc at 298 K*

Sample	[η] x10^{-2}, cm^{3}g^{-1} (toluene/EtAc)	Mx10^{-3}, g mol^{-1}	Dx10^7, cm^{2}s^{-1}	R, nm	w$_2$*	n**
1	(0.78/ 0.62) ±0.05	350	3 (toluene)	14	-	-
2	(0.80/0.58) ±0.05	-	3 (fast mode)/ 0.3 (slow mode) (toluene)	14/ 240	-	-
3	(0,35/0.26) ±0.02	120	6.2 (EtAc)	8.3	-	-
4	(0.34 /0.27) ±0.02	-	6.0 (fast mode)/ 0.5 (slow mode) (EtAc)	8.5/ 100	0.02	100
5	(0.33/0.26) ±0.02	-	6.2 (fast mode)/ 0.5 (slow mode) (EtAc)	8.3/ 100	0.08	108
6	(0.17/0.12) ±0.02	50	13 (fast mode)/ 0.5 (slow mode) (EtAc)	3.8 /100	0.05	220
7	(0.18/0.15) ±0.02	-	14 (fast mode) /0.6 (slow mode) (EtAc)	3.5 /83	0.07	205

* w$_2$ is weight fraction of ZrO$_2$ containing hybrid macromolecules in the modified PMMA sample;

** n is equivalent number of PMMA molecules attached to the surface of ZrO$_2$ in PMMA-4-7 synthesized in toluene solution.

The transition from the unimodal to the bimodal particle distribution was typical for all studied unmodified-modified PMMA pairs synthesized in solution and for the PMMA-1-PMMA-2 pair synthesized in bulk. However, for the modified PMMA-2, the distribution peak corresponding to the "slow" particles was very wide as compared with that coresponding to the modified PMMA samples synthesized in solution, i.e, the hybrid particles in the PMMA-2 composition obviously show wider spread in sizes.

Both modes of the relaxation time distributions for the modified PMMA samples are of the diffuse nature, which is seen in Fig. 5 that shows the inverse relaxation time $1/\tau$ as a function of squared sine of half-scattering angle for the PMMA-4 sample. The dependences presented in Fig. 5 are linear and traverse the origin of coordinates. From the slopes of these dependences, the translational diffusion coefficients D were determined according to Eq. (4). The diffusion coefficients D were independent of solute concentration. The concentration-averaged D values and hydrodynamic radii R of the particles calculated by Eq. (5) are presented in Table 2.

As follows from the DLS data (Table 2), the composite PMMA samples contain particles of hydrodynamic sizes corresponding to the size of molecules in the unmodified analogues, and much larger particles of sizes by an order of magnitude and more (in the case of bulk synthesis) exceeding the size of the linear PMMA macromolecules. The large-sized particles should contain ZrO_2 nanocrystals.

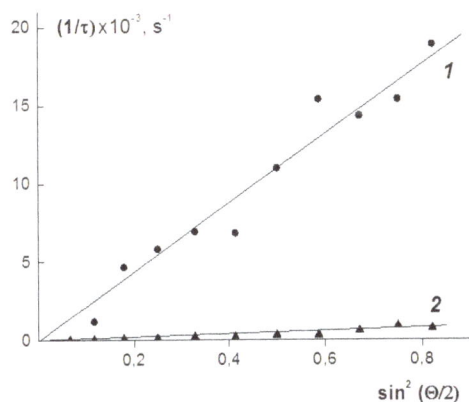

Figure 5. *Dependence of the inverse relaxation time values $1/\tau$ versus scattering angle as $\sin^2(\theta/2)$ for the fast mode (1) and slow mode (2) of corresponding distribution function received by DLS for modified sample PMMA-4 in EtAc at solute concentration 0. 698 $x10^{-2}$ g cm^{-3}.*

It can be readily inferred that, for *in situ* bulk polymerization of MMA, no systemic pattern could be expected in arrangement of polymer molecules covalently bound to ZrO_2 through TMSPM. Therefore, the zirconium-containing particles in the PMMA-2 sample may be in the form of irregular networks with ZrO_2 nanocrystals in its nodes. Obviously, such formations should be of significantly larger size than free PMMA molecules. The irregularity and large number of chemical bonds between PMMA and TMSPM may be responsible for the shape of the absorption spectrum of modified PMMA-2 (Fig. 3a) discussed

above. During synthesis of the modified PMMA in toluene solution, the formation of the hybrid particles of the "core-shell" type with ZrO_2 as the core is more likely.

Conclusion on the structural difference of hybrid particles in PMMA-2 (bulk polymerization) and PMMA-4-7 (polymerized in solution) is based on the results of TEM study. Figure 6 shows the HAADF-STEM images of the specially prepared thin films of zirconium dioxide modified PMMA-2 and PMMA-4 (see, please, experimental part). As easy to observe, the both polymers demonstrate similar images corresponding to homogeneous distribution of ZrO_2 in PMMA matrix in the scale of 1000 nm (Fig.6 (a, c)). More detail images in the scale of 50 nm and 200 nm (Fig. 6 (b, d)) show that zirconia particles in polymer matrix are larger in size and consist of several nanocrystalls in PMMA-2 film when a majority of zirconia particles of ~20 nm in size are locked separately in the film of PMMA-4.

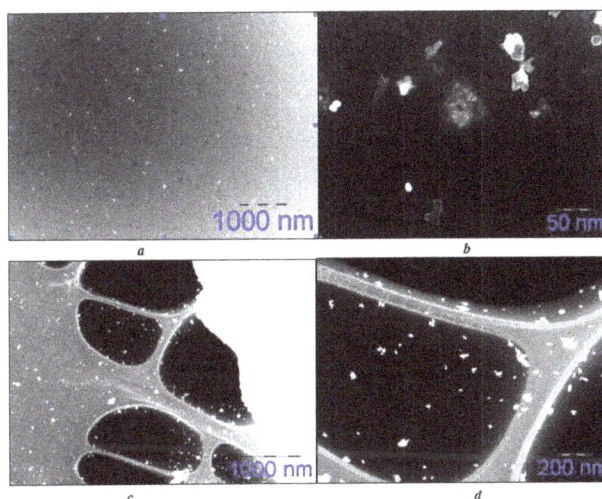

Figure 6. *Transmission electron microscope images of thin films of ZO_2-containing samples PMMA-4 synthesized in solution (a, b), and PMMA-2 synthesized by in situ bulk polymerization (c, d), at the scale variation. Black background corresponds to an open area of the carbon-film-coated copper grid.*

The fact that a size of the large particles in PMMA-4-7 synthesized in solution are only by an order of magnitude larger than a free (not connected with ZrO_2) molecule size attests in favor of this statement. Apparently, this may occur due to the growth of a sufficiently large number of the PMMA chains in the direction normal to the ZrO_2 surface. In this case, the macromolecules bound to the nanocrystal will be unable to take the inherent Gaussian coil conformation and will be forced to stretch [24]. Consequently, the size of the hybrid particles essentially differ from those of individual macromolecules in the PMMA-4—7 solutions (Table 2).

For the bimodal distributions of the diffusion coefficients of the particles in the modified PMMA-4—7 solutions, we determined the limits of the peak area ratio (S_1/S_2) at the zero scattering angle $q\rightarrow0$ (Fig. 7). Using Eq. (7a) for the toluene solutions and Eq. (7b) for EtAc solutions, the (m_1/m_2) ratios were obtained.

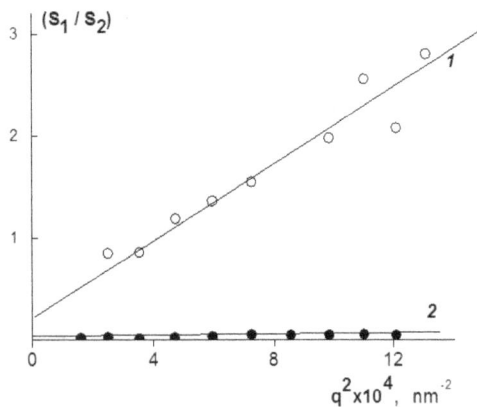

Figure 7. *Dependence of peak area ratio (S_1/S_2) of the diffusion coefficient distribution function versus the squared scattering vector q^2 for modified samples PMMA-2 (1) and PMMA-4 (2) in EtAc at solute concentration 0.8 $x10^{-2} g\ cm^{-3}$.*

The total amount of the solute in polymer solution $m=m_1+m_2$ is always known, which, with due account of the additional data on the (m_1/m_2) ratio, makes it possible to determine the weight fraction $w_2= m_2/(m_1+m_2)$ of the polymer-inorganic particles in the composition of the modified PMMA samples. Thereby obtained estimates of the w_2 value are listed in Table 2.

Correlation can be noted between the wt.% content of ZrO_2 nanocrystals in the samples presented in Table 1 and the calculated weight fraction w_2 of the hybrid particles listed in Table 2. As can be seen, the weight fraction of the hybrid particles in the modified PMMA appears fairly small, which properly explains the viscometry insensitivity to their presence in the composite polymers.

Assuming that polymerization degree of the free PMMA molecules and those linked to ZrO_2 through TMSPM should not strongly differ, let us estimate the equivalent number n of the polymer molecules bound to ZrO_2 in the modified samples synthesized in solution.

The M_1 value of the molecular component and (m_1/m_2) ratio being known, Eq. (6) provides the estimation of the mean molecular mass M_2 of the hybrid particles. For the PMMA-6, 7, we estimated the M_1 value using the Mark-Kuhn equation for PMMA in toluene [25] and the experimental value of the intrinsic viscosity of PMMA-6 (Table 2). The M_1 value for PMMA molecules in the content of PMMA-6, 7 was amounted to $50x10^3$ g mol^{-1}. By subtracting the mass of the zirconia core and contribution of TMSPM-modifier from M_2, the mean equivalent number of the PMMA molecules in the hybrid particles of the modified polymers synthesized in solution can be calculated.

The n value was estimated subject to the condition that the ZrO_2 nanocrystals are spherical in form, 20 nm in diameter, and their density is 6 g cm^{-3} [8,11]. The total mass of the bound TMSPM modifier molecules was calculated for the mean density of 10 molecules per nm^2 [9, 10]. The n value was calculated by the equation $n = (M_2 - 25.12x10^{-18}$ g $- M_{TMSPM}x12.5x10^3)/M_1$, converting the polymer molecular mass to grams. Thus received n values are listed in the last column in Table 2.

It should be noted that the n value is different for PMMA-4, 5 and PMMA-6,7, that is, for the samples synthesized under varied conditions (see Table 1). This fact is the evidence of the effect of temperature and concentration of initiator in the reaction mixture on the number of PMMA molecules covalently associated with ZrO_2, since the density of the bound TMSPM molecules per unit surface of nanocrystals was not varied. In this case, it appeared that the size of the hybrid particles in the PMMA-4,5 and PMMA-6, 7 samples are close to one another. The latter confirms the above inference that the polymer chains are forced to acquire the elongated conformations when being attached to ZrO_2 nanocrystals through TMSPM. Hence, the extent of their elongation may increase as the density of the attachments increases.

Note also that the masses of the inorganic and organic parts of the hybrid particles in PMMA-4-7 are practically equal to each other. Therefore, only those modified polymers are completely soluble which contain ZrO_2 nanocrystals with covalently bound polymer whose total mass is close to the mass of the nanocrystals proper. Apparently, this specificity provides high solubility of the hybrid particles in organic solvents, and it is an important factor for creating uniform distribution of inorganic particles in a polymer matrix.

4. Conclusions

In conclusion, it can be stated that the conditions of MMA polymerization in a presence of surface-modified ZrO_2 nanocrystals affect the type of the formed polymer-inorganic particles in a final PMMA composite. During MMA polymerization in toluene solution, mainly the "core-shell" hybrid particles are formed with ZrO_2 as the core. In a process of *in situ* bulk polymerization of MMA, the hybrid particles of the irregular cross-linked structure are formed of considerably larger size than those formed during polymerization in solution. Synthesis of a composite PMMA material in solution may be preferable if compared with that in bulk, because a higher yield of a polymer containing soluble hybrid polymer-inorganic particles can be obtained under these conditions.

Acknowledgements

The authors are grateful to Interdisciplinary resource center for nanotechnologies of St. Petersburg State University where TEM imaging was performed and also grateful to St. Petersburg State University Center for Diagnostics of functional materials for medicine, pharmacology and nanoelectronics for the dynamic light scattering measurements.

References

[1] D. Vollath, "Nanomaterials: An introduction to Synthesis, Properties and Applications", 2d ed., New York: Wiley, 2008, 386 pp.

[2] S. Pavlidou and C.D. Papaspyrides, "A review on polymer–layered silicate nanocomposites", Progress in Polymer Science, 2008, vol. 33, pp. 1119-1198.

[3] K. Friedrich, S. Fakirov and Z. Zhang, "Polymer Nanocomposites: From Nano- to Macro-scale", New York: Springer, 2005, 367 pp.

[4] V.E. Yudin and V.M. Svetlichnyi, "Effect of the structure and shape of filler nanoparticles on the physical properties of polyimide composites", Russian J. General Chem. , 2010, vol. 80, pp. 2157-2169.

[5] S. M. Khaled, R. Sui, A. Paul, P.A. Charpentier and A.S. Rizkalla, "Synthesis of TiO2–PMMA Nanocomposite: Using Methacrylic Acid as a Coupling Agent", Langmuir, 2007, vol. 23, pp. 3988-3995.

[6] P. Obreja , D. Cristea, V.S. Teodorescu, A. Dinescu, A.C. Obreja, F. Comanescu and R. Rebigan, " Preparation and patterning of nanoscale hybrid materials for micro-optics", 2010, vol. 12, pp. 2007-2013.

[7] A.N. Bugrov, E.N. Vlasova, M.V. Mokeev, E.N. Popova, E.M. Ivankova, O.V. Almjasheva and V.M. Svetlichnyi, "Distribution of zirconia nanoparticles in the matrix of poly(4,4'-oxydiphenylenepyromellitimide)", Polym.Sci. Ser. B., 2012, vol. 54, pp. 486-495.

[8] S. Shokoohi, A. Arefazar and R. Khorsokhavar, "Silane Coupling Agents in Polymer-based Reinforced Composites: A Review", J. Reinf. Plast. Comp.2008, vol. 27, pp. 473-485.

[9] Y. Hu, S. Zhou and L. Wu, " Surface mechanical properties of transparent poly(methyl methacrylate)/zirconia nanocomposites prepared by in situ bulk polymerization", Polymer, 2009, vol. 50, pp.3609-3616.

[10] T. Otsuka and Y. Chujo, "Poly(methyl methacrylate) (PMMA)-based hybrid materials with reactive zirconium oxide nanocrystals", Polymer J., 2010, vol. 42, pp. 58-65.

[11] A.N. Bugrov, "Polymer-inorganic composites based on carbo- and heterochain polymers modified by ZrO2 nanoparticles", PhD Thesis, St.Petersburg , 2013.

[12] O.V. Pozhidaeva, E.N. Korytkova and D.P. Romanov, " Formation of ZrO2 Nanocrystals in Hydrothermal Media of Various Chemical Compositions". Russian J. General Chem. 2002, vol. 72, pp. 849-853.

[13] A. Ya. Malkin and A.E. "Diffusion and viscosity of polymers. Measurements methods" Moscow: Chemistry, 1979, 304 pp.

[14] V.N. Tsvetkov, "Rigid-chain polymers: Hydrodynamic and Optical properties in solution", New York: Consultants Bureau, 1989, 397 pp.

[15] B. J. Berneand R. Pecora, "Dynamic Light Scattering", New-York : Courier Dover Publication, 2000, 376 pp.

[16] http://photocor.com/dynals/

[17] N.V. Agrinskaja, E.G. Guk, I.A. Kudrjavtsev and O.G. Ljublinskaja, "Spectral study of polydiacethylene-THD in PMMA matrix", Phys. Solid State, 1995, vol. 37,pp. 969-975.

[18] S. Okabe, S. Sugihara, S. Aoshima, and M. Shibayama, "Heat-Induced Self-Assembling of Thermosensitive Block Copolymer.Rheology and Dynamic Light Scattering", Macromolecules, 2003, vol.36, pp. 4099-4106.

[19] M. Shibayama, T. Karino and S. Okabe, " Distribution analysis of multy-model dynamic light scattering data", Polymer, 2006, vol. 47,pp. 6446-6456.

[20] C. Wu, M. Siddiq and K. F. Woo," Laser Light-Scattering Characterization of a polymer Mixture Made of Individual Linear Chain and Clusters", Macromolecules, 1995, vol.. 28, pp. 4914-4919.

[21] Y. Zhang, C. Wu, Q. Fang and Y.-X. Zhang, "A Light-Scattering Study of the Aggregation Behavior of Fluorocarbon-Modified Polyacrylamides in Water", Macromolecules, 1996, vol. 29, pp. 2494-2497.

[22] C. Wu, S. Bo, M. Siddiq, G. Yang and T. Chen, "Laser Light-Scattering Study of Novel Thermoplastics. 2. Phenolphthalein Poly(ether sulfone)" Macromolecules, 1996, vol. 29, pp. 3157-3160.

[23] E.A. Litmanovich and E.M. Ivleva, "The problem of bimodal distributions in dynamic light scattering: Theory and experiment", Polym. Sci. Ser. A., 2010, vol. 52, pp. 671-678.

[24] A.A. Mercurieva, T.M. Birshtein and V. M. Amoskov , "Theory of Liquid-crystalline ordering in polymer brushes" , Macromol. Symp., 2007, vol. 252, pp. 90-100.

[25] S.N. Chinai, J.D. Matlack, A.L. Resnick and R.J. Samuels, "Poly(methyl methacrylate): Dilute solution properties by viscosity and light scattering", J. Polym. Sci., 1955, vol.17, pp. 391-401.

12

Evaluation of Atomic, Physical, and Thermal Properties of Bismuth Oxide Powder: An Impact of Biofield Energy Treatment

Mahendra Kumar Trivedi[1], Rama Mohan Tallapragada[1], Alice Branton[1], Dahryn Trivedi[1], Gopal Nayak[1], Omprakash Latiyal[2], Snehasis Jana[2, *]

[1]Trivedi Global Inc., Henderson, USA
[2]Trivedi Science Research Laboratory Pvt. Ltd., Bhopal, Madhya Pradesh, India

Email address:
publication@trivedisrl.com (S. Jana)

Abstract: Bismuth oxide (Bi_2O_3) is known for its application in several industries such as solid oxide fuel cells, optoelectronics, gas sensors and optical coatings. The present study was designed to evaluate the effect of biofield energy treatment on the atomic, physical, and thermal properties of Bi_2O_3. The Bi_2O_3 powder was equally divided into two parts: control and treated. The treated part was subjected to biofield energy treatment. After that, both control and treated samples were investigated using X-ray diffraction (XRD), thermogravimetric analysis (TGA), Fourier transform infrared (FT-IR) spectroscopy, and electron spin resonance (ESR) spectroscopy. The XRD data exhibited that the biofield treatment has altered the lattice parameter (-0.19%), unit cell volume (-0.58%), density (0.59%), and molecular weight (-0.57%) of the treated sample as compared to the control. The crystallite size was significantly increased by 25% in treated sample as compared to the control. Furthermore, TGA analysis showed that control and treated samples were thermally stable upto tested temperature of 831°C. Besides, the FT-IR analysis did not show any significant change in absorption wavenumber in the treated sample as compared to the control. The ESR study revealed that g-factor was increased by 13.86% in the treated sample as compared to the control. Thus, above data suggested that biofield energy treatment has altered the atomic and physical properties of Bi_2O_3. Therefore, the biofield treated Bi_2O_3 could be more useful in solid oxide fuel cell industries.

Keywords: Bismuth Oxide, Biofield Energy Treatment, X-ray Diffraction, Differential Scanning Calorimetry, Thermogravimetric Analysis, Fourier Transform Infrared Spectroscopy

1. Introduction

Bismuth oxide (Bi_2O_3) is known for its optical and electrical properties such as dielectric permittivity, refractive index, large energy band gap, photoconductivity and photoluminescence [1]. Due to these properties, Bi_2O_3 play a vital role in the various fields such as optoelectronics, gas sensors and optical coatings. Bi_2O_3 has five polymorphs *i.e.* α- Bi_2O_3 (monoclinic), β- Bi_2O_3 (tetragonal), γ- Bi_2O_3 (BCC), δ- Bi_2O_3 (Cubic), and ε- Bi_2O_3 (triclinic) [2]. Among these phase, α- Bi_2O_3 and δ- Bi_2O_3 are the stable phases, while rest other phases are metastable. Furthermore, δ- Bi_2O_3 exists in the form of face centered cubic crystal structure [3]. The δ-Bi_2O_3 is known for its high conductivity among all other phases, which make it best material in solid oxide fuel cell. Although its application as

oxide ion conductor is limited, because it is only stable in the narrow temperature range. Recently, the stability of δ- Bi_2O_3 is reported to be increased by various kind of dopants such as Er_2O_3 [4] and Y_2O_3 [5] etc. Verkerk *et al.* reported that erbia-stabilized bismuth oxides are among the best solid oxide oxygen ion conductors [6]. In numerous research, the stabilization of the δ- Bi_2O_3 was enhanced by doping with 20 to 50% lanthanide ions [7-9]. However, all these process are either required costly equipment setup or high temperature devices. Thus, it is important to study an alternative approach *i.e.* biofield energy treatment, which could modify the Bi_2O_3 with respect to its atomic, physical, thermal properties. Recently, the biofield energy treatment has gained significant attention, due to its ability to alter the physical, atomic, and structure properties of metals [10, 11] and ceramics [12, 13].

Furthermore, a human has the capability to harness the energy from the environment/Universe and transmit it to any object around the Globe. The object(s) receive the energy and respond into a useful way that is called biofield energy, and this process is known as biofield energy treatment. Besides, the National Center for Complementary and alternative medicine (NCCAM) has recommended uses of CAM therapies (*e.g.* healing therapy) in the healthcare sector [14]. Nevertheless, Mr. Trivedi's unique biofield energy treatment (The Trivedi Effect®) had altered the atomic, physical and thermal characteristics in several ceramics oxides [15,16]. Thus, after considering the excellent outcomes with biofield energy treatment on ceramics and the industrial applications of Bi_2O_3, this work was undertaken to evaluate the effect of this treatment on the atomic, physical and thermal properties of the Bi_2O_3 using X-ray diffraction (XRD), thermogravimetric analysis (TGA), Fourier transform infrared (FT-IR) spectroscopy and electron spin resonance (ESR) spectroscopy.

2. Materials and Methods

The Bi_2O_3 powder was procured from Sigma Aldrich, USA. The procured powder was equally divided into two parts and coded as control and treated. The control part was remained the same and the treated part was in sealed pack, handed over to Mr. Trivedi for biofield energy treatment under standard laboratory conditions. Mr. Trivedi provided the treatment through his energy transmission process to the treated sample without touching the sample. After that, the control and treated samples were characterized using XRD, TGA, FT-IR and ESR techniques.

2.1. XRD Study

The XRD analysis of control and treated Bi_2O_3 was accomplished on Phillips, Holland PW 1710 X-ray diffractometer system. The X-ray of wavelength 1.54056 $\times10^{-10}$ m was used. From the XRD diffractogram, the peak intensity counts, d value (Å), full width half maximum (FWHM) ($\theta°$), relative intensity (%) values were obtained. The PowderX software was used to compute the lattice parameter and unit cell volume of the control and treated Bi_2O_3 samples. The crystallite size (D) was calculated by using Scherrer equation as following:

$$D = k\lambda/(bCos\theta)$$

Here, b is full width half maximum (FWHM) of XRD peaks, k=0.94, and λ =1.54056 Å.

The percentage change in crystallite size was calculated using following formula:

$$\% \text{ change in crystallite size} = [(D_t-D_c)/D_c] \times100$$

Where, D_c and D_t are crystallite size of control and treated powder samples respectively.

2.2. Thermal Analysis

The thermal analysis of Bi_2O_3 powder was done using

TGA-DTG. For that, Mettler Toledo simultaneous TGA-DTG instrument was used. The samples were heated from room temperature to 900°C with a heating rate of 10°C/min under nitrogen atmosphere.

2.3. FT-IR Spectroscopy

The FT-IR analysis of control and treated Bi_2O_3 samples were carried out on Shimadzu's FT-IR (Japan) with frequency range of 4000-500 cm^{-1}. The analysis was accomplished to evaluate the effect of biofield treatment on dipole moment, force constant and bond strength in chemical structure.

2.4. ESR Spectroscopy

The ESR analysis of control and treated Bi_2O_3 samples were performed on Electron Spin Resonance (ESR), E-112 ESR Spectrometer of Varian USA. In this experiment, X-band microwave frequency (9.5 GHz), having sensitivity of 5 x 1010, ΔH spins was used.

3. Results and Discussion

3.1. XRD Study

The XRD is a quantitative and non-destructive technique, which have been widely used to study the crystal structure parameters of a compound. Figure 1 shows the XRD diffractogram of control and treated Bi_2O_3 samples. It can be observed that the control sample showed the crystalline peaks at Bragg angle (2θ) 27.08°, 27.68°, 32.70°, 32.93°, 34.70°, 37.24°, 46.03°, and 52.08°. However, the treated sample showed the peaks at Bragg's angle 27.17°, 27.79°, 32.82°, 33.03°, 34.84°, 37.39°, 46.09°, and 52.17°. It indicated that the XRD peaks were shifted toward higher angles in the treated sample as compared to control, after biofield energy treatment. It is reported that the reduction in lattice parameter and unit cell volume lead to shifting of the XRD peaks toward higher angles [17]. The XRD data of the control and treated samples were analyzed using PowderX software. The crystal structure parameters such as lattice parameter, unit cell volume, density, and molecular weight were computed and presented in Table 1. The data showed that the lattice parameter of treated sample was decreased from 5.6596 Å to 5.6487Å. Kumar *et al.* reported that the XRD peaks can shift to the higher side if larger radii atoms are replaced by smaller radii atoms [18].

Nevertheless the unit cell volume of treated Bi_2O_3 powder was decreased by 0.58% as compared to control. Thus, the decrease in lattice parameter and unit cell volume were supported by shifting of XRD peaks toward higher angles. Hence, based on shifting of XRD peaks and reduction in the lattice parameter, it is assumed that the biofield treatment might induce compressive stress in treated Bi_2O_3 powder and this might be responsible for the internal strain in treated Bi_2O_3. Ekhelikar *et al.* reported that the lattice parameter of Bi_2O_3 unit cell was reduced from 5.560 Å to 5.540 Å when the doping composition of Y_2O_3 was increased from 10 to 20% in Bi_2O_3 and increased the stability of δ- Bi_2O_3 [19]. Thus, it is assumed that the decrease in lattice parameter of Bi_2O_3 after biofield treatment might increase the stability of δ- Bi_2O_3. Moreover, the reduction in unit cell volume led to the increase in density of treated Bi_2O_3 powder by 059% as compared to the control. The molecular weight of the treated Bi_2O_3 powder was reduced by 0.57% as compared to the control. Besides, the crystallite size of treated Bi_2O_3 was increased from 85.10 nm (control) to 106.39 nm after biofield treatment. It indicated that the crystallite size was significantly increased by 25% as compared to the control. It is possible that the neighboring crystalline plane reoriented themselves in the same plane and increased the crystallite size. Li *et al.* reported that the

crystallite size of Bi_2O_3 containing compound was increased with increased in sintering time [20]. It is also mentioned that the increase in crystallite size caused an increase in ionic conductivity in Bi_2O_3. Thus, based on this, it is assumed that the increase in crystallite size in treated Bi_2O_3 may lead to increase the ionic conductivity. It could be due to the reduction of crystallite boundaries in treated Bi_2O_3 as compared to control since an increase in crystallite size decrease the crystallite boundaries. Therefore, the increase in ionic conductivity and stability of Bi_2O_3 could play a major role in the enhancement of efficiency of the solid oxide fuel cell.

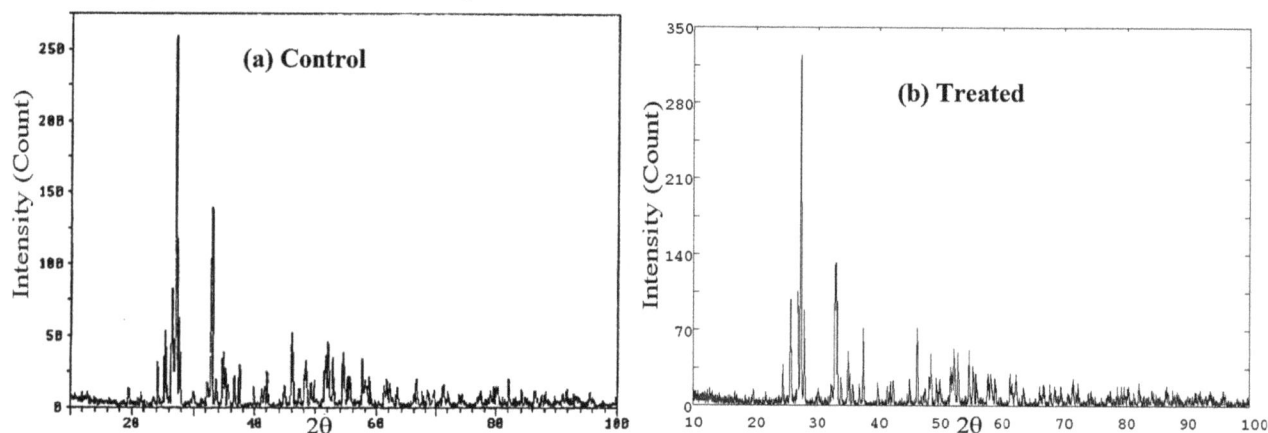

Fig. 1. *X-ray diffractogram of bismuth oxide powder.*

Table 1. *Effect of biofield energy treatment on lattice parameter, unit cell volume density, molecular weight, and crystallite size of bismuth oxide powder.*

Group	Lattice parameter (Å)	Unit cell volume ($\times 10^{-23}$ cm^3)	Density (g/cc)	Molecular weight (g/mol)	Crystallite size (nm)
Control	5.6596	18.13	8.6097	470.04	85.10
Treated	5.6487	18.02	8.6599	467.32	106.39
% Change	-0.19	-0.58	0.59	-0.57	25

3.2. Thermal Analysis

The analysis result of TGA are presented in Table 2. The data exhibited that both the control and treated sample were started weight loss at temperature around 50°C. The control sample lost around 0.22% upto temperature 335°C. After that, the control sample stared to gain the weight and which led to increase the weight by 0.2% upto temperature 821°C and so on. It indicated that control sample was thermally stable. However, the treated sample started to lose its weight at 50°C that continued till 660°C. In this process, the sample lost around 1.7% of its initial weight. The weight loss in temperature upto 335°C in control and treated sample could be due to the elimination of water from the samples. Klinkova *et al.* had studied the thermal behavior of Bi_2O_3, where it was reported that the sample continue to show weak weight loss upto 600°C due to the removal of oxygen. It was also mentioned that the formula unit was changed to $Bi_2O_{2.902}$ after heating of the sample upto 600°C [21]. Furthermore, the weight loss observed in control and treated samples were less than 1.7% which may be due to the loss of oxygen or water, thus, it indicated that both samples were thermally stable.

Table 2. *TGA analysis of bismuth oxide powder.*

Parameter	Control	Treated
Onset temperature (°C)	50	50
Endset temperature (°C)	335	660
Percent weight loss (%)	0.22	1.7

3.3. FT-IR Spectroscopy

The FT-IR spectra of control and treated Bi_2O_3 samples are presented in Figure 2. The band observed at around 3439 cm^{-1} in both control and treated samples could be due to O-H stretching vibrations indicating the presence of the water molecule. In addition, the band was observed in the range 400-700 cm^{-1} *i.e.* 440 and 506 cm^{-1} in control and treated sample could be the characteristics vibrations of Bi-O bond [22]. In addition, Wang *et al.* also reported the Bi-O stretching vibrations at 515cm^{-1} [23]. Thus, the FT-IR data did not show any significant alteration in absorption wavenumbers of treated sample as compared to the control.

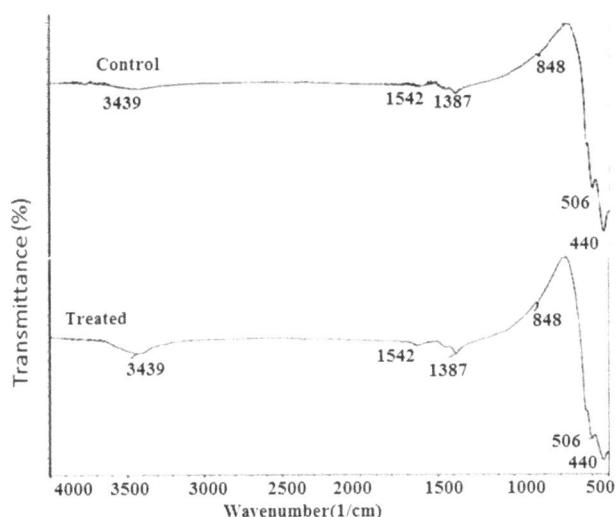

Fig. 2. FT-IR spectra of bismuth oxide powder.

3.4. ESR Spectroscopy

The analysis result of control and treated Bi_2O_3 samples using ESR are illustrated in Table 3. The data showed that the g-factor was increased from 2.0054 (control) to 2.2833 in treated Bi_2O_3 sample. It suggested that the g-factor was increased by 13.86% in treated sample as compared to the control. Also, the width and height of the ESR signal were significantly increased by 311.1 and 1188.9% respectively, as compared to the control. Thus, the increase in ESR signal width and height indicated that the interaction of electron with neighboring elements in treated sample probably altered after biofield treatment. It is assumed that the biofield energy, which probably transferred through treatment, possibly acted at atomic level to cause these modifications.

Table 3. Effect of biofield energy treatment on the ESR properties of bismuth oxide.

Group	g-factor	ESR signal width	ESR signal height
Control	2.0054	90	1.41×10^{-4}
Treated	2.2833	370	1.81×10^{-3}
Percent Change	13.86	311.1	1188.9

4. Conclusions

The XRD data revealed that the lattice parameter was reduced in treated sample as compared to the control. The decrease in lattice parameter may lead to enhance the stability of δ- phase of Bi_2O_3 in treated sample as compared to the control. Also, the increase in crystallite size upto 25% suggest that the ionic conductivity of treated Bi_2O_3 might increase after biofield treatment. The TGA study showed the stability of control and treated Bi_2O_3 samples upto the tested temperature of 900°C. Besides, the ESR spectra study revealed that the signal width and height were significantly increased by 311.1 and 1188.9% respectively, as compared to the control. Thus, overall study concludes that biofield

treatment has altered the atomic and physical properties of Bi_2O_3. Therefore, the treated Bi_2O_3 could be more beneficial in solid oxide fuel cell as compared to the control.

Acknowledgments

Authors would like to acknowledge Dr. Cheng Dong of NLSC, Institute of Physics, and Chinese academy of sciences for permitting us to use Powder-X software for analyzing XRD results. The authors would also like to thank Trivedi Science, Trivedi Master Wellness and Trivedi Testimonials for their support during the work.

References

[1]   Oudghiri-Hassani H, Rakass S, Al Wadaani FT, Al-ghamdi KJ, Omer A, et al.(2015) Synthesis, characterization and photocatalytic activity of α-Bi_2O_3 nanoparticles. J Taibah Univ Sci 9: 508-512.

[2]   Fan HT, Pan SS, Teng XM, Ye C, Li HG, et al. (2006) δ-Bi_2O_3 thin films prepared by reactive sputtering: Fabrication and characterization. Thin Solid Films 513:142-147.

[3]   Kayali R, Kasikci M, Durmus S, Ari M (2011) Investigation of electrical, structural and thermal stability properties of cubic $(Bi_2O_3)_{1-x-y}(Dy_2O_3)_x(Ho_2O_3)_y$ ternary system. Fuel Cells 1219-1234.

[4]   Vinke IC, Seshan K, Boukamp BA, Vries KJ de, Burggraaf AJ (1989) Electrochemical properties of stabilized δ-Bi_2O_3.Oxygen pump properties of Bi_2O_3-Er_2O_3 solid solutions. Solid State Ionics 34: 235-242.

[5]   Battle PD, Catlow CRA, Heap JW, Moroney LM (1986) Structural and dynamical studies of δ-Bi_2O_3 oxide ion conductors: I. The structure of $(Bi_2O_3)_{1-x}$ $(Y_2O_3)_x$ as a function of x and temperature. J Solid State Chem 63: 8-15.

[6]   Verkerk MJ, Keizer K, Burggraaf AJ (1980) High oxygen ion conduction in sintered oxides of the Bi_2O_3-Er_2O_3 system. J Appl Electrochem 10: 81-90.

[7]   Tanabe H, Fukushima S (1986) Cathodic polarization characteristics of the oxygen electrodes/stabilized Bi_2O_3 solid electrolyte interface. Electrochem Acta 31: 801-809.

[8]   Esaka T, Iwahara H, Kunieda H (1982) Oxide ion and electron mixed conduction in sintered oxides of the system Bi_2O_3-Pr_6O_{11}. J Appl Electrochem 12: 235-240.

[9]   Battle PD, Catlow CRA, Chadwick AV, Cox P, Greaves GN, et al. (1987) Structural and dynamical studies of δ-Bi_2O_3 oxide ion conductors: IV. An EXAFS investigation of $(Bi_2O_3)_{1-x}$ $(M_2O_3)_x$ for M = Y, Er, and Yb. J Solid State Chem 69: 230-239.

[10] Trivedi MK, Tallapragada RM, Branton A, Trivedi D, Nayak G, et al. (2015) Potential impact of biofield treatment on atomic and physical characteristics of magnesium. Vitam Miner 3: 129.

[11] Trivedi MK, Nayak G, Patil S, Tallapragada RM, Latiyal O, et al. (2015) An evaluation of biofield treatment on thermal, physical and structural properties of cadmium powder. J Thermodyn Catal 6: 147.

[12] Trivedi MK, Nayak G, Patil S, Tallapragada RM, Latiyal O, et al. (2015) Impact of biofield treatment on atomic and structural characteristics of barium titanate powder. Ind Eng Manage 4: 166.

[13] Trivedi MK, Patil S, Nayak G, Jana S, Latiyal O (2015) Influence of biofield treatment on physical, structural and spectral properties of boron nitride. J Material Sci Eng 4: 181.

[14] Barnes PM, Powell-Griner E, McFann K, Nahin RL (2004) Complementary and alternative medicine use among adults: United States, 2002. Adv Data 343: 1-19.

[15] Trivedi MK, Nayak G, Patil S, Tallapragada RM, Latiyal O (2015) Studies of the atomic and crystalline characteristics of ceramic oxide nano powders after bio field treatment. Ind Eng Manage 4: 161.

[16] Trivedi MK, Nayak G, Patil S, Tallapragada RM, Latiyal O (2015) Impact of biofield treatment on physical, structural and spectral properties of antimony sulfide. Ind Eng Manage 4:165.

[17] Schwertmann U, Cornell RM (2007) Iron oxides in the laboratory: preparation and characterization. John Wiley & Sons.

[18] Kumar P, Kar M (2014) Effect of structural transition on magnetic and dielectric properties of La and Mn co-substituted $BiFeO_3$ ceramics. Mater Chem Phys. 148: 968-977.

[19] Ekhelikar S, Bichile GK (2004) Synthesis and structural characterization of $(Bi_2O_3)_{1-x}$ $(Y_2O_3)_x$ and $(Bi_2O_3)_{1-x}$ $(Gd_2O_3)_x$ solid solutions. Bull Mater Sci 27: 19-22.

[20] Li R, Zhen Q, Drache M, Rubbens A, Estournès C, et al (2011) Synthesis and ion conductivity of $(Bi_2O_3)_{0.75}(Dy_2O_3)_{0.25}$ ceramics with grain sizes from the nano to the micro scale. Solid State Ionics 198: 6-15.

[21] Klinkova LA, Nikolaichik VI, Barkovskii NV, Fedotov VK (2007) Thermal stability of Bi_2O_3. Russ J Inorg Chem 52: 1822-1829.

[22] Mallahi M, Shokuhfar A, Vaezi MR, Esmaeilirad A, Mazinani V (2014) Synthesis and characterization of bismuth oxide nanoparticles via sol-gel method. AJER 3: 162-165.

[23] Wang SX, Jin CC, Qian WJ (2014) Bi_2O_3 with activated carbon composite as a supercapacitor electrode. J Alloy Compd 615: 12-17.

13

The Casimir Effect as a Pure Topological Phenomenon and the Possibility of a Casimir Nano Reactor – A Preliminary Conceptual Design

Mohamed S. El Naschie

Dept. of Physics, University of Alexandria, Alexandria, Egypt

Email address:

Chaossf@aol.com

Abstract: A preliminary conceptual design of a new free energy nano reactor is presented. The design is based on the following: A basically topological interpretation of the Casimir effect is given as a natural intrinsic property of the geometrical topological structure of the quantum-Cantorian micro spacetime. This new interpretation compliments the earlier conventional interpretation as vacuum fluctuation or as a Schwinger source and links the Casimir energy to the so called missing dark energy density of the cosmos. We start with a general outline of the theoretical principle and basic design concepts of a proposed Casimir dark energy nano reactor. In a nutshell the theory and consequently the actual design depends crucially upon the equivalence between the dark energy density of the cosmos and the faint local Casimir effect produced by two sides boundary condition quantum waves. This Casimir effect is then colossally amplified as a one sided quantum wave pushing from the inside with nothing balancing it from the non-existent outside. In view of the present theory, this is essentially what leads to the observed accelerated expansion of the cosmos. As in any reactor, the basic principle in the present design is to produce a gradient so that the excess energy on one side flows to the other side. Thus in principle we will restructure the local topology of space using material nanoscience technology to create an artificial local high dimensionality with a Dvoretzky theorem like volume measure concentration. Without going into the intricate nonlinear dynamics and technological detail, it is fair to say that this would lead us to pure, clean, free energy obtained directly from the topology of spacetime. Needless to say the entire design is based completely on the theory of quantum wave dark energy proposed by the present author for the first time in 2011 in a conference held in the Bibliotheca Alexandrina, Egypt and a little later in Shanghai, Republic of China. The quintessence of the present theory is easily explained as the ϕ^3 intrinsic Casimir topological energy where $\phi = (\sqrt{5} - 1)/2$ produced from the zero set ϕ of the quantum particle when we extract the empty set quantum wave ϕ^2 from it and find $\phi - \phi^2 = \phi^3$ by restructuring space via conducting but uncharged plates similar to that of the classical Casimir experiments but with some modification. Our proposed preliminary design of the reactor follows in a natural way from the above.

Keywords: Casimir Effect, Dark Energy, E-Infinity, Cantorian Spacetime, Nano Reactor Avant Project, Free Energy

1. Introduction

Apart of presenting phase one of an avant projet for a nano Casimir-dark energy reactor (see Figs. 1-4), the present paper has two different messages to communicate, a scientific one centered around the quantum vacuum as a source of energy [1-58] and a socio-economical, political message that we must invest in this new revolutionary source of energy [59-61]. The idea of zero point energy and the fluctuation of vacuum may seem at first glance to be more science fiction than science fact. However there are, and since quite some time, a host of hard core experimental evidence that the vacuum may be more real ad fundamental than most of what we habitually consider the materialistic reality of physical phenomenon [1-73]. We just need to mention in this context the Lamb shift, Schwinger correction [62-66] and the van der Waals forces to realize how physical and real the vacuum is [22-24][28]. Nonetheless, and we do not think it is a minority opinion, nothing could be more impressive and inspiring as the Casimir effect (Fig. 1) [22-24]. This effect is a natural consequence and fundamental aspect of quantum field theory. There are at least two fundamental interpretations of this miraculous effect [63,64]. The first is loosely connect to boundary conditions and the zero point quantum vacuum fluctuation which may be the common way of looking at the Casimir effect within the working physicists community. The second, which may be more theoretical and fundamental, is to see Casimir as a source in the mold of J. Schwinger's way of thinking and not far from the Casimir operators of quantum field theory [62,63,66]. Thus we could look upon the Casimir effect as a cousin of Hawking's negative energy fluctuation around a black hole or as Unruh's temperature for an accelerated, observed in a Rindler wedge, universe. Alternatively we could follow Schwinger's ideas and see it as something related to a fundamental mathematical scenario such as the Banach-Tarski theorem advanced for the first time in the cosmology of the big bang by the present author [67,68].

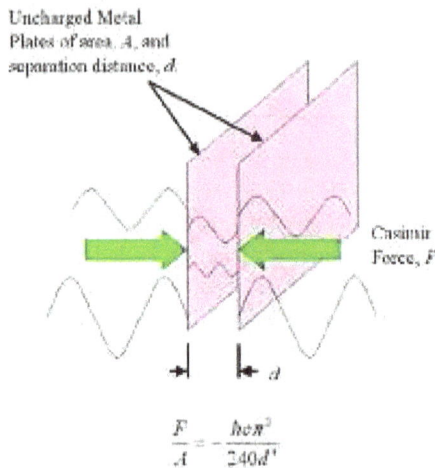

Fig. 1. *The Casimir effect as the basic idea behind the present work [22-24][62-66] is best visualized and understood using a hydrodynamical analogy. Suppose we put two walls in relatively shallow water near to the shore with the reasonable objective of building a small boats harbour. Clearly within the two walls constituting the harbour, there will only be moderate water wavelets. Outside however we have the open sea with its relatively strong waves that would have the effect of pushing the walls towards each other being much stronger than the waves between the two walls. In this analogy the open sea is the vacuum of fractal-Cantorian spacetime while the waves are the quantum waves of quantum field theory and finally the two walls are our two Casimir plates as shown in the figure.*

In the present paper however we opted for a rather different point of viewing the Casimir effect as a natural topological necessity of a Cantorian spacetime fabric which was woven from an infinite number of zero Cantor sets and empty Cantor sets [37]. The zero set is taken following von Neumann-Connes dimensional function to model the quantum particles while the empty set models the quantum wave. Following this road we come we come to realize that the Casimir latent energy is nothing but the universal fluctuation ϕ^3 which gives birth to the core of Cantorian-fractal spacetime by inversion $1/\phi^3 = 4 + \phi^3 = 4.23606797$ where $\phi = \left(\sqrt{5} - 1\right)/2$ [52-53]. This is nothing but the difference between the Hausdorff dimension of the particle

zero set ϕ and the wave empty set ϕ^2. The result not surprisingly is almost equal to double the value found using imaginative modification of the classical Casimir experiment by Zee [62] who found the dimensionless Casimir energy to be $\pi/24 \simeq 0.1308$ [62]. Using E-infinity methodological reasoning, the exact value of Zee in the limit must be the ratio of the dimensionality of a Calabi-Yau transfinite manifold $6 + k = 6.18033889$ and the transfinite dimension of bosonic string theory, i.e. 26.18033989. That means

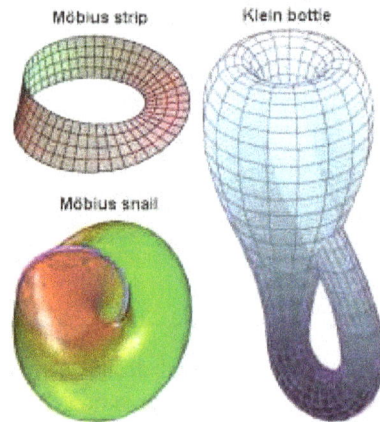

Fig. 2. *The Möbius strip is a well known one sided surface which mathematicians call one dimensional non-orientable manifold. The Klein bottle is also a well known generalization of a Möbius strip representing a bottle which does not have a distinct inside and an outside. The generalization does not stop here and there are categories of Kleinian groups [29, 32] that live in higher dimensionalities representing a generalization of Klein's bottle. We conjectured in the present work that the boundary of the boundary of the universe is such a one sided structure so that all these little Casimir effects ramify at the one sided boundary of the universe and becomes this force causing the accelerated cosmic expansion which we attribute to the conjectured missing dark energy.*

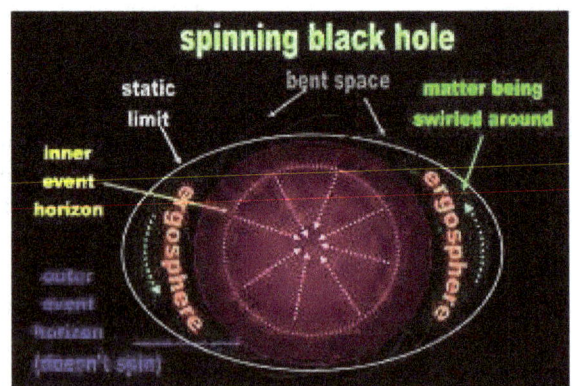

Fig. 3. *Kerr's black hole [73] could be viewed as an elementary particle with all the ordinary energy concentrated inside the singularity [26-28][58]. With E-infinity theory this amounts to $\phi^5/2 \simeq 1/22$ of the total unit energy of $E = mc^2$ when setting $m = c = 1$ where $\phi = (\sqrt{5} - 1)/2$. The rest, which is $5\phi^2/2 \simeq 21/22$, is the dark energy and dark matter portion and represents a quantum wave energy density related to the internal and external Kerr horizons. That way $E = mc^2$ acquires a quantum mechanical interpretation as $E = (mc^2/22) + mc^2 (21/22)$ for a Schwartzchild non-spinning mini black hole-elementary particle. However we can go one step further because the energy between the two horizons could be divided into dark matter energy density equal 5/22 and pure dark energy 16/22. Thus at the end E = is divided into three parts $E = mc^2/22 + mc^2 (5/22) + mc^2 (16/22) = mc^2$.*

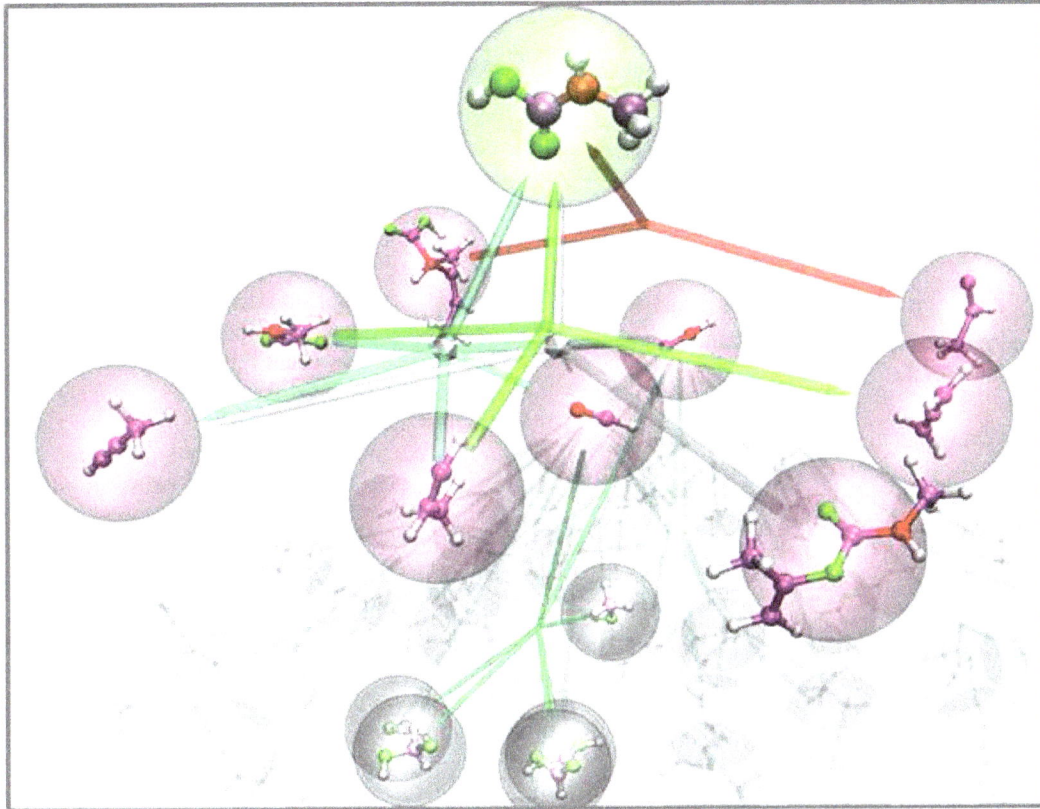

Fig. 4a. *The three figures 4a, 4b and 4c are basically a scientist-engineer and artist impression of how a Casimir-dark energy nano reactor could be designed. It is the Avant projet phase one of our reactor. In Fig. 4a we give the lay out of a reactor tree made of modular units on all scales like a fractal.*

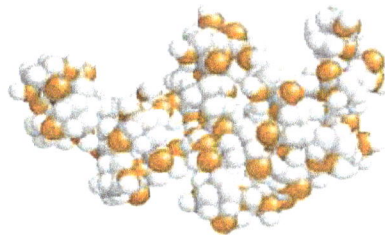

Fig. 4b. *In this figure we have a bird's eye view of our fractal tree [72,73] where the branches cannot be seen except as contact points of a sphere packing arrangement like lattice animals [50-53]. The connection to our Cantorian E-infinity spacetime becomes evident when we remember that a non-compactified holographic boundary of the Kleinian curve type has 336 units or degrees of freedom. In the fractal, i.e. compactified case we have $16k \simeq 3$ more where $k = \phi^3(1-\phi^3) = 2\phi^5$, ϕ^5 is Hardy's quantum entanglement and $\phi = (\sqrt{5}-1)/2$. Climbing down from super space $D = 8$ and super string space $D = 10$ we just need to divide $336 + 16k = 338.8854382$ by $(8)(10) = 80$ and find our core E-infinity Cantorian space Hausdorff dimension $4 + \phi^3 = 4.236067899$ [1-3].*

Fig. 4c. *In this figure we give the basic idea of how every unit of Fig. 4a and Fig. 4b represent in effect a nano universe with a nano holographic horizon and how we in principle could extract the 95.5 per cent energy as per our dark energy theory and in conformity with Dvoretzky's theorem [3-6]. The actual design of the devise shown here is not of direct relevance for the limited purpose of the present discussion and should for the moment be understood symbolically.*

$(6+k)/(26+k)=\phi^3/2$. Needless to say, the division by 2 is due to the subdivision of the 'vacuum' of E-infinity theory and is analogous to dividing Hardy's entanglement $P(H)=\phi^5$ by 2 to obtain the density of the ordinary measurable energy of the cosmos $E(O) = (\phi^5/2)mc^2$. The dimensional quantity analogous to mc^2 for the Casimir effect is trivially clear to be $\hbar c$ where c is the speed of light and $\hbar$ is the Planck quantum. From this new topological interpretation it becomes obvious that Casimir ϕ^3 is the counterfactual or global part of Hardy's entanglement $P(H)=(\phi^3)(\phi^n)$ where n is the number of quantum particles and is found for n = 0. It is therefore closely related to the Unruh temperature where n = 1, the Immirzi parameter n = 3 as well as Hardy's generic quantum-topological entanglement n = 2. These insights are not only simple mathematical insights. It goes far beyond that and suggests hat Casimir energy and dark energy are two sides of the same coin, differing only with regard to exo and endo boundary conditions [6,7] which will be made clearer in the main body of the present work. Second, by manipulating the local dimensionality of spacetime using an elaborate and complex set up of Casimir plates system we could build a nano universe and extract its dark energy concentrated at its boundary (fig. 4). The way to do this economically may be five, ten or more years of experimental work using the modern developments of cutting edge nanotechnology [1]. Never the less, the promise of near to infinite, clean, free energy is a goal worth any effort and the financial risks are minimal compared to the possible gains, so let the present modest steps be the first into this new world of a nano, Casimir-dark energy reactor.

2. Preliminary Remarks

Based on his E-infinity Cantorian spacetime theory [1-21], it was recently argued by the author that the Casimir effect is a local manifestation of the quantum wave while dark energy is the global manifestation of the same [1]. The only difference is that of the details of the boundary conditions [1,2]. It was further reasoned by the author that the universe as a whole has a one sided boundary akin to that of higher dimensional Möbius (Fig. 2) band and consequently the "local" Casimir effect ramifies at this one sided boundary located at infinity to produce the negative gravity pressure of the conjectured dark energy [1-3]. In other words, three rather mysterious physical notions are tied together and explained in terms of each other. At the top resides the quantum wave [4], which is not a mathematical artifact [4-8] but according to E-infinity theory of dark energy, a real physical entity fully described by the empty set fixed by Connes-El Naschie bi dimension $(-1,\phi^2)$ where $\phi=1(\sqrt{5}+1)$ [1-8]. On the other hand the gradient caused by different wave energy density in different bounded regions of space compared to the unbounded outside of the same space is behind the Casimir forces which in the limit can be show to be equal to the difference between the quantum zero set

$(0,\phi)$ and the wave empty set $(-1,\phi^2)$ leading to $\phi-\phi^2=\phi^3$ topological energy pressure [1, 5-8]. Finally at the edge of the universe there is only internal Casimir quantum wave pressure not balanced by outside pressure which is the dark energy concentration of 96 percent as per the consequences of Dvoretzky's theorem and the present author's dissection of Einstein's $E = mc^2$ to $E(O) = mc^2/22$ for ordinary energy of the quantum particle and $E(D) = mc^2(21/22)$ for the dark energy of the quantum wave (see Fig. 3) [7,8].

The sources of the ideas contained in the present work go back to many years ago when we attempted to improve on the traditional fast and slow fission reactors using the modern mathematics of fractals and nonlinear dynamics [9-17]. The second source is our recent reinterpretation of Einstein's $E = mc^2$ and finally the third source is the unexpected results of the earlier mentioned Dvoretzky's theorem of Banach spaces [3,6]. However in the final analysis building an actual reactor could not be possible, not even in principle, without first a sound theory [1-56] and second the combination of modern nanotechnology and state of the art Casimir effect experimentation [18-24]. In addition a reasonable amount of imaginative thinking similar to that of the man who is famed for inventing the 20th century is also recommended [49,57].

To keep the present paper short and yet to cover the large amount of the needed prerequisites we opted for a condensed presentation coupled to a large number of references. We recommend to start by reading Ref. 1 and Ref. 60, then it is a personal choice of how to proceed after that.

3. Möbius Strip in Higher Dimensions As The Boundary of the Holographic Boundary of the Universe

The holographic boundary theory goes back to the pioneering work of 'tHooft and Susskind [25-27]. On the other hand the principle that the boundary of a boundary is zero goes back to the out of the box thinking of J.A. Wheeler [28]. Pushing their ideas further still, it became obvious to the present author that the boundary of the holographic boundary is not only a zero limit set but actually a hierarchy of empty and emptier still sets ramifying at a most general form of a one sided higher dimensional Möbius band [28-33]. This limit set resembles a fundamental polyhedron group or better still, a Schottky-Kleinian group (Fig. 2) [29-33] which changes the topology of our conventional Casimir experiment to that of a sphere with internal Casimir pressure inflating the balloon-like universe and makes it expand into the surrounding "nothingness" fixed by the well known E-infinity formula $d_c^{(-\infty)} = \phi^\infty = 0$ where $\phi=1/(\sqrt{5}+1)$ [34]. From the preceding elementary reasoning it is clear that Casimir-effect and dark energy have the same cause, namely the topology of a Banach-spacetime like manifold and the only difference is the difference of local exophysics and global endophysics and the respective associated boundary conditions [1,2]. There is already a vast body of literature on the subject published in the last three years alone by the present author and his associates [1-56]. However what we are aiming at in the present paper is to point out the way to move from theory to useful, practical application of which nothing could be more important and pressing than building a free energy reactor, based on real science rather than wishful thinking. Thus we will combine the dreams of visionaries like N. Tesla with hard nosed modern mathematics and physics which were not yet available in the time of Tesla [49]. As we said we are of course in a far better position than Tesla because we can fall back on modern results both theoretical and experimental in black holes and fractal research [71-73].

4. Nano Engineering

There has been no want of imaginative experimental set ups for measuring, testing and visualizing the Casimir effect since it was proposed by Dutch physicist, H. Casimir [22-24]. In recent years nanotechnology invaded all scientific fields and played a significant role in Casimir effect experiments. Thanks to E-infinity we now know that the true physical-mathematical connection between dark energy and the Casimir effect. A natural consequence of this discovered reality of the quantum wave, is rendering it a relatively simple task to find a way to harness dark energy or Casimir energy. Of course this "simple" is extremely difficult but no longer impossible. As mentioned earlier on, we are aided in our quest for a Casimir reactor by the many wonderful results obtained in fractal sciences and the geometry of black holes [71-73].

We can start with a highly complex sub-structuring of space using nano tubes and nano particles and create that way nanosphere packing modeling the moonshine conjecture that relates superstrings to other fields of theoretical physics. We presently have, in embryonic form, the main idea of constructing a nano universe and extracting dark energy from its nano boundary of its holographic boundary. Our program to actually extract energy from such a nano reactor may still need five or more years but the road is marked and reasonably clear. It is only at the edge of the universe that 96% of the energy resides as dark energy. However we could create many nano universes from which its 96% energy concentration could be extracted without actually reaching to the boundary of our universe (see Figs. 2-4) [3-8].

5. Imaginative Concepts of Planned Laboratory Work

In noncommutative geometry as well as E-infinity theory, the Penrose universe plays a significant role as a generic concrete model for both theories [50-53]. On the other hand Penrose universe or Penrose fractal tiling is basically a quasi-crystal mathematical model with the forbidden 5-fold symmetry [53,54]. This form of matter not found naturally on earth, was produced experimentally by the great Israeli engineer D. Schechtman, who after facing a long period of fierce opposition from high profile scientists, for instance Nobel Laureate Linus Pauling, was rehabilitated and bestowed with a Nobel Prize. The 5 fold symmetry could be

thought of theoretically as five Kaluza-Klein dimensions and using nano particles and nano tubes combinations we could build in the lab a nano holographic universe [5-8] akin to our own from which energy could be experimented with and extracted. For sure it will be a journey in uncharted seas with many trials and errors but sooner or later we will find out the right road to a Casimir dark energy nano reactor [1,22]. There are other conceivable ways of producing artificial nano universes with high dimensionality for Dvoretzky's theorem to be applicable. For instance we could use Ji-Huan He's ten dimensional polytope [42] as a skeleton to grow on it a hierarchy of nano particles using the methods applied in the clustering of diffusion limited aggregation. In other words, we can let our scientific imagination run free but checked with E-infinity mathematical rigor and nanotechnological facts.

6. The Topological E-Infinity Interpretation of The Casimir-Dark Energy Density

It may come as a pleasant mild surprize that exact limits could easily be established for Casimir-dark energy using nothing more than the topology of our E-infinity Cantorian spacetime [56,60]. We can do this in a variety of ways which are essentially tautologies leading to the same basic conclusion in the limit. Thus we could view the energy density of the space outside the two Casimir plates as that of Einstein's $E = mc^2$ density, i.e. $\gamma(\text{Einstein})=1$. Inside the plate the energy density in the limit could only be a statical, quasi potential energy of the quantum particle, i.e. $E = mc^2/22$ and consequently $\gamma(0)=1/22$. It follows then that the net pressure of the Casimir plates must be $1-(1/22)=21/22$ which is, in the meantime rather well known, as the dark energy density of spacetime. A second way to interpret the same situation and reach the same result is to argue that within the Casimir plates there is no "space" except for the empty set with a Hausdorff dimension ϕ^2 where $\phi=1/(\sqrt{5}+1)$. Outside on the other hand we have the zero set. The difference is a net $\phi-\phi^2=\phi^3$ which is the universal fluctuation of spacetime and simply the reciprocal value of its Hausdorff dimension $(1/\phi^3)=4+\phi^3$ [3][52,56]. Finally we could see the situation as the difference of the completely empty set in the limit, i.e. zero between the Casimir plates and the spacetime fluctuation ϕ^3 [60]. That way the Casimir effect could be set in the limit equal to ϕ^3 and may easily be seen to be a relative to the Immirzi parameter ϕ^6 and the Unruh temperature ϕ^4 apart of Hardy's entanglement ϕ^5, i.e. a member of a generalized quantum-topological entanglement family [60].

7. Intermediate Discussion

It would be a gross error to place the present nano reactor proposal within the context of science fiction. There is definitely a trivial element of speculation and trial and error but that is all. Exploding stars and galaxies are scientific facts. Consequently to presume that these are only topological defects in to near infinitely large spacetime is not outlandish nor science fiction [54-56]. In fact the near identity of the Casimir effect and dark energy and the fact that both originate from the quantum wave aspect of quantum mechanics clearly shows to any open minded scientific thinker that to pursue clean free energy is not a scientific 'crackpot' idea but a real and reachable aim. I ask the sensitive reader to forgive me for using the ugly word 'crackpot' which is not a proper English word but merely slang which invaded the scientific English language like a virus. The 4.5% of ordinary energy in the universe is nothing but the multiplicative volume of a five dimensional K-K zero set while the 95.5% dark energy is the additive volume of the same 5D Kaluza-Klein empty set [34]. Seen that way we think that making humanity free from oil and traditional sources of energy is a higher and moral aim worth investing heavily in for what is a million or even billion dollar research grant funding compared to the three trillion dollar Iraq war [59]. In fact the highly enlightened rules of the United Arab Emirates are already looking towards a future free of oil based energy [61]. It was Nobel Laureate in Economics, Prof. J. Stiglitz who calculated with Prof. L. Bilmes the true cost of the Iraq war for the USA. The staggering three trillion dollars do not actually include the loss and destruction for the economy of the entire world. The author dares to say with a tongue in cheek, that the mere sight of only one trillion dollars funding for our nano Casimir-dark energy reactor is sufficient to make this reactor spontaneously pop out of spacetime like virtual particles!

The author, who was born and raised in the Middle East with its unrivalled rich history and unparalleled chequered present day politics feels morally obliged to call all the governments of the region to participate in a new dawn of science and life.

8. An Avant Projet For A Nano Casimir-Dark Energy Reactor Phase I

Building a reactor is predominantly more of an engineering task than it is a scientific one. The present reactor is definitely an exception because it is extremely unusual. In a sense our reactor is emulating the very act of creation, inspired by it and supported by science and hard facts. However in such a situation we need to combine all what we have as a human and not only our ability for logical deduction and classification. Besides our scientific brain power we need our engineering as well as artistic imagination and to embark upon a road of illumination by simple trial and error similar to that which was taken to decipher the genome [69] and defeat cancer [70]. In Figs. 1

to 4 we have in all six pictures to guide us along the aforementioned trail to a real Casimir-dark energy powered reactor. We have purposely used the French word Avant projet used in the building industry all over the Middle East to signal that it is the beginning of a huge engineering project and that so far we have solved the theoretical part and moved towards the first phase of physical implementation which the reader will surely understand must be protected by an appropriate patent which we hope will be in place in the not too distant future. For the moment we just mention Refs. [71-73] on fractals and the work of R. Kerr to fill the gaps.

9. Conclusion

In the present work we have made some substantial progress in the road towards the dream of spacetime free energy via two real experimentally and observationally documented facts, namely the Casimir effect as well as the accelerated expansion of the cosmos attributed to dark energy. We were able to show here that $E = mc^2$ of Einstein cannot only be dissected to $E = E(O) + E(D) = mc^2$ where $E(O) = mc^2/22$ is the ordinary energy of the quantum particle and $E(D) = mc^2(21/22)$ is the dark energy density of the quantum wave but it can be split down into three parts, namely ordinary energy as well as dark matter energy $E(DM) = mc^2(5/22)$ and pure dark energy density $E(PD) = mc^2(16/22)$ so that at the end we have $E(O) + E(DM) + E(PD) = mc^2$ exactly as Einstein showed us many years ago although neither he nor anyone else suspected for a second that $E = mc^2$ encapsules in it so many quantum secrets of the universe. To do that we needed the work of many scientists including Kerr and his solution as well as the work of Casimir, Penrose, Perlmuter and many others to put the present picture together piece by piece like a jigsaw puzzle. One of our most important insights gained here is undoubtedly the Möbius-like boundary of the boundary of our universe which led us to realize that dark energy expansion of the universe is a global manifestation of the local Casimir forces. Another equally important insight is the relation between Cantorian-quantum spacetime and branching polymers and clusters like lattice animals [72-73] which inspired us to build mini universes to be assembled using nanotechnology into an economically viable nano reactor based on the Casimir-dark energy equivalence principle outlined in this paper. Seen from our view point it is only a matter of funding and time before humanity can enjoy free spacetime Casimir energy reactors.

References

[1]   Mohamed S. El Naschie: Three quantum particles Hardy entanglement from the topology of Cantorian-fractal spacetime and the Casimir effect as dark energy – A great opportunity for nanotechnology. American Journal of Nano Research and Applications, 2015, 3(1), pp. 1-5.

[2] Mohamed S. El Naschie: Casimir-like energy as a double Eigenvalue of quantumly entangled system leading to the missing dark energy density of the cosmos. International Journal of High Energy Physics, 2014, 1(5), pp. 55-63.

[3] Mohamed S. El Naschie: The measure concentration of convex geometry in a quasi Banach spacetime behind the supposedly missing dark energy of the cosmos. American Journal of Astronomy & Astrophysics, 2014, 2(6), pp. 72-77.

[4] M. Slezak: Quantum wave function gets real. New Scientist, 7 February, 2015, pp. 14.

[5] Mohamed S. El Naschie: Dark energy and its cosmic density from Einstein's relativity and gauge fields renormalization leading to the possibility of a new 'tHooft quasi particle. The Open Astronomy Journal, 2015, 8, pp. 1-17.

[6] Mohamed S. El Naschie: Banach spacetime-like Dvoretzky volume concentration as cosmic holographic dark energy. International Journal of High Energy Physics, 2015, 2(1), pp. 13-21.

[7] Mohamed S. El Naschie: From $E = mc^2$ to $E = mc^2/22$ – A short account of the most famous equation in physics and its hidden quantum entanglement origin. Journal of Quantum Information Science, 2014, 4, pp. 284-291.

[8] Mohamed S. El Naschie: The hidden quantum entanglement roots of $E = mc^2$ and its genesis to $E = mc^2/22$ plus $mc^2(21/22)$ confirming Einstein's mass-energy formula. American Journal of Electromagnetics and Applications, 2014, 2(5), pp. 39-44.

[9] Mohamed S. El Naschie: From implosion to fractal spheres. A brief account of the historical development of scientific ideas leading to the trinity test and beyond. Chaos, Solitons & Fractals, 1999, 10(1), pp. 1955-1965.

[10] Mohamed S. El Naschie and S. Al Athel: Estimating the Eigenvalue of fast reactors and Cantorian space. Chaos, Solitons & Fractals, 2000, 11, pp. 1957-1961.

[11] M.S. El Naschie: On Nishina's estimate of the critical mass for fussion and early nuclear research in Japan. Chaos, Solitons & Fractals, 2000, 11(11), pp. 1809-1818.

[12] M.S. El Naschie: Remarks on Heisenberg's Farm-Hall lecture on the critical mass of fast neutron fission. Chaos, Solitons & Fractals, 2000, 11(8), pp. 1327-1333.

[13] M.S. El Naschie and A. Hussein: On the Eigenvalue of nuclear reaction and self-weight buckling. Chaos, Solitons & Fractals, 2000, 11, pp. 815-818.

[14] M.S. El Naschie: Elastic buckling loads and fission critical mass as an Eigenvalue of a symmetry breaking bifurcation. Chaos, Solitons & Fractals, 2000, 11, pp. 631-629.

[15] M.S. El Naschie: On the Zel'dovich-Khuriton critical mass for fast fission. Chaos, Solitons & Fractals, 2000, 11, pp. 819-824.

[16] M.S. El Naschie: On the Eigenvalue of transport reaction involving fast neutrons. Chaos, Solitons & Fractals, 2000, 11, pp. 929-934.

[17] M.S. El Naschie: Heisenberg's critical mass calculations for an explosive nuclear reaction. Chaos, Solitons & Fractals, 2000, 11, pp. 987-997.

[18] M.S. El Naschie: Chaos and fractals in nano and quantum technology. Chaos, Solitons & Fractals, 1998, 9(10), pp. 1793-1802.

[19] M.S. El Naschie: Nanotechnology for the developing world. Chaos, Solitons & Fractals, 2006, 30(4), pp. 769-773.

[20] M.S. El Naschie: The political economy of nanotechnology and the developing world. International Journal of ElectrospunNanofibrers and Applications, 2007, 1(1), pp. 41-50.

[21] M.S. El Naschie: Some tentative proposals for the experimental verification of Cantorian micro spacetime. Chaos, Solitons & Fractals,1998, 9(1/2), pp. 143-144.

[22] H. Johnston: Physicists Solve Casimir Conundrum. Physicsworld.com. July 18, 2012.

[23] S.Rencroft and J. Swain: What is the Casimir effect? Scientific American, June 22, 1998.

[24] P.Wongjun: Casimir dark energy, stabilization and the extra dimensions and Gauss-Bonnet term. The European Physical Journal C, 2015, 75(6).

[25] M.S. El Naschie: A review of application and results of E-infinity. International Journal of Nonlinear Science & Numerical Simulation, 2007, 8(1), pp. 11-20.

[26] L. Smolin: The strong and the weak holographic principles. Nuclear Physics B, 2001, 601(1-2), pp. 209-247.

[27] M.S. El Naschie: Holographic dimensional reduction center manifold theorem and E-infinity. Chaos, Solitons & Fractals, 2006, 29(4), pp. 816-822.

[28] C. Misner, K. Thorne and J.A. Wheeler: Gravitation. Freeman, New York, 1973.

[29] M.S. El Naschie: Kleinian groups in E-infinity and their connection to particle physics and cosmology. Chaos, Solitons & Fractals, 2003, 16, pp. 637-649.

[30] M.S. El Naschie: A guide to the mathematics of E-infinity Cantorian spacetime theory. Chaos, Solitons & Fractals, 2005, 25, p. 935-964.

[31] M.S. El Naschie: The concepts of E-infinity: An elementary introduction to the Cantorian-fractal theory of quantum physics. Chaos, Solitons & Fractals, 2004, 22, pp. 495-511.

[32] M.S. El Naschie: Complex vacuum fluctuation as a chaotic 'limit' set of any Kleinian group transformation and the mass spectrum of high energy particle physics via spontaneous self organization. Chaos, Solitons & Fractals, 2003, 17, pp. 631-638.

[33] M.S. El Naschie: Modular groups in Cantorian E-infinity high energy physics. Chaos, Solitons & Fractals, 2003, 16, pp. 353-366.

[34] M.S. El Naschie: On certain 'empty' Cantor sets and their dimensions. Chaos, Solitons & Fractals, 1994, 4(2), pp. 293-296.

[35] Ji-Huan He et al: Twenty six dimensional polytope and high energy spacetime physics. Chaos, Solitons & Fractals, 2007, 33(1), pp. 5-13.

[36] M.S. El Naschie: Is quantum space a random Cantor set with a golden mean dimension at the core? Chaos, Solitons & Fractals, 1994, 4(2), pp. 177-179.

[37] M.S. El Naschie: Mathematical foundation of E-infinity via Coxeter and reflection groups. Chaos, Solitons & Fractals, 2008, 37, pp. 1267-1268.

[38] M.S. El Naschie: Banach-Tarski theorem and Cantorian micro spacetime. Chaos, Solitons & Fractals, 1995, 5(8), pp. 1503-1508.

[39] M.S. El Naschie: On the initial singularity and the Banach-Tarski theorem. Chaos, Solitons & Fractals, 1995, 5(7), pp. 1391-1392.

[40] M.S. El Naschie: COBE satellite measurement, hyper spheres, superstrings and the dimension of spacetime. Chaos, Solitons & Fractals, 1998, 9(8), pp. 1445-1471.

[41] M.S. El Naschie: Infinite dimensional Branes and the E-infinity toplogy of heterotic superstrings. Chaos, Solitons & Fractals, 2001, 12, pp. 1047-1055.

[42] M.S. El Naschie: Ji-Huan He's ten dimensional polytope and high energy particle physics. International Journal of Nonlinear Science & Numerical Simulation, 2007, 8(4), pp. 475-476.

[43] M.S. El Naschie: Hyper-dimensional geometry and the nature of physical spacetime. Chaos, Solitons & Fractals, 1999, 10(1), pp. 155-158.

[44] D. Finkelstein: Quantum sets and Clifford algebras. International Journal of Theoretical Physics, 1982, 21(6/7).

[45] M.S. El Naschie: Derivation of the threshold and absolute temperature Ei = 273.16k from the topology of quantum spacetime. Chaos, Solitons & Fractals, 2002, 14, pp. 1117-1120.

[46] M.S. El Naschie: Quarks confinement via Kaluza-Klein theory as a topological property of quantum-classical spacetime phase transition. Chaos, Solitons & Fractals, 2008, 35, pp. 825-829.

[47] M.S. El Naschie: On a class of general theories for higher energy particle physics. Chaos, Solitons & Fractals, 2002, 14, pp. 649-668.

[48] Ji-Huan He: Hilbert cube model for fractal spacetime. Chaos, Solitons & Fractals, 2009, 42, pp. 2754-2759.

[49] R. Lomas: The Man Who Invented The Twentieth Century, Nicola Tesla, Forgotten Genius of Electricity. Headline Books, London 1999.

[50] M. Helal, L. Marek-Crnjac and Ji-Huan He: The three page guide to the most important results of M.S. El Naschie's research in E-infinity quantum physics. Open Journal of Microphysics, 2013, 3, pp. 141-145.

[51] M.S. El Naschie: A review of E-infinity theory and the mass spectrum of high energy physics. Chaos, Solitons & Fractals, 2004, 19(1), pp. 209-236.

[52] M.S. El Naschie: The theory of Cantorian spacetime and high energy particle physics (An informal review). Chaos, Solitons & Fractals, 2009, 41(5), pp. 2635-2646.

[53] L. Marek-Crnjac and Ji-Huan He: An invitation to El Naschie's theory of Cantorian spacetime and dark energy. International Journal of Astronomy and Astrophysics, 2013, 3, pp. 464-471.

[54] Mohamed S. El Naschie: A resolution of cosmic dark energy via quantum entanglement relativity theory. Journal of Quantum Information Science, 2013, 3, pp. 23-26.

[55] Mohamed S. El Naschie: What is the missing dark energy in a nutshell and the Hawking-Hartle quantum wave collapse. International Journal of Astronomy & Astrophysics, 2013, 3, pp. 205-211.

[56] Mohamed S. El Naschie: Topological-geometrical and physical interpretation of the dark energy of the cosmos as a 'halo' energy of the Schrodinger quantum wave. Journal of Modern Physics, 2013, 4, pp. 591-596.

[57] F.D. Peat: In Search of Nikola Tesla. Ashgrove Publications, London & Bath, 1983.

[58] L. Susskind and James Lindesay: The Holographic Universe. World Scientific, Singapore, 2005.

[59] J. Stiglitz and L. Bilmes: The Three Trillion Dollar War: The True Cost of The Iraq Conflict. Allen-Lane, Penguin Books, London, 2008.

[60] M.S. El Naschie: A unified Newtonian-relativistic quantum resolution of supposedly missing dark energy of the cosmos and the constancy of the speed of light. International Journal of Modern Nonlinear Theory & Application, 2013, 2, pp. 43-54.

[61] Caline Malek: Abu Dhabi Crown Prince details UAE leaders' vision of future without oil. The National Newspaper, UAE, February 10th, 2015 (http://www.thenational.ae/uae/government/abu-dhabi-crown-prince-details-uae-leaders-vision-of-future-without-oil?utm_content='%20vision%20of%20future%20without%20oil).

[62] A. Zee: Quantum Field Theory In A Nutshell. Princeton University Press, Princeton, USA, 2003.

[63] B. Duplantier and V. Rivasseau (Editors): Vacuum Energy-Renormalization. Birkhauser, Basel, Switzerland, 2003.

[64] P.W. Milonni: The Quantum Vacuum. Academic Press, Boston, USA, 1994.

[65] V.A. Parsegian: van der Waals Forces. Cambridge University Press, Cambridge, UK, 2006.

[66] K. Huang: Fundamental Forces of Nature. World Scientific, Singapore, 2007.

[67] L. M. Wapner: The Pea And The Sun. A.K. Peters Ltd., Wellesley, MA, USA, 2005.

[68] M.S. El Naschie: Banach-Tarski Theorem and Cantorian Micro Spacetime. Chaos, Solitons & Fractals, 5(8), 1995, pp. 1503-1508.

[69] P.Y. Agvis: Decoding the genome: A modified view. Nucleic Acids Research, 32(1), 2014, pp. 223-238.

[70] W.W. Xiao and C.N. Qian: Our dream of defeating cancer. Chinese Journal of Cancer, 33(3), 2014, pp. 125-132.

[71] D. Stauffer and H.E. Stanley: From Newton to Mandelbrot. Springer, Berlin, 1989.

[72] A. Bunde and S. Havlin (Editors): Fractals in Science. Springer, Berlin, 1995.

[73] V. Froler and A. Zelnikov: Introduction to Black Hole Physics. Oxford University Press, Oxford, 2013.

14

The Studing of Silver Nanoparticle Effect on the Copper Bioleaching Output from Low Grade Sulfidic Ores

Jamshid Raheb[1, *], Sorur SHaroknyan[2], Fatemeh Nazari[1], Yasin Rakhshany[3]

[1]National Institute of Genetic Engineering and Biotechnology, Tehran, Iran
[2]Department of Environment, Science and Research Branch, Islamic Azad University, Tehran, Iran
[3]Department of Microbiology, Islamic Azad University North Tehran Branch, karaj, Iran

Email address:
jamshid@nigeb.ac.ir (J. Raheb)

Abstract: The extraction of metals from ores causes various environmental pollutions. Since Iran is located on the so-called 'copper belt' and holds a significant share of the world's copper mines and resources, reduction of pollution from these mines can have an important effect on the overall reduction of pollution. Copper processing methods include pyrometallurgy and hydrometallurgy. Pyrometallurgy is mainly used in high grade mines whereas hydrometallurgy process is used in lower grade mines. In low grade copper sulfide mines, hydrometallurgy processes are used which use a lot of energy to covert mineral deposits into oxide forms which are then leached using sulfuric acid, or are extracted using bioleaching process. In acidic leaching, a lot of environmental pollution is created. Bioleaching process is an environmentally-friendly method which is mainly used in mines where the common physicochemical methods are not profitable. In this study, we have tried to increase the efficiency of bioleaching process by adding silver nanoparticle in order to increase the popularity of this method. For this purpose, initially the indigenous bacteria were separated from the ores and after adoption to silver, the bacteria were used in bioleaching tests. Three concentrations of silver component were used for the bioleaching tests. The results were compared to cases where no bacteria and no silver compounds were used, which showed significant increase in copper extraction efficiency. In the next step, the optimum concentration of silver was used in the percolation column. In this stage, four columns were set up for 'with bacteria and silver', 'with silver', 'with bacteria' and 'without bacteria and silver'. Results show that the column with bacteria and silver produced the highest efficiency of copper extraction.

Keywords: Silver Nanoparticle, Bioleaching, Copper, Ores

1. Introduction

The minerals present on the surface the earth's crust according to their importance, are being considerable by the man, and trying to find new mining areas and using more from these present mining areas is continuing, too. As simple as possible, we can say that without these mineral materials, our lives will not possible. According to increasing progress in using and consuming from these materials in different and various industries, Nowadays economics minerals are taken into account as one of the most important and essential foundations in every country[1].

The investing a researcher divided the history of mankind indifferent periods according to these minerals such as:

Rock age, Bronzeage and Iron age. Cheap minerals frequency is being created necessary field for establishing the industrial civilization and man societies general welfare is related to consume per capita these mineral productions[2].

Current living of mankind is dependent in to these new technology and industries and the mankind is not able to separate itself from the all sources which are providing all the living devices. The environment is a set of very huge and complicated different parameters and elements which have being created by the organisms gradual evolution of any living and the elements of making the earth surface, and then it has been influence on human activities and it is influenced by these activities. By doing mining operations, the surrounding environment has exposed to different changes and if the man often doesn't pay attention on controlling and monitoring these effects، may cause environment pollution[3].

The most and serious environment problem is related to having acidic mine sewages, which have caused harmful effect on ground and underground waters. Most mines are sulfide minerals especially pyrite minerals.

Having oxidation of these minerals cause to become condensed, especially pyrite and cause the formation of AMD.AMD is become specified by characteristics such as high Iron and Sulfate content and low pH [4]. Copper usually is found in the mineral form. The other minerals such as Azurite, Malachite and Bernite similar to other sulfites such as Chalcopyrite ($CuFeS_2$), Kovolin(CuS), Kalcozine (Cu_2S) or other oxides such as Copyrite (Cu_2o) are from the Copper sources [5]. The compounds such as Fleming solution which are used in chemistry and Copper sulfate which is used as toxin and is used as water filtration is another form of consuming Copper, today, too[6]. Nowadays, Bioleaching dumplings are very low cost techniques for Copper recovery from main mass which there are no other low cost techniques like these.

In addition to economical success Bioleaching dumplings, very little operations are needed for preparing these dumplings [7]. Chalcopyrite is the most important of the Copper source. About %70 of known sources are Chalcopyrite [8]. In the most mines in the world, masses of sulfite stone minerals are dissolved because of water penetration, and green solutions (containing Iron ion)and blue solutions (containing Copper ion) are sweat from those. Rodolf and Heilbroner discovered in 1922 that bacteria can extract metals from sulfide stones which have low purity and assay. By investigating and studying this phenomenon, the investigator were suspected to the performances of some bacterial which are being in these waters, and in next investigation, they proved that, these biological elements, when they have suitable environment for living, they receive their required food and energy from the Iron and Sulfur oxidation which are got from the sulfide minerals of these quarry.

In the next investigation, different kinds of these bacteria were explored, and each one of them in special physical and chemical situation (such as pH and temperature) could continue their living and performances[9]. The investigations show that adding other ions such as Bismuth, Plumb (Pb), Cobalt (Co), Magnesium (Mg), Zinc (Zn) and Nickel (Ni) don't have any influence on extracting Copper. Silver is a key element in Bioleaching and in extracting Copper from low purity and assay mines.

Hiroyoshi in his investigations find out that the suffering of biological leaching of chalcopyrite is related to its potential oxidation and reduction, and it is determined by Ferric iron [10]. Bioleaching of the metals which means extracting and exploiting the metals from the mines, is done by the neutral elements such as microorganisms.

During this procedure, an indissoluble metal (which is usually a metal such as Ni, Cu or Zn) is changed into a dissoluble from (which is metal sulfate) [11]. Present of the last extracted rich residue resources which are contained a lot of metals that are not justifiable economically[12]. The hyolro metallurgical procedures are included: leaching, extracting and exploiting solvent, Electrovining and Sendimentology

are in recent decades, by worldwide several big mine companies, in order to extract low purity and assay resources which are not possible and economical by the piro mythological procedures, are being more attractive[13]. Bioleaching procedure for extracting the metals such as Copper, Nickel, Kobalt, Zinc and Gold and etc. using the low purity and assay sulfite resources according usual physicochemical methods and technologies are not profitable. Extracting metal from the mines is done by using natural procedures such as microorganisms. These microorganisms have the best growth in the pH(1.8-2) and temperature degree (20-45°c) environment.

Bioleaching in contrast by usual physicochemical procedures, has the following advantages:

- All the microbiological process have very high coordination with the environment
- These procedures don't need too much energy and are profitable economically.
- They don't produce sulfur dioxide (sio_2) and other harmful gasses.
- In the case of quarry with low purity and assay, these procedures are usable. Most of these quarries are not recyclable or not profitable.
- Main bacteria of Bioleaching, have common physiological features. All of them chemo lithotrophic, and they can use from Iron or other nonorganic sulfur resources as electron donor. As the sub-production of sulfur oxidation is Sulfuric acid therefore, these organisms are acidophilus, and they grow at the pH 1/5 or 2. This acidophilus feature is correct about Bio minis organisms which they only oxidize the Iron. Although these bacteria can use of other receiver electrons which are non-oxygenic, But they commonly can grow better in aeration solutions [14].
- Microorganisms, in their structures, they need to a small amount of metals which their supplying resources of these metals are eco-environment. Most of bacteria, they deposit large amount of using metals into their inner cells, in order to use them when the amount of these metals are decreased in their environment, because they need these resources to supply their structural needs.

It is possible that the metal influence the phosphorylation oxidative and permeability of the cell membrane. We can see this condition in Vanadium and Mercury[15]. Bacterial active oxidation of Iron in soil and mines is done at the usual temperature, 10°c. Extracting desirable of metal from mine sulfides by these bacteria, is happened at the temperature degrees between 25-45°c.

This process stopped completely, because of bacterial decreasing in proteins and enzymes in the temperature degree 55°c and in the temperature degree above 55°c, only the chemical oxidation of the stone mine is happened.

Thiobacillusferroxidans, is done the metal extracting operation in the temperature degree between 10-40°c.

Because of protoplasmic special compound of organism, the least temperature for metals bacterial extraction in every region, is related to the special condition of that region [6].

2. Materials and Methods

2.1. Isolation of Bacteria

First of all, the surface stones are collected from the mine and for bacterial separate ,is sent in genetic researcher lab. In order to create a natural environment for Bioleaching bacteria the stone samples collected from the mine are put in big plastic dishes and they are sprayed by 1/5 normal sulfuric acid every 4 hours a day.

These samples are kept in similar green house at the temperature degree 30°c. Extracting chromosomal DNA is done by the DNA FAST Kit which is made by National Institute of Genetic Engineering and Biotechnology.

In order to detect those separated bacteria, the gens of the 16 RNA region are duplicated by the PCR set. That PCR production, after doing the electrophoresis, It is send for doing sequence test, and during the sequencing tests, present nucleotide sequences in the duplicated DNA, are determined or sequenced. Those obtained sequences are compared by the existing sequences which are in the genetic data base by the Blast program, and then by using these data Blast, the bacteria are detected by point of molecule.

2.2. Adaptation of Bacteria

Having more than certain amount of heavy metals are poisonous for bacteria, and they prevent the growth of bacteria. Then we must pay attention to the amount of them, and we must adapted the bacteria to silver in order to it haven't any problem for the growth of bacteria and Bioleaching tests.

In order to adapt the bacteria, first of all we use the 0.005g in 100cc culture solution and after have been completed the growth of bacteria, The bacteria sub cultured and 0.01g silver was added to culture and in every time, bacteria sub culturing was added about 0.005g silver to the previous culture. At the end the amount of 0.03g of silver has been detected. In order to prepare 9k bacteria culture ,we added 500 ml distilled water in an 1000 ml flask and then the following substances are added, respectively too.

K_2HPO_4(0.34 g), $Ca(NO_3)_2$ (0.02 g), KCL(0.2 g), $MgSO_4.7H_2O$(0.43 g), $(NH_4)_2SO_4$(1.5 g), $FeSO_4.7H_2O$ (25 g).Then after adding these substances, flask is received to 1000 ml.

2.3. The Bacteria Growth Test

The growth of bacteria is examined by adding the silver. Thus growths of all added substances in adaptation are measured separately. This work is done by measurement every day and then the bacteria growth diagram is drawn which is included bacteria plus seven different amounts of silver nitrate.

2.4. Iron Oxidation Tests

In order to determine the Iron oxidation, at the first standard solutions of 10 Fenantronin ($1H_2O$), Acetatsodium and Feroamonium sulfate ($6H_2O$) is prepared. Then by the help of Feroamoniumsulphate, the standard diagram of Iron is drawn. 50 μ of the sample which is contained bacteria and 5cc of Fenantroine and 4cc Acetat sodium are poured into falkon 50 and after 20 minutes, its OD is read in the wave length of 509 nanometer by the help of spectrophotometer.

2.5. The First Bioleaching Tests

In order to study the release Cu and Fe by the separated bacteria, we were used 12 flask with the volume 100cc. Three amounts of 0.01, 0.02, 0.03silver nanoparticles, is selected for doing these tests. Every one of these tests were contained 4 flask and they contained 0.01, 0.02, 0.03silver nanoparticles and 10 g of Nochonestone mines and 5cc *Thiobacillus ferroxidans* bacteria in the 95cc cultural solution which was homogeneous completely. Every one of three tests by having different amounts of silver which are prepared in 4 similar dishes, are removed in the first and second and third week. Three remained flasks of three amounts which were in the twenty eighth day or fourth week, are removed, too. The sediments are separated from the solutions by filter paper after the complete washing of the sediments, then sampling of the remained solution is done and these samples are sent to Atomic Energy Organization for analyzing ICP (Atomic spectrometer). The evidence tests which are included twelve above tests with similar amounts silver, are done without adding any more bacteria. Other evidence tests which are included tests of without silver and by adding bacteria and the tests that any without bacteria and silver are done, too.

2.6. Tests of Percolation Column

After doing shaking flask tests, these tests are done in the higher level volume and in the semi-test condition and in the percolation column. In these columns is prepared the suitable environment condition for the growth of bacteria and amount of 0.03 silver is added to the columns of containing silver. The column tests were containing four columns with 7cm the diameter. The weight of the stones by the diameter of 400-500 micron particles were 2400 g. Too, amount of 3700cc distilled water is added to the system bins and the pH of solution is lowered by 3cc of acid and spraying is done by the speed of 5 lh/m^2 in every column. The pH is measured every day and if it goes higher than 2, then it is lower by adding twice Sulfuric acid. For preparing the first column, amount of 25cc bacteria in 475cc distilled water is dissolved, and this solution is added to the column from the above of it. The bacteria are stayed in the same position for 5 days in order to adhere or join on the stone. Then 3700cc distilled water and 58.25cc silver solution %1 is added to the column. The second column is contained the same amount of bacteria without adding any silver and the third column is contained silver without adding any bacteria. In order to compare the results, in the forth column, we don't add any of the bacteria and silver because of we want to measure the influence of every factors.

3. Results and Discussion

After separating, the bacteria are observed by the fluorescent microscope and then they are observed rod-shaped bacteria (Fig 1).

Fig. 1. thebacterias are observed by the fluorescent micros cope.

The result of electrophoresis of the PCR products, on the agarose gel %1 has shown the length of 500bp bound. The compared result of this sequence with the other available data in the world wide bank has shown that this bacteria is *Thiobacillus ferroxidans* (Fig 2).

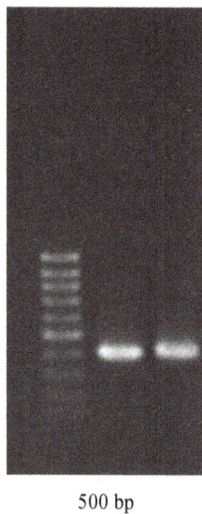

500 bp

Fig. 2. The result of electrophoresis of the PCR products, on the agarose gel %1.

3.1. The Tests of Bacteria Growth

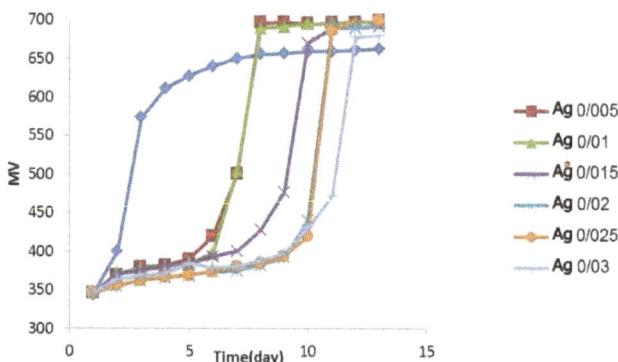

Fig. 3. The tests of bacteria growth.

In the growth tests had shown that whatever the more amount of silver was added, the time of starting the growth of bacteria was more delayed (Fig. 3).

3.2. Iron Oxidation Tests

Iron oxidation tests had shown that in compare with the standard bacteria DSM 583, it had more quickly growth. The result had been shown in diagram, too (Fig. 4).

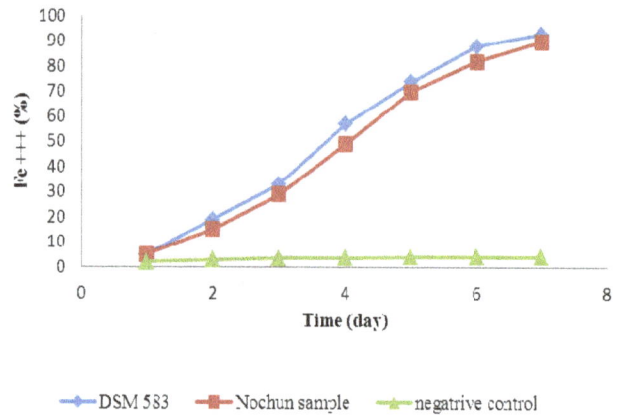

Fig. 4. Iron oxidation tests.

3.3. The Early Bioleaching Tests

The following diagram is shown the tests in three amount of silver without any bacteria. As it has been shown in the diagram, by adding more the amount of silver, the efficiency of releasing copper is increasing more, too (Fig. 5).

As it has been shown in the diagram, in the Bioleaching tests, by adding silver, the efficiency is increasing more, considerably. The first diagram which is related to the Bioleaching test without adding silver (Fig. 6) and the next diagram is related to the adding three amounts of silver (Fig. 7).

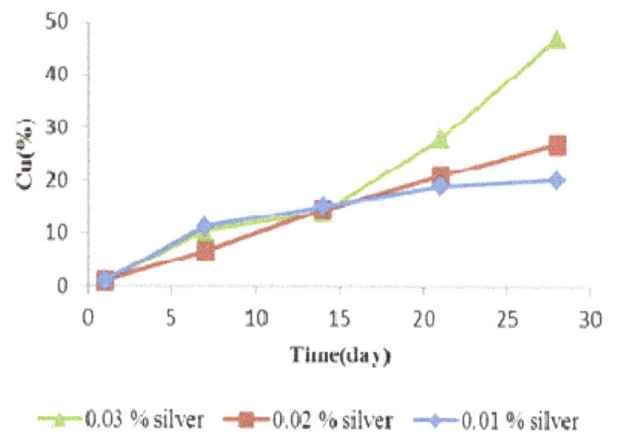

Fig. 5. The early Bioleaching tests.

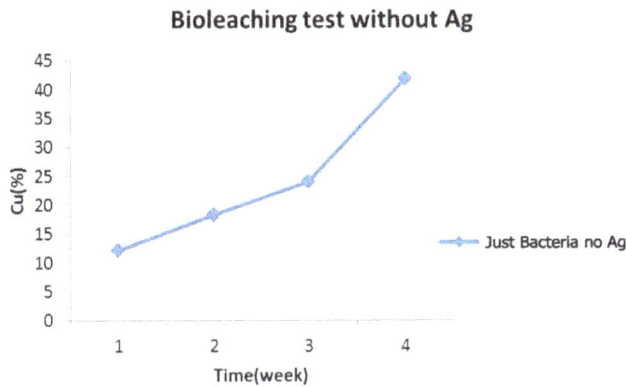

Fig. 6. *The early Bioleaching tests.*

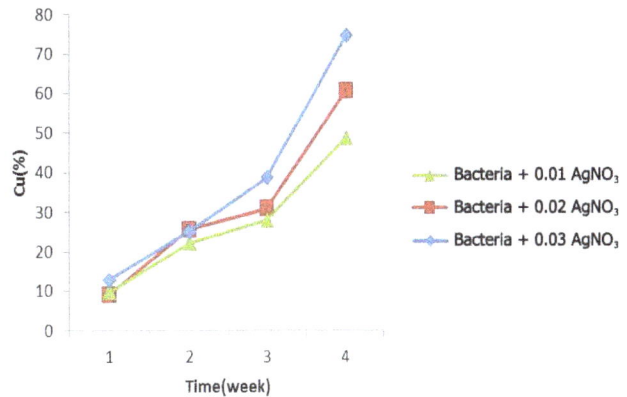

Fig. 7. *The early Bioleaching tests.*

3.4. The Column Results

In every column the previous result, are repeated and by adding the silver, releasing the Copper is increased, too (Fig. 8).

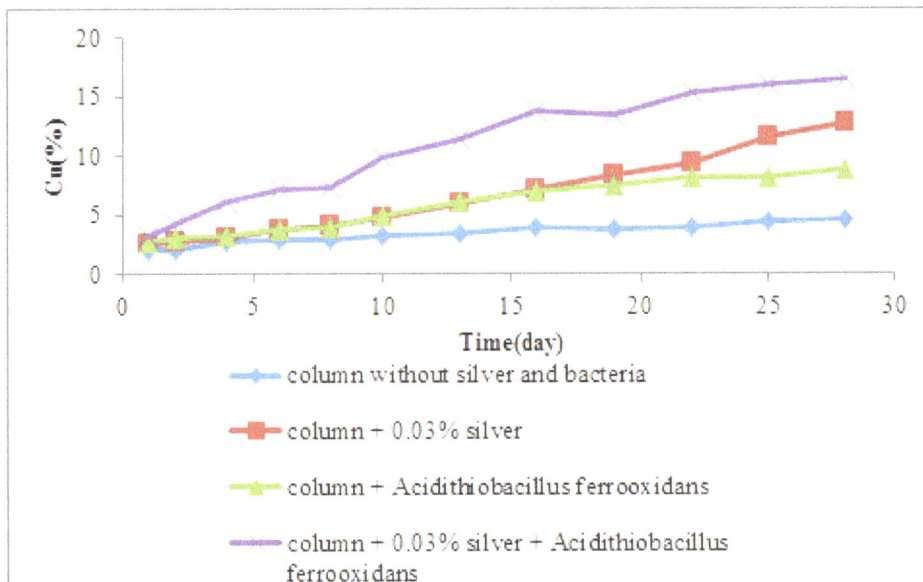

Fig. 8. *The column results.*

4. Discussion

Noadays, the procedure of ecoleaching (Bioleaching) has a special position in extracting metal technologies from mines, and many investigations in this field are doing now. Bioleaching (ecoleaching) is a suitable and economical technology for taking care of special mineral stones, according to the laboratory scale and semi-industry which is doing now, the noticeable amount of Bioleaching units in comparison with the developing countries (such as Iran) are establishing and working now. The reason of this subject is related to the simple operations and early cost investment for operating these units. In accordance with the fact that Iran has noticeable resources of valuable metals such as Copper and gold, the investigation in this field is very important. Having finding new microbial species by separating from mineral stones and studying the efficient parameters in the

Bioleaching procedure, these stones will cause to progress this technology in the country. Therefore, first of all in this investigation, collecting the samples and species of *Thiobacillus* were separated suitably from Nochon mines, and after detecting those molecular factors, the effective factors on leaching procedure were evaluated, and finally the leaching in percolation system were evaluated, too. Comparing the exploiting methods of metals, especially Copper extraction, which has more mine in our country, has shown that the Bioleaching method has little damages on eco-environment, because this method doesn't need to consume more energy and those harmful gasses such as sulfure dioxide which are produced by the other methods, are not produced. The necessity which causes the detection of domestic mine bacteria was this, that the activity of every bacteria in its local region is more than non-local bacterias. The present of *Thiobacillus ferroxidans* and controlling of this bacteria in Bioleaching procedure, cause to produce Acid higher than usual, and it has good influence on this economical procedure. This method is used for those stones which the exploiting of Copper from them is not profitable by previous methods because of low purity and assay, this method has coordination with the eco-environment and it is cost-effective. But this method needs too much time. Extraction of Copper from those composed Hypes is lasted several months, and for this reason this need is felt that some preparation activities must be considered to increase the efficiency of this method. One of these ways which were used in this project, was adding the silver compounds. The results showed that the released Copper in the presence of silver is increased, and by the noticeable achievements which were obtained in the laboratory site, the column tests for doing Bioleaching at a high and semi-industrial level were performed. The results were repeated in columns, too and they showed that adding silver cause to increase efficiency of the Bioleaching and then cause to increase the releasing the Copper.

References

[1] Najafi M. the series of articles from the first seminar about mine and related sciences, Azad university press of Tabas 1385.

[2] Rahimi N. procedures in using coal in power plants, the series of articles about strategic of the energy1376; vol. 1.

[3] Shokofeh N. Protecting of eco-environment in mines, eco-environment protection organization publicate1382.

[4] Orei and Tarigi. Benefitting of Iran coal mines Amir Kabir investigation and scientific publication1378; year 11, No. 41.

[5] Neale J. Bioleaching technology in minerals processin. Mintek, Biotechnology Division 2006; p. 110-115.

[6] Davis B, Carol S Nicolle J and Paul R .Ferrous iron oxidation and leaching of copper ore with halotolerant bacteria in ore columns. Hydrometallurgy 2008; 94(1–4): p. 144-147.

[7] Lundgren D G and Silver M. Ore leaching by bacteria. Annu. Rev.Microbial 1980; 34: p. 263-283.

[8] Munoz DB and Dreisinger WC. Silver- catalyzed bioleaching of low-grade copper ores, Part III Column reactors, Hydrometallurgy 2007; 88 P:35-57

[9] Dee Jay F, Debby F, Karen SM and Daphne LS. Evaluation of a Fluorescent Lectin-Based Staining Technique for Some Acidophilic Mining Bacteria. Applied and enviromental microbiology 2000; 5(66): p. 2208-2210.

[10] Barrie D and Jonhnson N O. Effect of temperature on the bioleaching of chalcopyrite concentrates containing different concentration of silver. Hydrometallurgy ; 2008 Vol 94 p:42-47.

[11] Foucher S, Battaglia-Brunet F, Hugues P, Clarens M, Godon JJ and Morin D. Evolution of the bacterial population during the batch bioleaching of a cobaltiferous pyrite in a suspended-solids bubble column and comparison with a mechanically agitated reactor. Hydrometallurgy 2003; 71(1–2): p. 5-12.

[12] Rawlings DE. Heavy metal mining using microbes. Annu Rev Microbiol 2002; 56: p. 65-91.

[13] Chen Z and Zhong L 2002. Growth Kinetics of Thiobacilli Strain HSS and Its Application in Bioleaching Phosphate Ore. Industrial & Engineering Chemistry Research 2002; 41(5): p. 1329-1334.

[14] Ac̦ıkel a, Ü,.Erșana M. and Sag˘ Ac Y. Optimization of critical medium components using response surface methodology for lipase production by Rhizopus delemar. food and bioproducts processing. Microbiol 2010; 8 8: p. 31–39.

[15] Rohwerder T and Gehrke T. Bioleaching review part A: progress in bioleaching: fundamentals and mechanisms of bacterial metal sulfide oxidation. Appl Microbiol Biotechnol 2003; 63(3): p. 239-48.

[16] Gardner M.N, Deane S M. and Rawlings D E. Isolation of a New Broad-Host-Range IncQ-Like Plasmid, pTC-F14, from the Acidophilic Bacterium Acidithiobacillus caldus and Analysis of the Plasmid Replicon. Journal of Bacteriology. 2001; 183(11): p. 3303-3309.

[17] Kelly D P and Wood A P. Reclassification of some species of Thiobacillus to the newly designed genera Acidithiobacillus gen gov .,Halothiobacillus gen. gov. and and Thermithiobacillus gen. gov. International journal of systemic and evolutionary microbiology 2000; 50: p. 511-516.

[18] Watling H R . The bioleaching of sulphide minerals with emphasis on copper sulphides - A review, Hydrometallurgy 2006; 84(1–2): p. 81-108.

[19] Watling HR. The bioleaching of sulphide minerals with emphasis on copper sulphides - A review, Hydrometallurgy 2006; 84(1–2): p. 81-108.

[20] Xia Jin-lan, Peng An-an He, huan Yang yu, LIU, Xue-duan QIU and Guan-zhou.. A new strain Acidithiobacillus albertensis BY-05 for bioleaching of metal sulfides ores. Trans.Nonferrous Met.Soc.China 2007; 17: p. 168-175.

15

The Cantorian Monadic Plasma behind the Zero Point Vacuum Spacetime Energy

Mohamed S. El Naschie

Dept. of Physics, University of Alexandria, Alexandria, Egypt

Email address:

Chaossf@aol.com

Abstract: Stimulated by the recent work on quarks-gluons plasma we present E-infinity theory in the form of Cantorian monadic plasma and proceed from there to a general explanation of the Casimir effect and dark energy as a zero point vacuum energy which could be utilized via advanced nanotechnology to build a clean energy reactor with near to unlimited capacity.

Keywords: Quark-Gluon Plasma, E-Infinity, Cantorian Spacetime, Monadic Cantorian Plasma, Highly Structured Rings, Casimir Effect Dark Energy, Zero Point Vacuum Energy

1. Introduction

In a lucidly written and beautifully illustrated Scientific American feature article by R. Ent, T. Ullrich and Raju Venugopalan they wrote [1]:

"When the cosmos was young, it was too hot for atoms or even stable protons and neutrons to form. Quarks and gluons buzzed around freely in a roiling swarm. Accelerators on Earth recently succeeded in replicating this state, called a quark-gluon plasma by smashing atomic nuclei together at near light speed. By studying the plasma as it cools, physicists can learn not just about the behaviour of quarks and gluons but also about the early evolution of our universe"

Here the present author may add that this is as near as mainstream physics [1-3] can come to a similar view point to that of our Cantorian spacetime [4-9] and our approach leading to the ordinary and dark energy density of the cosmos and that in a superb agreement with measurement and cosmic observations [8].

The present work aims at putting the highly intuitive picture of quark –gluon plasma [1] into a relation to Cantorian spacetime which obviously has the advantage of a stringent mathematical formulation of inert transfinite set theory [9]. Thus in a sense we are translating the rather mainstream but highly intuitive picture of the said plasma [1] to the language of self similar random Cantor sets and its associated highly structured golden mean rings of E-infinity theory [4-9]. The road we are taking here stretches from quantum set to the theories of extra spacetime dimensions and noncommutative geometry [10-15].

*) This paper is dedicated, in deep belated respect, to the memory of the imaginative inventor N. Tesla who died in 1943.

In anticipation of the next section we should mention that in the Cantorian E-infinity space picture, elementary particles and spacetime are made of each other via the zero set representing the first image of the quantum particle monad and the quantum wave monad which is the cobordism, i.e. the surface of the quantum particle monad [9]. The common cobordism of particle and wave then arises naturally as a random multi fractal with the same expectation topology equal to the cobordism of the quantum wave [4-9]. This multi-fractal is thus nothing else but the quantum Cantorian monad which gives rise to Cantorian quantum spacetime itself [4-6] (see also Fig. 1). This situation may seem superficially complex but it is not once it is elaborated upon via von Neumann-Connes dimensional function of noncommutative geometry [4-9], which we do next.

This is basically a two dimensional projection in which each of the larger balls (circles) are a zero set $(0;\phi)$ representing the quantum particle while the surface (circumference) represents the empty set $(-1,\phi^2)$ which in turn represents the quantum wave. This wave is then surrounded by an infinite hierarchy of smaller (fractal) spheres (surfaces) which may be seen as the emptier set $(-2,\phi^3)$, i.e. the surface of the empty set quantum wave. Remarkably the average set of all zero and empty sets is an expectation value equal $\langle -2 \, ; \, \phi^3 \rangle$. In other words $\langle -2 \, ; \, \phi^3 \rangle$ is our quantum spacetime which is the cobordism of the quantum wave which in turn is the cobordism of the quantum particle floating and propagating with the help of its wave in our Cantorian E-infinity spacetime. It is likewise remarkable that ϕ^3 is simultaneously equal to the topological Casimir force as well as the topological mass of the ordinary energy of spacetime. Thus all matter and energy manifestations in our cosmos are essentially a manifestation of the ero point energy of the vacuum of spacetime. To obtain Einstein maximal energy density we just need to find first the topological energy density by adding Kaluza-Klein D = 5 to ϕ^3 of the spacetime vacuum and find the fractal Kaluza-Klein dimension $5 + \phi^3$ then multiply this with the average Cantorian interval speed of light c = ϕ squared. The result is

$(5 + \phi^3)\ \phi^2 = 2$. Inserting in Newton's kinetic energy one finds $E(Einstein) = \dfrac{1}{2}m(v \to c)^2(2) = mc^2$ exactly as should be. The preceding explanation amounts to a paradigm shift in physics where the totally empty vacuum of spacetime is taken as fundamental and everything else is derivable from it. To prove this point was a dream of Serbian American inventor N. Tesla who died in 1943 as well as Soviet physicist A. Zakharof. In fact in his later years Nobel Laureate J. Schwinger was a champion of cold fusion which comes very near to our present concept of a Casimir-nano energy reactor.

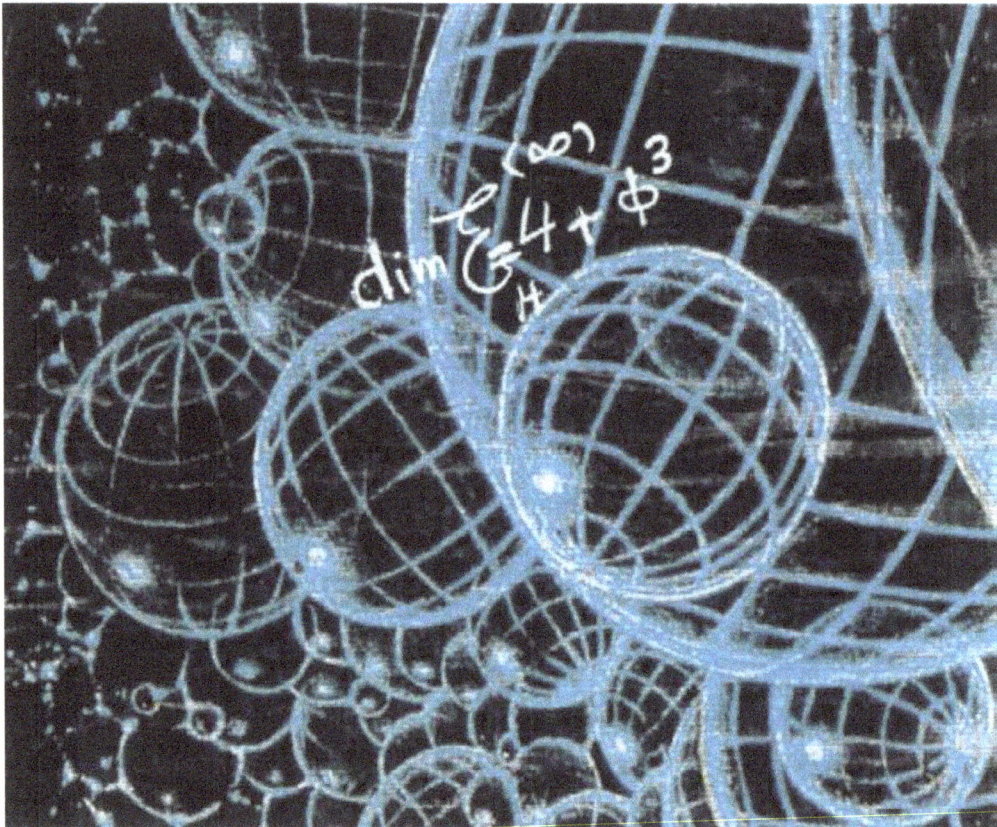

Fig. 1. *Cantorian spacetime of E-infinity theory which is considered here to model our actual spacetime may be envisaged advantageously as in this artist impression.*

2. The Dimensional Function

Following A. Connes [7,9,10] one finds that the dimensional function of a Klein-Penrose compactified space [10,12] which is representative of the holographic boundary of our superstring |E8 E8| exceptional Lie groups universe [4,7], [10-13] is given by [10]

$$D = a + b\phi \qquad (1)$$

where $a, b \in Z$ and $\phi = (\sqrt{5} - 1)/2$. Thus for the zero set we have a = 0 and b = ϕ and therefore

$$D = \phi. \qquad (2)$$

Consequently we have

$$D(\text{zero set}) = (0\ ;\ \phi) \qquad (3)$$

Similarly for the empty set we have a = 1 and b = −1 which leads to

$$D = 1 - \phi$$
$$= \phi^2 \qquad (4)$$

and

$$D(\text{empty set}) = (-1 ; \phi^2) . \qquad (5)$$

We move next to the monadic interpretation of this result which is basically an extension of the pioneering work of D. Finkelstein on quantum sets [15] and A. Connes on noncommutative geometry [10].

3. The Monadic Quantum Mechanics and Quantum Spacetime

Quantum mechanics was initially developed, and contrary to relativity, without any reference to spacetime [12]. However in our E-infinity theory as well as noncommutative geometry and superstrings and M-theory, spacetime is fundamental [4-8], [11-13]. It was reasoned in various previous publications that the monad of the quantum particle is the zero set $D(0) = (0, \phi)$ while the quantum wave monad is basically the surface of $D(0)$, namely by Menger-Urysohn deductive dimensionality theory and von Neumann-Connes theory $D(-1) \equiv (-1, \phi^2)$ [4-]. Now we conclude that the surface of the quantum wave or empty set monad must be an emptier set $D(-2) = (-2, \phi^3)$. This is remarkable because the average Hausdorff dimension of a quasi multi-fractal made of the zero set and all empty sets, that is to say ϕ^n for n = 2, 3, 4,.... ∞ is exactly ϕ^3 because of the well known fact that [4-8]

$$\langle d_c \rangle = \frac{1}{d_c^{(0)}(1 - d_c^{(0)})} = 1/\phi^3 = 4 + \phi^3 \qquad (6)$$

as well as [4-8]

$$\langle n \rangle = \sum_0^\infty \phi^n = 4 + \phi^3 \qquad (7)$$

and thus

$$1/\langle d_c \rangle = 1/4 + \phi^3 = \phi^3 \qquad (8)$$

where $d_c^{(0)} = \phi$ is the zero set Hausdorff dimension in the bijection version of von Neumann-Connes dimensional function [4,7,10] (see Fig. 1). The intuitive interpretation of this result will be discussed next.

4. The Spacetime Cantorian Plasma Picture

From the preceding analysis one could develop the following spacetime picture (Fig. 1) and we restrict it to a two dimensional projection of our space with the zero set particle represented by a ϕ circle living in an empty set represented by ϕ^2 containing ϕ and finally ϕ^2 may be seen as floating in an infinitely large circle representing an average emptier set ϕ^3 which is our Cantorian quantum

space [4-7]. This is in a sense a Cantorian plasma corresponding in a fuzzy intuitive way to the quark-gluon plasma [1]. Next we look at the application of the previous analysis to dark energy [8].

5. From Cantorian Monadic Plasma to Ordinary and Dark Energy Density of the Cosmos

Noting that multiplicative Hausdorff "volume" of the zero set in Kaluza-Klein five dimensional space is given by [14]

$$\text{Vol}^{(5)}(\text{particle}) = (\phi)^5 , \qquad (9)$$

it is easily reasoned that the corresponding energy density

$$E = \frac{1}{2}mv^2 \qquad (10)$$

may easily be understood in this case as [14]

$$E(O) = \left(\frac{1}{2}\right)\left(vol^{(5)}(particle)\right)(v \to c)^2 \qquad (11)$$
$$= (mc^2)(\phi^5/2) \simeq mc^2/22$$

Consequently the dark energy is the difference between the maximal density E(Einstain) = mc^2 and E(O) which means [14]

$$E(\text{Dark}) = mc^2(21/22). \qquad (12)$$

This could also be obtained directly from the additive volume of the empty set quantum wave [14]

$$vol^{(5)}(wave) = (5)(\phi^2) \qquad (13)$$

and consequently one finds in a similar way [14]

$$E(D) = \frac{1}{2}m(v \to c)^2 = mc^2(5\phi^2/2) \simeq mc^2(21/22) \qquad (14)$$

exactly as before [7,8]. Next we discuss the application of our theory to the Casimir effect and more general cases of zero point vacuum energy.

6. The E-Infinity Cantorian Plasma and the Zero Point Vacuum Energy

The Casimir effect is probably the most theoretically and experimentally documented manifestation of the zero point quantum vacuum energy. We take this Casimir effect very seriously which was linked to dark energy via a higher dimensional Möbius-like boundary of the holographic boundary [8]. As a result of this view it is not difficult to see that ϕ^3 is essentially the intrinsic Casimir effect and result from the wave pressure difference between the outside topological pressure ϕ of the two Casimir plates giving a net topological Casimir dimensionless force of [8]

$$P_T(Casimir) = \phi - \phi^2 = \phi^3 \qquad (15)$$

exactly as found in earlier publications [8].

7. Discussion of a Paradigm Shift towards Empty Spacetime Vacuum

In the present paper we addressed many subjects that were wrapped together into a unity. This may be in some sense a paradigm shift taking Einstein's insight of gravity being more or less the curvature of spacetime to its ultimate conclusion. Let us explain these main points in successive order:

A. The most abundant matter in the universe is the non-materialist void around everything which we call empty space or real vacuum and which Einstein generalized to a geometrical manifold called spacetime.

B. Spacetime is not just a stage where particles or particles and waves can perform their act nor is it a Pirandello play where audience, actors and Director interact together but far more than that. Cosmic spacetime on the quantum scale as well as on the Hubble scale is a unique theatre where everything is a by product of spacetime. In less prosaic terms, spacetime produces the quantum particle and the quantum wave and in turn the quantum particle and quantum waves produce spacetime and that in the most stringent mathematical meaning and not only as a metaphor.

C. From our previous work and analysis we know that the photon has a fractal weight equal to $\phi = (\sqrt{5}-1)/2$. This harmonizes fully with our Cantorian interval universe where the topological velocity of the light is exactly equal to ϕ. This topological ϕ is not only only the Hausdorff dimension of a cosmos made of a single random Cantor set, but it is also the topological velocity of light and the fractal weight of the photon.

D. Witten's M-theory is eleven dimensional. A fractal self-similar M-theory is consequently an eleven dimensional 'cube' inside another eleven dimensional cube and so on, indefinitely. This leads to a dimension equal to 11 plus ϕ^5 where ϕ^5 is nothing but Hardy's quantum entanglement which was proven to be accurate both theoretically as well as experimentally. In other words the probability of finding a 'quantum' point in $11 + \phi^3$ is clearly the reciprocal value, namely $P = 1/(11+\phi^5) = \phi^5$. On the other hand the probability of not finding any is clearly equal $1 - P = 1 - \phi^5$. This shows that the average probability is simply the definite probability of one of the two outcomes of any measurement, namely $1/2\left[\phi^5 + (1-\phi^5)\right] = 1/2$. Remembering our analysis of ordinary energy density and dark energy density of the cosmos we see that $1-(\phi^5/2) = 5\phi^2/2$ is the topological dimensionless dark energy of the cosmos leading to $E(D) = (2\phi^2/2)(mc^2) \cong mc^2(21/22)$

and the ordinary energy density

$$E(O) = (\phi^5/2)mc^2 \cong mc^2/22 .$$

The sum is of course E(Einstein) = mc2.

E. From the fact that E(O) = mc^2 $\phi^5/2$ and $\phi^5 = \phi^3\phi^2$ where $\phi^2 = c^2$ we see that ϕ^3 is in fact the topological mass and thus the Casimir latent topological energy is the mass of, if you want, the topological Higgs field given as ordinary measurable mass or ordinary measurable energy which on this deep level are one and the same thing. Extending the same idea we see that the topological mass of the dark energy is 5 and consequently the topological dark energy is (5)(ϕ^2). The average of both leads us again to Einstein's formula.

F. We surmised long ago that gravity is due to the passing of fractal time. Add this to the concept that there can be a Casimir force due to a gravity field and we end with a unified picture of all fundamental forces and the origin of all types of energies being spacetime vacuum and nothing else.

G. Noting that cyrstalines increase the Casimir attraction we note that Fullerene nano particles can do this even better because the flat faces resemble mini Casimir channels and are currently our conjectured candidate for a nano Casimir reactor pending some actual experimental trials.

The above conclusions are presented once more in a condensed pictural and mathematical fashion in Fig. 1 and its lengthy legend.

8. Conclusion

Self similar quasi-crystal fractals manifest themselves to us on the largest as well as the smallest scales in nature. The cosmic fractal spacetime plasma and the quarks-gluon plasma are but an expression of this scale relativity theory [1-9]. These ideas are not entirely new and were considered before by many authors including Nottale, Ord, Finkelstein and Penrose [4,7,10,12]. What is new is our ability to calculate these vital physical effects connected to the big picture. Our model is really the simplest that there is because it starts with the basics of mathematics which is in turn the base of physics, namely set theory. Only three sets are needed in our present theory, the zero set, the empty set and the emptier still set. That way we have three concentric circles. The first is the particle with 0 and ϕ as bi-dimensions, the second is the surrounding circle which is the empty set with -1 and ϕ^2 as their dimensions and then the infinitely large surrounding circle to average expectation bi-dimensions for infinitely many sets or multi-fractal given by $\langle -2 \rangle$ and $\langle \phi^3 \rangle$ expectation values. That means our Cantorian plasma picture leads to a spacetime picture of a particle wave set floating in a space made of the surface of the particle wave in monadic form. Putting this "flat" universe in five dimensions leads us to the correct ordinary and dark energy density of the cosmos as well as to understanding the zero point vacuum energy to the extent that we can start thinking seriously about harnessing it using nanotechnology [8].

References

[1]　R. Ent, T. Ulrich and R. Venugopalau: The glue that binds us. Scientific American, 312(5), May 2015, pp. 32-39.

[2]　M. Plumer, S. Raha and R. Weiner: How free is the quark-gluon plasma. Nukl. Phys. A, 418, 1984, pp. 549-557.

[3]　W.A. Zajc: The fluid nature of quark-gluon plasma. Nuclear Physics A, 805, 2008, p. 283-249.

[4]　M.S. El Naschie: A review of E-infinity theory and the mass spectrum of high energy physics. Chaos, Solitons & Fractals, 2004, 19(1), pp. 209-236.

[5]　M.S. El Naschie: Superstrings, knots and noncommutative geometry in E-infinity space. International Journal of Theoretical Physics, 37(12), 1998, pp. 2935-2951.

[6]　Mohamed S. El Naschie: Quantum Entanglement as a Consequence of a Cantorian Micro Spacetime Geometry. Journal of Quantum Information Science, 1(2), 2011, pp. 50-53.

[7]　M.S. El Naschie: The theory of Cantorian spacetime and high energy particle physics (An informal review). Chaos, Solitons & Fractals, 41(5), 2009, pp. 2635-2646.

[8]　Mohamed S. El Naschie: Kerr black hole geometry leading to dark matter and dark energy via E-infinity theory and the possibility of a nano spacetime singularity reactor. Natural Science, 7(4), 2015, pp. 210-225.

[9]　Ji-Huan He, M.S. El Naschie: On the monadic nature of quantum gravity as a highly structured golden ring, spaces and spectra. Fractal Spacetime and Noncommutative Geometry in Quantum and High Energy Physics. 2(2), 2012, pp. 94-98.

[10]　A. Connes: Noncommutative Geometry. Academic Press, San Diego, USA, 1994.

[11]　J. Polchinski: String Theory. Cambridge University Press. Cambridge 1998.

[12]　R. Penrose: The Road to Reality. A complete guide to the laws of the universe. Jonathan Cape, London. (2004).

[13]　M. Kaku: Introduction to Superstrings and M-theory. Springer, New York (1999).

[14]　Mohamed S. El Naschie: Topological-Geometrical and Physical Interpretation of the Dark Energy of the Cosmos as a 'Halo' Energy of the Schrodinger Quantum Wave. Journal of Modern Physics, 4(5), 2013, pp. 591-596.

[15]　D.R. Finkelstein: Quantum relativity. A synthesis of the ideas of Einstein and Heisenberg. Springer, Berlin, Germany, 1996.

16

On a Casimir-Dark Energy Nano Reactor

Mohamed S. El Naschie

Dept. of Physics, University of Alexandria, Alexandria, Egypt

Email address:

Chaossf@aol.com

Abstract: The paper is a general outline of the theoretical principle and basic design concepts of a proposed Casimir dark energy nano reactor. In a nutshell the theory and consequently the actual design depends crucially upon the equivalence between the dark energy density of the cosmos and the faint local Casimir effect produced by two sides boundary condition quantum waves. This Casimir effect is then colossally amplified as a one sided quantum wave pushing from the inside against the Möbius-like boundary with nothing balancing it from the non-existent outside. In view of our theory, this is essentially what led to the observed accelerated expansion of the cosmos. As in any reactor, the basic principle in the present design is to produce a gradient so that the excess energy on one side flows to the other side. Thus in principle we will restructure the local topology of space using material nanoscience technology to create an artificial local high dimensionality with a Dvoretzky theorem like volume measure concentration. Without going into the intricate nonlinear dynamics and technological detail, it is fair to say that this would be pure, clean, free energy obtained directly from the topology of spacetime. Needless to say the entire design is based completely on the theory of quantum wave dark energy proposed by the present author for the first time in 2011 in a conference held in the Bibliotheca Alexandrina, Egypt and a little later in Shanghai, Republic of China.

Keywords: Casimir Effect, Dark Energy, E-Infinity, Cantorian Spacetime, Nano Reactor, Free Energy, Möbius Boundary

1. Introduction

Based on his E-infinity Cantorian spacetime theory [1-21], it was recently argued by the author that the Casimir effect is a local manifestation of the quantum wave while dark energy is the global manifestation of the same [1]. The only difference is that of the details of the boundary conditions [1,2]. It was further reasoned by the author that the universe as a whole has a one sided boundary akin to that of higher dimensional Möbius band and consequently the "local" Casimir effect ramifies at this one sided boundary located at infinity to produce the negative gravity pressure of the conjectured dark energy [1-3]. In other words, three rather mysterious physical notions are tied together and explained in terms of each other. At the top resides the quantum wave [4], which is not a mathematical artifact [4-8] but according to E-infinity theory of dark energy, a real physical entity fully described by the empty set fixed by Connes-El Naschie bi dimension $(-1, \phi^2)$ where $\phi = 1/(\sqrt{5}+1)$ [1-8]. On the other hand the gradient caused by different wave energy density in different bounded regions of space compared to the unbounded outside of the same space is behind the Casimir forces which in the limit can be show to be equal to the difference between the quantum zero set $(0, \phi)$ and the wave

empty set $(-1, \phi^2)$ leading to $\phi - \phi^2 = \phi^3$ topological energy pressure [1, 5-8]. Finally at the edge of the universe there is only internal Casimir quantum wave pressure not balanced by outside pressure which is the dark energy concentration of 96 percent as per the consequences of Dvoretzky's theorem and the present author's dissection of Einstein's E = mc² to E(O) = mc²/22 for ordinary energy of the quantum particle and E(D) = mc²(21/22) for the dark energy of the quantum wave [7,8].

The sources of the ideas contained in the present work go back to many years ago when we attempted to improve on the traditional fast and slow fission reactors using the modern mathematics of fractals and nonlinear dynamics [9-17]. The second source is our recent reinterpretation of Einstein's E = mc² and finally the third source is the unexpected results of the earlier mentioned Dvoretzky's theorem of Banach spaces [3,6]. However in the final analysis building an actual reactor could not be possible, not even in principle, without first a sound theory [1-56] and second the combination of modern nanotechnology and state of the art Casimir effect experimentation [18-24]. In addition a reasonable amount of imaginative thinking similar to that of the man who is famed for inventing the 20th century is also recommended [49,57].

To keep the present paper short and yet to cover the large amount of the needed prerequisites we opted for a condensed presentation coupled to a large number of references. We recommend to start by reading Ref. 1 and Ref. 60, then it is a personal choice of how to proceed after that.

2. What is the Boundary of the Universe

The holographic boundary theory goes back to the pioneering work of 'tHooft and Susskind [25-27]. On the other hand the principle that the boundary of a boundary is zero goes back to the out of the box thinking of J.A. Wheeler [28]. Pushing their ideas further still, it became obvious to the present author that the boundary of the holographic boundary is not only a zero limit set but actually a hierarchy of empty and emptier still sets ramifying at a most general form of a one sided higher dimensional Möbius band [28-33]. This limit set resembles a fundamental polyhedron group or better still, a Schottky-Kleinian group [29-33] which changes the topology of our conventional Casimir experiment to that of a sphere with internal Casimir pressure inflating the balloon-like universe and makes it expand into the surrounding "nothingness" fixed by the well known E-infinity formula $d_c^{(-\infty)} = \phi^\infty = 0$ where $\phi = 1/(\sqrt{5}+1)$ [34]. From the preceding elementary reasoning it is clear that Casimir-effect and dark energy have the same cause, namely the topology of a Banach-spacetime like manifold and the only difference is the difference of local exophysics and global endophysics and the respective associated boundary conditions [1,2]. There is already a vast body of literature on the subject published in the last three years alone by the present author and his associates [1-56]. However what we are aiming at in the present paper is to point out the way to move from theory to useful, practical application of which nothing could be more important and pressing than building a free energy reactor, based on real science rather than wishful thinking. Thus we will combine the dreams of visionaries like N. Tesla with hard nosed modern mathematics and physics which were not yet available in the time of Tesla [49].

3. The Role of Nanotechnology

There has been no want of imaginative experimental set ups for measuring, testing and visualizing the Casimir effect since it was proposed by Dutch physicist, H. Casimir [22-24]. In recent years nanotechnology invaded all scientific fields and played a significant role in Casimir effect experiments. Thanks to E-infinity we now know that the true physical-mathematical connection between dark energy and the Casimir effect. A natural consequence of this discovered reality of the quantum wave, is rendering it a relatively simple task to find a way to harness dark energy or Casimir energy. Of course this "simple" is extremely difficult but no longer impossible.

We can start with a highly complex sub-structuring of space using nano tubes and nano particles and create that way nanosphere packing modelling the moonshine conjecture which relates superstrings to other fields of theoretical physics. We presently have, in embryonic form, the main idea of constructing a nano universe and extracting dark energy from its nano boundary of its holographic boundary. Our program to actually extract energy from such a nano reactor may still need five or more years but the road is marked and reasonably clear. It is only at the edge of the universe that 96% of the energy resides as dark energy. However we could create many nano universes from which its 96% energy concentration could be extracted without actually reaching to the boundary of our universe [3-8].

4. Some Experimental Proposals in a Nutshell

In noncommutative geometry as well as E-infinity theory, the Penrose universe plays a significant role as a generic concrete model for both theories [50-53]. On the other hand Penrose universe or Penrose fractal tiling is basically a quasi-crystal mathematical model with the forbidden 5-fold symmetry [53,54]. This form of matter not found naturally on earth, was produced experimentally by the great Israeli engineer D. Schechtman, who after facing a long period of fierce opposition from high profile scientists, for instance Nobel Laureate Linus Pauling, was rehabilitated and bestowed with a Nobel Prize. The 5 fold symmetry could be thought of theoretically as five Kaluza-Klein dimensions and using nano particles and nano tubes combinations we could build in the lab a nano holographic universe [5-8] akin to our own from which energy could be experimented with and extracted. For sure it will be a journey in unchartered seas with many trials and errors but sooner or later we will find out the right road to a Casimir dark energy nano reactor [1,22]. There are other conceivable ways of producing artificial nano universes with high dimensionality for Dvoretzky's theorem to be applicable. For instance we could use Ji-Huan He's ten dimensional polytope [42] as a skeleton to grow on it a hierarchy of nano particles using the methods applied in the clustering of diffusion limited aggregation. In other words, we can let our scientific imagination run free but checked with E-infinity mathematical rigor and nanotechnological facts.

5. Quantification of Casimir-Dark Energy Using E-Infinity Theory

It may come as a pleasant mild surprize that exact limits could easily be established for Casimir-dark energy using nothing more than the topology of our E-infinity Cantorian spacetime [56,60]. We can do this in a variety of ways which are essentially tautologies leading to the same basic conclusion in the limit. Thus we could view the energy density of the space outside the two Casimir plates as that of Einstein's $E = mc^2$ density, i.e. γ(Einstein)=1. Inside the plate the energy density in the limit could only be a statical, quasi potential energy of the quantum particle, i.e. $E = mc^2/22$ and consequently $\gamma(0)=1/22$. It follows then that the net pressure of the Casimir plates must be $1-(1/22)=21/22$ which is, in the meantime rather well known, as the dark energy density of spacetime. A second way to interpret the same situation and reach the same result is to argue that within the Casimir plates there is no "space" except for the empty set with a Hausdorff dimension ϕ^2 where $\phi = 1/(\sqrt{5}+1)$. Outside on the other hand we have the zero set. The difference is a net $\phi-\phi^2 = \phi^3$ which is the universal fluctuation of spacetime and simply the reciprocal value of its Hausdorff dimension $(1/\phi^3) = 4 + \phi^3$ [3][52,56]. Finally we could see the situation as the difference of the completely empty set in the limit, i.e. zero between the Casimir plates and the spacetime fluctuation ϕ^3 [60]. That way the Casimir effect could be set in the limit equal to ϕ^3 and may easily be seen to be a relative to the Immirzi parameter ϕ^6 and the Unruh temperature ϕ^4 apart of Hardy's entanglement ϕ^5, i.e. a member of a generalized quantum-topological entanglement family [60].

6. Conclusion

Exploding stars and galaxies are scientific facts. Consequently to presume that these are only topological defects in to near infinitely large spacetime is not outlandish nor science fiction [54-56]. In fact the near identity of the Casimir effect and dark energy and the fact that both originate from the quantum wave aspect of quantum mechanics clearly shows to any open minded scientific thinker that to pursue clean free energy is not a scientific 'crackpot' idea but a real and reachable aim. I ask the sensitive reader to forgive me for using the ugly word 'crackpot' which is not a proper English word but merely slang which invaded the scientific English language like a virus. The 4.5% of ordinary energy in the universe is nothing but the multiplicative volume of a five dimensional K-K zero set while the 95.5% dark energy is the additive volume of the same 5D Kaluza-Klein empty set [34]. Seen that way we think that making humanity free from oil and traditional sources of energy is a higher and moral aim worth investing heavily in for what is a million or even billion dollar research grant funding compared to the three trillion dollar Iraq war [59]. In fact the highly enlightened rules of the United Arab Emirates are already looking towards a future free of oil based energy [61]. It was Nobel Laureate in Economics, Prof. J. Stiglitz who calculated with Prof. L. Bilmes the true cost of the Iraq war for the USA. The staggering three trillion dollars do not actually include the loss and destruction for the economy of the entire world. The author dares to say with a tongue in cheek, that the mere sight of only one trillion dollars funding for our nano Casimir-dark energy reactor is sufficient to make this reactor spontaneously pop out of spacetime like virtual particles!

References

[1]   Mohamed S. El Naschie: Three quantum particles Hardy entanglement from the topology of Cantorian-fractal spacetime and the Casimir effect as dark energy – A great opportunity for nanotechnology. American Journal of Nano Research and Applications, 2015, 3(1), pp. 1-5.

[2]   Mohamed S. El Naschie: Casimir-like energy as a double Eigenvalue of quantumly entangled system leading to the missing dark energy density of the cosmos. International Journal of High Energy Physics, 2014, 1(5), pp. 55-63.

[3]   Mohamed S. El Naschie: The measure concentration of convex geometry in a quasi Banach spacetime behind the supposedly missing dark energy of the cosmos. American Journal of Astronomy & Astrophysics, 2014, 2(6), pp. 72-77.

[4]   M. Slezak: Quantum wave function gets real. New Scientist, 7 February, 2015, pp. 14.

[5]   Mohamed S. El Naschie: Dark energy and its cosmic density from Einstein's relativity and gauge fields renormalization leading to the possibility of a new 'tHooft quasi particle. The Open Astronomy Journal, 2015, 8, pp. 1-17.

[6]   Mohamed S. El Naschie: Banach spacetime-like Dvoretzky volume concentration as cosmic holographic dark energy. International Journal of High Energy Physics, 2015, 2(1), pp. 13-21.

[7]   Mohamed S. El Naschie: From $E = mc^2$ to $E = mc^2/22$ – A short account of the most famous equation in physics and its hidden quantum entanglement origin. Journal of Quantum Information Science, 2014, 4, pp. 284-291.

[8]   Mohamed S. El Naschie: The hidden quantum entanglement roots of $E = mc^2$ and its genesis to $E = mc^2/22$ plus $mc^2(21/22)$ confirming Einstein's mass-energy formula. American Journal of Electromagnetics and Applications, 2014, 2(5), pp. 39-44.

[9]   Mohamed S. El Naschie: From implosion to fractal spheres. A brief account of the historical development of scientific ideas leading to the trinity test and beyond. Chaos, Solitons & Fractals, 1999, 10(1), pp. 1955-1965.

[10] Mohamed S. El Naschie and S. Al Athel: Estimating the Eigenvalue of fast reactors and Cantorian space. Chaos, Solitons & Fractals, 2000, 11, pp. 1957-1961.

[11] M.S. El Naschie: On Nishina's estimate of the critical mass for fussion and early nuclear research in Japan. Chaos, Solitons & Fractals, 2000, 11(11), pp. 1809-1818.

[12] M.S. El Naschie: Remarks on Heisenberg's Farm-Hall lecture on the critical mass of fast neutron fission. Chaos, Solitons & Fractals, 2000, 11(8), pp. 1327-1333.

[13] M.S. El Naschie and A. Hussein: On the Eigenvalue of nuclear reaction and self-weight buckling. Chaos, Solitons & Fractals, 2000, 11, pp. 815-818.

[14] M.S. El Naschie: Elastic buckling loads and fission critical mass as an Eigenvalue of a symmetry breaking bifurcation. Chaos, Solitons & Fractals, 2000, 11, pp. 631-629.

[15] M.S. El Naschie: On the Zel'dovich-Khuriton critical mass for fast fission. Chaos, Solitons & Fractals, 2000, 11, pp. 819-824.

[16] M.S. El Naschie: On the Eigenvalue of transport reaction involving fast neutrons. Chaos, Solitons & Fractals, 2000, 11, pp. 929-934.

[17] M.S. El Naschie: Heisenberg's critical mass calculations for an explosive nuclear reaction. Chaos, Solitons & Fractals, 2000, 11, pp. 987-997.

[18] M.S. El Naschie: Chaos and fractals in nano and quantum technology. Chaos, Solitons & Fractals, 1998, 9(10), pp. 1793-1802.

[19] M.S. El Naschie: Nanotechnology for the developing world. Chaos, Solitons & Fractals, 2006, 30(4), pp. 769-773.

[20] M.S. El Naschie: The political economy of nanotechnology and the developing world. International Journal of Electrospun Nanofibrers and Applications, 2007, 1(1), pp. 41-50.

[21] M.S. El Naschie: Some tentative proposals for the experimental verification of Cantorian micro spacetime. Chaos, Solitons & Fractals,1998, 9(1/2), pp. 143-144.

[22] H. Johnston: Physicists Solve Casimir Conundrum. Physicsworld.com. July 18, 2012.

[23] S. Rencroft and J. Swain: What is the Casimir effect? Scientific American, June 22, 1998.

[24] P. Wongjun: Casimir dark energy, stabilization and the extra dimensions and Gauss-Bonnet term. The European Physical Journal C, 2015, 75(6).

[25] M.S. El Naschie: A review of application and results of E-infinity. International Journal of Nonlinear Science & Numerical Simulation, 2007, 8(1), pp. 11-20.

[26] L. Smolin: The strong and the weak holographic principles. Nuclear Physics B, 2001, 601(1-2), pp. 209-247.

[27] M.S. El Naschie: Holographic dimensional reduction center manifold theorem and E-infinity. Chaos, Solitons & Fractals, 2006, 29(4), pp. 816-822.

[28] C. Misner, K. Thorne and J.A. Wheeler: Gravitation. Freeman, New York, 1973.

[29] M.S. El Naschie: Kleinian groups in E-infinity and their connection to particle physics and cosmology. Chaos, Solitons & Fractals, 2003, 16, pp. 637-649.

[30] M.S. El Naschie: A guide to the mathematics of E-infinity Cantorian spacetime theory. Chaos, Solitons & Fractals, 2005, 25, p. 935-964.

[31] M.S. El Naschie: The concepts of E-infinity: An elementary introduction to the Cantorian-fractal theory of quantum physics. Chaos, Solitons & Fractals, 2004, 22, pp. 495-511.

[32] M.S. El Naschie: Complex vacuum fluctuation as a chaotic 'limit' set of any Kleinian group transformation and the mass spectrum of high energy particle physics via spontaneous self organization. Chaos, Solitons & Fractals, 2003, 17, pp. 631-638.

[33] M.S. El Naschie: Modular groups in Cantorian E-infinity high energy physics. Chaos, Solitons & Fractals, 2003, 16, pp. 353-366.

[34] M.S. El Naschie: On certain 'empty' Cantor sets and their dimensions. Chaos, Solitons & Fractals, 1994, 4(2), pp. 293-296.

[35] Ji-Huan He et al: Twenty six dimensional polytope and high energy spacetime physics. Chaos, Solitons & Fractals, 2007, 33(1), pp. 5-13.

[36] M.S. El Naschie: Is quantum space a random Cantor set with a golden mean dimension at the core? Chaos, Solitons & Fractals, 1994, 4(2), pp. 177-179.

[37] M.S. El Naschie: Mathematical foundation of E-infinity via Coxeter and reflection groups. Chaos, Solitons & Fractals, 2008, 37, pp. 1267-1268.

[38] M.S. El Naschie: Banach-Tarski theorem and Cantorian micro spacetime. Chaos, Solitons & Fractals, 1995, 5(8), pp. 1503-1508.

[39] M.S. El Naschie: On the initial singularity and the Banach-Tarski theorem. Chaos, Solitons & Fractals, 1995, 5(7), pp. 1391-1392.

[40] M.S. El Naschie: COBE satellite measurement, hyper spheres, superstrings and the dimension of spacetime. Chaos, Solitons & Fractals, 1998, 9(8), pp. 1445-1471.

[41] M.S. El Naschie: Infinite dimensional Branes and the E-infinity toplogy of heterotic superstrings. Chaos, Solitons & Fractals, 2001, 12, pp. 1047-1055.

[42] M.S. El Naschie: Ji-Huan He's ten dimensional polytope and high energy particle physics. International Journal of Nonlinear Science & Numerical Simulation, 2007, 8(4), pp. 475-476.

[43] M.S. El Naschie: Hyper-dimensional geometry and the nature of physical spacetime. Chaos, Solitons & Fractals, 1999, 10(1), pp. 155-158.

[44] D. Finkelstein: Quantum sets and Clifford algebras. International Journal of Theoretical Physics, 1982, 21(6/7).

[45] M.S. El Naschie: Derivation of the threshold and absolute temperature E_i = 273.16k from the topology of quantum spacetime. Chaos, Solitons & Fractals, 2002, 14, pp. 1117-1120.

[46] M.S. El Naschie: Quarks confinement via Kaluza-Klein theory as a topological property of quantum-classical spacetime phase transition. Chaos, Solitons & Fractals, 2008, 35, pp. 825-829.

[47] M.S. El Naschie: On a class of general theories for higher energy particle physics. Chaos, Solitons & Fractals, 2002, 14, pp. 649-668.

[48] Ji-Huan He: Hilbert cube model for fractal spacetime. Chaos, Solitons & Fractals, 2009, 42, pp. 2754-2759.

[49] R. Lomas: The Man Who Invented The Twentieth Century, Nicola Tesla, Forgotten Genius of Electricity. Headline Books, London 1999.

[50] M. Helal, L. Marek-Crnjac and Ji-Huan He: The three page guide to the most important results of M.S. El Naschie's research in E-infinity quantum physics. Open Journal of Microphysics, 2013, 3, pp. 141-145.

[51] M.S. El Naschie: A review of E-infinity theory and the mass spectrum of high energy physics. Chaos, Solitons & Fractals, 2004, 19(1), pp. 209-236.

[52] M.S. El Naschie: The theory of Cantorian spacetime and high energy particle physics (An informal review). Chaos, Solitons & Fractals, 2009, 41(5), pp. 2635-2646.

[53] L. Marek-Crnjac and Ji-Huan He: An invitation to El Naschie's theory of Cantorian spacetime and dark energy. International Journal of Astronomy and Astrophysics, 2013, 3, pp. 464-471.

[54] Mohamed S. El Naschie: A resolution of cosmic dark energy via quantum entanglement relativity theory. Journal of Quantum Information Science, 2013, 3, pp. 23-26.

[55] Mohamed S. El Naschie: What is the missing dark energy in a nutshell and the Hawking-Hartle quantum wave collapse. International Journal of Astronomy & Astrophysics, 2013, 3, pp. 205-211.

[56] Mohamed S. El Naschie: Topological-geometrical and physical interpretation of the dark energy of the cosmos as a 'halo' energy of the Schrodinger quantum wave. Journal of Modern Physics, 2013, 4, pp. 591-596.

[57] F.D. Peat: In Search of Nikola Tesla. Ashgrove Publications, London & Bath, 1983.

[58] L. Susskind and James Lindesay: The Holographic Universe. World Scientific, Singapore, 2005.

[59] J. Stiglitz and L. Bilmes: The Three Trillion Dollar War: The True Cost of The Iraq Conflict. Allen-Lane, Penguin Books, London, 2008.

[60] M.S. El Naschie: A unified Newtonian-relativistic quantum resolution of supposedly missing dark energy of the cosmos and the constancy of the speed of light. International Journal of Modern Nonlinear Theory & Application, 2013, 2, pp. 43-54.

[61] Caline Malek: Abu Dhabi Crown Prince details UAE leaders' vision of future without oil. The National Newspaper, UAE, February 10th, 2015 (http://www.thenational.ae/uae/government/abu-dhabi-crown-prince-details-uae-leaders-vision-of-future-without-oil?utm_content='%20vision%20of%20future%20without%20 oil).

17

Morphological and Structural Properties of Silver Nanofilms Annealed by RTP in Different Atmospheres

P. D. Nsimama

Dar Es Salaam Institute of Technology, Department of Science and Laboratory Technology, Dar Es Salaam, Tanzania

Email address:

pnsimama@yahoo.com

Abstract: This study aims at investigating the influence of gas atmospheres on the dewetting properties of DC sputtered and rapid thermally annealed silver (Ag) nanofilms. The annealing temperature ranged from 400℃ to 600℃ and the gases studied were argon (Ar) and nitrogen (N_2). Scanning electron microscope (SEM) and focused ion beam (FIB) were employed for morphological studies, while the X-ray diffraction (XRD) technique was applied in the structural analysis of the films. The SEM and top-view FIB-SEM images of Ag films annealed in both atmospheres were characterized by irregular shaped holes. At fixed temperature, the films annealed in the N_2 atmospheres gave higher hole density and larger hole sizes than the film annealed in the Ar atmosphere. Additionally, the hole density decreased with the annealing time. For films annealed in the N_2 atmosphere, isolated dewetted particles were only obtained at 600℃ substrate temperature. The XRD patterns of all the films were characterized by Ag metallic peaks. No significant difference was observed among the films' crystal structures. The annealing atmospheres mainly influences the morphologies of Ag nanofilms.

Keywords: Ag, Dewetting, Annealing Atmosphere, FIB

1. Introduction

Thin metal films tend to disintegrate into an array of particles upon annealing. The process is referred to as solid state dewetting and is driven by surface, interface and strain energy minimization [1]. In particular, the structural and morphological properties of thin metal films deposited on non-metal surfaces have drawn more interest due to their potential applications in numerous electronic, magnetic and optical devices [2].

Generally dewetting begins with the formation of holes reaching the substrate surface. The holes then grow and develop a thickened rim due to a local curvature gradient at their edges and as the rim thickens the net curvature is reduced and edge retraction slows down. Rims break down via a fingering or pinch-off instability that lead to formation of lines that subsequently decay into isolated islands through a Rayleigh-like instability [2]. Once the isolated solid metal particles are formed, their size, shape and spacing evolution depends on the annealing temperature, time and atmosphere [3]. The annealing atmospheres play a great role in their dewetting mechanisms. This is due to the fact that in the presence of adsorbates, different diffusion paths may change [4]. The adsorbates can change the surface energy and its anisotropy; the anisotropy can affect the dewetting process by causing the texture and grain boundary character distribution changes in polycrystalline thin films [4]. Despite of the potential role of annealing atmosphere in the dewetting process of thin films, there are only few studies documented on the subject.

Anna et al. [4] investigated on the influence of annealing atmosphere on the dewetting mechanisms of magnetron sputtered gold (Au) thin films on c-plane oriented sapphire substrates. The annealing atmospheres studied were air and forming gas. They found that Au films annealed in forming gas exhibit higher surface energy anisotropy than those annealed in air. It was also observed in their microscopic images that Au films annealed in air translated into a more branched, tortuous morphology of the holes, signifying lower degree of surface energy anisotropy.

Sharma et al. [5] investigated on the influence of annealing atmospheres on the morphologies of cathodic sputtered Ag thin films. They used quartz substrates and the annealing was done in vacuum, He, O_2 and Ar atmospheres using a large mobile furnace. In all annealing atmospheres, hillock formation took place and was explained on the basis of thermal stress relaxation by diffusion creep. The largest size of hillocks was recorded by the sample that was annealed in the O_2 atmosphere and was attributed to the enhanced surface self-diffusion of silver atoms, which tends to be enhanced up to a factor of 100 in the presence of oxygen. It was also reported that, with the exception of oxygen atmosphere, there were no holes even after 20 hours. The island formation took place by the holes joining together and ultimately leading to agglomeration. The reduction in the surface energy of the islands acted as a driving force for the surface diffusion of silver atoms during agglomeration. The maximum surface area of the substrate not covered with film was about 65 % after complete agglomeration and was observed for a film of thickness 50 nm annealed at 470℃ for 2 hours.

In another study [6], the thermally evaporated Ag films coated on quartz substrates were annealed in vacuum and air. The dewetted surface was covered by holes, whose density decreased with the increase in the film thickness. They also reported from their work that the hole formation in the Ag films annealed in air was faster than in those annealed in vacuum, the result that was attributed to the increased self-diffusion of Ag films due to the presence of oxygen in air.

In a recent study, Jongpil [7] thermally annealed e-beam evaporated Ag thin films on Si (100) substrate using a vacuum tube at 400℃ in the hydrogen and oxygen atmospheres. He investigated the dominant dewetting mechanism through evaluation of temporal changes in the spatial distribution of holes. His results showed stronger spatial correlation in samples with a greater number of holes. The observed trend was attributed to the formation site of new holes, which were spatially increasing due to the presence of other holes. Additionally, Ag films annealed in the O_2 atmosphere had a larger number density of holes than those annealed in the H_2.

In this work we report on the influence of annealing atmospheres in the dewetting properties of Ag nanofilms prepared by DC magnetron sputtered technique. Two different atmospheres are considered, i.e. Ar and N_2. The sputtering technique has been chosen because of its simplicity and flexibility in the materials combination. The annealing was done using the rapid thermal process (RTP) at 400℃ for 30 minutes in both atmospheres and further at 500℃ and 600 ℃ in N_2 atmosphere. The RTP annealing is

considered to be more effective than the tube furnace annealing since the sample is heated at a much faster rate. We show that Ag films annealed in different atmospheres result into different dewetting properties.

2. Experimental Details

Ag films of 30 nm thickness each were deposited onto c-plane oriented sapphire ((0001) single crystal α-Al$_2$O$_3$) substrates at room temperature using *LA440S Von Ardenne Anlagentechnik GMBH* sputtering machine. Prior to deposition, the chamber was evacuated to a base pressure of 2×10^{-7} mbar. The DC power and argon flow rate were set at 200 W and 80 sccm respectively. The films were annealed using thermal process (*RTA, Jipelec Jetstar 100*) at 400℃ for 30 minutes in Ar and N_2 atmospheres. Then annealing in N_2 atmosphere was further done at 500 ℃ and 600 ℃ temperatures. While fixing the substrate temperature at 400℃, Ag films were annealed in the N_2 atmosphere for 10, 30 and 40 minutes. The details of the rapid thermal annealing process employed in the current work can be found elsewhere [8]. After annealing, the samples were cooled to room temperature in the furnace. The morphologies of films were investigated by high-resolution scanning electron microscopy (*SEM, Hitachi S-4800*). The machine is equipped with the energy dispersive (EDS) facility, which was used for elemental composition analysis of the films. The top view and cross-sectional FIB-SEM analysis of the samples was done by *Zeiss Auriga 60 DualBeam*. The electron and ion beam were employed to deposit carbon and Pt respectively to protect the thin film during FIB analysis. The particle distribution analysis of the annealed Ag nanofilms was done by using the online free software; ImageJ. The Grazing incidence X-ray diffraction (GIXRD) data were collected using *SIEMENS D 5000* theta-theta diffractometer machine with Cu Kα radiation of $\lambda = 1.5405$ nm.

3. Results and Discussion

3.1. SEM Results

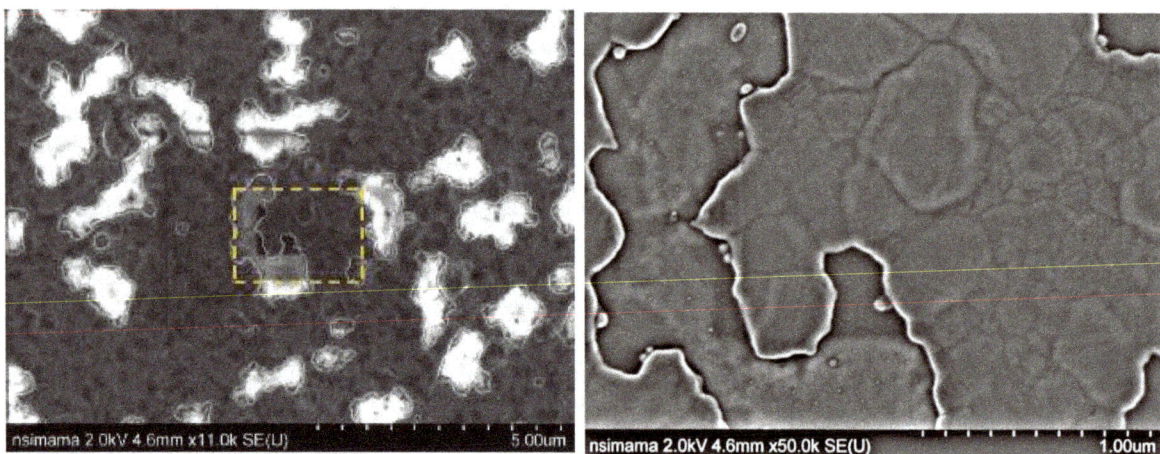

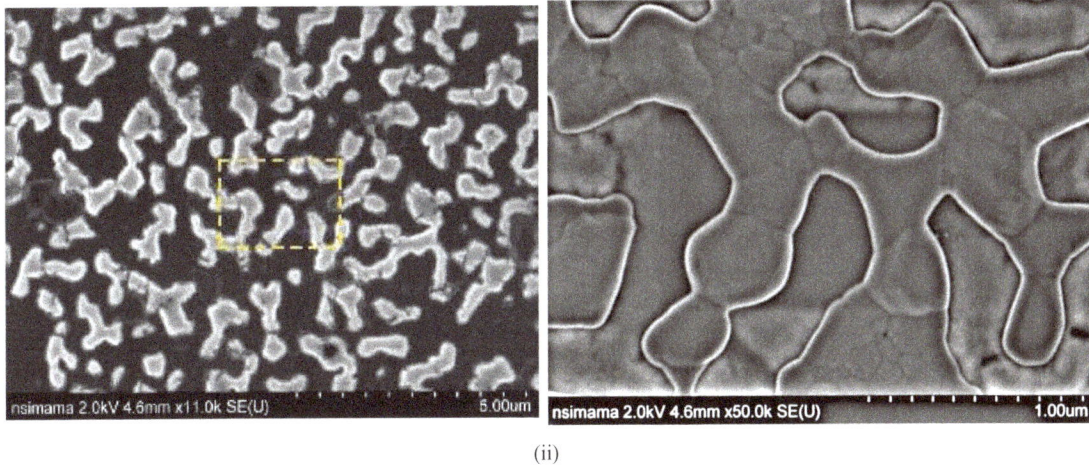

(ii)

Figure 1. *SEM images for Ag nanofilms annealed at 400 ℃ for 30 minutes in (i) Ar and (ii) N₂ atmosphere. The higher magnification images represent the dotted lined-rectangles.*

Figure 1 shows the lower (5 μm x 5 μm) and higher (1 μm x 1 μm) magnification scale SEM images for the Ag films annealed in the Ar and N₂ atmospheres at 400 ℃ for 30 minutes. The surface images of samples annealed in both Ar and N₂ atmospheres (Fig. 1 (i) and (ii)) consists of irregular holes. The Ag film annealed in the N₂ atmosphere has higher hole density than the film annealed in the Ar atmosphere.

The SEM images for Ag films annealed in the N₂ atmosphere at temperatures higher than 400℃, i.e., 500 and 600℃ for 30 minutes are shown in Figure 2. The surface of the sample annealed at 500℃ (Fig. 2 (i)), has a combination of isolated particles and some which are connected by thin necks. Thin necks are signs of isolations/breaking upon further annealing process due to the Winterbottom effect. Despite the formation of isolated particles, the surface seems to be covered by undewetted Ag films at such a temperature.

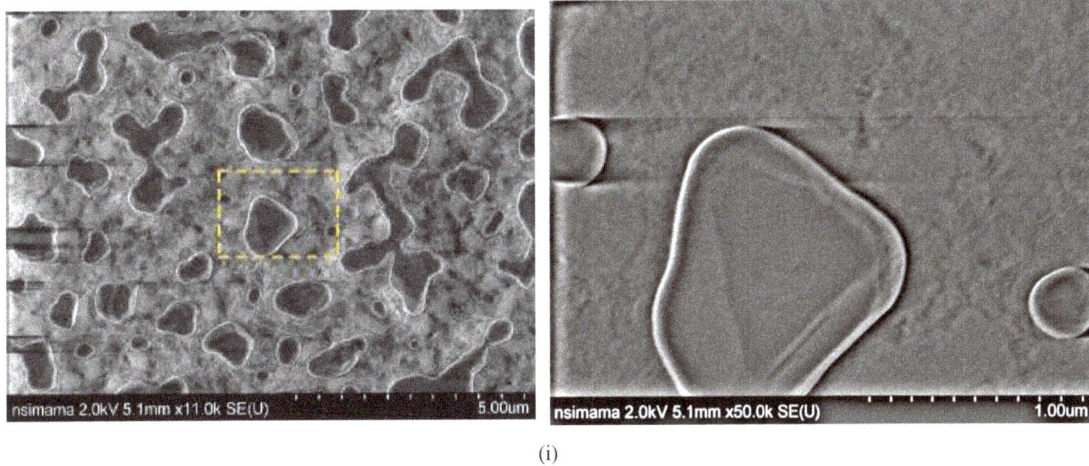

(i)

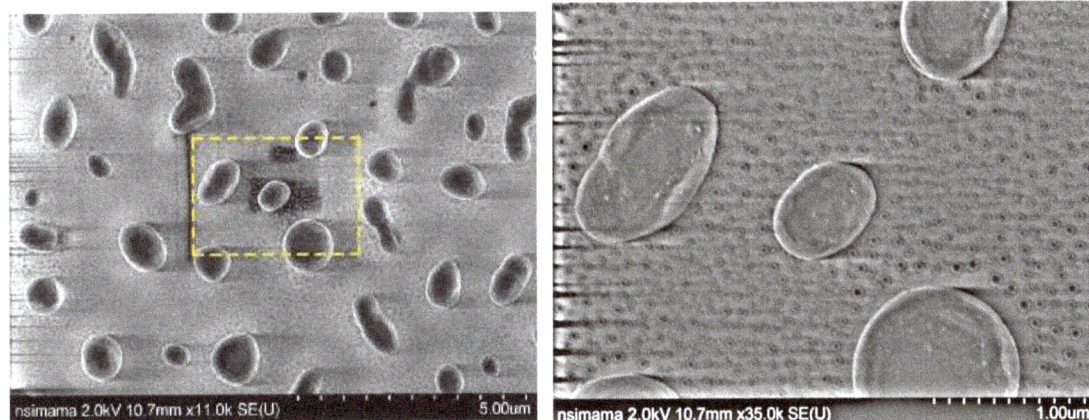

(ii)

Figure 2. *SEM images for Ag nanofilms annealed at (i) 500 ℃ (ii) 600 ℃ for 30 minutes in the N₂ atmosphere.*

The surface of the sample annealed at 600℃ (Fig. 2 (ii)), has only isolated particles of spherical and elliptical shapes. At this temperature, almost the whole Ag layer seems to have undergone dewetting leaving only the substrate (sapphire). This is substantiated by the EDS spectra shown in Figure 3 in which, the spectrum for an isolated particle (Fig. 3 (i)) is dominated by Ag peak while that of the dewetted area (bare surface) is dominated by Al and O peaks. The observed carbon (C) is resulting from the contaminations during the annealing process.

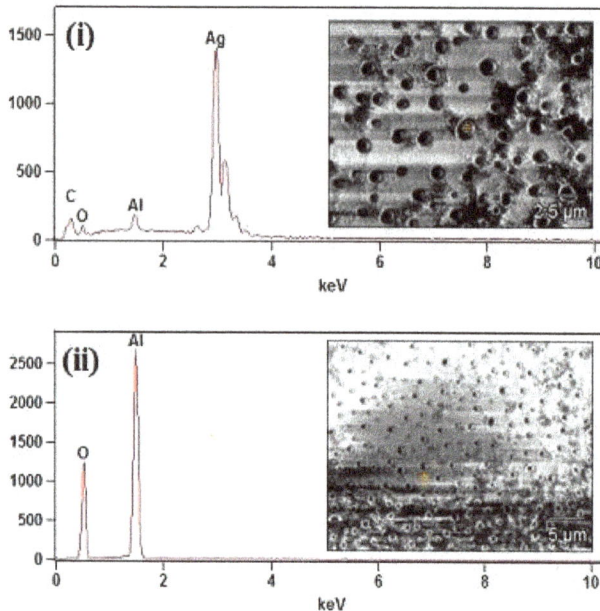

Figure 3. *The EDS elemental compositions for Ag film annealed at 600 ℃ in N_2 atmosphere for 30 minutes (i) on the particle and (ii) on the dewetted area. The symbol "⊞" indicates the analyzed area.*

3.2. Holes Distribution Analysis

The results in this section are presented according to the following sequence: the hole size distribution is discussed before the analysis of the dewetted areas for different annealing atmospheres. This analysis excludes samples annealed at temperatures higher than 400℃.

3.2.1. The Hole Sizes Distribution

The hole size distributions for annealed Ag films are shown in Figure 4 (i). Generally, the results show that Ag film annealed in the N_2 atmosphere records higher number of holes and bigger hole sizes than the film annealed in the Ar atmosphere. This implies faster dewetting of Ag films is obtained when the annealing is done in the N_2 atmosphere. It is worth noting that the majority of hole sizes for both films falls under the nano-range.

The variations of normalized average holes density and average hole size with the annealing atmospheres are shown in Figure 4(ii). The Ag films annealed in the N_2 atmosphere records superior normalized hole density and average hole size than the film annealed in the Ar atmosphere.

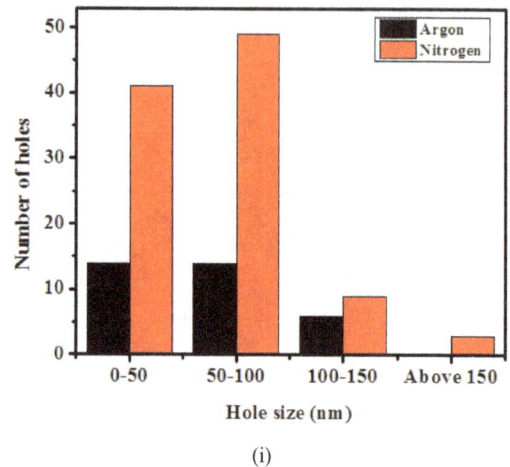

(i)

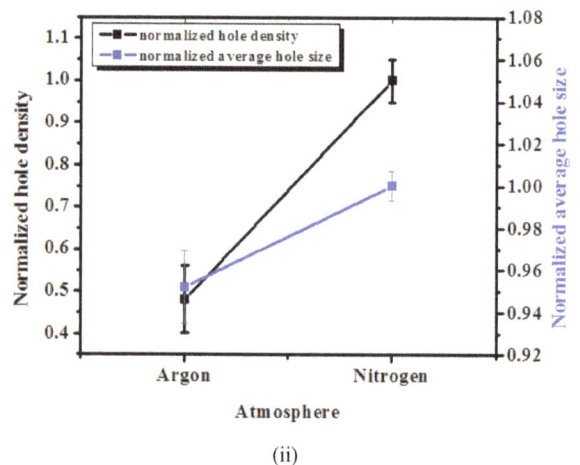

(ii)

Figure 4. *(i) The hole size distribution histogram for Ag films annealed in different atmospheres (ii) Variation of normalized hole density and normalized average hole size with the annealing atmospheres.*

3.2.2. The Dewetted Area

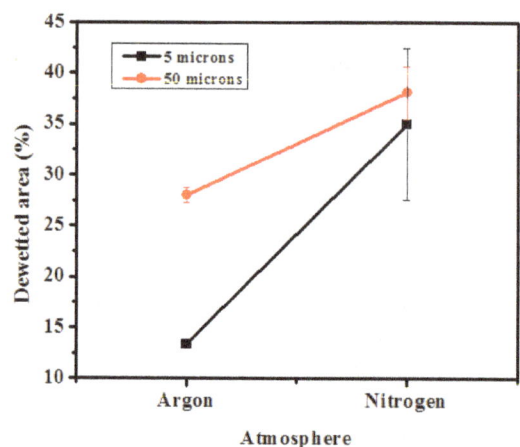

Figure 5. *Variation of dewetted areas with the annealing atmosphere; the 50 and 5 microns scale data from two different SEM image scales are compared.*

The analysis of dewetted areas for the samples was done by using both 5 microns and 50 microns scale SEM images to check the consistency of hole distribution data. The trend of percentage dewetted area for both magnification scales seem to be similar. Consistent with the results from Figure 4, the Ag film annealed in the N_2 atmosphere records higher dewetted area than the film annealed in the Ar atmosphere.

The diatomic molecule of N_2 is held tightly together by a strong triple bond between two nitrogen which requires a great amount of energy to break the bonds [9] minimizing the possibility of dissociation, which would have negatively influenced the diffusion process through attaching to the film surface. The same applies to the Ar atom, which is an inert gas. Therefore, the possible reason for slower and faster dewetting in the Ar and N_2 atmosphere respectively is to do with the atomic radii and masses of the two gases. The larger atomic radius (71 pm) and molecular mass (39.948 kg/mole) of Ar compared to those of N_2 (56 pm and 28.01 kg/mole) [10] might have interfered more with the free movement of Ag atoms in the surface diffusion process in the Ar atmosphere.

The changes of the dewetted areas, hole sizes and hole densities of the films with the annealing times for Ag nanofilms annealed at 400 ℃ in the N_2 atmosphere are summarized in Figure 6. The dewetted area and hole size increase with the increase in the annealing times due to the increase in the number of holes. On the other hand, holes tend to merge at higher annealing times leading to lower hole density.

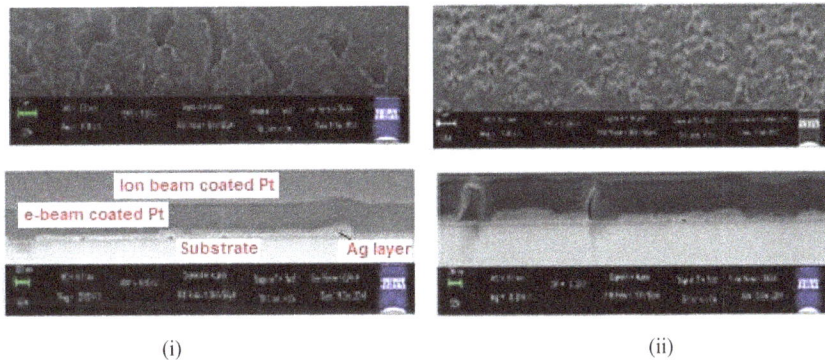

Figure 6. *The changes of dewetted areas, hole sizes and hole densities of Ag films with the annealing times; the temperature was fixed at 400°C.*

3.3. FIB-SEM Results

The top-view and cross sectional FIB-SEM images for Ag films annealed in the two atmospheres are shown in Figure 7. Comparing the top-view images one can notice that irregular shaped holes are observed on the surfaces of samples annealed in both Ar and N_2 atmospheres with the sample annealed in Ar having larger grains with well defined grain boundaries than the sample annealed in the N_2 atmosphere. The top-view FIB-SEM images are similar to those of the SEM images observed in Figure 1.

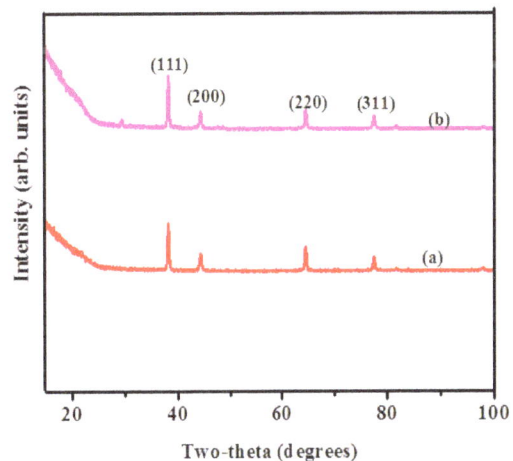

Figure 7. *Top view and cross-sectional view FIB-SEM images for Ag nanofilms annealed in (i) Ar and (ii) N_2 atmosphere.*

For the cross-sectional FIB-SEM images the arrangement of layers for each image is as labelled in Figure 7 (i). Both cross-sectional images have holes of varying sizes and spacing. The sample annealed in Ar atmosphere has the higher number of holes, which are uniformly spread across the Ag layer. Closely spaced holes are observed on the image of the sample annealed in N_2 atmosphere. The top-view FIB-SEM images look similar to SEM images from Figure 1.

3.4. XRD Results

The grazing incidence XRD patterns for the Ag samples annealed in different atmospheres are shown in Figure 8. There is no significance difference between the XRD patterns of the films, suggesting that the annealing atmospheres have no significant influence on the crystal structures of Ag films. Both samples are characterized by metallic Ag orientations, i.e. (111), (200), (220) and (311) [11].

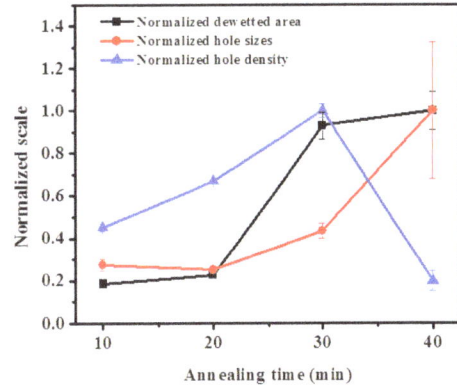

Figure 8. *GIXRD patterns for Ag nanofilms dewetted in different atmospheres; (a) Ar and (d) N_2*

4. Conclusions

We have successfully investigated the influence of Ar and N_2 annealing atmospheres on the dewetting properties of magnetron sputtered Ag films coated on sapphire substrates. The annealing atmosphere influences the dewetting properties of Ag films. Ag films annealed in the N_2 atmosphere dewet faster than the film annealed in the Ar atmosphere. Holes tend to merge at higher annealing times leading to lower hole density. At higher annealing temperature (600°C) in the N_2 atmospheres isolated spherical and elliptical shaped particles were obtained. The top-view FIB-SEM images showed irregular shaped holes similar to the SEM images. The samples annealed in both atmospheres display similar XRD patterns, mainly from metallic Ag peaks.

Acknowledgements

This work was funded by the Alexander von Humboldt Foundation, Bonn, Germany.

The author would like to acknowledge Prof. Peter Schaaf, Dr. Dong Wang and Herz Andreas for their useful advice; Dr. Rolf Grieseler and Miss Diana Rossberg all from the Chair, Materials for electronics and electrical Engineering, TU Ilmenau for their assistances in the sample preparation and FIB measurements respectively. Miss Anna Franz's help in the rapid thermal annealing is highly appreciated.

References

[1]   Claudia Manuela Müller, Ralph Spolenak, (2010), Acta Materialia, 58, 6035-6045.

[2]   F. Ruffino and M. G. Grimaidi, (2014), Vacuum, 99, 28-37.

[3]   O. Malyi and E. Rabkin, (2012), Acta Materialia, 60, 261-268.

[4]   Anna Kosinova, Oleg Covalenko, Leonid Klinger and Eugen Rabkin, (2015), Acta Materialia, 83, 91-101.

[5]   S. K. Sharma and J. Spitz, (1980), Thin Solid Films, 65, 339-350.

[6]   S. K. Sharma, S. V. M Rao and N. Kumar, (1986), Thin Solid Films, 142, L95- L98.

[7]   Jongpil Ye, (2014), Appl. Phys. Express, 7, 085601.

[8]   A. Herz, D. Wang, Th. Kups and P. Schaaf, J. Appl. Phys., (2014), 116, 044307.

[9]   Satoko Kuwano-Nakatani, Takeshi Fujita, Kazuki Uchisawa, Daichi Umetsu, Yu Kase, Yusuke Kowata, Katsuhiko Chiba, Tomoharu Tokunaga, Shigeo Arai, Yuta Yamamoto, Nobuo Tanaka and Mingwei Chen, (2015), Materials Transactions, 56, No. 4, pp. 468-472.

[10] http://www.webelements.com/

[11] X. H Yang, H. T. Fu, K. Wong, X. C Jiang, A. B. Yu, (2013), Nanotechnology, 24, 415601 (10 pp).

18

Progress in solid acid fuel cell electrodes

Aron Varga

Leibniz Insitute of Surface Modification, Permoserstraße 15, D-04318 Leipzig, Germany

Email address:

aron.varga@iom-leipzig.de

Abstract: Solid acid fuel cells represent a relatively new technology with the advantage of an intermediate operating temperature of 240°C and a solid state proton conducting electrolyte (CsH_2PO_4). Widespread commercial application has been hindered mainly by low performance and costly electrodes containing a high Pt loading. Here we review the recent progress and current status of solid acid fuel cell electrodes. Major efforts include creating nanostructured composites leading to much reduced Pt loadings while maintaining or even increasing performance. Furthermore, fundamental studies on Pt thin films, as geometrically controlled electrodes, have recently revealed the possibility of an electrochemical pathway through the two-phase boundary in addition to the classic three-phase boundary. Carbon nanotubes as electronic interconnects have been shown to dramatically improve Pt catalyst utilization and hence electrode performance. Major efforts are spent to search for alternative, non-precious metal catalysts.

Keywords: Solid Acid Fuel Cells, Electrodes, CsH_2PO_4, Pt, CNTs

1. Introduction

Fuel cells have been heralded as the energy technology of the future for many decades. Start-up companies, industrial giants, government supported pilot programs came and went periodically causing boom and bust periods. Despite multiple disappointments, the technology is just too elegant and attractive to be abandoned altogether, and here, a highly promising, new class of fuel cell is regarded. Especially in the light of an ever increasing necessity for efficient energy storage and conversion devices, fuel cells are becoming more important since they can most efficiently convert chemical energy to electrical and vice versa. Since their invention in the 19th century, multiple technologies have been developed, with most efforts spent on the portable and flexible low temperature technology based on polymer electrolytes (PEMFCs) and the stationary but highly efficient technology based on solid oxide materials (SOFCs). A relative newcomer in this field is the solid acid fuel cell (SAFC), based on the solid acid material CsH_2PO_4[1]. Their intermediate operating temperature of 240°C and a truly solid state electrolyte provide multiple technological advantages, such as the suitability of stainless steel interconnects, fuel flexibility and resistance to catalyst poisoning. However, widespread commercial application has been hindered by the need for relatively high loadings of precious metal Pt as the electrocatalyst and low electrode performance[2]. Here we review recent progress and current status of solid acid fuel cell electrodes.

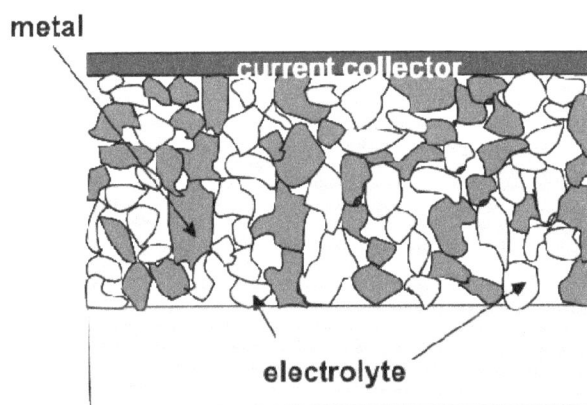

Figure 1. Schematic of a solid acid fuel cell electrode consisting of porous, interconnected Pt catalyst and CsH_2PO_4 electrolyte particles[3].

In general, SAFCs employ a pure ion conducting electrolyte (no electron-ion mixed conductivity as with some solid oxide fuel cell electrode materials) and Pt as the electrocatalyst. The electrochemical reactions occur where the simultaneous transport of ions, electrons and gas molecules is possible, i.e. the so-called triple phase boundary between the electrolyte, the catalyst and the gas phase, Fig 1[3]. Considering these restrictions, the ideal electrode consists of an interpenetrating, 3-dimensional, interconnected structure of the electrolyte and the electrocatalyst with high porosity for gas access.

As with other low and intermediate temperature fuel cells, the overpotential at the cathode is far greater than that at the anode and with electrolyte membranes as thin as 20 μm, the overpotential due to Ohmic ion transport resistance can be neglected[4].

2. Electrochemical Measurements of Electrodes

The most convenient method to measure the electrochemical performance of solid acid fuel cell electrodes is AC impedance spectroscopy in a single chamber, symmetric gas environment. Such a measurement does not require complicated and error-prone sealing but allows a detailed analysis of the electrode reaction kinetics as well as separation of the electrolyte response. However, such symmetric cell measurements are limited to hydrogen (hydrated to $pH_2O = 0.4$ atm to prevent dehydration of the electrolyte[5]) since Pt is in its oxidized form in thermodynamic equilibrium in an oxygen gas environment at ca. 240°C. Despite this challenge, the anode performance is often assumed to be indicative for cathode performance, since the same geometric features, namely the triple phase boundary, is rate limiting for both the oxygen reduction and the hydrogen oxidation reaction. A high performance anode is also a high performance cathode with this material system. During a symmetric cell measurement, the forward and the reverse electrochemical reaction is captured simultaneously and it is assumed that the electrochemical impedance contribution is equal in magnitude:

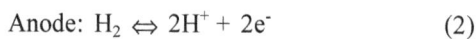

$$\text{Cathode: } 2H_2O \Leftrightarrow 4H^+ + 4e^- + O_2 \qquad (1)$$

$$\text{Anode: } H_2 \Leftrightarrow 2H^+ + 2e^- \qquad (2)$$

In order to uniquely separate the electrode impedance for a specific reaction, such as the hydrogen oxidation, a reference electrode is traditionally used. However, it has been shown by Adler[6] that in a solid state system, the placement of the reference electrode is extremely sensitive, potentially introducing a large error in the measurement. An elegant solution to this problem has been described by Sasaki et. al.[7], suggesting to geometrically restrict the working electrode such that its overpotential becomes dominant, in a reference electrode free configuration, Fig 2. The impedance of the much larger counterelectrode can be neglected. A quantitative relationship between the size ratio of the electrodes (working electrode radius: r_{we} and counter electrode radius: r_0) and the electrolyte thickness (t) where the electrode responses are well separated, was established through a computational and experimental study.

AC impedance spectroscopy of electrochemical cells with an asymmetric geometry permit, with a suitable size ratio of the working electrode and the counter electrode, the measurement of the cathodic reaction in a symmetric oxygen

environment. Thermochemical considerations reveal that at the SAFC operating temperature of 240°C and at 0.6 atm pO_2, the standard free energy of formation of PtO is ca. -22 kJ/mol.[8] Hence a voltage of at least 0.22 V across the electrode shifts the equilibrium of the oxidation reaction in favor of Pt. This implies that for an electrochemical cell with an asymmetric geometry (t = 3 mm, $r_0 = 1$ cm, and $r_{WE} = 1$ mm) an electric DC bias of 0.3 V is sufficient to access the electrochemical activity of Pt in its reduced form. The overpotential across the electrolyte film and the counter electrode, even when its specific activity as PtO is lower than that of Pt, is negligible.

3. Composite Solid Acid Fuel Cell Electrodes

3.1. Microstructured Powder Electrodes

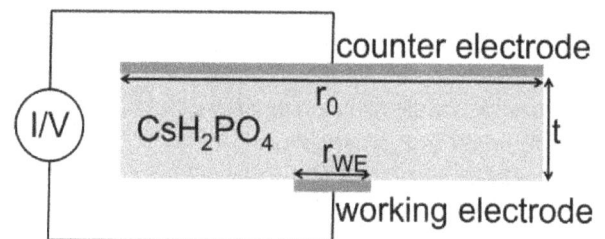

Figure 2. Schematic of an electrochemical cell with asymmetric geometry and symmetric gas configuration for AC impedance spectroscopy.

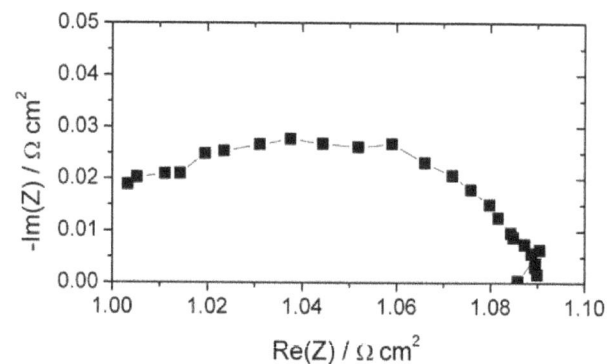

Figure 3. Nyquist plot of the electrode impedance for symmetric cell with powder electrodes ($CsH_2PO_4 : Pt : Pt/C = 3 : 3 : 1$ weight ratio) measured at 240°C in humidified hydrogen.

The first generation solid acid fuel cell electrodes consisted of a composite powder of CsH_2PO_4 microparticles, Pt and Pt on carbon nanoparticles, mixed with the pore former naphthalene [4,9]. An intimate mixture of the components is achieved with ultrasonication as a suspension in toluene and subsequent drying, ballmilling, or grinding with mortar and pestle. The membrane electrode assembly is obtained by co-pressing the electrode powder mix with a uniaxial cold-press with a layer of pure CsH_2PO_4 as the electrolyte at 34 MPa. Upon heating to the operating temperature of 240°C, the electrolyte particles partially sinter and ensure a continuous ion transport from the catalytically active centers to the electrolyte membrane. The large difference in the particle sizes and the mechanical properties of the electrolyte and the electrocatalyst, as well as the complete decomposition of naphthalene ensures that continuous pores remain for gas access. With a Pt loading of 7.7 mg/cm^2, the to-date lowest published electrode resistance of 0.06 Ω cm^2 was obtained, and here successfully reproduced, Fig. 3. Nevertheless, the Pt mass normalized activity of such electrode with 2.2 S/mg is low, implying that the spatial distribution of the catalyst is suboptimal. Not all catalyst particles are connected with the electrolyte or electronically connected to the current collector. From simple geometric considerations, an improved matching of the electrolyte and electrocatalyst particle size leads to improved catalyst distribution and hence utilization. It has been shown experimentally that mechanical mixtures of composite electrode powders with decreasing electrolyte particle sizes indeed result in lower electrode overpotentials[2].

3.2. Nanostructured Electrodes via Electrospray

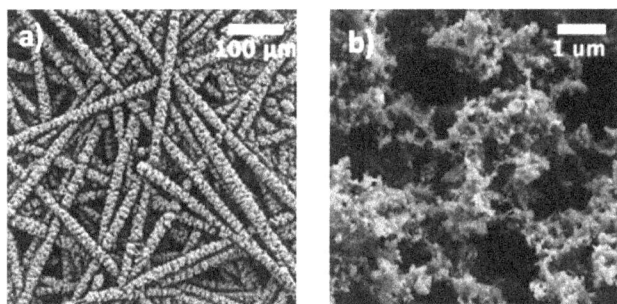

Figure 4. Scanning electron micrograph of nanocomposite SAFC electrode (CsH$_2$PO$_4$ and Pt nanoparticles at (a) low and (b) high magnification, deposited onto carbon paper via electrospray [10].

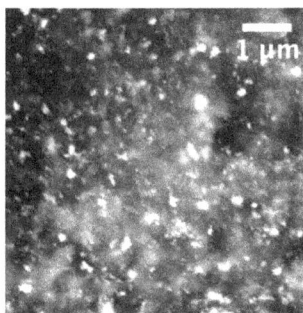

Figure 5. Scanning electron micrograph of nanocomposite SAFC electrode (CsH$_2$PO$_4$, Pt) obtained with a back scattered electron detector, showing isolated Pt particles as bright areas.

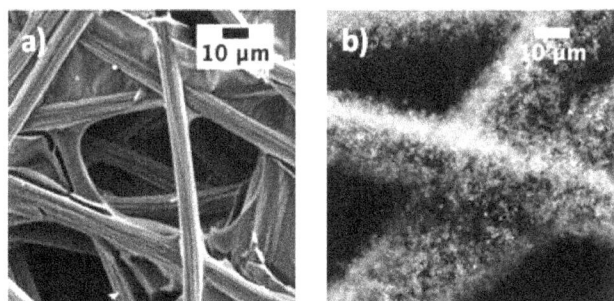

Figure 6. Scanning electron micrographs of (a) blank carbon paper and (b) CNT overgrown carbon paper.

Figure 7. Scanning electron micrographs of SAFC electrode deposited via spraydrying onto a CsH$_2$PO$_4$ electrolyte pellet (a) top-view, (b) side view.

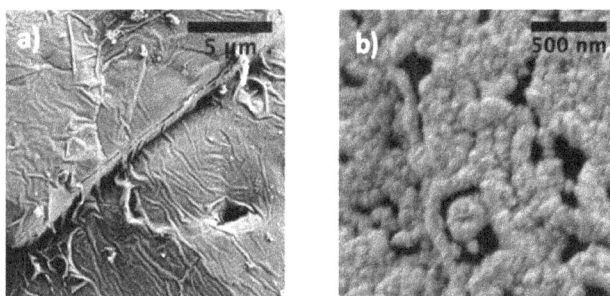

Figure 8. Scanning electron micrographs of SAFC electrode WITH Pt deposited via metal organic chemical vapor deposition (MOCVD) onto a CsH$_2$PO$_4$ electrolyte pellet at (a) low, (b) high magnification.

Substantial efforts have been spent on reducing the electrolyte particle size to the sub-micron regime. The aim is to increase the density of electrochemically active triple phase boundaries between the CsH$_2$PO$_4$ electrolyte, the Pt electrocatalyst nanoparticles (ca. 10nm diameter for commercially available standard catalysts) and the gas phase, by improving the catalyst distribution. The challenges for CsH$_2$PO$_4$ particle size reduction stem from the fact that CsH$_2$PO$_4$ is a hygroscopic material with a low melting point and thus is prone to agglomeration. As an in-situ nanoparticle fabrication method, without the need for post-synthesis particle processing, electrospray deposition does indeed allow the production of sub-micron CsH$_2$PO$_4$ particles. When directly deposited onto a substrate, an interconnected, highly porous electrolyte structure with a feature size of down to 100 nm can be created[10]. In addition, the electrocatalyst can be co-deposited when suspended in the electrospray precursor solution, Fig. 4. Further additives, such as the polymer polyvinylpyrrolidone (PVP) stabilize the otherwise unstable nanostructure even under fuel cell operating conditions. With 0.3 mg/cm^2, a substantial reduction of the catalyst loading has been achieved without sacrificing electrode performance, when compared to mechanically mixed electrodes[7].

However, the mass normalized activity of 2.2 S/mg Pt is similar to the values obtained with the mechanically mixed electrodes. Scanning electron micrographs taken with a backscattered electron detector show isolated Pt nanoparticles, Fig. 5, and hence indicate that the loss of electronic interconnectivity between the catalyst nanoparticles and agglomeration in the precursor solution is the main cause of the low mass normalized activity. In addition, a major drawback of electrospray synthesis is the low deposition rate of 5 mg/h.

To improve the electronic interconnectivity of Pt catalyst nanoparticles, without increasing the Pt loading, carbon nanotubes (CNTs) have been grown directly onto the standard current collector, carbon paper via chemical vapor deposition (CVD). The matching size scales of the current collector (CNT + carbon paper) and the feature size of the electrosprayed nanostructure result in a higher statistical likelyhood of a Pt nanoparticle to be connected with the CNT (20 – 40 nm diameter), compared with a bare carbon fibers (10 μm diameter) from the carbon paper, Fig. 6. The mass normalized activity thus was improved 3-fold to 6.6 S/mg and the stability was demonstrated with AC impedance measurements with a DC voltage bias [11].

3.3. Nanostructured Electrodes via Spraydrying

Spraydrying has been explored as a more easily scalable method compared to electrospray and hence technologically more relevant process, to obtain nanostructured solid acid fuel cell electrodes[12]. Here, an aerosol is generated via a vibrating mesh membrane with micron sized CsH_2PO_4 solution droplets. The droplets are carried by a heated gas stream to an electrophoretic deposition area. During flight, the solvent evaporates and the solute precipitates to form nanoparticles, with the size mainly depending on the concentration. Ultimately, a diffusion limited aggregate is deposited on a given substrate, such as carbon paper current collector, or prefabricated electrolyte pellet, with a feature size similar to electrosprayed structures, Fig. 7. A deposition rate of 165 mg/h can be achieved covering a surface area of 1750 cm^2. The Pt electrocatalyst was added via sputtering. The electrode impedance was 3.6 Ω cm^2 and a mass normalized activity of 13 S/mg was obtained.

4. Thin Film Electrodes

4.1. Sputtered Thin Film Electrodes

In contrast to high performance, nanocomposite structures, simple, geometrically controlled electrodes in the form of a thin films, deposited onto a polished electrolyte pellet, allow fundamental studies of the rate-limiting step of electrochemical reactions. Careful analysis of such electrodes by Louie and Haile[13], with systematically varying film thickness and outer diameter reveal that for sub-50 nm Pt films, obtained by DC magnetron sputtering, a significant contribution to the electrode performance is via the two-phase boundary between the electrocatalyst and the electrolyte. Here the reaction rate is not limited by the diffusion process of dissociated hydrogen across the metal thin film. In addition, a lower bound for the specific activity of the triple phase boundary was estimated to be 41 kΩ cm. A remarkable mass normalized activity of 19 S/mg and an area normalized electrode impedance of ca. 3.1 Ω cm^2 was measured with Pt films of 7.5 nm thickness. Due to its significantly higher hydrogen permeability, Pd is expected to play an important role in high performance thin film electrodes[14], such as Pt-Pd-Pt sandwich structure.

4.2. Thin Film Electrodes via MOCVD

Pt particles or thin films can be obtained via metal organic chemical vapor deposition (MOCVD), using $Pt(acac)_2$ as the precursor material[15]. Here, the metal organic compound is mixed with CsH_2PO_4 powder and heated to at least 150°C in a vacuum oven that has been evacuated and purged with nitrogen. Water that keeps CsH_2PO_4 from dehydration may also assist in the decomposition of the metal organic compound on the acidic material surfaces. Thus, conformal Pt coatings can be obtained in a single step, batch synthesis method, Fig. 8, providing a significant advantage over line of sight deposition methods, such as sputtering. Precise control of the film thickness is not difficult, as it can be adjusted either by the amount of $Pt(acac)_2$ mixed with CsH_2PO_4 powder or by the number of successive depositions. An area normalized impedance of 0.3 Ω cm^2 was measured with a Pt loading of 1.8 mg/cm^2 , i.e. with a mass normalized activity of ca. 1.9 S/mg.

5. Conclusion and Prospects

Solid acid fuel cells represent a highly attractive technology because of their intermediate operating temperature, considered as the "sweet spot" for fuel cells. About ten years of research and development has shown the tremendous potential in terms of performance and cost. With equivalent catalyst utilization as with PEMFCs, a much higher fuel cell power density can be expected due to the 240°C operating temperature for CsH_2PO_4 based electrolytes.

A summary of the main electrode developments is given in table 1. A combination of high surface area electrode and fine distribution of Pt catalyst should lead to an even higher mass normalized activity and lower area specific electrode resistance than the current state of the art. Compared to other fuel cell technologies, such as PEMFCs and SOFCs, the initial relatively competitive 417 W/cm^2 [16] fuel cell power density is projected to surpass current state of the art, upon optimizing the cathode performance. Assuming negligible electrode overpotential and a 10 μm electrolyte membrane, a 2500 W/cm^2 power density can be calculated when using hydrogen and oxygen as the fuel.

Furthermore, alternative, non-precious metal catalyst materials are needed to allow further necessary raw material cost reduction, before widespread technological applicability is possible. The challenge here is the general reactivity of many common metal and metaloxide candidate materials with the electrolyte CsH_2PO_4 and the generation of non-conducting, new phases as the reaction product. Here, functionalized, carbon based materials may provide an interesting research direction.

Acknowledgements

Financial support was provided by the ESF Forschergruppe "Applied and theoretical molecular electrochemistry as a key for new technologies in the area of energy conversion and storage".

Table 1. Comparison of solid acid fuel cell anodes measured between 238 and 250°C

Electrodes	Electrode resisitivity (Ohm cm^2)	Pt loading (mg/cm^2)	Mass normalized activity (S/mg)
Pt:Pt/C:CsH$_2$PO$_4$ (3 :1 : 3 wt) – mech. mix[9,17]	0.06	7.7	2.2
Pt : CsH$_2$PO$_4$ (1 : 2 wt) – mech. mix[7]	1.7	10	0.06
Pt : CsH$_2$PO$_4$ (1 : 2 wt) – electrosprayed[10]	1.5	0.3	2.2
Pt : CsH$_2$PO$_4$: CNT – electrosprayed[11]	0.5	0.3	6.6
10 nm Pt film, spraydried CsH$_2$PO$_{4[12]}$	3.6	0.021	13.2
7.5 nm Pt film, polished CsH$_2$PO$_4$[13]	3.1 ± 0.5	0.017	19

References

[1] S. M. Haile, D. A. Boysen, C. Chisholm, and R. Merle, Nature 410, 910 (2001).

[2] C. R. I. Chisholm, D. A. Boysen, A. B. Papandrew, S. K. Zecevic, S. Cha, K. A. Sasaki, Á. Varga, K. P. Giapis, and S. M. Haile, Interface Magazine 18, 53 (2009).

[3] M. Louie, California Institute of Technology, 2011

[4] T. Uda and S. M. Haile, Electrochemical and Solid-State Letters 8 (5), A245 (2005).

[5] A. Ikeda, S. M. Haile, Solid State Ionics 213, 63 (2012)

[6] S. B. Adler, Journal of The Electrochemical Society 149 (5), E166 (2002).

[7] K. A. Sasaki, Y. Hao, and S. M. Haile, Physical chemistry chemical physics : PCCP 11 (37), 8349 (2009).

[8] K. Ota and Y. Koizumi, in Handbook of Fuel Cells - Fundamentals, Technology and Applications, edited by W. Vielstich, H. Yokokawa, and H. A. Gasteiger (John Wiley and Sons, 2009), Vol. 5, pp. 243.

[9] S. M. Haile, C. R. I. Chisholm, K. Sasaki, D. A. Boysen, and T. Uda, Faraday Discussions 134, 17 (2007).

[10] Á. Varga, N. A. Brunelli, M. W. Louie, K. P. Giapis, and S. M. Haile, Journal of Materials Chemistry 20 (30), 6309 (2010).

[11] Á. Varga, M. Pfohl, N. A. Brunelli, M. Schreier, K. P. Giapis, and S. M. Haile, Physical chemistry chemical physics : PCCP 15 (37), 15470 (2013).

[12] R. C. Suryaprakash, F. Lohmann, M. Wagner, B. Abel, and A. Varga, RSC Adv. (2014).

[13] M. W. Louie and S. M. Haile, Energy & Environmental Science 4 (10), 4230 (2011).

[14] M. W. Louie, K. Sasaki, and S. M. Haile, ECS Transactions 13 (28), 57 (2008).

[15] A. B. Papandrew, C. R. I. Chisholm, R. A. Elgammal, M. M. Özer, and S. K. Zecevic, Chemistry of Materials 23 (7), 1659 (2011).

[16] T. Uda, D. A. Boysen, C. R. I. Chisholm, and S. M. Haile, Electrochemical and Solid-State Letters 9 (6), A261 (2006).

19

Tailored nano- and micrometer sized structures of gold-nanoparticles at polymeric surfaces via photochemical and kinetic control of the synthesis and deposition process

Christian Elsner[*], **Andrea Prager, Ulrich Decker, Sergej Naumov, Bernd Abel**

Leibniz Institute of Surface Modification, Chemical Department, Permoser Strasse 15, D-04318 Leipzig, Germany

Email address:

christian.elsner@iom-leipzig.de (C. Elsner)

Abstract: The goal of the present work is to elucidate complex nano- and micrometer surface modification of soft materials via photochemical and kinetic control of the synthesis and deposition process of gold-nanoparticles. The key to this technology is the synthesis of gold-nanoparticles from different $HAuCl_4$ precursor solutions with photons of a defined short wavelength emitted by Xe_2^* (172 nm) and XeCl* (308 nm) vacuum UV and UV-C excimer lamps. The size and plasmonic properties of the spherical nanoparticles are tailored by the application of different irradiation conditions. Additionally, with 172 nm irradiation porous nanomembranes are generated. Furthermore, the spatial and density controlled immobilization of nanoparticles on to solid supports such as paper and PES membranes is demonstrated leading to defined 2-dimensional structures in the micrometer range. The synthesis of high gold content structures on paper substrates allows for the rapid and simple generation of conductive paths in electronic circuits. The generated micro– and nanosystems are characterized by scanning electron and light microscopy, photoelectron spectroscopy, dynamic light scattering and UV/VIS spectroscopy. In order to shed light into the kinetic mechanism quantum chemical calculations are employed that help to identify preferred reaction paths of the photo-induced reduction of Au(III) to Au(0).

Keywords: Excimer Lamps, Printed Electronics, Conductive Structures, Membranes, UV

1. Introduction

Gold-nanoparticles are among the most extensively studied nanomaterials and have been widely used in several fields.[1] Their size- and shape-dependent optical properties, especially their absorption capacities in the visible region of light based on the surface plasmon resonance (SPR) phenomena of nanoparticles makes them suitable for reporter probes in chemical or biochemical sensors.[2-5] Furthermore, gold-nanoparticles have become attractive in heterogeneous catalysis, imaging agents, hyperthermia medium and LDI mass spectrometry.[6-8] Buttom-up as well as top-down processes for the generation of gold-nanoparticles by radiation techniques, which are clean, one-step, and easy tuneable processes different from conventional chemical ones have been proposed since many years. Radiolysis and photolysis of aqueous $HAuCl_4$-solutions, for instance conducted by the use of a low pressure mercury lamp ($\lambda_{max\,em.}$ = 254 nm) or KrF excimer laser ($\lambda_{max\,em.}$ = 248nm) as well as laser ablation techniques from solid targets are the most suitable approaches.[9-14] Glow discharge techniques are proposed for the generation of highly dispersive and catalytically active gold-nanoparticles in a short timescale and a low technical complexity.[15]

Several techniques exist for the immobilisation of gold-nanoparticles on to solid supports, mainly co-precipitation and impregnation.[16] In the first case the support and the gold-precursor are formed simultaneously, in the latter one the pores of the support are filled with the precursor solution. Thermal treatments, often carried out in air, convert these precursors to elemental, immobilised particles. Consequently, these conventional methods are only suitable for the modification of thermally stable materials, e.g. oxides, and spatial resolved modification as well as integration into a continuous production process e.g. for the manufacturing of nanoparticle modified web fabrics have serious problems.

In the present study we describe the use of VUV/UVC-excimer lamps for the synthesis of gold-nanoparticles. UV-light provided by the decay of excited dimers is characterised by a single dominant emission band and the absence of any thermal emissions. Depending on the wavelength of the emitted photons strong effects on the treated material can be anticipated, which is limited to a certain penetration depth, typically of a few hundred nanometers into the treated material. Chemical bond dissociation, ionisation and radical formation besides of excitation are the most prominent results of high energy photon interaction with molecules. The short-wavelength photons are especially suited for large area modification of temperature sensitive polymeric materials under normal ambient conditions and have been applied to low-temperature oxidation, photoetching, photodegradation and microstructuring of polymeric surfaces.[17] Moreover, excimer lamps were successfully employed to photodeposition and photoinitiator-free photo-induced free radical polymerisation as well as photoconversion reactions in thin coatings on polymeric substrates.[18-20]

Herein, aqueous solutions of $HAuCl_4$ were exposed to UV-light from Xe_2^*- ($\lambda_{max\ em.}$ = 172nm) and $XeCl^*$- ($\lambda_{max\ em.}$ = 308nm) excimer lamp sources. The obtained gold-nanomaterials were characterized according their size and shape using scanning electron microscopy (SEM) and dynamic light scattering (DLS). Their optical properties were elucidated by UV/VIS spectroscopy. In terms of a technical application the feasibility of the large scale radiation induced synthesis and immobilisation of gold-nanoparticles on web-fabrics was exemplarily demonstrated by the use of paper and membrane flexible substrates. Photoelectron spectroscopy was used to determine the amount of immobilzed particles.

2. Results and Discussion

2.1. Synthesis of Gold-Nanoparticles

Kurihara et al. have studied the process of gold-nanoparticle formation in water and water-in-oil emulsions by pulse radiolysis and laser photolysis and proposed a reduction schema for both systems.[25] The radiolytic generation of elemental gold proceeds via the reduction of Au(III) to Au(II) by solvated electrons (eq. 1'), the disproportionation of Au(II) to Au(III) and Au(I) (eq. 2'), and further reduction of Au(I) to elemental Au (eq. 3'). Recently, a revised multistep reduction mechanism has been proposed by the investigation of gold cluster formation in the presence of iso-propyl radicals which are known as scavengers of oxidising OH-radicals.[26] There, the initial step of reduction of Au(III) is achieved by alcohol radicals (eq. 1) and to a smaller extend by solvated electrons (eq. 1'). The formed Au(II) species disproportionate via the dissociation of a long lived dimer (eq. 2) rather than direct disproportionation (eq. 2'). Moreover, gold cluster formation did not proceed before

complete consumption of Au(III). Thus, it is supposed that Au(0) is directly involved in the reduction process of Au(III) after its formation via comproportionation (eq. 4). Consequently, the synthesis of gold-nanoparticles should be inhibited at a constant dose if the amount of Au(III) species is enhanced and exceeds a critical concentration. Based on the investigations of Kurihara et al., the photolytic generation is a multi-photon event and proceeds via the formation of a caged divalent gold complex after excitation, its dissociation and disproportionation followed by a further reduction of the formed monocation to elemental gold.

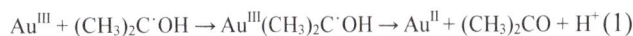

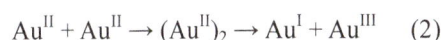

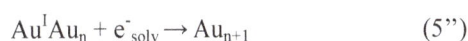

$$Au^{III} + (CH_3)_2C^{\cdot}OH \rightarrow Au^{III}(CH_3)_2C^{\cdot}OH \rightarrow Au^{II} + (CH_3)_2CO + H^+ (1)$$

$$Au^{III} + e^-_{solv} \rightarrow Au^{II} \qquad (1')$$

$$Au^{II} + Au^{II} \rightarrow (Au^{II})_2 \rightarrow Au^{I} + Au^{III} \qquad (2)$$

$$Au^{II} + Au^{II} \rightarrow Au^{I} + Au^{III} \qquad (2')$$

$$Au^{I} + (CH_3)_2C^{\cdot}OH \rightarrow Au^{I}(CH_3)_2C^{\cdot}OH \rightarrow Au^{0} + (CH_3)_2CO + H^+ \quad (3)$$

$$Au^{I} + e^-_{solv} \rightarrow Au^{0} \qquad (3')$$

$$Au^{0} + Au^{III} \rightarrow Au^{I} + Au^{III} \qquad (4)$$

$$nAu^{0} \rightarrow Au_n \qquad (5)$$

$$Au^{I} + Au_n \rightarrow Au^{I}Au_n \qquad (5')$$

$$Au^{I}Au_n + e^-_{solv} \rightarrow Au_{n+1} \qquad (5'')$$

$$Au^{I}Au_n + (CH_3)_2C^{\cdot}OH \rightarrow Au_{n+1} + (CH_3)_2CO + H \quad (5''')$$

$$(CH_3)_2CHOH + OH^{\cdot} \rightarrow (CH_3)_2C^{\cdot}OH + H_2O \qquad (6)$$

Dependingon the wavelength of the emitted photons from the excimer lamp sources the cleavage of chemical bonds, radical formation and initiation of radiolytic pathways is an option, especially if alcohols are constituents of the reaction mixture. Thus, we have investigated the formation of gold-nanoparticles using irradiation from a Xe_2^* excimer lamp in the presence and absence of t-butanol, 2-propanol, ethanol, methanol, and trifluoroethanol under comparable conditions (Figure 1).

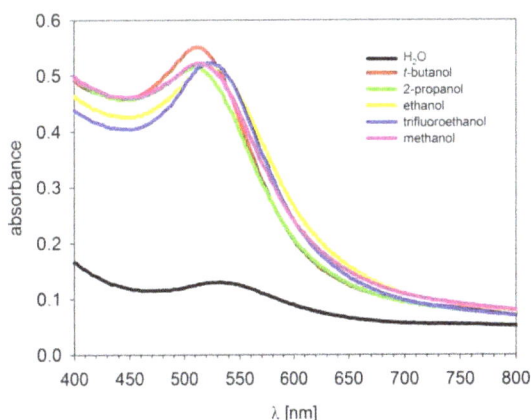

Figure 1. UV/VIS-absorption spectra of gold-nanoparticles obtained by Xe_2^* excimer irradiation of aqueous $HAuCl_4$ solutions in the presence of different alcohols. Condition I, lamp power: 100 %, irradiation time: 2 min.

The absorption spectra of the formed nanoparticles present bands with λ_{max} ranging from 512 nm (t-butanol) to 532 nm (H$_2$O). The differences in the absorbance intensity and the position of λ_{max} of the SPR-band can be attributed to differences in the amount and the size of the gold-nanoparticles, respectively.[27] Obviously, the addition of the alcohols promotes the nanoparticle formation. It was already stated that the addition of 2-propanol (but also of ethanol and methanol) has a positive influence on the radiolytic synthesis of gold-nanoparticles according eq. 6. However, we also observed a promotion of nanoparticle formation in the presence of t-butanol. The corresponding t-butyl radical has no reducing properties. Moreover, the application of photons from XeCl* excimer lamps which are not able to cleave chemical bonds above 4.2 eV bond dissociation energy (BDE) show a similar behaviour. Thus, the formation of reducing species and the initiation of a radiolytic pathway of gold-nanoparticle formation is presumably not the dominant process. We assume that the used alcohols have a positive influence on the reduction of Au(III) to Au(0) by the promotion of the electron transfer at different stages of the reduction process. Quantum chemical calculation show, that excited solvent molecules may act as electron donors and acceptors and may facilitate the reduction process (Figure 2). Briefly, the Au(III)Cl$_4^-$ anion is formed through the dissociation of the HAu(III)Cl$_4$ molecule (reaction (1)), where the H$_2$O molecule could act as a proton acceptor. The next step (reaction (2)) may be a direct formation of Au(III)Cl$_3$ through the abstraction of Cl$^-$. However, this reaction should be energetically unfavorable because of the high Gibbs free energy of the reaction (ΔG = 33.1 kcal mol^{-1}). An alternative way may be the formation of Au(III)Cl$_4$ through a possible electron transfer to the electronically excited solute molecule (CF$_3$CF$_2$OH or CHMe$_2$OH) as an electron acceptor, followed by the abstraction of the Cl radical. This reaction could proceed with lower ΔG (ΔG=16 kcal mol^{-1}). The next step (reaction (3)), namely the formation of the Au(II)Cl$_2$ molecule, could proceed either by the disproportionation reaction with the Au(III)Cl$_3$ molecule or by the absorption of the solvated electron (e$^-_{solv}$) by Au(III)Cl$_3$ followed by the abstraction of the Cl$^-$. However, the disproportionation reaction is energetically more favorable with ΔG = 15 kcal mol^{-1}. The next step (reaction (4)), namely the formation of the Au(I)Cl molecule, can proceed endergonic both through the disproportionation reaction with the Au(II)Cl$_2$ molecule and through the absorption of the solvated electron (e$^-_{solv}$) by Au(II)Cl$_2$, followed by the abstraction of the Cl$^-$. Both of these reactions are exergonic (ΔG = -9 and ΔG = -141 kcal mol^{-1}, respectively) and may possibly proceed spontaneously. The high reactivity of Au(II)Cl$_2$ molecule towards dissociation may be explained by the fact, that Au(II)Cl$_2$ molecule is actually a radical with unpaired number of electrons. It is also in agreement with the experimental finding, that the chemistry of gold is dominated by the oxidation states Au(I) and Au(III). The last step (reaction (5)), the formation of Au(0), could proceed in different ways as shown in

Figure 2. However, only two ways, namely: 1) the reaction of the solvated electron (e$^-_{solv}$) with Au(I)Cl followed by the

abstraction of the Cl$^-$ and 2) the electron transfer from excited solvent molecules to Au(I)Cl followed by the abstraction of the Cl$^-$, may be energetically probable. Additionally, the alternative pathway of formation of Au(0) through disproportionation (reaction (6), Figure 2) as proposed in the literature [28] seems to be energetically improbable because of the very high positive Gibbs free energy of the reaction. Thus, it should be mention, that excited solvent molecules may act as electron donors (e-transfer(I)) or acceptors (e-transfer(II)) and may facilitate the reduction process.

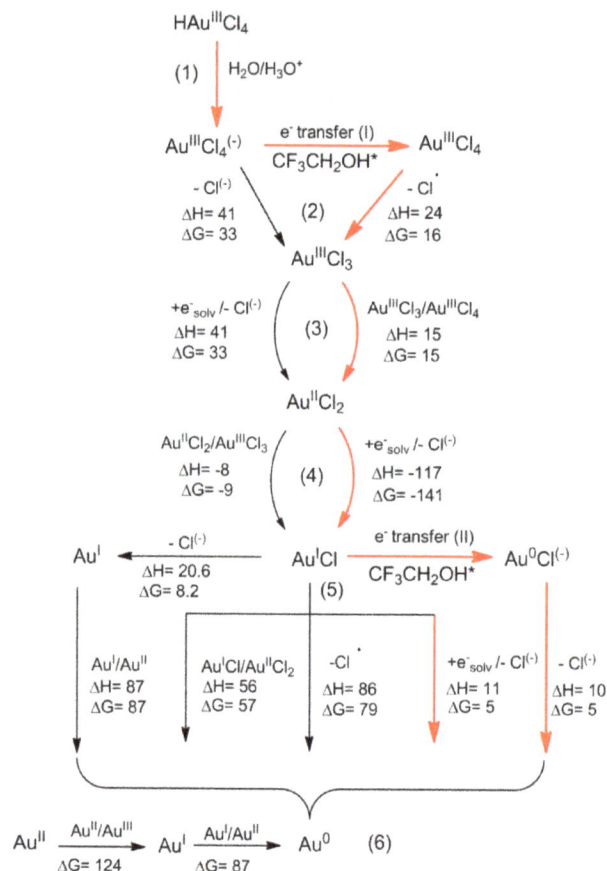

Figure 2. *Transformation scheme for the possible reaction pathways for the Au(0) formation starting from HAu(III)Cl$_4$. The most probable reaction pathway is marked in red. Here, ΔH and ΔG are the reaction enthalpy and Gibbs free energy of the reaction (kcal mol^{-1}) as calculated in water at M06-D/LACV3P+** */PBF level of theory. For further details see the text.*

For further investigations on the photo-induced synthesis of gold-nanoparticles aqueous HAuCl$_4$ solutions in the presence of 2-propanol were used. Depending on the applied dose and wavelength of the emitted VUV-photons different absorbance spectra were obtained for the gold-nanoparticles. A dose influence on λ_{max} of the SPR absorption was observed for the XeCl*-system, whereas for the Xe$_2^*$-system only a reduction of the absorbance was found (Figure 3). In the case of the XeCl*-system the λ_{max} of the SPR-bands were in the range of 530-580 nm. They were shifted towards longer wavelengths by the reduction of the lamp power (dose). For the Xe$_2^*$-system λ_{max} of the SPR absorption was around 526 nm independently of the applied dose. The red-shift of λ_{max} in the case of the XeCl*-excimer lamp irradiation suggests the formation of larger gold-nanoparticles. Indeed, DLS measurements revealed an increase of the particle sizes from 17 nm to 148 nm. The particle sizes after Xe_2*-excimer lamp irradiation were in the range of 12 nm to 28 nm. Figure 4 shows a plot of the particle sizes versus λ_{max} of the corresponding SPR-absorption in comparison to values obtained from the literature.

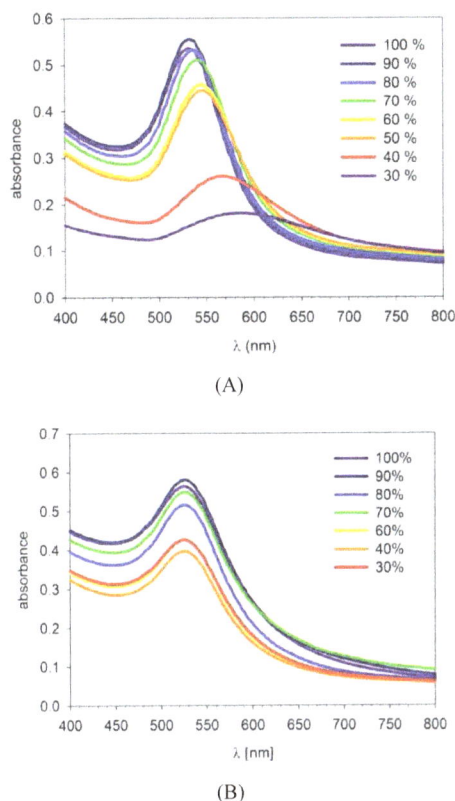

(A)

(B)

Figure 3. *UV/VIS-absorption spectra of gold-nanoparticles obtained by XeCl* *(A) and Xe₂* *(B) irradiation of aqueous HAuCl₄-solutions at different doses (lamp power). Condition II, lamp power: 100 % - 30 %, irradiation time: 2 min.*

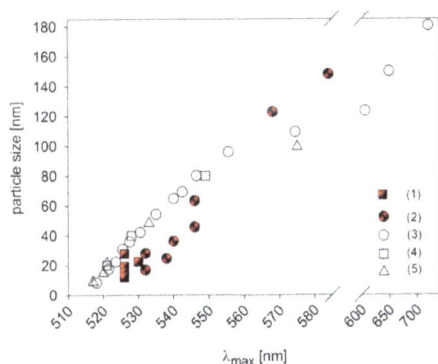

Figure 4. *Comparison of plots of gold-nanoparticle sizes vs. SPR-λ_max-values based on data obtained from the literature (3-5, 3: Bastus et al. [29], 4: Jain et al. [30], 5: Link et al. [2]) and synthesized herein by the irradiation of aqueous HAuCl₄-solutions using Xe₂* (1) and XeCl*- (2) excimer lamps. For conditions refer to Figure 3.*

They are in a good agreement; small deviations may be attributed to different crystallinity and shape of the nanoparticles based on the variable preparation methods. The differences of particle sizes obtained during the excimer-lamp irradiation by the Xe_2*- and XeCl*-system may be explained by the differences in the penetration depths of the photons into the reaction media and the availability of the generated nuclei/seeds and further elemental gold material for coalescence processes in this specific area. The formation of nanoparticles involves competing nucleation processes which proceed above the saturation concentration of elemental gold, and growth processes. The proportion of both determines the final size and shape. Accordingly, if the proportion of seeds to gold atoms suited for growth processes $Au_n/(Au(0)+Au(I))$ becomes smaller, larger particles were obtained. In general, this was observed in the case of XeCl*-excimer lamp irradiation. Comparatively, due to the short wavelength of Xe_2* derived photons their penetration depth into the reaction liquid is smaller and in general in the range of a few hundred nanometers. Thus, initial processes of gold reduction and particle formation proceed mainly nearby the liquid-gas interface. As a consequence of the absorption of the short wavelength photons in a limited solvent volume the saturation concentration of elemental gold is permanently exceeded and nuclei are formed which preferable take part in coalescence processes if they are impeded in that "Hot zone". Diffusion outside of the "hot zone" reduces the chance for further reduction and coalescence processes dramatically. Consequently, the particle sizes and SPR absorption are in a limited range.

2.2. Synthesis of Porous Gold-Nanomembranes

Increasing the amount of HAuCl₄ in the precursor solution, the irradiation with Xe_2*-excimer lamp derived photons results in an additional formation of a porous nanomembrane due to the preferred particle formation, coalescence and even conversion processes at the interfacial region. The membrane, which appears as a continuous metallic-golden film by the naked eye swimming on top of the reaction liquid, has a thickness of 70 nm (Figure 5). The pores in the range of a few micrometers are not homogeneously distributed over the complete surface area. Especially the border region is characterized by fractal structures and larger pores compared to the more dense structures in the centered regions of the membrane. The grainy, fractal structures suggest a multi-step growth process starting from inter-particle aggregation, domain formation, domain interconnection and the adsorption of further particles, clusters or atoms. To our knowledge, the formation of a metallic, a few tenth nanometer thick membrane by photon induced processes has never been described in the literature. This pronounces the efficiency of excimer lamps for transformation processes which are especially limited to substrate surfaces.

2.3. Immobilisation of Gold-Nanoparticles

Excimer lamps can be used to effectively treat large substrate areas under normal ambient conditions without any high demands on protective or shielding barriers. This is a clear advantage over other radiation based approaches, e.g. laser and low pressure plasma treatment or even electron beam or gamma irradiation. Since reactive species are only generated in a very thin surface area around 100 nm in depth bulk properties of the substrate will never been altered. As shown herein, the synthesis of gold-nanoparticles by high energy photons from excimer lamp sources proceeds in a short time frame. Thus, the approach seems to be suitable for the synthesis and immobilization of gold-nanoparticles on temperature sensitive web fabrics under roll-to-roll conditions. For this purpose we have used paper and membrane substrates because of their unique open porous structure which facilitates wetting of the precursor solution. Both substrates are considered as interesting and promising materials for the development of disposable diagnostic devices and they became more and more attractive in point-

of-care testing.[31, 32] Briefly, a paper and a membrane substrate were immersed with a precursor solution and the wet specimens were passed through the irradiation zone of a series of Xe$_2$*-excimer lamps using a conveyor system. The absence of thermal emissions prevents the evaporation of volatile liquids such as alcohols, which promote the generation of gold-nanoparticles. A pink colour after irradiation indicated the formation and immobilisation of gold-nanoparticles on the substrate even under non-stationary conditions. A detailed assessment by SEM confirmed the synthesis of gold-nanoparticles on paper and membranes (Figure 6). Although just not studied in detail, the kind of the substrate may play a crucial role in the formation process. Thus, we have observed spherical gold-nanoparticles with a diameter of 20 nm on paper and extremely tiny particles of a few nanometers on the membrane.

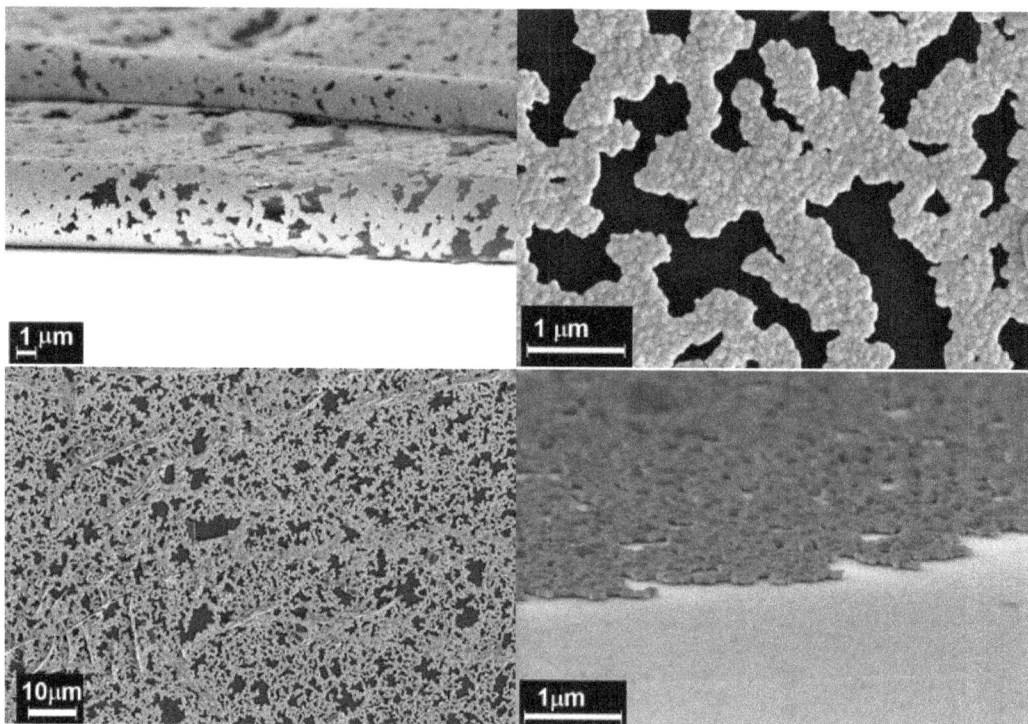

Figure 5. *SEM of porous gold-nanomembrane obtained by the Xe$_2$* excimer irradiation of aqueous HAuCl$_4$ solutions. The highly flexible and bendable membrane appears as an interconnected network of sintered nanoparticles. Condition III, lamp power: 100 %, irradiation time: 5 min.*

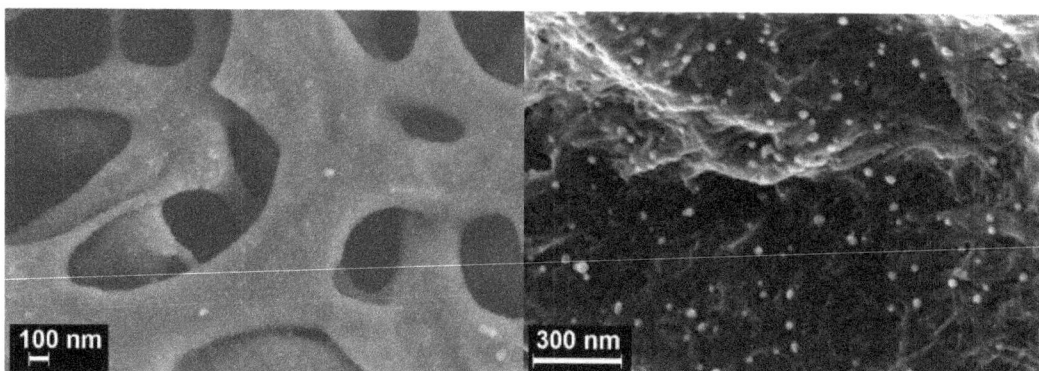

Figure 6. SEM of immobilized gold-nanoparticles on PES-membrane (left) and paper (right) substrates. Condition II.

As a continuation and extension of the experiments we were interested in the spatial resolved immobilisation of

nanoparticles on flexible substrates. Photopatterning of structures on the micrometer scale involving nanoparticles has potential for application to electronic, sensoric and MEMS devices. However, only a few methods for micropatterning of surfaces with nanoparticles exist, mostly based on laser writing.[33] Recently, it has been shown, that the chemistry of the surface can be altered for further structuring and deposition processes by the spatial resolved treatment with short wavelength photons using a photo mask.[34] Herein, we have employed irradiations through a polyimide photo mask for the rapid generation, immobilisation, and arrangement of gold-nanoparticles on paper substrates. The general principle and examples of generated micropatterns on paper substrates are shown in Figure 7.

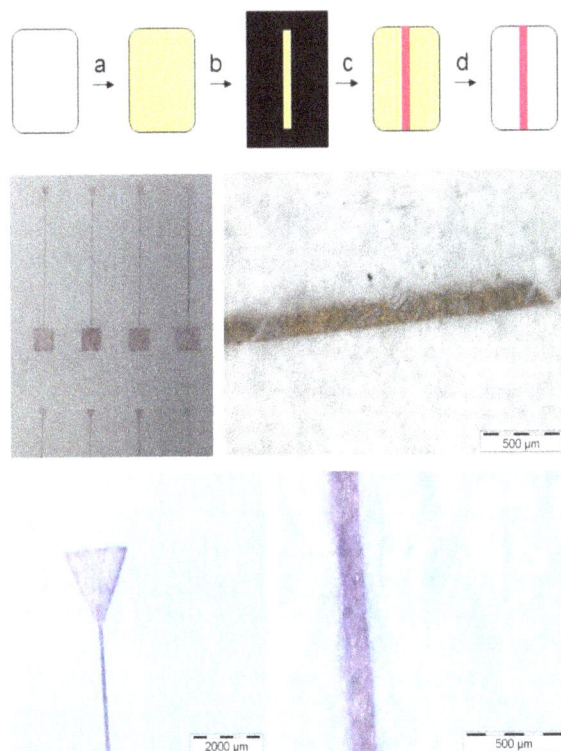

Figure 7. Schematic of the spatially resolved functionalization of flexible substrates with gold-nanoparticles and examples of 2-dimensional patterns on paper substrates. (a) Immersion of a absorptive flexible substrate with aqueous HAuCl₄-solution, (b) application of a photo mask and irradiation, (c) removal of the photo mask, (d) and washing of the substrate with water.

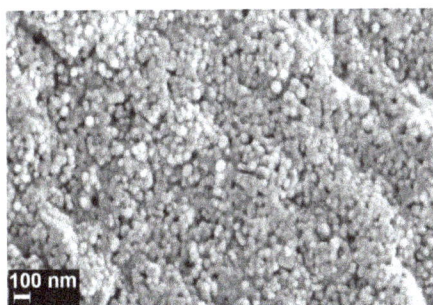

Figure 8. SEM of a dense gold-nanoparticle film on a paper substrate. The interconnected, presumably sintered aggregates are conductive. The chemical composition [atom-%] of the film was determined by XPS as follows: Au: 32, C: 48, O: 20. Condition IV, lamp power: 100 %, irradiation time: 2 min; two time application.

Thereby, the density of the formed nanoparticles can be altered from single isolated particles (Figure 7) up to the formation of interconnected grainy nanostructures (Figure 8) appearing as golden-coloured films by the alteration of the concentration of the $HAuCl_4$ in the precursor solution as previously described in the nanomembrane-section.

A detailed investigation was carried out by the determination of the chemical surface composition using photoelectron spectroscopy (XPS). The amount of gold on the surface can be adjusted by the $HAuCl_4$ concentration of the precursor solution as shown in Figure 9.

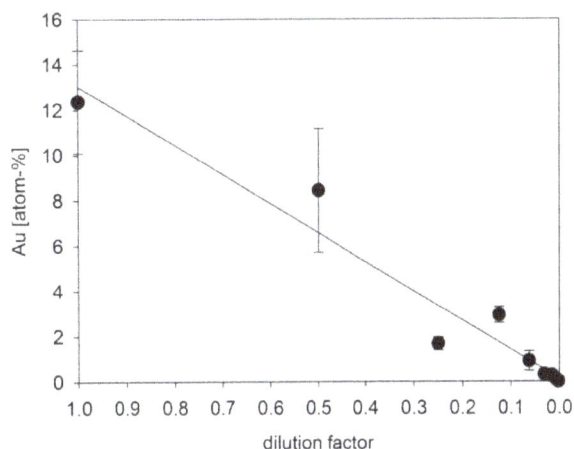

Figure 9. Atomic concentration of gold on the surface of a paper substrate as a function of the concentration of HAuCl₄ in the precursor solution. Condition 5, lamp power: 100 %, irradiation time: 2 min.

3. Conclusion

In conclusion we have demonstrated the rapid and controlled synthesis of gold-nanoparticles by VUV/UVC-photons of 172 nm and 308 nm wavelength generated by Xe_2*- and XeCl-excimer lamp systems and have applied the approach polymeric substrate surfaces. for the modification of sensitive polymeric substrates. The technique allows for an enhanced control of the synthesi process with respect to time, localization, and density of the formed nanoparticles. The kinetic mechanism for the reduction of the gold precursor has been understood with the help of quantum chemical calculations that enabled us to identify reaction paths of the photo-induced reduction of Au(III) to Au(0). We have developed asuited technology to easily generate complex nano- and micrometer-dimensional structures on soft polymeric surfaces via photochemical and kinetic control of the synthesis and deposition of gold-nanoparticles. Moreover, the immobilization of gold-nanoparticles on suited flexible substrates is not only limited to the lab scale and can in principle be up-scaled to a continuous roll-to-roll process. This may be an important feature for the application of metallic-nanoparticle based systems in the coating or printing industry and allows for alternative approaches of the design of new devices based on the unique properties of nanomaterials.

4. Experimental Section

4.1. Materials

All chemicals and solvents were obtained from commercial sources and used without further purification: 2-propanol (HiPerSolv, HPLC grade), VWR international; methanol, ethanol (Rotisolv, HPLC grad. grade), Carl-Roth GmbH, Karlsruhe, Germany; 2,2,2-trifluoroethanol (synthesis grade), MERCK KGaA, Hohenbrunn, Germany; $HAuCl_4$ (~ 30 wt.% in dilute HCl), Sigma-Aldrich, Germany. For all experiments Millipore-grade water was used. $HAuCl_4$-precursor solutions were prepared by mixing of different amounts (v/v) of water, alcohol, and $HAuCl_4$ (~ 30 wt.% in dilute HCl) as follows: Condition I: $HAuCl_4$-stock solution: 7.2 µl $HAuCl_4$ (~ 30 wt.% in dilute HCl) in 5 ml H_2O; working mixture: 275 µl $HAuCl_4$-stock solution, 250 µl H_2O, 25 µl alcohol, condition II: working mixture: 12.5 ml H_2O, 1.25 ml 2-propanol, 9 µl $HAuCl_4$ (~ 30 wt.% in dilute HCl), condition III: working mixture: 982 µl H_2O, 15.6 µl 2-propanol, 2.8 µl $HAuCl_4$ (~ 30 wt.% in dilute HCl), condition IV: working mixture: 98.2 µl H_2O, 1.56 µl 2-propanol, 2.8 µl $HAuCl_4$ (~ 30 wt.% in dilute HCl), condition V: A serial dilution (1:1) was prepared starting from a mixture of 400 µl H_2O, 20 µl 2-propanol, 100 µl $HAuCl_4$ (~ 30 wt.% in dilute HCl) by the addition of the dilution mixture consisting of 4 ml H_2O, 200 µl 2-propanol.

4.2. Excimer Lamp Set Up

Stationary approaches: Open $Xe_2{}^*$-excimer irradiation system (λ_{max} = 172 nm, lamp 172/630) and $XeCl^*$-excimer irradiation system (λ_{max} = 308 nm) from Heraeus Noblelight, Kleinostheim, Germany were used. All excimer irradiation experiments were carried out in a nitrogen atmosphere at room temperature and normal pressure. The dose was determined using a Flatlog-system V. 1.2 (Jenoptik Polymer Systems GmbH, Berlin, Germany) with calibrated measuring diodes (GaP: 172 nm, SiC: 308 nm, Figure 10). For the irradiation experiments 96-well microtiter plates (NunclonTM Surface, Nunc, Denmark) with a volume of 0.2 ml/well $HAuCl_4$-precursor solution were used. For non-stationary approaches $HAuCl_4$-precursor solution immersed substrates (paper, membrane) were passed through the irradiation zone of a series of 172 nm excimer lamps using a conveyor system at 2 m min^{-1} applying a dose of 1.7 J/cm^2.

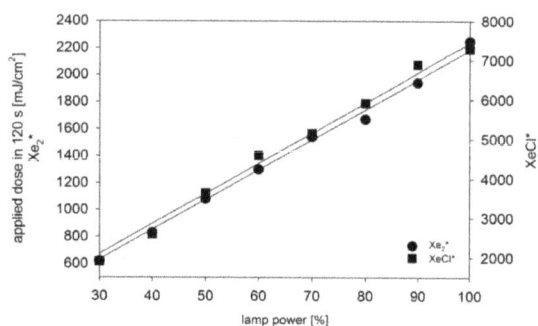

Figure 10. *Applied dose of Xe$_2$* and XeCl*-excimer lamps over a period of 120 s at an adjusted lamp power.*

4.3. Instruments

Absorbance spectra in the range of 400 – 800 nm were recorded on a TECAN infinite M 200 microtiter-plate reader using flat bottom micro titer plates and a volume of 0.1 ml/well (Nunclon™ Surface, Nunc, Denmark). DLS measurements were performed on a MALVERN Nanoseries Nano ZS zetasizer. Scanning electron micrographs were generated on an ULTRA 55 Carl Zeiss SMT.

4.4. Computational Methods

Density Functional Theory (DFT) calculations were carried out using the M06-D3 density functional. MO6 functional is parameterized for organometallic and noncovalent interactions (Zhao and Truhlar, 2008).[21] M06-D3 functional includes physically and chemically very important London dispersion interactions (Grimme et al., 2010).[22] The molecular geometries and energies of the all calculated molecules were calculated at the M06-D3/LACV3P+** level of theory as implemented in Jaguar 8.1 program (Jaguar 8.1, 2013). The LACV3P+** basis set uses the standard 6-311+G** basis set for light elements and the LAC pseudopotential (Wadt and Hay, 1985) for the heavier elements, such as Au in this case.[23] Frequency calculations were done at the same level of theory to characterize the stationary points on the potential surface and to obtain total enthalpy (H) and Gibbs free energy (G) at a standard temperature of 298.15 K using un-scaled vibrations. The reaction enthalpies (ΔH) and Gibbs free energies of reaction (ΔG) were calculated as the difference of the total enthalpy H and the Gibbs free energies G between the reactants and products, respectively. To take solvent effect on the structure and reaction parameters of studied molecules into account the calculation were done using Jaguar's dielectric continuum Poisson-Boltzmann solver, which fits the field produced by the solvent dielectric continuum to another set of point charges (Tannor et al.).[24]

Acknowledgements

The work was supported by the Federal Government of Germany and the Freistaat Sachsen.

References

[1] Daniel, M. C.; Astruc, D., *Chem Rev* 2004, *104* (1), 293-346. DOI 10.1021/cr030698+.

[2] Link, S.; El-Sayed, M. A., *Journal of Physical Chemistry B* 1999, *103* (21), 4212-4217. DOI 10.1021/jp984796o.

[3] Link, S.; El-Sayed, M. A., *Int. Rev. Phys. Chem.* 2000, *19* (3), 409-453. DOI 10.1080/01442350050034180.

[4] Watanabe, S.; Yamamoto, S.; Yoshida, K.; Shinkawa, K.; Kumagawa, D.; Seguchi, H., *Supramolecular Chemistry* 2011, *23* (3-4), 297-303. DOI Pii 936375251Doi 10.1080/10610278.2010.527977.

[5] Stewart, M. E.; Anderton, C. R.; Thompson, L. B.; Maria, J.; Gray, S. K.; Rogers, J. A.; Nuzzo, R. G., *Chemical Reviews* 2008, *108* (2), 494-521. DOI 10.1021/cr068126n.

[6] Boronat, M.; Corma, A., *Journal of Catalysis* 2011, *284* (2), 138-147. DOI DOI 10.1016/j.jcat.2011.09.010.

[7] Della Pina, C.; Falletta, E.; Rossi, M., *Chem Soc Rev* 2012, *41* (1), 350-69. DOI 10.1039/c1cs15089h.

[8] Haruta, M.; Date, M., *Applied Catalysis a-General* 2001, *222* (1-2), 427-437.

[9] Belloni, J.; Mostafavi, M.; Remita, H.; Marignier, J. L.; Delcourt, M. O., *New Journal of Chemistry* 1998, *22* (11), 1239-1255. DOI 10.1039/a801445k.

[10] Henglein, A., *Langmuir* 1999, *15* (20), 6738-6744. DOI 10.1021/la9901579.

[11] Sau, T. K.; Pal, A.; Jana, N. R.; Wang, Z. L.; Pal, T., *Journal of Nanoparticle Research* 2001, *3* (4), 257-261. DOI 10.1023/a:1017567225071.

[12] Esumi, K.; Matsuhisa, K.; Torigoe, K., *Langmuir* 1995, *11* (9), 3285-3287. DOI 10.1021/la00009a002.

[13] Watanabe, M.; Takamura, H.; Sugai, H., *Nanoscale Res. Lett.* 2009, *4* (6), 565-573. DOI 10.1007/s11671-009-9281-2.

[14] Biswal, J.; Ramnani, S. P.; Shirolikar, S.; Sabharwal, S., *Radiat. Phys. Chem.* 2011, *80* (1), 44-49. DOI 10.1016/j.radphyschem.2010.08.016.

[15] Liang, X.; Wang, Z. J.; Liu, C. J., *Nanoscale Res. Lett.* 2010, *5* (1), 124-129. DOI 10.1007/s11671-009-9453-0.

[16] Bond, G. C.; Louis, C.; Thompson, D. T., Preparation of Supported Gold Catalysts. In *Catalysis by Gold*, Hutchings, G. J., Ed. Imperial College Press: 2006; Vol. 6, pp 72-120.

[17] Kogelschatz, U., *Appl. Surf. Sci.* 1992, *54* (C), 410-423.

[18] Scherzer, T.; Knolle, W.; Naumov, S.; Mehnert, R., *Nucl. Instr. Meth. Phys. Res. B.* 2003, *208* (1-4), 271-276.

[19] Elsner, C.; Lenk, M.; Prager, L.; Mehnert, R., *Appl. Surf. Sci.* 2006, *252* (10), 3616-3624.

[20] Prager, L.; Dierdorf, A.; Liebe, H.; Naumov, S.; Stojanovic, S.; Heller, R.; Wennrich, L.; Buchmeiser, M. R., *Chem. Eur. J.* 2007, *13* (30), 8522-8529.

[21] Zhao, Y.; Truhlar, D. G., *Theoretical Chemistry Accounts* 2008, *120* (1-3), 215-241. DOI 10.1007/s00214-007-0310-x.

[22] Grimme, S.; Antony, J.; Ehrlich, S.; Krieg, H., *Journal of Chemical Physics* 2010, *132* (15). DOI 10.1063/1.3382344.

[23] Wadt, W. R.; Hay, P. J., *Journal of Chemical Physics* 1985, *82* (1), 284-298. DOI 10.1063/1.448800.

[24] Tannor, D. J.; Marten, B.; Murphy, R.; Friesner, R. A.; Sitkoff, D.; Nicholls, A.; Ringnalda, M.; Goddard, W. A.; Honig, B., *Journal of the American Chemical Society* 1994, *116* (26), 11875-11882. DOI 10.1021/ja00105a030.

[25] Kurihara, K.; Kizling, J.; Stenius, P.; Fendler, J. H., *Journal of the American Chemical Society* 1983, *105* (9), 2574-2579. DOI 10.1021/ja00347a011.

[26] Dey, G. R.; El Omar, A. K.; Jacob, J. A.; Mostafavi, M.; Belloni, J., *J. Phys. Chem. A* 2011, *115* (4), 383-391. DOI 10.1021/jp1096597.

[27] Huang, W. C.; Chen, Y. C., *Journal of Nanoparticle Research* 2008, *10* (4), 697-702. DOI DOI 10.1007/s11051-007-9293-8.

[28] Shang, Y. Z.; Min, C. Z.; Hu, J.; Wang, T. M.; Liu, H. L.; Hu, Y., *Solid State Sci.* 2013, *15*, 17-23. DOI 10.1016/j.solidstatesciences.2012.09.002.

[29] Bastus, N. G.; Comenge, J.; Puntes, V., *Langmuir* 2011, *27* (17), 11098-11105. DOI 10.1021/la201938u.

[30] Jain, P. K.; Lee, K. S.; El-Sayed, I. H.; El-Sayed, M. A., *Journal of Physical Chemistry B* 2006, *110* (14), 7238-7248. DOI 10.1021/jp057170o.

[31] Rozand, C., *Eur. J. Clin. Microbiol. Infect. Dis.* 2014, *33* (2), 147-156. DOI 10.1007/s10096-013-1945-2.

[32] Veigas, B.; Jacob, J. M.; Costa, M. N.; Santos, D. S.; Viveiros, M.; Inacio, J.; Martins, R.; Barquinha, P.; Fortunato, E.; Baptista, P. V., *Lab on a Chip* 2012, *12* (22), 4802-4808. DOI 10.1039/c2lc40739f.

[33] Spano, F.; Castellano, A.; Massaro, A.; Fragouli, D.; Cingolani, R.; Athanassiou, A., *Journal of Nanoscience and Nanotechnology* 2012, *12* (6), 4820-4824. DOI 10.1166/jnn.2012.4931.

[34] Elsner, C.; Naumov, S.; Zajadacz, J.; Buchmeiser, M. R., *Thin Solid Films* 2009, *517* (24), 6772-6776. DOI 10.1016/j.tsf.2009.05.041.

20

High resolution imaging of a multi-walled carbon nanotube with energy-filtered photoemission electron microscopy

Andreas Neff[1], Olga Naumov[1], Timna-Josua Kühn[2], Nils Weber[2], Michael Merkel[2], Bernd Abel[1], Aron Varga[1], Katrin R. Siefermann[1, *]

[1]Leibniz Institute of Surface Modification (IOM), Chemical Department, Permoser Strasse 15, 04318 Leipzig, Germany
[2]FOCUS GmbH, Neukirchner Strasse 2, 65510 Hünstetten, Germany

Email address:

katrin.siefermann@iom-leipzig.de (K. R. Siefermann)

Abstract: Photoemission electron microscopy (PEEM) is a powerful and well established tool in surface science. In recent years, PEEM has been increasingly applied to new terrain, such as imaging of complex nano-objects and functional molecular materials, as well as time-resolved experiments. When applying PEEM to such new terrain, information on the mechanisms causing contrast in the PEEM image is particularly valuable. Here, we present a PEEM study on a complex nano-object – an individual multi-walled carbon nanotube (CNT) – to shed light on the origin of PEEM contrast. The presented PEEM images of the nanotube are of unsurpassed resolution and feature intensity variations along the nanotube. Complementary scanning electron microscopy (SEM) and atomic force microscopy (AFM) measurements on the same nanotube reveal topography as the dominant cause for the contrast observed along the nanotube. Energy-filtered PEEM measurements demonstrate that the contrast between nanotube and substrate mainly originates from their different electronic structures. The measurements further demonstrate that energy-filtered PEEM has the potential to image electronic structure variations of complex nano-objects and materials on nanometer length scales.

Keywords: Photoemission Electron Microscopy, PEEM, Carbon Nanotube, CNT, High Resolution Imaging, Contrast Mechanisms

1. Introduction

Photoemission electron microscopy (PEEM) is a widely used type of emission microscopy. It images electrons emitted from a sample upon irradiation with light, typically UV-light, X-rays or lasers. PEEM has proven a powerful tool for the characterization of nanometer-sized materials and the high spatial resolution has led to considerable advances in this field [1-6]. With aberration correction, a spatial resolution of few nanometers may be achieved [7]. Using femtosecond laser pulses as excitation source, it is possible to investigate ultrafast phenomena on a nanometer scale, such as surface plasmon polaritons [8-10] and carrier-dynamics in solids [11]. Even the combination of PEEM with attosecond pulse trains has been realized [12].

Contrast in the PEEM image is the result of an interplay of work function differences, variation in the density of states, photoionization yields, and topography [13]. An additional source of contrast are local distortions of the electric field on the sample, as the sample itself is part of the electron optical system [14-17]. When PEEM is applied to image complex nano-objects on surfaces, the interplay of these various contrast mechanisms complicates interpretation of resulting images. Here, we present a comprehensive study on an individual multi-walled carbon nanotube (CNT) to reveal major contrast mechanisms.

To date, only few PEEM-studies have been performed on complex nano-objects on surfaces – such as carbon nanotubes (CNTs) [18–21] – despite increasing scientific interest in respective materials and their potential for industrial applications [22–24]. Suzuki et al. have investigated the work function difference of individual single-walled carbon nanotubes on a Si line pattern using X-rays as the excitation source [18,19]. Sangwan has employed a UV-PEEM as a tool for probing field effects in CNT-based transistors [20]. Recently, Bao's group has used a

UV-PEEM to study pod-like CNTs encapsulating iron nanoparticles [21]. UPS and XPS studies of carbon nanotubes have been performed by several groups [25–29] providing insights into their valence-band structure. Besides, CNT networks on an insulating SiO_2 substrate were imaged with scanning electron microscopy (SEM) to reveal contrast mechanisms for different electron energies [30,31].

In this work, we present a UV-PEEM image of an isolated multi-walled CNT which features an unsurpassed resolution. In particular, intensity variations along the nanotube are clearly resolved. In order to interpret this observation, we performed complementary scanning electron microscopy (SEM) and atomic force microscopy (AFM) experiments on the exact same nanotube. Additionally, we measured a photoelectron spectrum of a single CNT using an imaging energy filter (IEF) installed in the PEEM. Hereby, we show the possibility of obtaining quantitative information about the electronic structure of nanometer scale objects using photoemission electron microscopy.

2. Experimental

Multi-walled CNTs were grown on carbon paper (Toray TGP-H-120) by chemical vapor deposition as described elsewhere [32], with hydrogen, acetylene, ammonia, and argon as the precursor gases. Nickel nanoparticles served as the growth catalyst. As-grown CNTs were removed from the carbon paper substrate by ultrasonication for one hour in ethanol, forming a suspension. The suspension was drop-cast on a polished SiC substrate (6H, 0001-Orientation, MaTecK GmbH, Germany), previously rinsed with distilled water, acetone, and ethanol. The solvent was allowed to evaporate at ambient conditions. The sample was stored in the PEEM vacuum chamber (p ~ 5×10^{-9} mbar) for five days prior to the PEEM measurements.

PEEM images were obtained with a UV-PEEM (FOCUS GmbH), using an Hg-lamp with a photon spectrum of 4.9-5.2 eV as excitation source. An imaging energy filter (IEF) integrated in the PEEM allowed for acquisition of energy-resolved images. The IEF is a retarding field analyzer that essentially works as a high pass filter for photoelectrons. Energy-filtered measurements were performed in single event counting mode. For all measurements, the extractor voltage was set to 15 kV at a working distance of 1.8 mm from the sample surface.

The sample was further characterized with a scanning electron microscope (Ultra 55, Carl Zeiss SMT) at accelerating voltages of 1–3 kV. The atomic force microscope (Dimension Icon AFM, Bruker) used in this study was operated in tapping mode using a commercial silicon cantilever (k = 26 N/m).

3. Results and Discussion

Fig. 1a) and b) show PEEM images of the sample at two different magnifications. Both images are background and flat field corrected. Fig. 1a) shows an image with a field of view

(FoV) of 50 μm (acquisition time = 10 s). The red arrow points towards an isolated CNT. The CNT appears substantially brighter than the underlying SiC substrate. On the left side of Fig. 1a), a large bright structure is visible, which is attributed to debris from the carbon paper used as the CNT growth substrate.

Fig. 1b) shows a PEEM image of the same CNT, as highlighted with a red arrow in Fig. 1a), but obtained with the highest magnification (FoV = 1.3 μm). This image is averaged over 50 exposures of 10 seconds each. Several kinks along the CNT are visible and well resolved in the PEEM image. The thickness of the CNT is not significantly changing over the length of the tube. A Gaussian fit of a line profile perpendicular to the nanotube (yellow line in Fig. 1b) exhibits a FWHM of 38.3 ± 0.4 nm (Fig. 1c).

As the width of the CNT is on the same order as the resolution of the PEEM, it is difficult to determine the exact spatial resolution of this image. However, it is apparent that it is well below 40 nm. For CNTs, this PEEM image is the one with the highest resolution and the most details published to date. The resolution in our measurement is comparable to previously reported resolutions of 25 nm achieved with other samples with the FOCUS IS-PEEM [33,34].

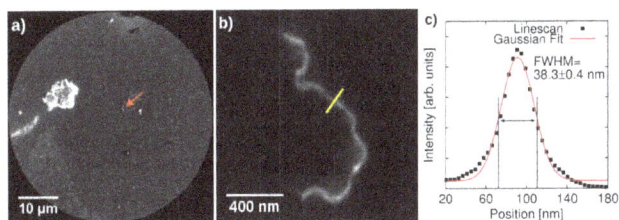

Figure 1. PEEM measurement showing a) a large field of view (50 μm) image of the sample with an isolated CNT indicated by the red arrow and b) the same CNT at highest magnification. c) Line profile of the intensity along the yellow line in b) together with a Gaussian fit.

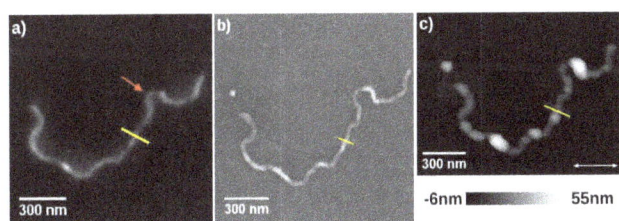

Figure 2. Comparison of a) PEEM b) SEM and c) AFM measurements of the same CNT. Images a) and b) have been rotated to allow for comparison. The width of the CNT at the indicated positions is 38 nm (PEEM) and 29 nm (SEM), respectively. The arrow in c) indicates the scan direction of the AFM. The height of the CNT in this measurement is 15 nm.

We attribute the unprecedented resolution and the strong contrast achieved here to two major points: (1) the resolution benefits from the overall flat topography of our sample, as both, the isolated nanotubes and detected debris are without excessive protrusions. We find that the presence of larger CNT agglomerates on the sample is detrimental for the resolution. (2) As we show later, the work function difference between CNTs and the SiC substrate, obtained from the energy-resolved PEEM measurement, exhibits a value as small as 0.35 eV. This is probably small enough to avoid formation of strong contact potentials, which would lead to increased field distortions, and in general to a lower resolution [14-17].

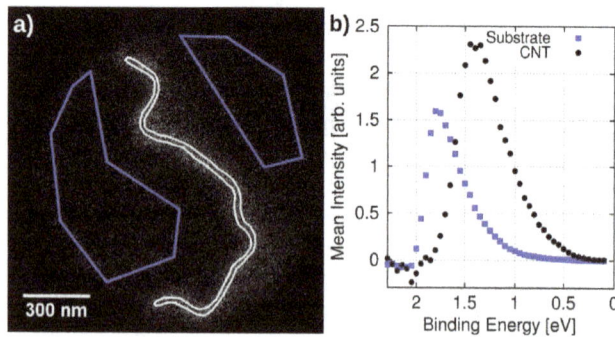

Figure 3. *a) Energy-filtered PEEM image at EB = 1.15 eV. b) Photoelectron spectra of the areas enclosed by the blue lines (SiC substrate) and the white line (CNT) in a).*

The high resolution presented here has not been achieved in respective work by other groups [18–21]. Reasons might be differences in sample preparation, as well as the usage of different combinations of substrate materials and excitation energies.

In order to interpret the variation in intensity along the CNT, we have recorded scanning electron microscopy (SEM) images and atomic force microscopy (AFM) images of the very same CNT previously imaged with PEEM and shown in Fig. 1b). All three images are shown in Fig. 2. The SEM image was obtained at an acceleration voltage of 1 kV and a working distance of 3.8 mm combining the signal of the SE2 (secondary electron detector sideways) and the InLense (secondary electron detector above objective lens) detector at a ratio of 1:1.

The PEEM (Fig. 2a) and SEM (Fig. 2b) images are strikingly similar. This underlines the high resolution achieved in the PEEM image and it indicates that field distortions – which are detrimental for a high resolution PEEM image – are small. The effect of a small field distortion is visible in the loop located in the upper part of the CNT (marked by the arrow). The size of this loop is slightly smaller in the PEEM image compared to the SEM image. The diameter of the CNT determined from the SEM image is 29 nm and thus slightly smaller than the one obtained in the PEEM measurement (FWHM = 38 nm), indicating the better resolution of the SEM. The variation in brightness along the CNT is remarkably similar in the PEEM and SEM image. Bright and dark sections are visible and found to be located at the exact same positions in both images. The origin of this intensity variation is found in the topography of the sample, as evidenced by the AFM image shown in Fig. 2c) and an SEM image obtained at a sample tilt of 30° shown in Fig. S2. Bright spots in the PEEM and SEM image thus correspond to positions at which the CNT is bent up and protruding away from the substrate, leading to local field distortions which in turn result in contrast formation due to change of electron trajectories [14-17,35,36]. We note that the bright spot above the left end of the CNT in Fig. 2b) and c) originates from the SEM

measurement, as the electron beam was standing still at that position for a few seconds. As the PEEM measurement was performed before, the spot is not visible in Fig. 2a).

In summary, the PEEM image has an unsurpassed resolution comparable to the respective SEM image. Intensity variations along the CNT are identical in PEEM and SEM image and directly correlate with the topography of the CNT imaged with AFM.

Fig. 3a) shows an energy-filtered PEEM image at a binding energy $E_B = 1.15$ eV. The image was obtained as part of a sequence of images for which the retarding field analyzer was scanned from $E_{kin,S} + \phi_S = E_{kin,A} + \phi_A = hv - E_B = 2.45$ eV to 4.95 eV in steps of 50 meV. $E_{kin,S}$ is the kinetic energy of the electrons at the sample surface, $E_{kin,A}$ the kinetic energy at the analyzer, ϕ_S the work function of the sample, $\phi_A = 4.05$ eV the work function of the analyzer and $hv \approx 5$ eV the photon energy. The retarding field analyzer works as a high pass filter. For each cutoff setting only electrons with a kinetic energy higher than the cutoff can pass the filter. Every image was recorded for 10 seconds and a total of 80 scans are shown (FoV = 1.3 μm). Event counting was used to improve signal to noise ratio. The images were differentiated to obtain energy-filtered images. The energy scale has been corrected for sample charging resulting from photoemission of electrons. The charging effect has been estimated to be 0.1 eV by evaluation of the shift of the spectrum for different photon intensities. To correct this effect, the spectra have been shifted by 0.1 eV to lower binding energies.

The intensity located within the area enclosed by the white line (CNT) and the blue lines (SiC substrate) was determined for each resulting differential image and plotted as a function of E_B in Fig. 3b). We note that the resulting spectra are dominated by secondary electrons that superimpose the electronic structure of the CNT and substrate, respectively. Spectra were also determined for three different sections within the nanotube, with the aim to compare the markedly bright spot in the lower half of the tube with the rest. Sections and respective spectra are shown in Fig. S3. All spectra were found to be identical within the experimental uncertainties. This finding supports our conclusion that the bright spot on the CNT observed in the PEEM image is solely caused by topography. The spectra presented in Fig. 3b) and S3 clearly demonstrate the power of energy-filtered PEEM to image electronic structure on a nanometer scale.

The sample work function is given by the high binding energy cutoff of the spectra, which is obtained by numerically calculating the inflection point of the respective edge. This yields values of $\phi_{CNT} = 3.43 \pm 0.05$ eV for the CNT and $\phi_{SiC} = 3.08 \pm 0.05$ eV for the SiC substrate. The error corresponds to the energy interval of the measurement.

Furthermore, CNT and substrate show a distinct difference in the signal onset on the low binding energy side of the spectra resulting from their different electronic structure.

A comparison of our work function measurement with values reported in literature ($\phi_{SiC} = 4.3$–4.89 eV [37–43], $\phi_{CNT} = 4.3$–4.9 eV [44–48]) reveals significant deviations. However, when comparing our work function values to respective literature values, several points have to be considered: (1) influence by the Schottky effect [49], (2) a contact potential between CNT and substrate, (3) doping of the CNT and (4) impact of contaminations. (1) The Schottky effect describes the influence of an electric field on the work function of materials. In the immersion lens of a PEEM, the sample itself is part of the objective lens, and thus exposed to an electric field perpendicular to the sample. In our experiment, this field is on the order of $E = 8 \times 10^6$ V/m. As an estimation, we treat the sample as a metal surface and obtain a lowering of the work function by $\Delta W = [e^3 E / (4\pi\varepsilon_0)]^{1/2} = 0.11$ eV. Deviations due to the semiconducting nature of SiC and carbon nanotubes should be small [50]. The actual work functions are thus about 0.1 eV higher than the values obtained from Fig. 3b). (2) The work function difference of CNT and SiC substrate leads to a contact potential as the Fermi levels equilibrate. In our measurements, the CNT exhibits a higher work function than the SiC substrate. Accordingly, electrons flow from the SiC substrate onto the CNT. As a consequence of this charging and the resulting additional potential, the photoelectron spectrum of the CNT shifts to higher kinetic energies (lower binding energies), while the spectrum of the SiC substrate shifts to lower kinetic energies (higher binding energies). The actual work function difference of the CNT and SiC substrate should thus be smaller than measured. However, this effect cannot explain the deviation from literature values. (3) The investigated CNTs were grown with ammonia gas present, with the aim of nitrogen-doping. Nitrogen-doping is known to reduce the work function of CNTs. For capped (5,5) CNTs, Wen et al. computed that a substitutional nitrogen atom reduces the work function by about 0.5 eV [51]. (4) Furthermore, CNT and substrate are most likely contaminated with adsorbed water molecules and substances used during sample preparation (e.g. ethanol, salts). These adsorbates are likely to be the most dominant contribution to the difference between the literature values and our measurement of the work functions. In particular, adsorbed water is known to reduce the work function and ionization energy of carbon nanotubes [52]. We note that substrate spectra recorded on different locations on the sample are found to differ by not more than 0.1 eV (cf. Fig. S4b).

In summary, our results demonstrate the potential of energy-filtered PEEM to image differences in electronic structure on a nanometer scale. The deviation of obtained work function values from the literature values is mainly attributed to contaminations on the sample investigated here.

4. Conclusions

We have conducted UV-PEEM studies on multi-walled carbon nanotubes with unsurpassed spatial resolution. Complementary SEM and AFM measurements reveal that the intensity variations along the CNT originate from local field distortions due to topography of the nanotube. An imaging energy filter installed in the PEEM allowed us to obtain photoelectron spectra from a single CNT and even from several regions along the CNT. These spectra recorded along

the CNT are identical within the experimental uncertainties, underlining that the PEEM contrast in this case is predominantly of topographic origin. With this, we shed light onto the importance of the different contrast mechanisms when imaging complex nano-objects with PEEM.

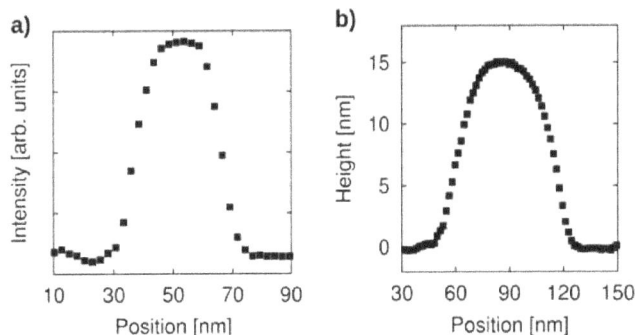

Figure S1. a) Intensity profile of SEM measurement (FWHM = 29 nm) and b) height profile of AFM measurement (FWHM = 65 nm, height = 15 nm) along yellow lines in Fig. 3b) and c). In b) background has been subtracted.

The work function determined from photoelectron spectra for the SiC substrate and the CNT is 0.9-1.8 eV lower than expected from literature. This is most likely due to adsorbates and residuals from sample preparation.

We conclude that photoemission electron microscopy (PEEM) has the potential to image complex nano-objects on surfaces with resolutions approaching those of a scanning electron microscope (SEM). In addition, energy-filtered PEEM bears great potential for fast imaging of electronic structure variation on nanometer length scales.

Acknowledgments

We thank Andrea Prager, Anika Gladytz, Ravikiran Chelur Suryaprakash and Dietmar Hirsch (Leibniz-Institute of Surface Modification) for SEM measurements. We are grateful for insightful discussion with Dietmar Hirsch (Leibniz-Institute of Surface Modification) and Martin Wolf (Fritz-Haber-Institute Berlin). We appreciate the support received from Dieter Polenz (FOCUS GmbH). CNT sample material was provided by Sossina M. Haile and Konstantinos P. Giapis (California Institute of Technology, Pasadena, USA). Nickel nanoparticles as CNT growth catalyst were provided by Nicholas Brunelli (Ohio State University). Andreas Neff was supported by a scholarship of the Beilstein-Institute, Germany. Katrin R. Siefermann was supported by the Robert-Bosch Foundation.

Appendix

SEM and AFM measurement

Fig. S1 shows line scans along the yellow lines shown in Fig. 2b) and c) of the SEM and AFM measurement. The size of the CNT in the AFM image differs considerably from PEEM (cf. line scan in Fig. 1c) and SEM images. While the CNT exhibits a height of about 15 nm, the FWHM has a value of 65 nm. The height of the nanotube in the AFM measurement is roughly half as large as the width in the SEM measurement. This is most likely due to a deformation of the CNT by the AFM tip as the measurement was performed in tapping mode. Similar deformations of multi-walled CNTs during AFM measurements have been detected by Yu et. al. [53]. When increasing the force applied by the AFM tip, they observed a decrease of CNT height by almost one half, which is consistent with the ratio of the SEM width and AFM height of our measurement. We note, however, that the sample itself as well as the AFM settings used in [53] differ from our measurement. Furthermore, it is known that van der Waals interactions of CNTs with a substrate can lead to a deformation of the spherical structure even in absence of an AFM tip [54,55].

Another reason for the large ratio between the width and the height of the AFM measurement is the width of the tip itself, which is much larger than the CNT. Thus, one would expect to measure a larger width than height of a perfectly round nanotube.

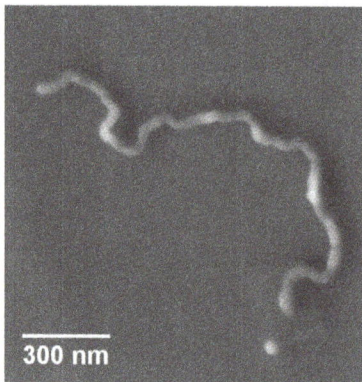

Figure S2. *SEM image of CNT at tilt of 30°. Combination of SE2 and InLens detector with ratio 1:1 (acceleration voltage: 3 kV).*

Fig. S2 shows an SEM image of the investigated CNT obtained at an acceleration voltage of 3 kV with a combination of SE2 and InLens detector signals at a ratio of 1:1. The sample was tilted by 30° to observe topography of the CNT.

Photoelectron Spectra

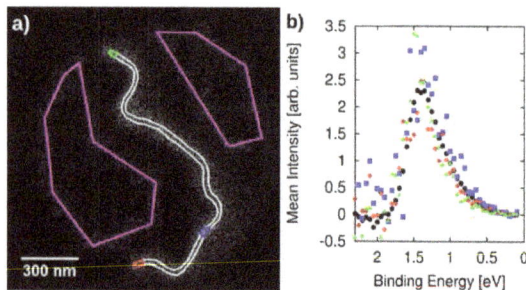

Figure S3. *a) Energy-filtered PEEM image at $E_B = 1.15$ eV. b) Photoelectron spectra of marked areas in a) (black circles: complete CNT, blue squares: bright spot in PEEM image, red diamonds: lower end of CNT, green triangles: upper end of CNT*

Fig. S3b) shows photoelectron spectra of several areas on

the CNT together with the respective areas in a). As the selected areas are much smaller than the whole CNT, the noise in the spectra increases. Within the accuracy of the measurement, no change of the electronic valence structure can be observed for the bright spot in the PEEM image or the end of the CNT. This is in line with the AFM measurement indicating that the PEEM contrast is, in this case, topographic.

Fig. S4b) shows a comparison of photoelectron spectra of two different substrate areas of the same sample. The magenta area in Fig. S3a) exhibits a work function of 3.10 ± 0.05 eV while the white area in Fig. S4a) shows a work function of 3.18 ± 0.05 eV. We attribute the discrepancy to different concentrations of adsorbates and residual matter on the substrate.

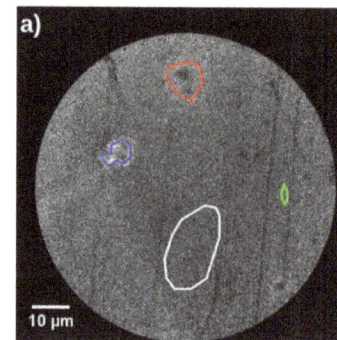

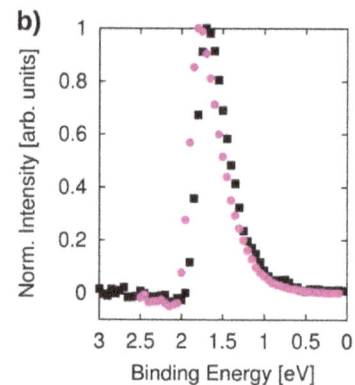

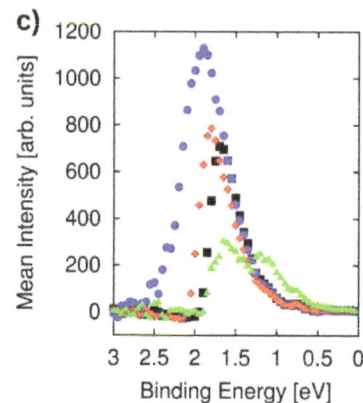

Figure S4. *a) Energy-filtered PEEM image at $E_B = 1.8$ eV. b) Comparison of two substrate areas. Magenta circles represent the spectrum for the respective area in Fig. S3a) and black squares for white area in Fig. S4a). c) Photoelectron spectra of respective areas indicated in a). Black squares represent spectrum for white area in a)*

Fig. S4c) shows photoelectron spectra of several areas marked in Fig. S4a). It reveals different electronic structure for different dirt particles on the substrate.

References

[1] A. Bailly, O. Renault, N. Barrett, L.F. Zagonel, P. Gentile, N. Pauc, et al., Direct quantification of gold along a single Si nanowire., Nano Lett. 8 (2008) 3709–3714.

[2] L. Douillard, F. Charra, Z. Korczak, R. Bachelot, S. Kostcheev, G. Lerondel, et al., Short range plasmon resonators probed by photoemission electron microscopy., Nano Lett. 8 (2008) 935–940.

[3] M. Hjort, J. Wallentin, R. Timm, a. a. Zakharov, J.N. Andersen, L. Samuelson, et al., Doping profile of InP nanowires directly imaged by photoemission electron microscopy, Appl. Phys. Lett. 99 (2011) 233113.

[4] S.J. Peppernick, A.G. Joly, K.M. Beck, W.P. Hess, Plasmonic field enhancement of individual nanoparticles by correlated scanning and photoemission electron microscopy., J. Chem. Phys. 134 (2011) 034507.

[5] L. Chelaru, M. Horn-von Hoegen, D. Thien, F.-J. Meyer zu Heringdorf, Fringe fields in nonlinear photoemission microscopy, Phys. Rev. B. 73 (2006) 115416.

[6] Y. Takamura, R. V Chopdekar, A. Scholl, A. Doran, J.A. Liddle, B. Harteneck, et al., Tuning magnetic domain structure in nanoscale La0.7Sr0.3MnO3 islands., Nano Lett. 6 (2006) 1287–1291.

[7] R. Könenkamp, R.C. Word, G.F. Rempfer, T. Dixon, L. Almaraz, T. Jones, 5.4 Nm Spatial Resolution in Biological Photoemission Electron Microscopy., Ultramicroscopy. 110 (2010) 899–902.

[8] R.C. Word, J.P.S. Fitzgerald, R. Könenkamp, Direct coupling of photonic modes and surface plasmon polaritons observed in 2-photon PEEM., Opt. Express. 21 (2013) 30507–30520.

[9] C. Lemke, T. Leißner, S. Jauernik, A. Klick, J. Fiutowski, J. Kjelstrup-Hansen, et al., Mapping surface plasmon polariton propagation via counter-propagating light pulses., Opt. Express. 20 (2012) 12877–12884.

[10] N.M. Buckanie, P. Kirschbaum, S. Sindermann, F.-J. Meyer zu Heringdorf, Interaction of light and surface plasmon polaritons in Ag islands studied by nonlinear photoemission microscopy., Ultramicroscopy. 130 (2013) 49–53.

[11] K. Fukumoto, K. Onda, Y. Yamada, T. Matsuki, T. Mukuta, S. Tanaka, et al., Femtosecond time-resolved photoemission electron microscopy for spatiotemporal imaging of photogenerated carrier dynamics in semiconductors., Rev. Sci. Instrum. 85 (2014) 083705.

[12] A Mikkelsen, J. Schwenke, T. Fordell, G. Luo, K. Klünder, E. Hilner, et al., Photoemission electron microscopy using extreme ultraviolet attosecond pulse trains., Rev. Sci. Instrum. 80 (2009) 123703.

[13] K. Siegrist, E.D. Williams, V.W. Ballarotto, Characterizing topography-induced contrast in photoelectron emission microscopy, J. Vac. Sci. Technol. A Vacuum, Surfaces, Film. 21 (2003) 1098.

[14] S.A. Nepijko, N.N. Sedov, C.H. Ziethen, G. Schönhense, M. Merkel, M. Escher, Peculiarities of imaging one- and two-dimensional structures in an emission electron microscope. 1. Theory, J. Microsc. 199 (2000) 124–129.

[15] S.A. Nepijko, N.N. Sedov, O. Schmidt, G. Schönhense, X. Bao, W. Huang, Imaging of three-dimensional objects in emission electron microscopy., J. Microsc. 202 (2001) 480–487.

[16] M. Lavayssière, M. Escher, O. Renault, D. Mariolle, N. Barrett, Electrical and physical topography in energy-filtered photoelectron emission microscopy of two-dimensional silicon pn junctions, J. Electron Spectros. Relat. Phenomena. 186 (2013) 30–38.

[17] V.K. Sangwan, V.W. Ballarotto, K. Siegrist, E.D. Williams, Characterizing voltage contrast in photoelectron emission microscopy., J. Microsc. 238 (2010) 210–217.

[18] S. Suzuki, Y. Watanabe, Y. Homma, S. Fukuba, S. Heun, A. Locatelli, Work functions of individual single-walled carbon nanotubes, Appl. Phys. Lett. 85 (2004) 127.

[19] S. Suzuki, Y. Watanabe, Y. Homma, S. Fukuba, A. Locatelli, S. Heun, Photoemission electron microscopy of individual single-walled carbon nanotubes, J. Electron Spectros. Relat. Phenomena. 144-147 (2005) 357–360.

[20] V.K. Sangwan, Carbon Nanotube Thin Film as an Electronic Material, PhD thesis, University of Maryland, 2009.

[21] D. Deng, L. Yu, X. Chen, G. Wang, L. Jin, X. Pan, et al., Iron encapsulated within pod-like carbon nanotubes for oxygen reduction reaction., Angew. Chemie. 52 (2013) 371–375.

[22] R.H. Baughman, A.A. Zakhidov, W.A. de Heer, Carbon nanotubes--the route toward applications., Science (80-.). 297 (2002) 787–792.

[23] W. Yang, P. Thordarson, J.J. Gooding, S.P. Ringer, F. Braet, Carbon nanotubes for biological and biomedical applications, Nanotechnology. 18 (2007) 412001.

[24] V. Popov, Carbon nanotubes: properties and application, Mater. Sci. Eng. R Reports. 43 (2004) 61–102.

[25] S. Suzuki, C. Bower, Y. Watanabe, O. Zhou, Work functions and valence band states of pristine and Cs-intercalated single-walled carbon nanotube bundles, Appl. Phys. Lett. 76 (2000) 4007.

[26] S. Suzuki, Y. Watanabe, S. Heun, Photoelectron spectroscopy and microscopy of carbon nanotubes, Curr. Opin. Solid State Mater. Sci. 10 (2006) 53–59.

[27] J.W. Chiou, C.L. Yueh, J.C. Jan, H.M. Tsai, W.F. Pong, I.-H. Hong, et al., Electronic structure of the carbon nanotube tips studied by x-ray-absorption spectroscopy and scanning photoelectron microscopy, Appl. Phys. Lett. 81 (2002) 4189.

[28] S. Suzuki, Y. Watanabe, T. Kiyokura, K. Nath, T. Ogino, S. Heun, et al., Electronic structure at carbon nanotube tips studied by photoemission spectroscopy, Phys. Rev. B. 63 (2001) 245418.

[29] S. Suzuki, Y. Watanabe, T. Ogino, S. Heun, L. Gregoratti, a. Barinov, et al., Electronic structure of carbon nanotubes studied by photoelectron spectromicroscopy, Phys. Rev. B. 66 (2002) 035414.

[30] W. Li, Y. Zhou, H.-J. Fitting, W. Bauhofer, Imaging mechanism of carbon nanotubes on insulating and conductive substrates using a scanning electron microscope, J. Mater. Sci. 46 (2011) 7626–7632.

[31] Y. Homma, S. Suzuki, Y. Kobayashi, M. Nagase, D. Takagi, Mechanism of bright selective imaging of single-walled carbon nanotubes on insulators by scanning electron microscopy, Appl. Phys. Lett. 84 (2004) 1750.

[32] Á. Varga, M. Pfohl, N. a Brunelli, M. Schreier, K.P. Giapis, S.M. Haile, Carbon nanotubes as electronic interconnects in solid acid fuel cell electrodes., Phys. Chem. Chem. Phys. 15 (2013) 15470–15476.

[33] J. Lin, N. Weber, A. Wirth, S.H. Chew, M. Escher, M. Merkel, et al., Time of flight-photoemission electron microscope for ultrahigh spatiotemporal probing of nanoplasmonic optical fields., J. Phys. Condens. Matter. 21 (2009) 314005.

[34] C. Ziethen, O. Schmidt, G.H. Fecher, C.M. Schneider, G. Schönhense, R. Frömter, et al., Fast elemental mapping and magnetic imaging with high lateral resolution using a novel photoemission microscope, J. Electron Spectros. Relat. Phenomena. 88-91 (1998) 983–989.

[35] S.A. Nepijko, N.N. Sedov, Resolution deterioration in emission and electron microscopy due to object roughness, Ann. Phys. 9 (2000) 441–451.

[36] J. Stöhr, S. Anders, X-ray spectro-microscopy of complex materials and surfaces, IBM J. Res. Dev. 44 (2000) 535–551.

[37] M. Wiets, M. Weinelt, T. Fauster, Electronic structure of SiC(0001) surfaces studied by two-photon photoemission, Phys. Rev. B. 68 (2003) 125321.

[38] V. Van Elsbergen, T. Kampen, W. Mönch, Surface analysis of 6H-SiC, Surf. Sci. 365 (1996) 443–452.

[39] T. Jikimoto, J.L. Wang, T. Saito, M. Hirai, M. Kusaka, M. Iwami, et al., Atomic and electronic structures of heat treated 6H–SiC surface, Appl. Surf. Sci. 130-132 (1998) 593–597.

[40] S. Kennou, An x-ray photoelectron spectroscopy and work-function study of the Er/α-SiC(0001) interface, J. Appl. Phys. 78 (1995) 587.

[41] S. Kennou, A. Siokou, I. Dontas, S. Ladas, An interface study of vapor-deposited rhenium with the two (0001) polar faces of single crystal 6H-SiC, Diam. Relat. Mater. 6 (1997) 1424–1427.

[42] J. Pelletier, D. Gervais, C. Pomot, Application of wide-gap semiconductors to surface ionization: Work functions of AlN and SiC single crystals, J. Appl. Phys. 55 (1984) 994.

[43] C. Benesch, M. Fartmann, H. Merz, k-resolved inverse photoemission of four different 6H-SiC (0001) surfaces, Phys. Rev. B. 64 (2001) 205314.

[44] Z. Xu, X.D. Bai, E.G. Wang, Z.L. Wang, Field emission of individual carbon nanotube with in situ tip image and real work function, Appl. Phys. Lett. 87 (2005) 163106.

[45] R. Gao, Z. Pan, Z.L. Wang, Work function at the tips of multiwalled carbon nanotubes, Appl. Phys. Lett. 78 (2001) 1757.

[46] P. Liu, Q. Sun, F. Zhu, K. Liu, K. Jiang, L. Liu, et al., Measuring the work function of carbon nanotubes with thermionic method., Nano Lett. 8 (2008) 647–651.

[47] H. Ago, T. Kugler, F. Cacialli, W.R. Salaneck, M.S.P. Shaffer, A.H. Windle, et al., Work Functions and Surface Functional Groups of Multiwall Carbon Nanotubes, J. Phys. Chem. B. 103 (1999) 8116–8121.

[48] M. Shiraishi, M. Ata, Work function of carbon nanotubes, Carbon N. Y. 39 (2001) 1913–1917.

[49] W. Schottky, Influence of structure-action, especially the Thomson constructive force, on the electron emission of metals, Phys. Zeitschrift. 15 (1914).

[50] G. Busch, J. Wullschleger, Der Schottky-Effekt an reinen Silizium-Oberflächen, Phys. Der Kondens. Mater. 12 (1970) 47–71.

[51] Q.B. Wen, L. Qiao, W.T. Zheng, Y. Zeng, C.Q. Qu, S.S. Yu, et al., Theoretical investigation on different effects of nitrogen and boron substitutional impurities on the structures and field emission properties for carbon nanotubes, Phys. E Low-Dimensional Syst. Nanostructures. 40 (2008) 890–893.

[52] C. Kim, Y.S. Choi, S.M. Lee, J.T. Park, B. Kim, Y.H. Lee, The Effect of Gas Adsorption on the Field Emission Mechanism of Carbon Nanotubes, J. Am. Chem. Soc. 124 (2002) 9906–9911.

[53] M. Yu, T. Kowalewski, R. Ruoff, Investigation of the radial deformability of individual carbon nanotubes under controlled indentation force, Phys. Rev. Lett. 85 (2000) 1456–1459.

[54] T. Hertel, R. Walkup, P. Avouris, Deformation of carbon nanotubes by surface van der Waals forces, Phys. Rev. B. 58 (1998) 13870–13873.

[55] R.S. Ruoff, J. Tersoff, D.C. Lorents, S. Subramoney, B. Chan, Radial deformation of carbon nanotubes by van der Waals forces, Nature. 364 (1993) 514–516.

21

Optical switching of azophenol derivatives in solution and in polymer thin films: The role of chemical substitution and environment

Yasser M. Riyad[1, 2, *]**, Sergej Naumov**[3]**, Jan Griebel**[1]**, Christian Elsner**[3]**, Ralf Hermann**[1]**, Katrin R. Siefermann**[3]**, Bernd Abel**[1, 3]

[1]Wilhelm-Ostwald-Institute for Physical and Theoretical Chemistry, Faculty of Chemistry and Mineralogy, University of Leipzig, Permoserstrasse 15, 04318 Leipzig, Germany
[2]Chemistry Department, Faculty of Science, Al-Azhar University, Nasr City, 11884, Cairo, Egypt
[3]Chemical Department, Leibniz Institute of Surface Modification, Permoserstrasse 15, 04318 Leipzig, Germany

Email address

yasser_riyad@yahoo.com (Y. M. Riyad)

Abstract: Design of polymer materials whose properties can be reversibly changed by illumination with light is a technology of particular scientific interest. Such materials contain molecular chromophors, which change their geometry and/or polarity upon absorption of light of a specific wavelength. The most prominent chromophores are azobenzene derivatives. Here, we present a systematic study on azobenzene derivatives in order to quantify the impact of chemical substitution and chemical environment on the dynamics of light-induced trans-cis isomerization (at 368 nm and 355 nm), thermal cis-trans relaxation, and light-induced cis-trans isomerization (at 434 nm). Systems under investigation were 4-hydroxyazobenzene (4-HAB) in acetonitrile (MeCN) solution and in a poly(methylmethacrylate) (PMMA) matrix. These two systems are compared to systems in which 4-HAB is esterified, namely 4-hydroxyazobenzene covalently bound (esterified) to PMMA matrix, and N-(tert-butoxycarbonyl)glycine-4- hydroxyazobenzene (Boc-Gly-4-HAB) in MeCN and in PMMA. Photoisomerization and thermal relaxation kinetics are monitored with UV-vis absorption spectroscopy and accompanied by quantum chemical calculations to shed light into the molecular origin of observed differences in switching properties. We find that the chemical environment (MeCN vs. PMMA) only has minor impacts (~10%) on trans to cis photoisomerization rates. Also, the impact of chemical environment on thermal cis to trans relaxation is small; with relaxation rates in PMMA beeing < 35% smaller compared to rates in MeCN solution. However, the thermal cis to trans relaxation rates of 4-HAB are clearly faster (factor > 400) than the rates of esterified systems. This difference is a clear result of the different substituents on the azobenzene moiety. Quantum chemical calculations suggest that the cis-configuration in the esterified systems is stabilized by an intramolecular H-bond between a carbonyl oxygen on the substituent and an H atom on the phenyl ring. In all systems, the cis to trans isomerization can be significantly accelerated by illumination with 434 nm light. For esterified systems, accelerations by factors of about 5700 – 15500 are observed. In the case of 4-hydroxyazobenzene covalently bound (esterified) to the PMMA matrix, complete light induced transfer from cis to trans is possible. In addition, it features a low thermal cis to trans isomerization rate and acceptable photoinduced trans to cis isomerization properties. With this, the material fulfills the basic requirements of a functional polymer material whose properties can be reversibly changed by illumination with light.

Keywords: Azobenzene Derivatives, Azophenol, Trans - Cis Isomerization, Photoisomerization, Thermal Relaxation, Chemical Environment, Chemical Substitution, Functional Polymer

1. Introduction

Using light for a variety of manufacturing, controlling, information processing and many other purposes is of particular interest in diverse fields. At the same time, the number of new applications involving light is increasing at an exceptionally high rate, in particular in the field of high precision applications. In this regard, photoresponsive systems have received increasing attention over the last years, owing to their broad applicability in a number of key fields such as optical memories,[1-5] molecular machines,[6-11] molecular switches,[12-15] photocontrol of biomolecules[16-20] and polymers[21-23], and surface relief gratings[24,25] etc.

One promising technology of particular scientific interest involves polymer materials whose properties can be reversibly changed by illumination with light. Polymeric materials can be controlled by light if they contain molecular chromophors. Suitable chromophores are able to change their geometry and/or polarity upon absorption of light of a specific wavelength. And advantageously, the initial state can be restored with light of a different wavelength. If the chromophore is chemically bound to a polymer network, this principle can be used to reversibly change properties of the respective polymer material.

The ideal chromophore for such applications features two stable states (e.g. conformations). Selective and efficient light induced transfer between the states is possible and light induced cycling between the states can be performed an unlimited number of times.

Among the photoswitches available today, azobenzene and its derivatives are of particular interest in this regard.[1-3,7,12-26]

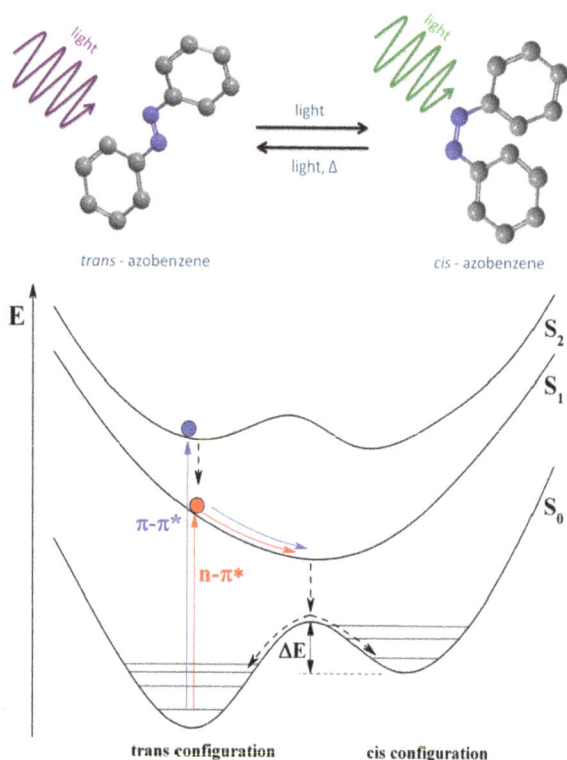

Figure 1. General scheme of the potential energy curves of the trans-cis isomerization process after excitation of the ππ and nπ* transition for azobenzene in solution, adapted from the results of calculations reported in Ref. 27.*

Azobenzene and its derivatives can be efficiently "switched" from their trans conformation to their cis conformation upon the absorption of light of a suitable wavelength (see Figure 1). In general, the trans conformation is the thermodynamical most stable conformation and the photoinduced switch to the cis conformation is characterized by high quantum yields. Furthermore, azobenzenes possess low photochemical fatigue and can be easily functionalized for a wide variety of photoactive materials. The photoinduced isomerization of azobenzene has been used to selectively change properties of azobenzene containing systems: absorption spectra, dielectric constants, refractive indices, oxidation/reduction potentials, phase transitions, and surface wettability.[28]

In general, the switching performance of azobenzene is influenced by chemical substituents and the chemical surrounding of the chromophore (e.g. solvents, polymer matrices, and temperature)[29-44] While a number of azobenzene-compounds with great switching properties is known from experiments in solution, it is not always clear how these properties change quantitatively when the azobenzene-molecules are incorporated into a polymer network. We have thus performed a systematic study in order to quantify the impact of a polymeric environment on the dynamics of light-induced trans-cis isomerization, thermal cis-trans relaxation, and light-induced cis-trans isomerization.

A particular focus of our study is on the thermal cis-trans relaxation, as the thermal stability of the cis-conformation (over days or even months) is a prerequisite for many applications e.g. in the field of optical storage devices.[45] In contrast, azobenzenes with a very fast thermal isomerization process (~ tens of milliseconds) are required for optical switching and real-time optical information processing.

In our systematic study, we investigated the switching properties of 5 systems (Figure 3):

(1) 4-HAB: 4-hydroxyazobenzene in acetonitrile (MeCN). In previous work, the thermal stability of cis-azobenzene was found to depend strongly on the chemical environment and the chemical substituents.[42,46,47] In particular, the thermal relaxation lifetime for cis-4-hydroxyazobenzene (4-HAB) at room temperature is very short (0.2 second) in protic ethanol solvent, whereas it is substantially longer in nonpolar toluene (31 min).[47] A limited number of studies has been devoted to the investigation of photochemical and thermal relaxation behaviors of 4-HAB and its derivatives in solution and in polymer matrices.[48]

(2) Boc-Gly-4-HAB: N-(tert-butoxycarbonyl)glycine-4-hydroxyazobenzene in MeCN. In this compound, the –OH group of 4-HAB is esterified. Compared to (1), differences in switching properties are thus a direct consequence of different chemical substituents.

(3) 4-HAB/PMMA: 4-hydroxyazobenzene embedded in poly(methyl methacrylate) (PMMA) matrix. The PMMA film contains individual 4-HAB molecules. The 4-HAB molecules are thus in a solid environment, whereas they are in liquid environment in (1).

(4) Boc-Gly-4-HAB/PMMA: N-(tert-butoxycarbonyl) glycine-4-hydroxyazobenzene embedded in PMMA matrix.

(5) 4-HAB-PMMA: 4-hydroxyazobenzene covalently bound to PMMA matrix. As in (3) and (4), the –OH group of 4-HAB is esterified.

With this systematic study we shed light onto the impact of chemical structure and chemical environment on the switching properties of azobenzene derivatives. The kinetics of photoinduced and thermal isomerization of systems (1)-(5) were monitored by UV-vis spectroscopy. Our experimental results were flanked by quantum chemical calculations to gain insights into the molecular origin of the observed differences in switching properties. The combination of experimental and theoretical investigations provides crucial information for the rational design of novel photoactive materials.

2. Experimental Section

Materials. All chemicals were purchased from Sigma-Aldrich and used as received. 4-Hydroxyazobenzene (4-HAB) was used after recrystallization from ethanol. Details on the syntheses and analytical data are given in the Appendix.

Synthesis of Boc-Gly-4-HAB (2). The coupling of N-(tert-butoxycarbonyl)glycine and 4-hydroxyazobenzene was performed using standard peptide coupling reagents like N,N,N',N'-tetramethyl-O-(benzotriazol-1-yl)uronium tetrafluoroborate and 1-hydroxybenzotriazole hydrate. The solvent was N,N-dimethylformamide and N-ethyldiisopropylamine was used as base. The reaction yielded 67% of an orange powder. Samples (1) and (2) were prepared by dissolving 4-HAB and Boc-Gly-4-HAB in MeCN, respectively. Concentrations were 20   mol dm-3.

Synthesis of 4-HAB/PMMA (3) and Boc-Gly-4-HAB/PMMA (4): The films were prepared by dissolving PMMA powder and 4-HAB or Boc-Gly-4-HAB in toluene. The solution was concentrated until most of the solvent was removed. Then, the solution was coated onto a stainless steel plate with a spatula. The films were removed after the solvent was completely evaporated. The obtained 4-HAB/PMMA and Boc-Gly-4-HAB/PMMA films were 0.3 mm and 0.5 mm thick, respectively, with 0.1 wt% azochromophores.

Synthesis of 4-HAB-PMMA (5): 4-(phenyldiazenyl)phenyl methacrylate was synthesized from 4-hydroxyazobenzene and methacryloyl chloride with triethylamine as base and diethylether as solvent according to literature procedure.[49] The reaction yielded 66% of an orange powder. 4-(phenyldiazenyl)phenyl methacrylate, methyl methacrylate, and AIBN (azobisisobutyronitrile) were mixed in a beaker and put in a domestic microwave oven. The reaction mixture was irradiated with a power of about 350 W in intervals of 90, 30 and 20 s, interrupted by 120 s breaks to avoid sputtering. The highly viscous mixture was poured in methanol and the precipitate was washed and dried at room temperature. For the film preparation, the polymer was solved in toluene and heated on a plate. The viscous mixture was coated of a stainless steel plate after most of the solvent was evaporated. The films were removed after evaporation of toluene. The

obtained film thickness was 0.5 mm with 0.03 wt% azochromophore (4-(phenyldiazenyl)phenyl methacrylate).

Physico-Chemical Characterization. The products were characterized by ¹H-NMR and ¹³C-NMR using a BRUKER Advance Ultra Shield 600 MHz spectrometer with a 5 mm BBO probe head. The chemical shifts (δ) are given in parts per million (ppm) and referenced to the solvent signal. IR measurements were carried out on a FTS 6000 spectrometer from BIO-RAD using the ATR modus. The glass transition temperatures were measured with a DSC 8500 PERKIN ELMER (heating rate 10 K·min⁻¹, inert gas nitrogen). The determination of the molar mass was made via size exclusion chromatography (SEC) using dimethyl formamide as mobile phase with 0.33 ml·min⁻¹ flow rate on a GRAM column (Company Polymer Standards Service, 4.6 mm diameter). The MS analysis was performed with a BRUKER „autoflex speed". The samples were crystallized on a pre-spotted AnchorChip (PAC) target with α-cyano-4-hydroxycinnamic acid (HCCA) matrix from a 1 μl acetonitrile/trifluoroacetic acid (TFA) (80:0.1) solution.

Photochromic experiments. The kinetics of photoinduced isomerization and thermal relaxation processes of the systems under investigation were followed by measuring UV-vis absorption spectra with a UV 2101 PC (Shimadzu) UV–vis spectrometer. Liquid samples (1) and (2) were measured in a 10 mm cuvette; solid films (3), (4), and (5) were inserted into a 1 mm cuvette. All experiments were performed at room temperature. A 450 W Xenon lamp (XBO 450, Osram) with a 5 cm water-filter and suitable interference filter (355 nm, 368 nm, 434 nm) was used for irradiation of the samples. We note that the bandwidth (FWHM) of the filters is 10 nm. The light intensity at the sample was 9 mW cm⁻² for 355 nm, and 12 mW cm⁻² for 368 nm and 434 nm.

Quantum Chemical Calculations. Quantum chemical calculations were done using the Density Functional Theory (DFT) B3LYP method[50,51] (Jaguar version 8.3 program)[52]. The structures of studied molecules were optimized at B3LYP/6-31(d,p) level both in gas phase and in solvent (acetonitrile). The interactions between the molecule and the solvent were evaluated at the same level of theory by Jaguar's Poisson-Boltzmann solver (PBF)[53] which fits the field produced by the solvent dielectric continuum to another set of point charges. The frequency analysis was made at the same level of theory to characterize the stationary points on the potential surface and to obtain thermodynamic parameters such as total enthalpy (H) and Gibbs free energy (G) at 298 K. The reaction enthalpies (ΔH) and Gibbs free energies of reaction (ΔG) were calculated as the difference of the calculated total enthalpies H and Gibbs free energies G between the reactants and products respectively. The electronic structure of studied molecules, namely molecular orbitals, Mulliken atomic charges, spin density distribution and energy of excited states were further analyzed using Gaussian 03 program.[54] The electronic transition spectra were calculated using time-dependent density functional theory (TDDFT)[55] both in the gas phase and with the PBF solvation model as implemented in Jaguar version 8.3 program package with the full linear response approximation.[56]

3. Results and Discussion

3.1 UV-Vis Absorption Spectra of Samples (1) - (5) are Presented in Figure 2

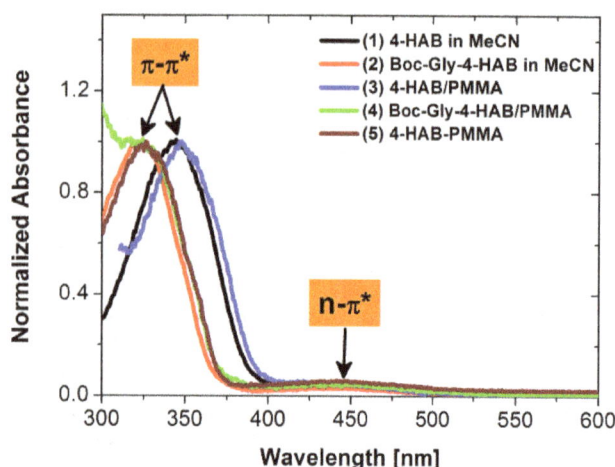

Figure 2. UV-vis absorption spectra of samples (1) – (5)

The spectra show the characteristic absorption bands of trans-azobenzene compounds, indicating that the azobenzene chromophores are in their thermodynamically most stable trans-configuration. The spectra exhibit a weak absorption band in the visible region, centered at about 440 nm, which corresponds to a symmetry-forbidden $n \rightarrow \pi^*$ transition. The strong absorption band centered at about 350 nm for samples (1) and (3), and at about 325 nm for (2), (4), and (5) corresponds to the symmetry allowed $\pi \rightarrow \pi^*$ transition. PMMA strongly absorbs below ~ 300 nm and prohibits evaluation of azobenzene-absorption bands below 300 nm for polymer samples (3)-(5). The spectral data of the investigated systems are given in Table 1. The spectral parameters of 4-HAB in MeCN agree well with the data in literature.[47] The $\pi \rightarrow \pi^*$ absorption band of 4-HAB in polar aprotic MeCN and the respective band in nonpolar PMMA matrix are similar. In the PMMA sample, a slight red shift of the band is detected, which we attribute to the different polarity of the environment.

In agreement with this finding, the spectra of Boc-Gly-4-HAB in MeCN and PMMA are similar.

Comparison of 4-HAB samples (1) and (3) with the esterified samples (2), (4), and (5) reveals the effect of different chemical substitution on the azobenzene moiety. The esterified samples exhibit a blue shift of the $\pi \rightarrow \pi^*$ band by about 20 nm. This shift is attributed to the electron-withdrawing character of the (–O–CO–) substituent.

3.2 Photoinduced Trans to cis Isomerization

In case the absorption spectrum of trans- and cis-configuration is significantly different, the kinetics of trans-cis isomerization can be monitored with UV-vis absorption spectroscopy. In the present experiments, we illuminated the samples with UV light at 368 nm and 355 nm (355 nm data shown in appendix). The experiments were performed as follows: A sample was irradiated for a certain amount of time and then inserted into the spectrometer to record a UV-vis absorption spectrum. Irradiation of the sample was afterwards continued and another spectrum was measured. The time needed for recording the spectrum, was kept to a minimum (approx. 45 s) to minimize inaccuracies from thermal cis-trans relaxation. The resulting series of spectra are shown in Figure 3. Irradiation of sample (1) 4-HAB in MeCN at 368 nm, results in a decrease of the absorption band at 350 nm and an increase of the absorption band at 440 nm. After a total irradiation time of 480 s, no further change in the UV-vis absorption spectrum upon further irradiation is observed (see Figure 3). This indicates that a photostationary state (PSS) is reached. The absorption spectrum in the PSS is distinctly different from the initial trans-spectrum indicating that photoinduced isomerization at 368 nm allows an almost complete transfer of trans- into cis isomer. We note that the irradiation wavelength of 368 nm was specifically selected to be on the red wing of the $\pi \rightarrow \pi^*$ transition, where absorption of the cis-isomer is small, allowing quantitative light-induced transition from trans to cis. Furthermore, the series of absorption spectra is characterized by the presence of two isosbestic points at 296 nm and 404 nm. This confirms the existence of only two absorbing species, i.e. the trans- and cis-isomers.

Table 1. Spectral properties of samples (1) – (5). Values for Cis-isomers were extracted from the PSS-spectra of the photoinduced trans to cis isomerization experiments.

		λ_{max} [nm] (ε [M^{-1} cm^{-1}])			
		Trans-		Cis-	
		$\pi \rightarrow \pi^*$	$n \rightarrow \pi^*$	$\pi \rightarrow \pi^*$	$n \rightarrow \pi^*$
1	4-HAB	346 (51000)	436 (2410)	308 (14900)	436 (4200)
2	Boc-Gly-4-HAB	325 (32000)	440 (950)	285 (10820)	438 (1850)
3	4-HAB/PMMA	348	436	-----	436
4	Boc-Gly-4-HAB/PMMA	325	438	-----	438
5	4-HAB-PMMA	327	440	-----	440

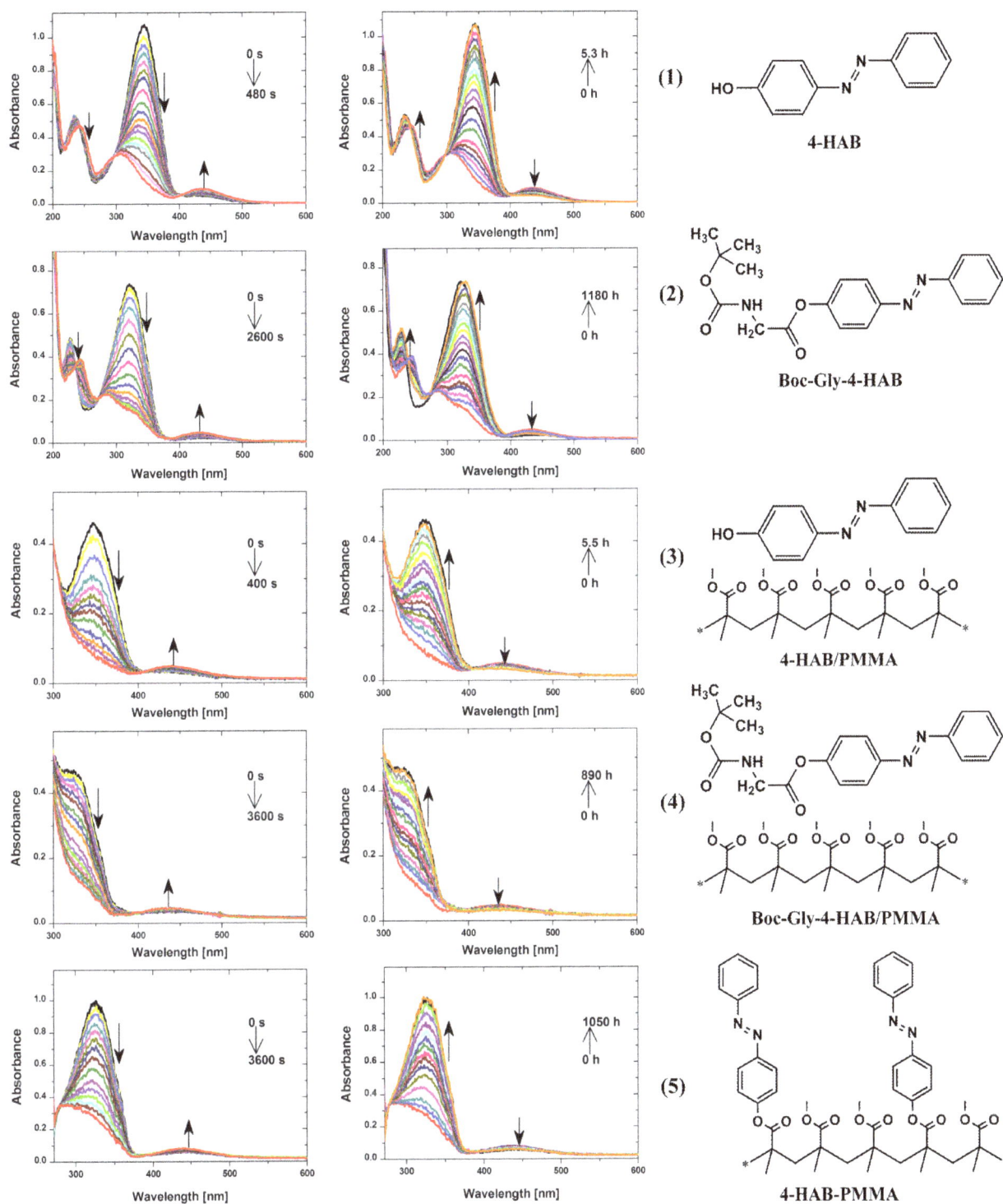

Figure 3. *Chemical structure of samples (1) – (5), and the variation of UV-vis spectra of sample (1) – (5) after illumination with 368 nm (left column) and thermally recovered (right column) at room temperature.*

Table 2. *Rate constants for photoinduced trans to cis isomerization (k_{tc}) at 368 nm and 355 nm, thermal cis to trans isomerization (k_b) and photoinduced cis to trans isomerization (k_{ct}) at 434 nm.*

		$k_{tc}^{368\,nm}$ [a] [s^{-1}]	$k_{tc}^{355\,nm}$ [a] [s^{-1}]	$k_{ct}^{434\,nm}$ [a] [s^{-1}]	k_b [s^{-1}]
1	4-HAB	1.4×10^{-2}		2.1×10^{-2}	8.9×10^{-4}
2	Boc-Gly-4-HAB	2.3×10^{-3}	1.4×10^{-3}	1.4×10^{-2}	0.9×10^{-6}
3	4-HAB/PMMA	1.6×10^{-2}	1.3×10^{-2}	1.8×10^{-2}	5.8×10^{-4}
4	Boc-Gly-4-HAB/PMMA	2.3×10^{-3}	1.6×10^{-3}	8.5×10^{-3}	1.1×10^{-6}
5	4-HAB-PMMA	2.1×10^{-3}	1.5×10^{-3}	7.4×10^{-3}	1.3×10^{-6}

[a] The error is ± 10%

The series of spectra obtained for sample (2) Boc-Gly-4-HAB in MeCN shows a similar behavior (Figure 3). Isosbestic points are found at 270 nm and 385 nm. The PSS is reached after a total illumination time of 2600 s. This indicates a slower kinetic of light-induced trans to cis isomerization compared to sample (1) 4-HAB for irradiation at 368 nm.

For the polymer film samples (3)-(5), the PSS is reached after a total illumination time of 400 s for sample (3) 4-HAB/PMMA, and after 3600 s for sample (4) Boc-Gly-4-HAB/PMMA, and sample (5) 4-HAB-PMMA.

In order to quantify these differences, photoisomerization kinetics of samples (1) - (5) were evaluated. According to equation (1), rate constants k_{tc} are obtained via the change in absorption at the wavelength at which the initial maximum of the $\pi \rightarrow \pi^*$ transition band is centered. These wavelengths are listed in Table 1 (values for the trans isomer).

$$\ln[(A_0\text{-}A_{pss})/(A_t\text{-}A_{pss})] = k_{tc}\, t \qquad (1)$$

A_0, A_t and A_{pss} are the absorbances at this wavelength at times 0, t, and infinite (=PSS), respectively.[57] The plots of $\ln[(A_0\text{-}A_{pss})/(A_t\text{-}A_{pss})]$ vs. illumination time (t) are shown in Figure 4.

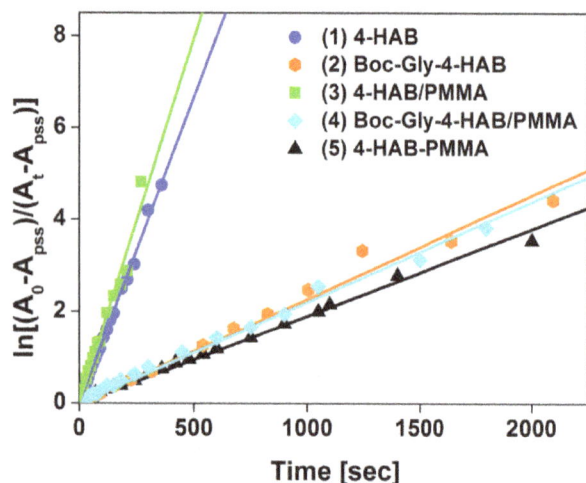

Figure 4. *Kinetics of photoinduced trans to cis isomerization of systems (1) – (5) with 368 nm. Linear functions (solid lines) are fitted to the data, and rate constants (k_{tc}) are obtained from the slopes of the linear functions.*

For all samples, the data in these plots is well represented by a linear function, indicating first-order kinetics. Rate constants (k_{tc}) are obtained from the slope of the linear function and summarized in Table 2.

We note that these rate constants are specific to the irradiation wavelength of 368 nm. According to equation (2) the rate constants (k_{tc}) for different systems depend on several parameters: [57]

$$k_{tc} = 2.303\ I\lambda\ (\varepsilon_t\Phi_{tc} + \varepsilon_c\Phi_{ct}) + k_b \qquad (2)$$

where Φ_{tc} and Φ_{ct} represent the quantum yields of the trans-cis and cis-trans photochemical reactions, ε_t and ε_c represent the molar extinction coefficients of the trans and cis isomers at the

illumination wavelength, I_λ is the intensity of the irradiation wavelength, and k_b is the rate constant for the thermal cis to trans isomerization. According to equation (2), the observed differences in rate constants in our experiments originate from two points:

(a) Differences in the molar extinction ε at the irradiation wavelength of 368 nm.

(b) Differences in quantum yields Φ related to the different chemical substituents on the azobenzene moiety and the chemical environment.

As k_b is small in all investigated systems at room temperature, its influence on the observed k_{tc} is small (see thermal relaxation section 2.2 and results in Table 2).

Table 2 demonstrates that rate constants for systems (1) 4-HAB in MeCN and (3) 4-HAB/PMMA are identical within experimental accuracy. Also identical are the rate constants for system (2) Boc-Gly-4-HAB in MeCN and (4) Boc-Gly-4-HAB/PMMA. As the molar extinction ε is not expected to change significantly with chemical environment (MeCN vs. PMMA), these results suggest that the quantum yields (Φ) are not influenced by the chemical environment.

A similar trend was observed by Paik et al.[30] and Sin et. al.[35] in experiments on polymers containing azobenzene side chains. The azobenzene moiety exhibited approximately similar photoisomerization kinetics in dilute solutions and in polymer films above and below the polymers glass transition temperature, respectively.

Another interesting finding is that the rate constants obtained for systems (5) 4-HAB-PMMA, (2), and (4) are identical within the accuracy of the experiment. Assuming that the molar extinction ε at the excitation wavelength is similar for these systems, this indicates that the far structure of the substituent, beyond the ester function directly attached to the azobenzene system, does not influence the photoisomerization kinetics.

We further find that rate constants of systems (2), (4), and (5) are one order of magnitude smaller than those of systems (1) and (3). This finding might be the result of different molar extinction coefficients at the excitation wavelength of 368 nm. We note that for systems (2), (4), and (5) this excitation wavelength is on the red edge of the $\pi \rightarrow \pi^*$ absorption band. However, additional experiments at 355 nm (see appendix and Table 2) indicate that also the quantum yields Φ are different for the two types of chemical substituents on the azobenzene moiety: -OH (in (1) and (3)) and ester (in (2), (4), and (5)).

3.3 Thermal Cis to Trans Isomerization

The kinetics of thermal cis to trans isomerization were monitored as follows: The samples were irradiated at 368 nm, according to the experiments described above, until the photostationary state was reached. Samples were then kept in the dark (at room temperature) and the thermal cis to trans relaxation was followed by recording UV-vis spectra. The resulting series of spectra for systems (1) - (5) are shown in Figure 3 (right column). Comparison to the start spectra of photoinduced trans to cis isomerization experiments (Figure 3, middle column) demonstrates that complete recovery of the initial signal occurs. Only the final spectrum of (2) Boc-Gly-4-HAB in MeCN solution slightly deviates from the initial trans spectrum. We attribute this to inaccuracies in data recording, originating from the long time over which the sample has been measured during relaxation experiments.

Figure 3 reveals significant differences in the timescales of relaxation. While relaxation in systems (1) and (3) is completed within a few hours, complete relaxation in systems (2), (4), and (5) takes over one month.

We quantify the thermal cis to trans isomerization kinetics by evaluating the series of spectra obtained for systems (1) to (5) according to equation (3). Similar to the case of photoinduced trans to cis isomerization, rate constants are obtained via the change in absorption at the wavelengths at which the maximum of the $\pi \rightarrow \pi^*$ transition band (of the trans isomer) is centered. For all systems, these wavelengths are listed in Table 1 (values for trans isomer).

$$\ln[(A_\infty - A_0)/(A_\infty - A_t)] = k_b\, t \qquad (3)$$

A_0 and A_t are the absorbances at this wavelength at times 0 and t, respectively (here $A_0 = A_{pss}$). A_∞ is the absorbance at infinite times, meaning at times for which no further change in the spectrum is observed. k_b is the rate constant of thermal cis to trans isomerization. The plots of $\ln[(A_\infty - A_0)/(A_\infty - A_t)]$ vs. illumination time (t) are shown in Figure 5. For all samples, the data in these plots are well represented by a linear function, indicating first-order kinetics. Rate constants (k_b) are obtained from the slope of the linear function and summarized in Table 2.

In our experiment, we find that the cis isomer of system (1) 4-HAB in MeCN has a thermal relaxation lifetime of 19 min. This lifetime is significantly longer than the lifetime (< 1 min) reported by Kojima et al.[47] for 4-HAB in MeCN. In constrast to Kojima et al. we have performed our experiments in MeCN, which contains < 0.01 wt% of water contamination. We find that the presence of water significantly decreases the thermal relaxation lifetime of 4-HAB. Indeed, in MeCN with a water content > 0.01 wt% we were unable to detect any change in the UV-vis spectrum upon irradiation of the sample at 368 nm and 355 nm, respectively. This indicates that the thermal relaxation lifetime is significantly below 1 minute, and thus not detectable with our experimental setup. The same result was observed for 4-HAB in methanol, indicating that the presence of polar protic molecules enables fast thermal cis to trans isomerization, possibly via the formation of hydrogen bound complexes of 4-HAB with protic solvents. Kojima et al.'s explanation of the fast relaxation by intermolecular hydrogen bonding between 4-HAB and MeCN seems less likely in the light of our experimental findings.

Comparison of thermal cis to trans relaxation lifetimes of systems (1) - (5) leads to the following findings: System (1) 4-HAB in MeCN has a lifetime of 19 min which is slightly shorter than the lifetime of system (3) 4-HAB in PMMA with a lifetime of 29 min. Besides, system (2) Boc-Gly-4-HAB in MeCN exhibits a lifetime of about 250 hours, slightly shorter than (4) Boc-Gly-4-HAB in PMMA

with ~310 hours. This finding suggests that the thermal relaxation rate is only slightly influenced by the chemical environment. This is consistent with findings by Sin et. al.[35]. They investigated azobenzene-containing polymers in solution and in polymer matrix at room temperature and find that isomerization kinetics are similar in both cases. Our observation that relaxation rates in PMMA matrix are slightly smaller compared to MeCN may thus rather be the result of their different polarities than their different viscosities.

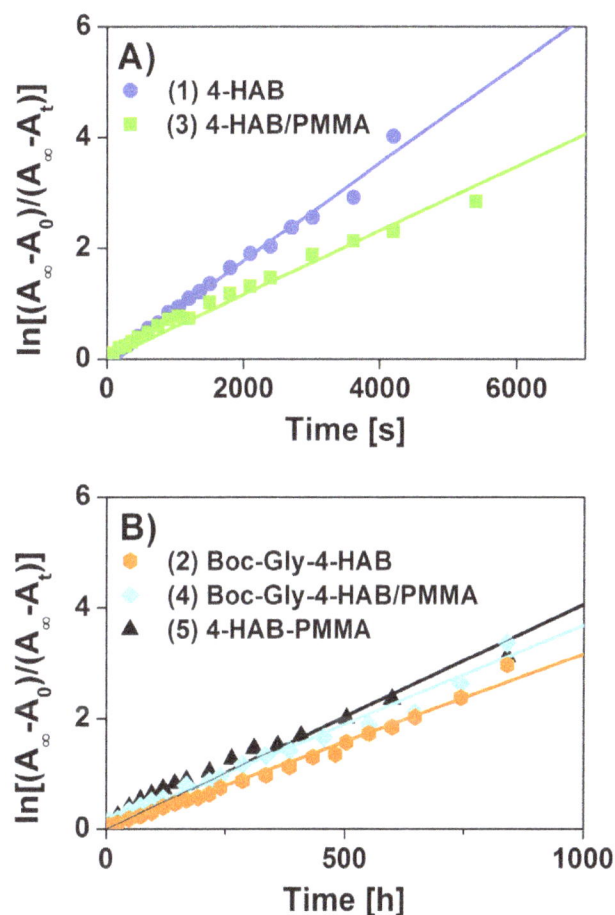

Figure 5. *Kinetics of thermal cis to trans isomerization of (A): systems (1) + (3), and (B): systems (2) + (4) + (5). Linear functions (solid lines) are fitted to the data, and rate constants (k_b) are obtained from the slopes of the linear functions.*

The thermal cis to trans relaxation lifetime of system (5) 4-HAB-PMMA is found to be about 210 hours, and thus comparable to the lifetimes of systems (2) and (4). These lifetimes are > 400 times longer than lifetimes for systems (1) and (3). This difference is a clear result of the different substituents on the azobenzene moiety. We have performed quantum chemical calculations (section 2.5) in order to understand how these substituents influence the stability of the cis-isomer.

3.4 Photoinduced Cis to Trans Isomerization

The thermal stability of cis isomers of systems (1) - (5) is high enough to allow experiments on the light induced cis to trans isomerization. In these experiments, the samples were irradiated with visible light at 434 nm in order to induce cis-trans isomerization. Figure 6 shows the obtained spectral series for systems (1) - (5). For system (5) 4-HAB-PMMA we observe a complete transfer of cis to trans, as the final spectrum is identical with the initial trans spectrum of the sample. For systems (1)-(4), illumination with 434 nm results in a photostationary state. In the case of systems (1), (2) and (3) the PSS still contains significant amounts of cis isomer. Rate constants kct were obtained with the procedure described in section 2.2 (see Figure 7) and listed in Table 2. In all systems, the cis to trans isomerization can be significantly accelerated by illumination with 434 nm light. In the case of system (5), where complete light induced transfer from cis to trans is observed, the rate constant kct is more than a factor of 5500 higher compared to the thermal cis to trans relaxation rate kb.

2.5 Quantum Chemical Calculations

The structures of studied systems (1)+(3) 4-HAB, (2)+(4) Boc-Gly-4-HAB and (5) 4-HAB-PMMA were optimized both in trans and cis forms. We use a structure containing three PMMA units to model 4-HAB-PMMA. In agreement with results of calculations with different methods on the trans-azobenzene structure58, the optimized trans structures of studied molecules have both phenyl rings slightly distorted from planarity. Accordingly, the phenyl rings in trans-4-HAB are rotated by $1.1°$ relative to the N=N-C plane. The distortion from planarity is slightly larger for trans-4-HAB-PMMA ($1.0°$ and $2.5°$ for both sites of N=N bond) and for trans-Boc-Gly-4-HAB ($1.2°$ and $3.7°$).

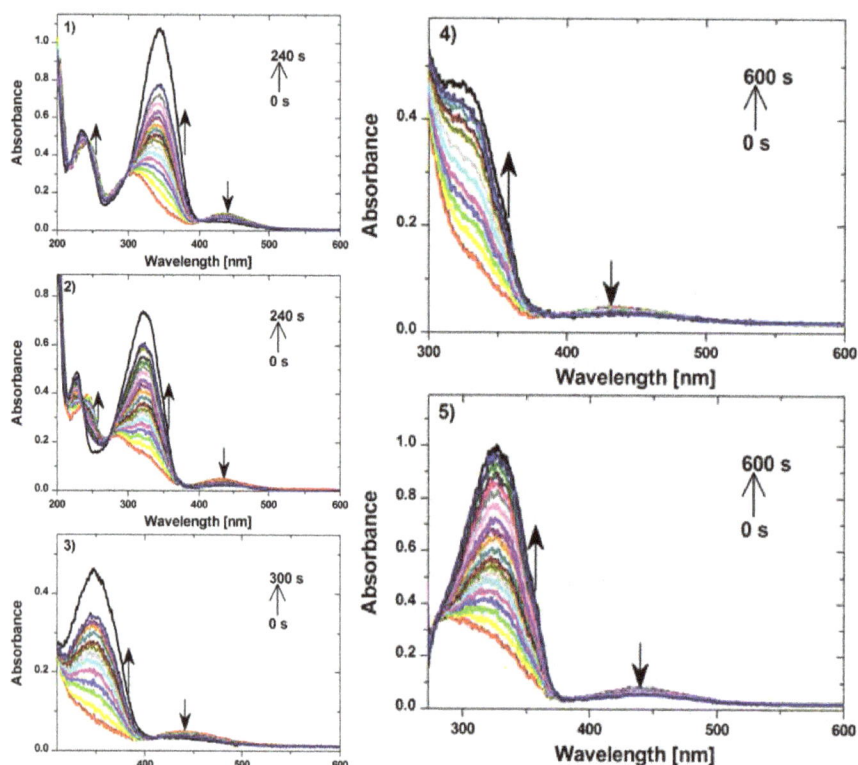

Figure 6. Variation of UV-vis. spectra of sample (1) – (5) after illumination with 434 nm light.

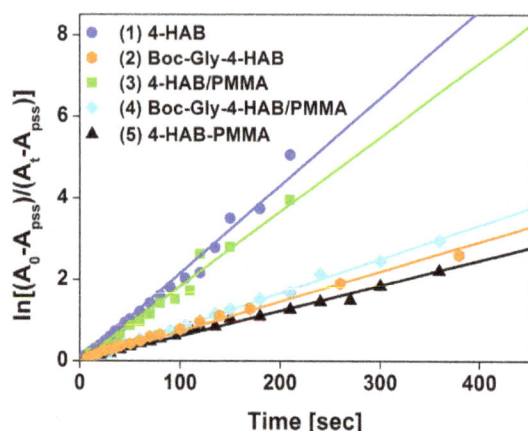

Figure 7. Kinetics of photoinduced cis to trans isomerization of systems (1) – (5) with 434 nm light. Linear functions (solid lines) are fitted to the data, and rate constants (k_ct) are obtained fom the slopes of the linear functions.

In all optimized cis-structures phenyl rings are strongly distorted from planarity. In the case of Boc-Gly-4-HAB and 4-HAB-PMMA, we find additional, energetically favored structures (denoted as cis*).

These structures feature a strong intramolecular hydrogen-bond between a carbonyl oxygen from the substituents and a H-atom on the phenyl ring. The optimized stable structures of 4-HAB-PMMA and Boc-Gly-4-HAB are shown in Figure 8

According to reported studies[48,59,60] the ground-state thermal isomerization of azobenzenes proceeds via rehybridization of one of the azo nitrogens rather than by rotation about the -N=N- bond. In this work calculations of activation energy for thermal cis-trans isomerization were performed for the systems 4-HAB and Boc-Gly-4-HAB, as well as for unsubstituted azobenzene (u-AB) for comparison.

Results of calculations are shown in Figure 9.

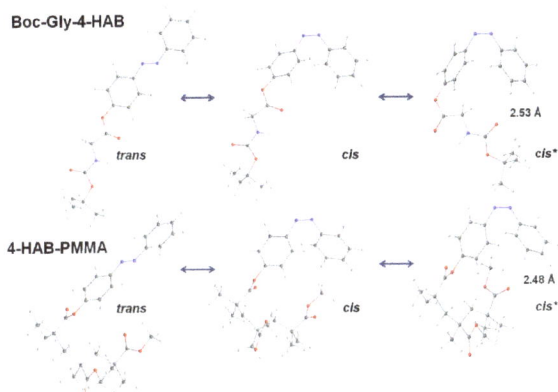

Figure 8. *Optimized structures (trans and cis forms) of Boc-Gly-4-HAB and structure of 4-HAB-PMMA bonded to model polymer PMMA (3 units). Here cis*structure forms intramolecular H-bond between carbonyl oxygen and H atom from phenyl ring.*

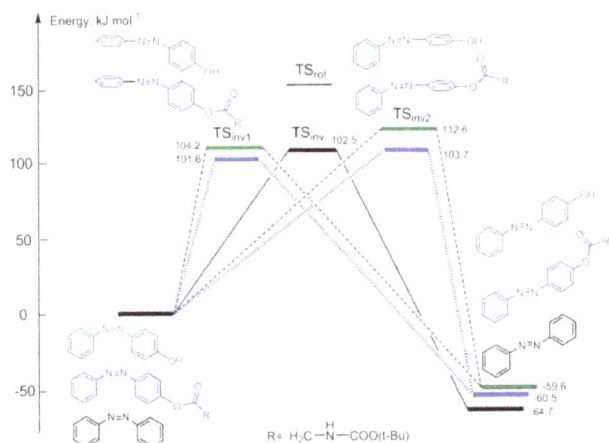

Figure 9. *Energy scheme for the ground-state thermal cis-trans isomerization calculated starting from optimized cis-isomer. Here TS$_{inv1}$ – transition state for the inversion on the site of free phenyl and TS$_{inv2}$ – on the site connected to substituent. TS$_{rot}$ – transition state for rotation about the -N=N- bond is energetically unfavorable and is not discussed here.*

The cis-trans isomerization barrier through inversion on -N=N- calculated by us for u-AB was 102.1 kJ mol^{-1}, which is in good agreement with the literature data.[48,68] In the case of 4-HAB and Boc-Gly-4-HAB molecules the activation barrier depends on the site of the -N=N- bond. As can be seen, the activation barrier is slightly higher for the inversion on the site connected to substituent. Thus, it is expected, that the thermal cis-trans isomerization through inversion should rather proceed on the site of the unsubstituted phenyl.

To get any idea on the large difference on the rate of the thermal cis-trans isomerization between free 4-HAB and both large molecules namely Boc-Gly-4-HAB and 4-HAB-PMMA bonded to model polymer PMMA, electronic structures, charge distributions and excitation energies were compared. For 4-HAB and Boc-Gly-4-HAB electron distributions of the two highest occupied molecular orbitals (HOMO and HOMO-1) and the lowest unoccupied molecular orbital

(LUMO), which are mostly involved into formation of S$_1$ and S$_2$ excited states are shown in Figure 10.

As can be seen, the large substituent in Boc-Gly-4-HAB does not have any essential effect on the electronic structure.

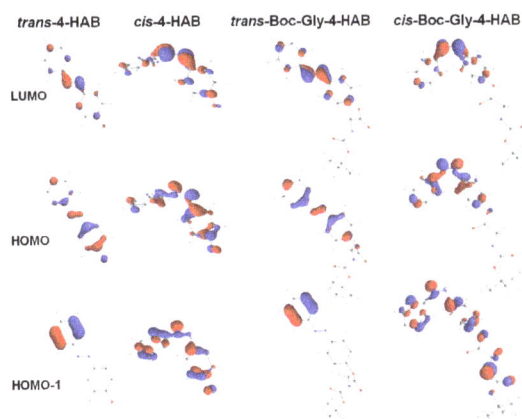

Figure 10. *Electron distributions from molecular orbitals mostly involved into formation of S$_1$ and S$_2$ excited states.*

The calculated excitations energies for all studied molecules are given in Table 3. The energies of excited states are in reasonable agreement with experimental spectra. The experimentally observed huge difference in the rate of the thermal cis-trans isomerization between systems (1)+(3) and (2)+(4)+(5) can thus not readily be explained by changes in electronic parameters caused by the different substituents. However, these differences may be explained by the possibility of intramolecular H-bond formation between a carbonyl oxygen on the substituent and an H atom on the phenyl ring, which is only possible in the cis conformation of systems (2), (4), and (5). The respective structures are denoted cis* and shown in Figure 8. The calculated Gibbs free energies ΔG of such intramolecular hydrogen-bond in the case of Boc-Gly-OAB and 4-HAB-PMMA molecules were 13.4 and 9.2 kJ mol^{-1} respectively. These values agree well with the typical strength of a O····H hydrogen bond (between 8.4 and 22 kJ mol^{-1}).[61] Apparently, such an H-bond hinders cis-trans isomerization via the inversion mechanism and may be the reason for the observed large thermal stability of the cis-form of systems (2), (4), and (5).

Table 3. *Excitation energies and oscillator strength (in parenthesis) of the investigated systems as calculated with time-dependent density functional theory (TDDFT) at B3LYP/6-31G(d,p) level.*

| | | | λ[nm] | |
			trans	*cis*
1	4-HAB (in MeCN)	S$_1$ (nπ*)	468 (0.0001)	469 (0.065)
		S$_2$ (ππ*)		325 (0.151)
		S$_3$ (ππ*)	345 (0.801)	299 (0.055)
		S$_4$ (ππ*)	235 (0.090)	262 (0.082)
2	Boc-Gly-4-HAB (in MeCN)	S$_1$ (nπ*)	479 (0.0003)	471 (0.062)
		S$_2$ (ππ*)		320 (0.072)
		S$_3$ (ππ*)	346 (0.969)	307 (0.129)
		S$_4$ (ππ*)	224 (0.057)	246 (0.099)
5	4-HAB-PMMA	S$_1$ (nπ*)	484 (0.0001)	480 (0.040)
		S$_2$ (ππ*)	342 (0.988)	305 (0.082)

3. Conclusions

We presented a systematic study on azobenzene derivatives in order to quantify the impact of chemical substitution and chemical environment on the dynamics of light-induced trans-cis isomerization, thermal cis-trans relaxation, and light-induced cis-trans isomerization. Systems under investigation were 4-hydroxyazobenzene (4-HAB) in acetonitrile (MeCN) solution and in poly(methylmethacrylate) (PMMA) matrix. These two systems are compared to systems in which 4-HAB is esterified, namely 4-hydroxyazobenzene covalently bound (esterified) to PMMA matrix, and N-(tert-butoxycarbonyl)glycine-4- hydroxyazobenzene in MeCN and in PMMA.

We find that different chemical environments (MeCN vs. PMMA) do not significantly influence the UV-vis absorption spectrum of azobenzene derivatives. However, the absorption spectra of 4-HAB are clearly different from the spectra of esterified systems.

The same trend is found for photoinduced trans to cis isomerization at 368 nm: chemical environment only has a minor impact on photoisomerization kinetics. Our results further indicate that the far structure of the substituent – in our case beyond the ester function directly attached to the azobenzene system – does not influence trans to cis photoisomerization kinetics.

In line with these findings, also the influence of chemical environment on the thermal cis to trans relaxation rate is found to be small. However, the relaxation rates of 4-HAB are clearly faster (factor > 400) than the rates of esterified systems. This difference is a clear result of the different substituents on the azobenzene moiety. Quantum chemical calculations indicate that the cis-configuration in the esterified systems is stabilized by an intramolecular H-bond between the carbonyl oxygen of the substituent and an H atom of the phenyl ring. Further experiments will be required to unambiguously prove this hypothesis, and to evaluate whether it can be used as a design principle for azobenzene-based materials with a low thermal cis to trans isomerization rate.

We further found indications that the presence of polar protic molecules (water or methanol) greatly enhances the thermal cis to trans isomerization rate of 4-HAB. Possibly via the formation of hydrogen bound complexes of 4-HAB with water and methanol, respectively.

In all systems, the cis to trans isomerization can be significantly accelerated by illumination with 434 nm light. In the case of system (5) 4-HAB covalently bound (esterified) to PMMA matrix, complete light induced transfer from cis to trans is observed. The rate constant k_{ct} is a factor of about 5700 higher compared to the thermal cis to trans relaxation rate k_b. In addition, it features a low thermal cis to trans isomerization rate and acceptable photoinduced trans to cis- isomerization properties. In summary, this renders it a promising starting point for the rational design of functional polymer materials whose properties can be reversibly changed by illumination with light.

4. Appendix

Synthesis of 4-(phenyldiazenyl)phenyl methacrylate

5.0 g (25.2 mmol) azobenzene was dissolved in 50 ml diethylether and 6.5 ml triethylamine was added. 3.5 ml (38.2 mmol) methyl methacryloyl chloride was added during 30 minutes with a dropping funnel under vigorous stirring at room temperature. The reaction mixture was kept at room temperature with a water bath. After 5.5 hours stirring 50 ml chloroform was added and additionally 30 minutes stirred. Then, 50 ml water was added and stirred till the clouding disappeared. The two phases were separated and the aqueous phase was extracted two times with 20 ml chloroform. The combined organic phases were washed with hydrochloric acid, water, $NaHCO_3$-solution and finally water and dried with sodium sulfate. The solvent was removed under reduced pressure. The crude product was recrystallized from ethanol/n-pentane (1/1, v/v) to yield 4.4 g (66%) orange powder. ^{1}H-NMR (600 MHz, acetone-d_6)δ (ppm): 2.06 (dd, 3 H), 5.88 (m, 1 H), 6.35 (m, 1 H), 7.41 (d, 2H), 7.61 -7.54 (m, 3 H), 7.94 (d, 2 H), 8.01 (d, 2 H); ^{13}C-NMR (125 MHz, acetone-d_6) δ (ppm): 18.9; 124.1; 125.2; 128.5; 130.7; 132.7; 137.2; 151.5; 153.8; 154.7; 166.4; IR (FT-ATR, cm^{-1}): 3105, 3058, 2975, 2954, 2922 (w), 1729 (s), 1193 (m), 1120 (s).

Synthesis of Boc-Gly-4-HAB

876 mg (5 mmol) N-(tert-Butoxycarbonyl)glycine, 892 mg (4.5 mmol) 4-Hydroxyazobenzene, 1.605 g (5 mmol) N,N,N',N'-Tetramethyl-O-(benzotriazol-1-yl)uronium tetrafluoroborate, and 675 mg (5 mmol) 1-Hydroxybenzotriazole hydrate were dissolved in 10 ml N,N-Dimethylformamide. 1.742 ml (10 mmol) N-Ethyldiisopropylamine was added and the reaction mixture was stirred overnight at room temperature followed by the addition of water and ethyl acetate. The two phases were separated and the organic phase was washed 4 times with water and dried with sodium sulfate. After evaporation of the solvent the crude product was re-crystallized from ethyl acetate/n-hexane to yield 1.06 g (3 mmol, 67 %) of an orange powder. MALDI-TOF-MS, m/z: 377.9 [M+Na]$^+$, ^{1}H-NMR (600 MHz, DMSO-d_6)δ (ppm): 1.44 (s, 9 H), 4.02 (d, 2 H), 6.94-7.97 (10 H); ^{13}C-NMR (150 MHz, DMSO-d6) δ (ppm): 28,0; 42,2; 78,4; 122,4; 122,5; 123,7; 129,3; 131,4; 149,6; 151,8; 152,4; 155,8; 169,0.

Polymerization of MMA with 4-(phenyldiazenyl)phenyl methacrylate: In a typical synthesis: 250 mg (0.95 mmol) 4-(phenyldiazenyl)phenyl methacrylate, 50 mg AIBN were dissolved in 10 ml (94 mmol) methyl methacrylate in a 100 ml beaker. The beaker was placed in a domestic microwave oven (PRIVILEG 8020) and the following time program at 350 W microwave power was performed: 90 – 30 – 20 s with 120 s breaks. The resulting viscous mixture was poured into 200 ml methanol. The precipitate was washed with methanol and dried at room temperature. The obtained polymer was the starting material for the film preparation. The best films for the UV/VIS experiments were obtained by dilution the abovementioned polymer with pure PMMA getting a final azobenzene concentration of about 0.03 wt%.

Table A1. Glass transitions and molecular weights of the polymer samples

	Properties	T_G / °C	M_N / Da	M_W / Da
3	4-HAB/PMMA	94	44720	105655
4	Boc-Gly-4-HAB/PMMA	80	43592	107437
5	4-HAB-PMMA	62	35021	127300

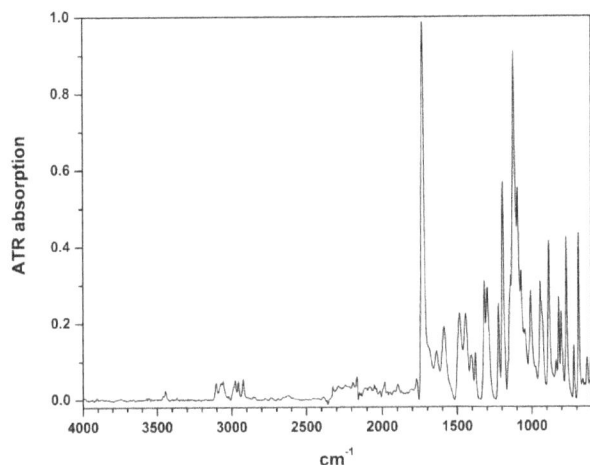

Figure A1. ATR-IR spectrum of 4-(phenyldiazenyl)phenyl methacrylate.

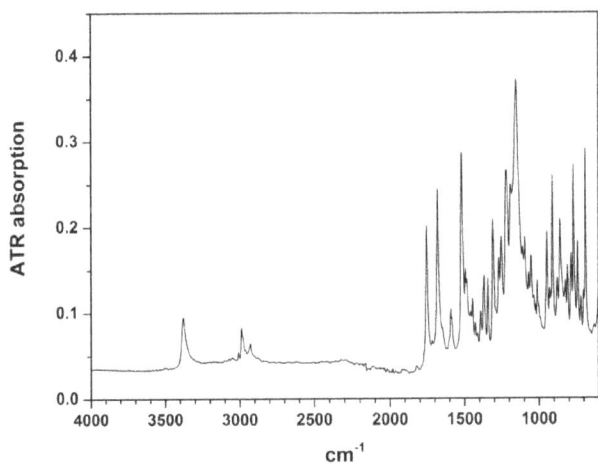

Figure A2. ATR-IR spectrum of Boc-Gly-4-HAB.

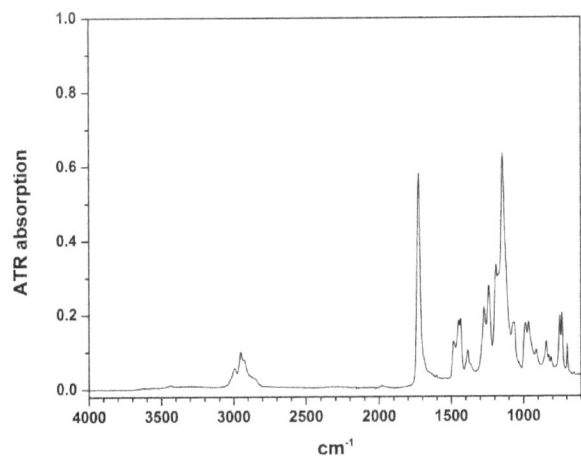

Figure A3. ATR-IR spectrum of 4-(phenyldiazenyl)phenyl methacrylate incorporated in PMMA.

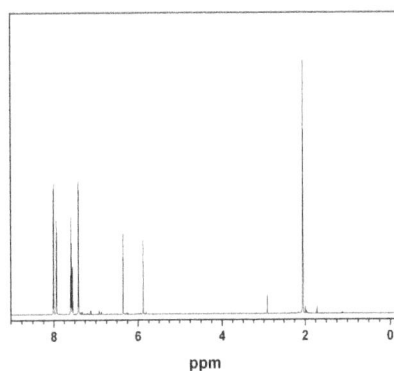

Figure A4. ^1H NMR spectrum of 4-(phenyldiazenyl)phenyl methacrylate, acetone-d_6.

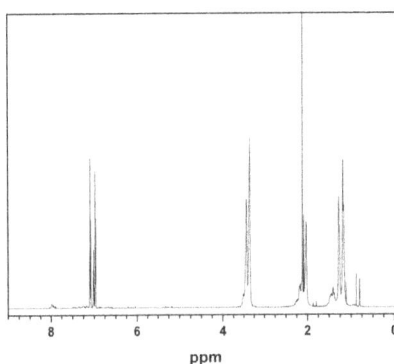

Figure A5. ^1H NMR spectrum of 4-(phenyldiazenyl)phenyl methacrylate covalently linked in PMMA, toluene-d_8.

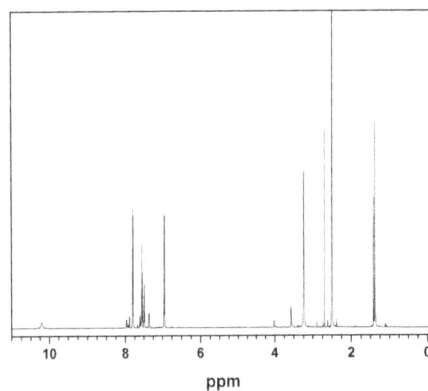

Figure A6. ^1H NMR spectrum of Boc-Gly-4-HAB, DMSO-d_6.

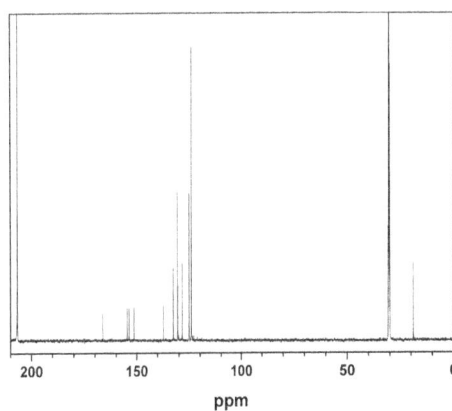

Figure A7. ^{13}C NMR spectrum of 4-(phenyldiazenyl)phenyl methacrylate, acetone-d_6.

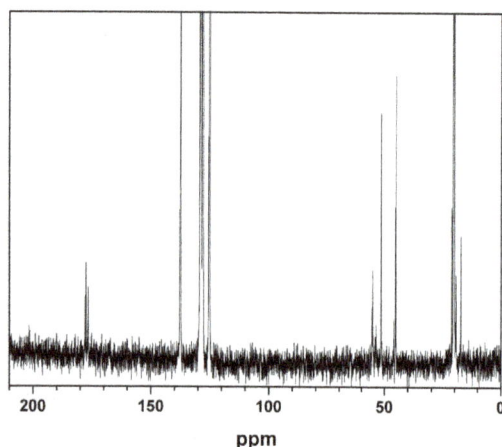

Figure A8. ^{13}C *NMR spectrum of 4-(phenyldiazenyl)phenyl methacrylate covalently linked in PMMA, toluene-d_8.*

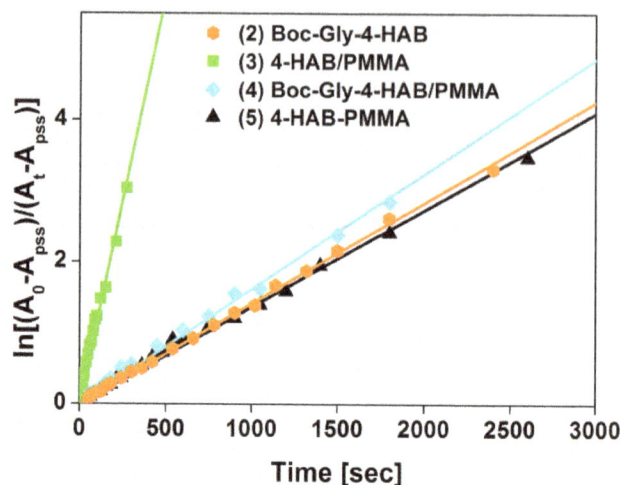

Figure A10. *Kinetics of photoinduced trans to cis isomerization of systems (2)-(5) with 355 nm light. Linear functions (solid lines) are fitted to the data and rate constants (k_{tc}) are obtained from the slopes of the linear functions.*

References

[1]   Z. F. Liu, K. Hashimoto, A. Fujishima. Photoelectrochemical information storage using an azobenzene derivative. *Nature* 1990, *347*, 658-660.

[2]   T. Ikeda, O. Tsutsumi. Optical switching and image storage by means of azobenzene liquid-crystal films. *Science* 1995, *268*, 1873-1875.

[3]   J. W. Brown, B. L. Henderson, M. D. Kiesz, A. C. Whalley, W. Morris, S. Grunder, H. Deng, H. Furukawa, J. I. Zink, J. F. Stoddart, O. M. Yaghi. Photophysical pore control in an azobenzene-containing metal-organic framework. *Chem. Sci.* 2013, *4*, 2858–2864.

[4]   J. E. Green, J. W. Choi, A. Boukai, Y. Bunimovich, E. Johnston-Halperin, E. DeIonno, Y. Luo, B. A. Sheriff, K. Xu, Y. Shik Shin, H.-R. Tseng, J. F. Stoddart, J. R. Heath. A 160-kilobit molecular electronic memory patterned at 10(11) bits per square centimetre. *Nature* 2007, *445*, 414–417.

[5]   E. Orgiu, N. Crivillers, M. Herder, L. Grubert, M. Pätzel, J. Frisch, E. Pavlica, D. T. Duong, G. Bratina, A. Salleo, N. Koch, S. Hecht, P. Samori. Optically switchable transistor via energy-level phototuning in a bicomponent organic semiconductor. *Nat. Chem.* 2012, *4*, 675–679.

[6]   B. Lewandowski, G. De. Bo, J. W. Ward, M. Papmeyer, S. Kuschel, M. J. Aldegunde, P. M. E. Gramlich, D. Heckmann, S. M. Goldup, D. M. D'Souza, A. E. Fernandes, D. A. Leigh. Sequence-Specific Peptide Synthesis by an Artificial Small-Molecule Machine. *Science* 2013, *339*, 189–193.

[7]   E. R. Kay, D. A. Leigh, F. Zerbetto. Synthetic molecular motors and mechanical machines. *Angew. Chem. Int. Ed.* 2007, *46*, 72–191.

[8]   D. S. Marlin, D. G. Cabrera, D. A. Leigh, A. M. Z. Slawin. An allosterically regulated molecular shuttle. *Angew. Chem. Int. Ed.* 2006, *45*, 1385-1390.

[9]   D. A. Leigh, J. K. Y. Wong, F. Dehez, F. Zerbetto. Unidirectional rotation in a mechanically interlocked molecular rotor. *Nature* 2003, *424*, 174-179.

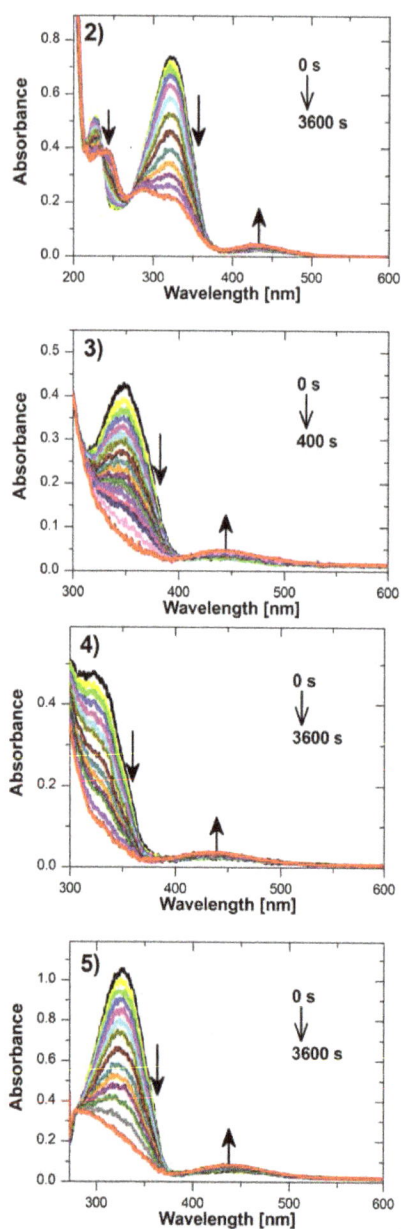

Figure A9. *Variation of UV-vis spectra of sample (2) – (5) after illumination with 355 nm light.*

[10] N. Koumura, R. W. J. Zijlstra, R. A. van Delden, N. Harada, B. L. Feringa. Light-driven monodirectional molecular rotor. *Nature* 1999, *401*, 152-155.

[11] A. M. Brouwer, C. Frochot, F. G. Gatti, D. A. Leigh, L. Mottier, F. Paolucci, S. Roffia, G. W. H. Wurpel. Photoinduction of fast, reversible translational motion in a hydrogen-bonded molecular shuttle. *Science* 2001, *291*, 2124-2128.

[12] N. Liu, Z. Chen, D. R. Dunphy, Y.-B. Jiang, R. A. Assink, C. J. Brinker, Photoresponsive nanocomposite formed by self-assembly of an azobenzene-modified silane. *Angew. Chem. Int. Ed.* 2003, *42*, 1731-1734.

[13] C. Zhang, M.-H. Du, H.-P. Cheng, X.-G. Zhang, A. E. Roitberg, J. L. Krause. Coherent electron transport through an azobenzene molecule: A light-driven molecular switch. *Phy. Rev. Lett.* 2004, *92*, 158301/1-158301/4.

[14] W. R. Browne, B. L. Feringa. Making molecular machines work. *Nature Nanotech.* 2006, *1*, 25-35.

[15] S. Muramatsu, K. Kinbara, H. Taguchi, N. Ishii, T. Aida. Semibiological molecular machine with an implemented and logic gate for regulation of protein folding. *J. Am. Chem. Soc.* 2006, *128*, 3764-3769.

[16] I. Willner, S. Rubin. Control of the structure and functions of biomaterials by light. *Angew. Chem. Int. Ed.* 1996, *35*, 367-385.

[17] L. Ulysse, J. Cubillos, J. Chmielewski. Photoregulation of cyclic peptide conformation. *J. Am. Chem. Soc.*, 1995, *117*, 8466-8467.

[18] H. Asanuma, X. Liang, T. Yoshida, A. Yamazawa, M. Komiyama. Photocontrol of triple-helix formation by using azobenzene-bearing oligo(thymidine). *Angew. Chem. Int. Ed.* 2000, *39*, 1316-1318.

[19] S. Spörlein, H. Carstens, H. Satzger, C. Renner, R. Behrendt, L. Moroder, P. Tavan, W. Zinth, J. Wachtveitl. Ultrafast spectroscopy reveals subnanosecond peptide conformational dynamics and validates molecular dynamics simulation. *Proc. Natl. Acad. Sci. U. S. A.* 2002, *99*, 7998-8002.

[20] X. Liang, H. Asanuma, M. Komiyama. Photoregulation of DNA triplex formation by azobenzene. *J. Am. Chem. Soc.* 2002, *124*, 1877-1883.

[21] G. S. Kumar, D. C. Neckers. Photochemistry of azobenzene-containing polymers. *Chem. Rev.* 1989, *89*, 1915-1925.

[22] T. Hugel, N. B. Holland, A. Cattani, L. Moroder, M. Seitz, H. E. Gaub. Single-molecule optomechanical cycle. *Science* 2002, *296*, 1103-1106.

[23] N. B. Holland, T. Hugel, G. Neuert, A. Cattani-Scholz, C. Renner, D. Oesterhelt, L. Moroder, M. Seitz, H. E. Gaub. Single molecule force spectroscopy of azobenzene polymers: Switching elasticity of single photochromic macromolecules. *Macromolecules*, 2003, *36*, 2015-2023.

[24] A. Natansohn, P. Rochon, Photoinduced motions in azo-containing polymers. *Chem. Rev.* 2002, *102*, 4139-4175.

[25] A. Priimagi, A. Shevchenk, Azopolymer-based micro- and nanopatterning for photonic applications. *J. Polym. Sci. Part B: Polym. Phys.* 2014, *52*, 163–182.

[26] E. Merino, M. Ribagorda, Control over molecular motion using the cis-trans photoisomerization of the azo group. *Beilstein. J. Org. Chem.* 2012, *8*, 1071-1090.

[27] S. Monti, G. Orlandi, P. Palmieri. Features of the photochemically active state surfaces of azobenzene. *Chem. Phys.* 1982, *71*, 87-99.

[28] H. Bouas-Laurent, H. Durr. Organic photochromism. *Pure Appl. Chem.* 2001, *73*, 639–665.

[29] P. D. Wildes, J. G. Pacifici, G. Irick, D. G. Whitten, Solvent and substituent effects on thermal isomerization of substituted azobenzenes. Flash spectroscopic study. *J. Am. Chem. Soc.* 1971, *93*, 2004–2008.

[30] C. S. Paik, H. Morawetz, Photochemical and thermal isomerization of azoaromatic residues in side chains and backbone of polymers in bulk. *Macromolecules*. 1972, *5*: 171–177.

[31] J. M. Nerbonne, R. G. Weiss, Elucidation of thermal-isomerization mechanism for azobenzene in a cholesteric liquid-crystal solvent. *J. Am. Chem. Soc.* 1978, *100*, 5953–5954.

[32] H. J. Haitjema, Y. Y. Tan, G. Challa, Thermal isomerization of azobenzene- based acrylic-monomers and (co)polymers with dimethylamino substituents in solution, influence of addition of (poly)acid, copolymer composition, spacer length, and solvent type. *Macromolecules*, 1995, *28*, 2867–2873.

[33] C. H. Wang, R. G. Weiss, Thermal *cis→trans* isomerization of covalently attached azobenzene groups in undrawn and drawn polyethylene films. Characterization and comparisons of occupied sites. *Macromolecules*, 2003, *36*, 3833–3840.

[34] D. Acierno, E. Amendola, V. Bugatti, S. Concilio, L. Giorgini, P. Iannelli, S. P. Piotto, Synthesi and characterization of segmented liquid crystalline polymers with the azo group in the main chain. *Macromolecules* 2004, *37*, 6418-6423.

[35] S. L. Sin, L. H. Gan, X. Hu, K. C. Tam, Y. Y. Gan, Photochemical and thermal isomerization of azobenzene-containing amphiphilic diblock copolymers in aqueous micellar aggregates and in film. *Macromolecules* 2005, *38*, 3943-3948.

[36] S. Furumi, K. Ichimura, Effect of para-substituents on azobenzene side chains tethered to poly(methacrylate)s on pretilt angle photocontrol of nematic liquid crystals. *Thin Solid Films* 2006, *499*, 135-142.

[37] P. Sierocki, H. Mass, P. Dragut, G. Richardt, F. Vögtle, L. De Cola, F. A. M. Brouwer, J. I. Zink, Photoisomerization of azobenzene derivatives in nanostructured silica. *J. Phys. Chem. B* 2006, *110*, 24390-24398.

[38] N. A. Wazzan, P. R. Richardson, A. C. Jones, Cis-Trans isomerisation of azobenzenes studied by laser-coupled NMR spectroscopy and DFT calculations. *Photochem. Photobiol. Sci.* 2010, *9*, 968-974.

[39] A. A. Beharry, O. Sadovski, G.A. Woolley. Azobenzene photoswitching without ultraviolet Light. *J. Am. Chem. Soc.* 2011, *133*, 19684-19687.

[40] U. Georgi, P. Reichenbach, U. Oertel, L.M. Eng, B. Voit. Synthesis of azobenzene-containing polymers and investigation of their substituent-dependent isomerisation behavior. *React. Funct. Polym.* 2012, *72*, 242-251.

[41] P. J. Coelho, C. M. Sousa, M. C. R. Castro, A. M. C. Fonseca, M. M. M. Raposo, Fast thermal cis-trans isomerization of heterocyclic azo dyes in PMMA polymers. *Opt. Mater.*, 2013, *35*, 1167-1172.

[42] J. Garcia-Amoos, D. Velasco, Understanding the fast thermal isomerization of azophenols in glassy and liquid-crystalline polymers. *Phys. Chem. Chem. Phys.*, 2014, *16*, 3108-3114.

[43] J. Garcia-Amoros, D. Velasco. Recent advances towards azobenzene-based light-driven real-time information-transmitting materials. *Beilstein J. Org. Chem.* 2012, *8*, 1003-1017.

[44] J. Garcia-Amoros, D. Velasco. *In Responsive Materials and Methods (Advanced Materials Series)*, ed. A. Tiwari and H. Kobayashi, WILEY-Scrivener Publishing LLC, New Jersey, 2013, chapter 2.

[45] Y. Kishimoto, J. Abe. A fast photochromic molecule that colors only under UV light. *J. Am. Chem. Soc.* 2009, *131*, 4227-4229.

[46] J. Garcia-Amoros, A. Sanchez-Ferrer, W. A. Massad, S. Nonell, D. Velasco. Kinetic study of the fast thermal cis-to-trans isomerisation of para-, ortho- and polyhydroxyazobenzenes. *Phys. Chem. Chem. Phys.* 2010, *12*, 13238-13242.

[47] M. Kojima, S. Nebashi, K. Okawa, N. Kurita. Effect of solvent on cis-to-trans isomerization of 4-hydroxyazobenzene aggregated through intermolecular hydrogen bonds. *J. Phys. Org. Chem.* 2005, *18*, 994-1000.

[48] H. M. D. Bandara, S. C. Burdette. Photoisomerization in different classes of azobenzene. *Chem. Soc. Rev.* 2012, *41*, 1809–1825

[49] A. Altomare, C. Carlini, F. Ciardelli, R. Solaro. Photochromism of 4-Acryloxybenzene/(-)-methyl acrylate copolymers. *J. Polym. Sci., Polym. Chem. Ed.* 1984, *22*, 1267-1280.

[50] A. D. Becke. Density-functional thermochemistry .4. A new dynamical correlation functional and implications for exact-exchange mixing. *J. Chem. Phys.*, 1996,*104*, 1040-1046.

[51] C. T. Lee, W. T. Yang; R. G. Parr. Development of the colle-salvetti correlation-energy formula into a functional of the electron density. *Phys. Rev. B* 1988, *37*, 785-789.

[52] Jaguar, version 8.3, Schrodinger, Inc., New York, NY, 2014.

[53] D. J. M. Tannor, B. Murphy, R. Friesner, R. A. Sitkoff, D. Nicholls, A. Ringnalda, W. A. M. Goddard, III, B. Honig. Charge-distribution and solvation energies from ab-Initio quantum mechanics and continuum dielectric theory. *J. Am. Chem. Soc.* 1994, *116*, 11875-11882.

[54] M. J. Frisch, G. W. Trucks, H. B. Schlegel, G. E. Scuseria, M. A. Robb, J. R. Cheeseman, V. G. Zakrzewski, J. A. Montgomery, Jr. R. E.Stratmann, J. C. Burant, et al. Gaussian 03, Revision A.11, Gaussian, Inc., Pittsburgh PA, 2003.

[55] R. A. Friesner, New methods for electronic structure calculations on large molecules. New methods for electronic-structure calculations on large molecules. *Ann. Rev. Phys. Chem.* 1991, *42*, 341-367.

[56] W. T. Pollard, R. A. Friesner. Efficient Fock matrix diagonalization by a Krylov-space method. *J. Chem. Phys.* 1993, *99*, 6742-6750.

[57] L. Angiolini, D. Caretti, L. Giorgini, E. Sabatelli, A. Altomare, C. Carlini, R. Solaro, Synthesis, chiroptical properties and photoresponsive behaviour of optically active poly[(*S*)-4-(2-methacryloyloxypropanoyloxy)azobenzene]. *Polymer* 1998, *39*, 6621–6629.

[58] N. K. S. Tanaka, S. Itoh. Ab initio molecular orbital and density functional studies on the stable structures and vibrational properties of trans- and cis-azobenzenes. *J. Phys. Chem. A* 2000, *104*, 8114-8120.

[59] A. A. Blevins, G. J. Blanchard. Effect of positional substitution on the optical response of symmetrically disubstituted azobenzene derivatives. *J. Phys. Chem. B* 2004, *108*, 4962-4968.

[60] L. Wang, C. Yi, H. Zou, J. Xu, W. Xu. Theoretical study on the isomerization mechanisms of phenylazopyridine on S_0 and S_1 states. *J. Phys. Org. Chem.* 2009, *22*, 888–896.

[61] C. Reichardt, *Solvents and Solvent Effects in Organic Chemistry*, VCH Verlagsgesellschaft, Weinheim, 1990.

22

Heat Transfer Enhancement Using Nanofluids

Mohammed Saad Kamel, Raheem Abed Syeal, Ayad Abdulameer Abdulhussein

Department of Mechanical Techniques, Al-Nasiriyah Technical Institute, Southern Technical University, Thi-Qar, Iraq

Email address:

mr.mohd1986@yahoo.com (M. S. Kamel), raheemsyeal@gmail.com (R. A. Syeal), ayad_abdulameer@yahoo.com (A. A. Abdulhussein)

Abstract: Nanofluids are a fluids containing nanometer-sized particles, called nanoparticles. These fluids are Suspension of nanoparticles in conventional fluids. Nanofluids have been the subject of intensive study worldwide since pioneering researchers recently discovered the anomalous thermal behavior of these fluids. The enhancement of heat transfer using nanofluids have been used as one of the passive heat transfer techniques in several heat transfer applications. It is considered to have great potential for heat transfer enhancement and are highly suited to application in heat transfer processes like microelectronics, fuel cells, pharmaceutical processes, and hybrid-powered engines, engine cooling/vehicle thermal management, domestic refrigerator, chiller, heat exchanger, and in boiler flue gas temperature reduction. This review covers the enhancement of heat transfer by using nanofluids and potential applications of nanofluids. This paper presents an updated review of the heat transfer applications of nanofluids to develop directions for future work because the literature in this area is spread over a wide range of disciplines, including heat transfer, material science, physics, chemical engineering and synthetic chemistry.

Keywords: Nanofluids, Nanoparticles, Enhancement of Heat Transfer, Friction Factor

1. Introduction

Nanofluids have attracted much attention recently because of their potential as high performance heat transfer fluids in electronic cooling and automotive. Performance of heat transfer equipment can be improved with studies related to a significant increase in heat flux and miniaturization. In many industrial applications such as power generation, microelectronics, heating processes, cooling processes and chemical processes, water, mineral oil and ethylene glycol are used as heat transfer fluid. Effectiveness and high compactness of heat exchangers are obstructed by the lower heat transfer properties of these common fluids as compared to most solids. It is obvious that solid particles having thermal conductivities several hundred times higher than these conventional fluids, as seen from *Figure 1*. To improve thermal conductivity of a fluid, suspension of ultrafine solid particles in the fluid can be a creative idea. Different types of particles (metallic, non-metallic and polymeric)can be added into fluids to form slurries. Due to the fact that sizes of these suspended particles are in the millimeter or even micrometer scale, some serious problems such as the clogging of flow channels, erosion of pipelines and an increase in pressure drop can occur. Moreover, they often suffer from rheological and instability problems. Especially, the particles tend to settle rapidly. For that reason, though the slurries have better thermal conductivities but they are not practical. [1].

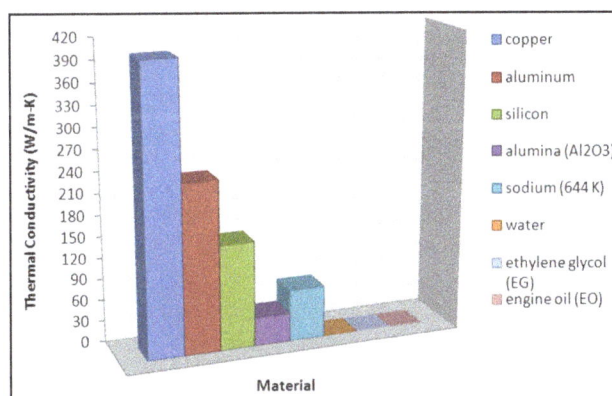

Figure 1. Thermal conductivities of various solids and liquids [2].

The present review provides a comprehensive overview of the attractive research progress made in the area of nanofluids.

It also summarizes the experimental, theoretical, and computational developments in this field.

2. Preparation Methods for Nanofluids

Various methods have been tried to produce different kinds of nanoparticles and nanosuspensions. There are two primary methods to prepare nanofluids: A two-step method in which nanoparticles or nanotubes are first produced as a dry powder. The resulting nanoparticles are then dispersed into a fluid in a second step and Single-step nanofluid processing methods have also been developed and there are a novel methods also mentioned in this section.

2.1. Two-Step Method

This method is the most widely used for preparing nanofluids. Nanoparticles, nanofibers, nanotubes, or other nanomaterials used in this method are first produced as dry powders by chemical or physical methods. Then, the nano-sized powder will be dispersed into a fluid in the second processing step with the help of intensive magnetic force agitation, ultrasonic agitation, high-shear mixing, homogenizing, and ball milling. Two-step method is the most economic method to produce nanofluids in large scale, because nanopowder synthesis techniques have already been scaled up to industrial production levels. Due to the high surface area and surface activity, nanoparticles have the tendency to aggregate. The important technique to enhance the stability of nanoparticles in fluids is the use of surfactants. However, the functionality of the surfactants under high temperature is also a big concern, especially for high-temperature applications. Due to the difficulty in preparing stable nanofluids by two-step method, several advanced techniques are developed to produce nanofluids, including one-step method. In the following part, we will introduce single-step method.

2.2. One-Step Method

The nanoparticles may agglomerate during the drying storage, and transportation process, leading to difficulties in the following dispersion stage of two-step method. Consequently, the stability and thermal conductivity of nanofluid are not ideal. In addition, the production cost is high. To reduce the agglomeration of the nanoparticles, one-step methods have been developed. There are some ways for preparing nanofluids using this method including direct evaporation condensation [21, 22], chemical vapour condensation [23], and single-step chemical synthesis.

2.3. Other Novel Methods

Wei et al developed a continuous flow micro fluidic micro reactor to synthesize copper nanofluids. By this method, copper nanofluids can be continuously synthesized, and their microstructure and properties can be varied by adjusting parameters such as reactant concentration, flow rate, and additive. CuO nanofluids with high solid volume fraction (up to 10 vol%)

can be synthesized through a novel precursor transformation method with the help of ultrasonic and microwave irradiation [24]. The precursor $Cu(OH)_2$ is completely transformed to CuO nanoparticle in water under microwave irradiation. The ammonium citrate prevents the growth and aggregation of nanoparticles, resulting in a stable CuO aqueous nanofluid with higher thermal conductivity than those prepared by other dispersing methods. Phase-transfer method is also a facile way to obtain monodisperse noble metal colloids [25]. Phase transfer method is also applied for preparing stable kerosene based Fe_3O_4 nanofluids. Oleic acid is successfully grafted onto the surface of Fe_3O_4 nanoparticles by chemisorbed mode, which lets Fe_3O_4 nanoparticles have good compatibility with kerosene [26]. In a water cyclohexane two-phase system, aqueous formaldehyde is transferred to cyclohexane phase via reaction with dodecylamine to form reductive intermediates in cyclohexane. The intermediates are capable of reducing silver or gold ions in aqueous solution to form dodecylamine-protected silver and gold nanoparticles in cyclohexane solution at room temperature. Feng et al. used the aqueous organic phase transfer method for preparing gold, silver, and platinum nanoparticles on the basis of the decrease of the PVP's solubility in water with the temperature increase [27].

3. Advantages of Nanofluids

Nanofluids cause drastic change in the properties of the base fluid so, the following benefits are expected to get on.
- Due to nano size particles, pressure drop is minimum.
- Higher thermal conductivity of nanoparticles will increase the heat transfer rate.
- Successful employment of nanofluid will lead to lighter and smaller heat exchanger.
- Heat transfer rate increases due to large surface area of the nanoparticles in the base fluid.
- Nanofluids are most suitable for rapid heating and cooling systems.
- Due to nano size particles, fluid is considered as integral fluid.
- Good mixture nanofluids will give better heat transfer.
- Microchannel cooling without clogging. Nanofluids are not only a better medium for heat transfer in general but they are also ideal for microchannel applications where high heat loads are needed.
- Cost and energy saving. Successful employment of nanofluids will result in significant energy and cost savings because heat exchange systems can be made smaller and lighter.

4. Literature Survey in Recent Years

Many researcher tried to study the effect of nanoparticles on base fluids like water, ethylene glycol and oil and shows the enhancement that be get on because of the higher thermal conductivities for those fluids so there are some experimental and numerical studies have be done in recent years. *Table 1, 2 and 3* show some studies that presented in recent years.

Table 1. Some studies in 2013.

Authors	Type of investigation	Funding / results
C. Yang et. al [3]	Theoretical	It has been found that Nusselt number has optimal bulk mean nanoparticle volume fraction value for alumina water nanofluids, whereas it only increases monotonously with bulk mean nanoparticle volume fraction for titanium water nanofluids.
Y. Raja Sekhar et. al [4]	Experimental	The results found that The Nusselt number and friction factor increases with increase of particle concentration. But, friction factor decreases with increase of Reynolds number of flow where as the Nusselt number increases. Using nanofluid with a high heat exchange can help in reduce the size of the heat exchanger or without increasing the size of the heat exchanger efficiency of the system can be improved. Further, using twisted tapes and nanofluids in the pipe flows is advantageous since it is visible from the results that the energy gained with heat exchange is more than the energy spent on pumping power. It is clear from the results that heat transfer enhancement in a horizontal tube increases with Reynolds number of flow and nanoparticle concentration.
Adnan M. Hussein et. al [5]	Numerical	Data measured showed that thermal conductivity and viscosity increase with increasing the volume concentration of nanofluids with maximum deviation 19%and 6%, respectively. Simulation results concluded that the friction factor and Nusselt number increase with increasing the volume concentration. On the other hand, the flat tube enhances heat transfer and decreases pressure drop by 6% and −4%, respectively, as compared with circular tube.
Faris Mohammed Ali et. al [6]	Numerical	The results show that, the thermal conductivity and thermal diffusivity enhancement of nanofluids increases as the particle size increases. Thermal conductivity and thermal diffusivity enhancement of Al_2O_3 nanofluids was increase as the volume fraction concentration increases. This enhancement attributed to the many factors such as, ballistic energy, nature of heat transport in nanoparticle, and interfacial layer between solid/fluids.
S. Zeinali Heris et. al [7]	Experimental	Experiments show that considerable enhancement of heat transfer coefficient is achieved and this enhancement is up to 27.6% at 2.5% volume fraction of nanoparticles comparing to the base fluid (water), also it has been noticed that convective heat transfer coefficient increases with the increment of nanoparticles concentration in nanofluid especially at high flow rates. The decrement of wall temperature observed using nanofluid.

Table 2. Some studies in 2014.

Authors	Type of investigation	Funding/ results
A. Azari et. al [8]	Experimental and numerical	Experimental and simulation results showed that the thermal performance of nanofluids is higher than that of the base fluid and the heat transfer enhancement increases with the particle volume concentration and Reynolds number.
Hassanain Ghani Hameed et. al [9]	Numerical	Results found The nanoparticles within the liquid enhance the thermal performance of the heat pipe by reducing the thermal resistance and temperature difference by 0.168 K/W and 5.06 K respectively. While increasing the maximum heat load and the capillary pressure by 96 W and 192.46 Pa respectively. All these results at input heat of 30 W and nanoparticles concentration of 5 Vol. %. The results of wall temperature distribution for the heat pipe have been compared with the previous study for the same problem and a good agreement has been achieved.
Mohamed H. Shedid [10]	Numerical	Results of numerical simulations are compared and showed an enhancement of Nusslet number as Peclet number grows with increasing concentration ratio of Al_2O_3 and TiO_2 nanoparticles.
Layth W. Ismael et. al [11]	Experimental	The experimental results emphasized the enhancement of the thermal conductivity due to the nanoparticles presence in the fluid greater than microfluids, also shown the effect of the particle size and concentration on the thermal conductivity. It has been recognized that the addition of highly conductive particles can significantly increase the thermal conductivity of heat – transfer fluids. Particles in the micro and nano – size range have attracted the most interest because of their enhanced stability against sedimentation and, as a result, reduction in potential for clogging a flow system. Furthermore the results showed that, the obtained thermal conductivities doubtlessly revealed that size and type particles was a key factor affecting conductive heat transport in suspensions.
Hooman Yarmand et. al [12]	Numerical	The numerical results indicate that SiO_2-water has the highest Nusselt number compared to other nanofluids while it has the lowest heat transfer coefficient due to low thermal conductivity. The Nusselt number increases with the increase of the Reynolds number and the volume fraction of nanoparticles.
Adnan M. Hussein et. al [13]	Experimental	Results showed that the heat transfer in car radiator increases with increasing of nanofluid volume fraction by using TiO_2 and SiO_2 nanoparticles dispersed in water as a base fluid.
Sami D. Salman et. al [14]	Numerical	The results show that the heat transfer enhancement increases with an increase in the volume fraction of the CuO nanoparticle.

Table 3. *Some studies in 2015.*

Authors	Type of investigation	Funding/ results
Hsien-Hung Ting et. al [15]	Numerical	The numerical results show that the heat transfer coefficients and Nusselt numbers of Al_2O_3/water nanofluids increase with increases in the Peclet number as well as particle volume concentration. The heat transfer coefficient of nanofluids is increased by 25.5% at a particle volume concentration of 2.5% and a Peclet number of 7500 as compared with that of the base fluid (pure water).
Rabah Nebbati et. al [16]	Numerical	The results of thermal and hydrodynamic fields show that nanofluids can provoke an increase in the average and local Nusselt numbers, a decrease of bottom surface local temperature and a slight decrease of the shear stress on the wall, when compared to predictions using constant properties and nanoparticles free water.
Dr. Khalid Faisal Sultan [17]	Numerical	The numerical results show that as the solid volume fraction increases, the heat transfer is enhanced for all values of Rayleigh number. This enhancement is more significant at high Rayleigh number. The lowest heat transfer was obtained for TiO_2 (50 nm) due to domination of conduction and large nanoparticles. whereas Ag (20 nm), Cu (30 nm) – distilled water nanofluids has the highest heat transfer, respectively.
Dr. Khalid Faisal Sultan [18]	Experimental	The measured results show that silver with oil nanofluid gives maximum heat transfer enhancement compared with oxide zirconium nanofluid used. The presence of Ag and ZrOR2R nanoparticles attributes to the generation of strong nano convection current and better mixing also, The heat transfer coefficient and pressure drop is increased by using nanofluids (Ag + oil, ZrO_2 + oil) instead of the base fluid (oil).
Abdolbaqi Mohammed Khdher et. al [19]	Numerical	Results found that the heat transfer rates and wall shear stress increase with an increase of nanofluid volume concentration. In addition, the results of viscosity and thermal conductivity of the nanofluids show a significant increment with the increase of volume fractions. Therefore, optimal particle volume fraction is considered in enhancing the performance of nanofluid in an engineering system.
Bayram Sahin et. al [20]	Experimental	It was found that the particle volume concentrations higher than 1%vol. were not appropriate with respect to the heat transfer performance of the CuO-water nanofluid. No heat transfer enhancement was observed at Re = 4.000. The highest heat transfer enhancement was achieved at Re = 16.000 and Φ = 0.005.

5. Application of Nanofluids

Nanofluids can be used to improve heat transfer and energy efficiency in a variety of thermal systems. That's means can be used as a cooling fluids in many application and there are some common application:

1. Engine cooling
2. Nuclear cooling system
3. Cooling of electronic circuit
4. Refrigeration
5. Enhancement of heat transfer exchange
6. Thermal storage
7. Biomedical application
8. Cooling of microchips
9. In defense and space application
10. Transportation
11. Petroleum industry
12. Inkjet printing
13. Environmental remediation
14. Surface coating
15. Fuel additives
16. Lubricant

6. Conclusion

The present papers gives a review about the enhancement of heat transfer by using nanofluids by many authors that preformed an experimental and numerical investigations related to heat transfer enhancement using nanofluids. So, we need to understanding the fundamentals of heat transfer and wall friction from this review because has a significant importance for developing nanofluids for a wide range of heat transfer applications and we can concluded the following:

1. Heat transfer rate is directly proportional to the Reynolds number and peclet number of Nanofluid.
2. Increasing volume Concentration of nanoparticles increases the pressure drop of Nanofluids.
3. Spherical shaped nanoparticles increases the heat transfer rate of Nanofluids compared with other shaped nanoparticles.
4. The fine grade of Nanoparticles increases the heat transfer rate but it's having poor stability.
5. Increasing size of nanoparticales (diameter of NP) led to decreasing in heat transfer because area per unit volume decreases.

References

[1]   A. S. Dalkilic et.al ' Forced Convective Heat Transfer of Nanofluids - A Review of the Recent Literature' Current Nanoscience, 2012, 8, 949-969.

[2]   Xiang-Qi Wang et.al 'A review on nanofluids - part i: theoretical and numerical investigations' 'Brazilian Journal of Chemical Engineering' Vol. 25, No. 04, pp. 613 - 630, October - December, 2008.

[3] C. Yang, W. Li, A. Nakayama, 'Convective heat transfer of Nanofluids in a concentric annulus' 'International Journal of Thermal Sciences' 71 (2013) 249-257.

[4] Y. Raja Sekhara, K. V. Sharmab, R. Thundil Karupparaja, C. Chiranjeevia 'Heat Transfer Enhancement with Al2O3 Nanofluids and Twisted Tapes in a Pipe for Solar Thermal Applications' 'International Conference On DESIGN AND MANUFACTURING, IConDM 2013'Procedia Engineering 64 (2013) 1474 – 1484.

[5] Adnan M. Hussein, K. V. Sharma, R. A. Bakar, and K. Kadirgama.'The Effect of Nanofluid Volume Concentration on Heat Transfer and Friction Factor inside a Horizontal Tube' Hindawi Publishing Corporation 'Journal of Nanomaterials' Volume 2013, Article ID 859563, 12 pages.

[6] Faris Mohammed Ali, W. Mahmood Mat Yunus and Zainal Abidin Talib "Study of the effect of particles size and volume fraction concentration on the thermal conductivity and thermal diffusivity of Al2O3 nanofluids'"International Journal of Physical Sciences' Vol. 8(28), pp. 1442-1457, 30 July, 2013.

[7] S. ZeinaliHeris, Taofik H. Nassan, S. H. Noie, H. Sardarabadi, M. Sardarabadi "Laminar convective heat transfer of Al2O3/water nanofluid through square cross-sectional duct" 'International Journal of Heat and Fluid Flow' 2013.

[8] A. Azari, M. Kalbasi and M. Rahimi "CFD AND EXPERIMENTAL INVESTIGATION ON THE HEAT TRANSFER CHARACTERISTICS OF ALUMINA NANOFLUIDS UNDER THE LAMINAR FLOW REGIME" 'Brazilian Journal of Chemical Engineering' Vol. 31, No. 02, pp. 469 - 481, April-June, 2014

[9] Hassanain Ghani Hameed, Proof. Dr. Abudl-Muhsin A. Rageb "NUMERICAL SIMULATION OF THERMAL PERFORMANCE OF CONSTANT CONDUCTANCE CYLINDRICAL HEAT PIPE USING NANOFLUID" Al-Qadisiya Journal For Engineering Sciences, Vol. 7, No. 4, 2014.

[10] Mohamed H. Shedid" Computational Heat Transfer for Nanofluids through an Annular Tube" "Proceedings of the International Conference on Heat Transfer and Fluid Flow" Prague, Czech Republic, August 11-12, 2014.

[11] Layth W. Ismael, Dr. Khalid Faisal. Sultan" A comparative Study on the Thermal Conductivity of Micro and Nano fluids by Using Silver and Zirconium Oxide" Al-Qadisiya Journal For Engineering Sciences, Vol. 7, No. 2, 2014.

[12] Hooman Yarmand, Samira Gharehkhani, Salim Newaz Kazi, Emad Sadeghinezhad, and Mohammad Reza Safaei " Numerical Investigation of Heat Transfer Enhancement in a Rectangular Heated Pipe for Turbulent Nanofluid" Hindawi Publishing Corporation, Scientific World Journal. Volume 2014, Article ID 369593, 9 pages.

[13] Adnan M. Hussein , R. A. Bakar, K. Kadirgama, K. V. Sharma" Heat transfer augmentation of a car radiator using nanofluids " Heat and Mass Transfer journal, November 2014, Volume 50, Issue 11, pp 1553-1561.

[14] Sami D. Salman, Abdul Amir H. Kadhum, Mohd S. Takriff, and Abu Bakar Mohamad "Heat Transfer Enhancement of Laminar Nanofluids Flow in a Circular Tube Fitted with

Parabolic-Cut Twisted Tape Inserts" Hindawi Publishing Corporation, Scientific World Journal Volume 2014, Article ID 543231, 7 pages.

[15] Hsien-Hung Ting and Shuhn-Shyurng Hou" Investigation of Laminar Convective Heat Transfer for Al2O3-Water Nanofluids Flowing through a Square Cross-Section Duct with a Constant Heat Flux" Materials 2015, 8, 5321-5335.

[16] Rabah Nebbati, Mahfoud Kadj" Study of forced convection of a nanofluid used as a heat carrier in a microchannel heat sink" "International Conference on Technologies and Materials for Renewable Energy, Environment and Sustainability, TMREES15", 2015.

[17] Khalid Faisal Sultan "Numerical Solution of Heat Transfer and Flow of Nanofluids in Annulus With Fins Attached on the Inner Cylinder" Journal of Babylon University/Engineering Sciences/ No.(2) / Vol.(23): 2015.

[18] Dr. Khalid Faisal Sultan "Augmentation of Heat Transfer Through Heat Exchanger With and Without Fins by Using Nano fluids" Journal of Engineering and Development, Vol. 19. No. 4, July 2015, ISSN 1813- 7822.

[19] Abdolbaqi Mohammed Khdher, Rizalman Mamat, Nor Azwadi Che Sidik "The Effects of Turbulent Nanofluids and Secondary Flow on the Heat Transfer through a Straight Channel" Recent Advances in Mathematical and Computational Methods, 2015.

[20] Bayram Sahin, Eyuphan Manay and Eda Feyza Akyurek " An Experimental Study on Heat Transfer and Pressure Drop of CuO-Water Nanofluid" Hindawi Publishing Corporation, Journal of Nanomaterials, Volume 2015, Article ID 790839, 10 pages.

[21] J. A. Eastman, U. S. Choi, S. Li, L. J. Thompson, S. Lee, Proceedings of the Materials Research Symposium(Nanophase and Nanocomposite Materials II), 1997.

[22] J. A. Eastman, S. U. S. Choi, S. Li, G. Soyez, L. J. Thompson, R. J. DiMelfi, J. Metastab. Nanocryst. 2(1998) 629.

[23] V. V. Srdic, M. Winterer, A. Mller, G. Miehe, H. Hahn, J. Am. Ceram. Soc. 84 (2001) 277.

[24] H. T. Zhu, C. Y. Zhang, Y. M. Tang, and J. X. Wang, "Novel synthesis and thermal conductivity of CuO nanofluid," Journal of Physical Chemistry C, vol. 111, no. 4, pp. 1646–1650, 2007.

[25] Y. Chen and X. Wang, "Novel phase-transfer preparation of monodisperse silver and gold nanoparticles at room temperature," Materials Letters, vol. 62, no. 15, pp. 2215– 2218, 2008.

[26] W. Yu, H. Xie, L. Chen, and Y. Li, "Enhancement of thermal conductivity of kerosene-based Fe3O4 nanofluids prepared via phase-transfer method," Colloids and Surfaces A, vol. 355, no. 1–3, pp. 109–113, 2010.

[27] X. Feng, H. Ma, S. Huang et al., "Aqueous-organic phasetransfer of highly stable gold, silver, and platinum nanoparticles and new route for fabrication of gold nanofilms at the oil/water interface and on solid supports," Journal of Physical Chemistry B, vol. 110, no. 25, pp. 12311–12317, 2006.

23

Green biosynthesis of gold nanoparticles and biomedical applications

Tuhin Subhra Santra[1], Fan-Gang Tseng[1, 2, 3], Tarun Kumar Barik[4]

[1]Department of Engineering and Systems Science, National Tsing Hua University, Hsinchu, Taiwan
[2]Institute of Nanoengineering and Microsystems (NEMS), National Tsing Hua University, Hsinchu, Taiwan
[3]Division of Mechanics, Research Center for Applied Sciences, Academia Sinica, Taipei, Taiwan
[4]Department of Applied Sciences, Haldia Institute of Technology, Haldia, West Bengal, India

Email address:
tarun.barik2003@gmail.com (T. K. Barik), santra.tuhin@gmail.com (T. S. Santra)

Abstract: Nanotechnology is an emerging field of science and technology with numerous applications in biomedical fields and manufacturing new materials. To extract gold nanoparticles with different techniques, green biosynthesis is in under exploration due to its cost effective ecofriendly preparation with controllable shape, size and disparity, tremendous physical and chemical inertness, optical properties related with surface plasmon resonance, surface modification, surface bio-conjugation with molecular probes, excellent biocompatibility and less toxicity. This review article presents the overview of green biosynthesis of gold nanoparticles (AuNP) and their recent biomedical applications.

Keywords: Biosynthesis, Gold Nanoparticles (AuNP), Biomedical Applications

1. Introduction

Nanomaterials are defined as zero or one dimensional materials with approximate size 1-100 nm. Nanoparticles are objected with three dimensions at nanoscale level. Engineered nanoparticles can be formed with different shape, size and surface chemistry, which influence optical, electronic, thermal and mechanical properties of the materials for their wide range of applications in nanotechnology. Mostly, high surface to volume ratio of nanoparticles is the key factor to enhance unique material properties. Among the variety of nanoparticles with their applications, metalic nanoparticles (like gold and silver nanoparticles) are playing most prominent role in biology and medicine [1-3]. Gold nanoparticles are enormously used for different applications such as optoelectronic devices, ultrasensitive chemical, biological sensors, catalysts, separation science, biomedical applications like drug delivery, cancer treatment, DNA, RNA analysis, gene therapy, antibacterial agent etc. [4-10]. Fig. 1 shows different recent important applications of green biosynthesized gold nanoparticles (AuNP). The synthesis of colloidal gold nanoparticles has been extensively studied for long time [11]. In 1951, Turkevich et al. suggested the synthesis of gold nanoparticles (AuNP) by reduction of Au^{3+}

ions to Au^0 with the use of citric acid. This method can also be stabilized to form monodispersed nanoparticles and it could be exchanged to other ligand [12].

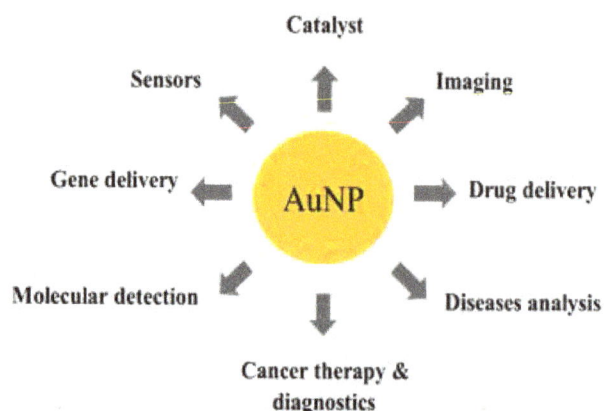

Fig. 1. Different important applications of green biosynthsized gold nanoparticles (AuNP).

In 1994, Brust et al. suggested the production of gold nanoparticles, where sodium borhydride was used as reducing agent and citric acid was immediately replaced by selected mercaptan. This method produced monodispersed nanoparticles which was easy to disperse in organic solvent and to reisolate as pure powders [13]. Using this method, different gold nanoparticles with their modified properties such as reactivity and solubility, through changing molecular structure of thiolates on the particle surface can be produced [14-17]. However, production of gold nanoparticles with these techniques is not ecofriendly due to toxic mercaptans and organic solvents, which greatly limit to its application in biomedical field especially for clinical purpose. Thus, "green chemistry" ensures clean, non-toxic and environment-friendly methods to produce nanoparticles with well-defined shape, and controllable size [18]. The green biosynthesis of nanoparticles have more advantages over chemical or physical methods such as it has significant application in biomedical fields due to its excellent chemical stability, biocompatibility, cost effectiveness, easy preparation, optical properties related with surface plasmon resonance, convenient surface bioconjugation with molecular probes and low toxicity [19-23]. To extract nanoparticles by green synthesis mechanism, three essential features are environmentally acceptable solvent system, eco-friendly reducing and capping agents. The green biosynthesis technique are synthetic route to use relatively non-toxic chemicals for the preparation of environment friendly stable functionalize nanoparticles with the use of non-toxic solvents such as water, plant extracts, biological systems etc.

2. Green Biosynthesis of Gold Nanoparticles (AuNP)

In recent years, biosynthesized metallic nanoparticles using plant extract has been received more attention due to simple and viable alternative against chemical and physical methods with their potential applications in nanomedicine.

To prepare gold nanoparticles (AuNP), initially plant leafs are collected and completely dried with expose of sunlight. After that aqueous plant leaf extract need to prepare by mixing of 100 mL deionized (DI) water with 10 g dried leaf powder in a flask and boiled for 10-20 minutes. Then leaf extract need to add into metallic salt solution (1 mM chloroauric acid (HAuCl$_4$) solution) with 60 ^{0}C to 80 ^{0}C temperature and

finally the color will change after 15-20 minutes which indicates the formation of gold nanoparticles (AuNP) [24-27]. This bioreduction of metal salt to metal nanoparticles are highly stable without impurities. Gardea-Torresdey et al. reported firstly the formation of biosynthesized gold nanoparticles using living plants [28]. They have used Alfalfa plants, which were grown in an AuCl$_4$ rich environment and found the nucleation and growth of gold nanoparticles (AuNP) inside the plant extract. The nanoparticles are in crystalline in nature with minimum 4 nm size, while large coalesced nanoparticles ranging from 20-40 nm. Dwivedi et al. reported biosynthesis of gold nanoparticles using Chenopodium album leaf extract. They have used aqueous leaf extract as mild reducing agent for nanoparticles synthesis. The biosynthesized gold nanoparticles were in quasi-spherical shapes within 10-30 nm range [29]. Fig. 2 shows the photograph of Chenopodium album leaf, transmission electron microscopic (TEM) image of leaf extract synthesized gold nanoparticles (AuNP) and their particle size distribution. The stability of the nanoparticles was determined at different pH with zeta potentiometer without adding any stabilizing agents [29].

Shiv Shankar et al. have synthesized stable gold nanoparticles (AuNP) from geranium leaves (Pelargonium graveolens) with variable size including rod, flat sheet and triangle. The shapes of particles are predominantly decahedral and icosahedral with 20-40 nm in sizes and their transmission electron microscopy (TEM) results suggest that they are multiply twinned particles (MTPs) [30]. Afterward, they synthesized thin, flat, single-crystalline gold nanotriangles (AuNP) from lemongrass plant extract, when it reacted with aqueous chloroaurate ions and then the process involve rapid reduction assembly and room-temperature sintering, resulting the formation of spherical shape "liquid like" gold nanoparticles (AuNP) [31]. The authors also extract gold nanotriangles by using lemongrass plant and found their potential application in infrared-absorbing optical coating. Fig. 3 shows transmission electron microscopic (TEM) image of gold nanoparticles (AuNP) with different shapes and sizes synthesized using lemongrass extract [32].

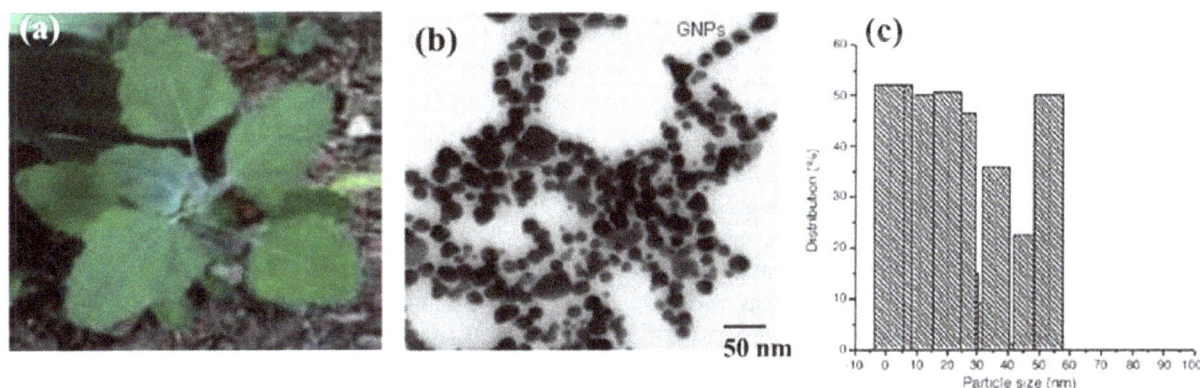

Fig. 2. (a) Photograph of Chenopodium album leaf, (b) transmission electron microscopy image of gold nanoparticles (AuNP or GNPs) and (c) histograms of particles size distribution. Permission to reprint obtained from Elsevier [29].

Fig. 3. Transmission electron microscopy (TEM) image of gold nanoparticles (AuNP) by the reduction of 5 mL of 10-3 M aqueous HAuCl₄ solution with (a) 0.2, (b) 0.3, (c) 0.5 and (d) 1.0 mL of lemongrass extract. Permission to reprint obtained from American Chemical Society (ACS) [32].

Armendariz et al. have shown gold nanoparticles (AuNP) formation with controllable size by Avena sativa biomass. They found Au(III) ions were bound to oat biomass in a pH-dependent manner and observed gold nanoparticles with fcc tetrahedral, decahedral, hexagonal, icosahedral multitwinned, irregular and rodlike shapes. The particles size influence by pH reaction, where larger particles (approximately 25-85 nm) are formed at pH 2 and smaller particles (approximately 20 nm) are observed at pH 3 and 4 [33]. Ghule et al. have reported microscale size triangular gold prisms are synthesized using bengal gram beans (Cicerarietinum L.) extract [34]. The extracellular transport of biomolecules and proteins from protein rich gram beans mediate the reduction of aqueous Au^{3+} ions and direct growth of triangular prisms. By varying the composition of gram bean extract and aqueous Au^{3+} solution, they control the morphology of gold nanoparticles [34].

Gold nanoparticles (AuNP) from sundried Cinnamomumcamphora leaf are extracted by Huang et al. using bioreduction process. The chloroauric acid (HAuCl₄) and dried powder of Cinnamomumcamphora leaf was used to synthesis the nanoparticles. Initially the leaf powder was added into 50 ml aqueous HAuCl₄ solution in the content of 100 ml conical flask at room temperature. Then the flask rotates with 150 rpm at 30 °C in a dark place resulting biomass reaction with solution and finally precipitate at the bottom of the flask within 1 hour. The suspended solution above the precipitate was collected for TEM observation. Using this technique, spherical gold nanoparticles (AuNP) were produced with different shapes ranging from 20-100 nm [27]. Again, Badrinarayanan et al. extracted gold nanoparticles (AuNP) by using Coriander leaf extract as reducing agent [35]. Here, the reduction of gold ions by Coriander leaf extract results the formation of stable morphological gold nanoparticles (size range 6.75-57.91 nm) with spherical, triangular, truncated triangular and decahedral shapes etc. The rate of reaction is rapid (12 h only) to synthesis nanoparticles by this method, when compared to microbes-mediated synthesis (24–120 h) [36, 37]. Armendariz et al. have extracted gold nanoparticles (AuNP) by the interaction of Au(III) ions with oat and wheat biomasses using

cetyltrimethylammonium bromide (CTAB) or sodium citrate at pH 4. The extraction was occurred under mild condition of pH with only sonication of the samples. The sizes of the extracted gold nanoparticles using CTAB were less than 20 nm in diameter [38]. Inbakandan et al. have shown the synthesis of gold nanoparticles (AuNP) from gold precursor using the extract derived from the marine sponge, Acanthellaelongata (Dendy, 1905) belonging to the primitive phylum porifera. To produce gold nanoparticles, the marine sponge extract were added to 10.3 M HAuCl₄ aqueous solution at 45 °C with continuous string for 4 hours. The particles were monodispersed and spherical in shapes with size range 7-20 nm. However, the average diameters of maximum particles were 15 nm [39]. Elavazhagan et al. have synthesized silver and gold nanoparticles by the use of an aqueous leaf extract of Memecylonedule (Melastomataceae). Their scanning electron microscopy (SEM) image analysis shows that aqueous gold ions when exposed to M. edule leaf broth, gold nanoparticles (AuNP) are formed with size range 20-50 nm. However, for TEM analysis shows that formation of gold nanoparticles with triangular, circular, and hexagonal shapes with size range 10-45 nm [40]. Arunachalam et al. investigated the effect of phytochemicals present in Memecylonumbellatum leaf extract during formation of stable silver and gold nanoparticles, and found the existence of saponins, phenolic compounds, phytosterols, quinines etc. They shows that most of the phytochemicals present in the plant extract and play an important role to form silver and gold nanoparticles. In their investigation, the sizes of silver and gold nanoparticles were 15-20 nm and 15-25 nm, respectively [41]. Yasmin et al. shows the fast synthesis of gold nanoparticles (AuNP) by using a medicinal plant (Hibiscus rosa-sinensis) extract and microwave heating. To form gold nanoparticles, different conditions were optimized by varying of plant extract concentration, gold salt solution concentration, microwave heating time and power of microwave hating. The average diameter of stable spherical nanoparticles was 16-30 nm [42].

Except plant extract, gold nanoparticles (AuNP) can be synthesized using different bacteria such as Pseudomonas stulzeri, Escherichia coli, Vibrio cholera, Pseudomonas aeruginosa, Salmonell styplus, Staphylococcus currens etc. [43-49]. The bacterial synthesis of gold nanoparticles is also ecofriendly and costs effective due to its environmental compatibility, lower energy consumption etc., when compared with other physical and chemical synthesis processes. The fungi are also extremely good candidate for synthesis of gold nanoparticles (AuNP). Different reported fungi used for this purpose are Aspergillusfumigatus, Fusariumoxysporum, Penicilliumbrevicompactum, Fusariumsemitectum, Penicilliumfellutanum, Cladosporiumcladosporioides, Volvariellavolvacea etc. [50].

3. Applications

Biosynthesized gold nanoparticles have tremendous applications in different fields such as biological and chemical sensors, heavy metal ion detection, catalysts, separation science, electrical coatings etc. [51-56]. The synthesis of gold nanoparticles (AuNP) using plant extract, is advantageous over biological process by eliminating the elaborate process to maintain cell culture and also suitable for large scale nanoparticle synthesis [57]. Due to high biocompatibility, chemical stability, convenient surface bioconjugation with molecular probes, excellent surface plasmon resonance and low toxicity, biosynthesized gold nanoparticles have diverse biomedical applications including drug delivery, cancer treatment, DNA-RNA analysis, gene therapy, sensing and imaging, antibacterial agent etc. [58-60]. Gold nanoparticles have tremendous optical and electronic properties, as a result it can act as biosensors to detect biomolecules. Zheng et al. reported that, biosynthesized Au-Ag alloy nanoparticles by yeast cells can be applied for electrochemical vanillin sensor. They reported that Au–Ag alloy nanoparticles modified glassy carbon electrode was able to enhance the electrochemical response of vanillin for at least five times. In ideal condition, the peak current of vanillin can linearly increase with concentration range of 0.2–50 µM with a low detection limit of 40 nM. Using this sensor, vanillin can be detected successfully from vanillin beam and vanillin tea [61]. On the other hand, another group of scientist reported a novel nonenzymatic amperometric biosensor of hydrogen peroxide (H_2O_2) by using one-pot green synthesis to prepare a self-assembled membrane of reduced graphene oxide–gold nanoparticles (RGO–AuNP) nanohybrids at liquid–air interface [62]. The Brownian motion, electrostatic interaction between RGO and AuNP and the encapsulation of AuNP in the hybrid membrane influence the formation of RGO–AuNP hybrid membrane. The RGO–AuNP hybrid membranes are very stable in various organic and inorganic solvents. This H_2O_2 biosensor has wide linear range 0.25-22.5 mM, low detection limit 6.2 µM (S/N = 3), high selectivity and long-term stability. Hu et al. have reported green-synthesized gold nanoparticles decorated graphene sheets for label-free electrochemical impedance DNA hybridization biosensing [63]. In their work initially the graphene sheets were functionalized with 3,4,9,10-perylene tetracarboxylic acid (PTCA). PTCA molecules can separate graphene sheets and introduced more negative –COOH, which can potentially beneficial for the decoration of graphene with gold nanoparticles (AuNP). Then the amine-terminated ionic liquid (NH_2-IL) was applied for reduction of $HAuCl_4$ to gold nanoparticles. The DNA probes immobilized via electrostatic interaction and adsorption effect due to graphene sheet and NH_2-IL protected AuNP. For label free DNA detection, electrochemical impedance value increases after DNA probes immobilization. This sensor can successfully detect the sequence of pol gene of human immunodeficiency virus 1. Due to high surface to volume ratio, gold nanoparticles have very high surface plasmon resonance and can detect biomolecules. Kuppusamy et al. have detected HCG hormone in pregnant women urine sample using biosynthesized gold nanoparticles synthesized using C. nudiflora plant extract [64]. It can be used to detect HCG hormone on both pregnancy positive and negative urine sample. Initially, 500 µl of AuNP solution was mixed with same volume of the test sample and used it for assays. Then the solution was tested using a

pregnancy test strip. When gold nanoparticles in the urine sample changed color into pink indicated the pregnancy, while the gray color indicated the absence of pregnancy. Authors claim this method is 100% accurate for pregnancy diagnosis and it can be used as alternative method for urine pregnancy test. Sayed et al. used biosynthesized monodispersed gold nanoparticles (AuNP) for cytotoxic assay test, biodistribution and bioconjugation with the anticancer drug doxorubicin. They produced monodispersed gold nanoparticles using thermophilic fungus Humicola spp. by green synthesis mechanism. They found Humicola spp. can reduce the precursor solution ($HAuCl_4$) at just 50 °C to form uniform spherical morphology with high stability gold nanoparticles with size 18-24 nm. The nanoparticles are capped by natural proteins and can be directly attached with multiple-receptors such as LHRH, EGFR and EpCAM without targeting agent involvement. These nanoparticles can also bind with integrins and VEGFs for the development of novel anti-angiogenesis strategy for wide range of tumor treatment [65]. Thus the gold nanoparticles (AuNP) can use for drug delivery and cancer treatment. Malathi et al. proposed green synthesis of gold nanoparticles using chitosan as a reducing/capping agent for controlled delivery. The authors designed biocompatible carrier prepared by using single oil-in water (O/W) emulsion for controlled release of hydrophobic drugs. The drug loaded with spherical nanoparticles of size 50 nm, while the average nanoparticles size was 2-3 nm. They also investigated controlled release of rifampicin (RIF) by in vitro studies by using phosphate buffer saline (PBS) at pH=7.4. The encapsulated drug can release at 37 ^{0}C with 71% loading efficiency. They again investigated the antibacterial activity of RIF loaded nanoparticles by Gram +ve (Bacillus subtils) and Gram -ve (Pseudomonas aeruginosa) bacteria and drug loaded nanocarrier for treating cancer diseases [66]. Fazal et al. have shown anisotropic gold nanoparticles synthesis by using green synthesis for photothermal cancer therapy [67]. The anisotropic gold nanoparticles were synthesized using an aqueous route with cocoa extract which served as both reducing and stabilizing agent. The sizes of nanoparticles are approximately 150-200 nm, which shows good biocompatibility with A431, MDA-MB231, L929, and NIH-3T3 cell lines in vitro experiment with concentration of 200 µg/mL. The successful photothermal ablation was tested with epidermoid carcinoma A431 cancer cells upon irradiation with a femtosecond laser pulse of wavelength 800 nm at low power density (6 W/cm^2). This report also claims, first time green synthesized anisotropic and cytocompatible gold nanoparticles are successfully able to phototheramal therapy without using any capping agents. Krishnaraj et al. have shown in vitro cytotoxic effect of biosynthesized silver and gold nanoparticles against MDA-MB-231, human breast cancer cells [68]. The various silver and gold nanoparticles with concentrations ranging from 1-100 µg/ml were used for acridine orange and ethidium bromide (AO/EB) dual staining, MTT, caspase-3 and DNA fragmentation assays. The nanoparticles with concentration 100 µg/ml showed cytotoxic effects and the apoptotic with human breast cancer cells which confirmed using caspase-3 activation and DNA fragmentation assays. Hampp et al. have shown the adhesion of biosynthesized gold nanoparticles for breast cancer detection and treatment [69]. The well developed, spherical, homogeneous gold nanoparticles were synthesized using a common soil bacterium, Bacillus megaterium. The authors showed the adhesion forces between biosynthesized AuNP and breast cancer cells were almost six times greater than adhesion forces between biosynthesized AuNP and normal breast cells by using atomic force microscopy (AFM). They also reported that the adhesion force between biosynthesized AuNP and breast cancer cells were three times greater than chemically synthesized AuNP and breast cancer cells. According to their results, biosynthesized AuNP conjugated to breast-specific antibodies (AuNP-Ab conjugates) and breast cancer cells were five times greater than adhesion forces between unconjugated AuNP and breast cancer cells. These results might be useful for the development of nanostructures for targeted detection and breast cancer treatment. Craig et al. have shown functionalization of gold nanoparticles for cancer imaging. The mAb-F19-conjugated gold nanoparticles were prepared and used to label human pancreatic adenocarcinoma. Initially gold nanoparticles were coated with dithiol bearing hetero-bifunctional PEG (polyethylene glycol), and cancer-specific mAb F19. These bioconjugated nanoparticles are completely stable and used to label sections of healthy and cancerous human pancreatic tissue [70]. Mukherjee et al. have shown potential diagnostic and therapeutic applications of one-step *in situ* biosynthesized gold nanoconjugates (2-in-1 system) in cancer treatment. The gold nanobioconjugates (AuNPs-OX) were extracted using Olax Scandens leaf. From TEM observation, the gold nanoparticles were in spherical (5–15 nm), few rod shape (18–55 nm), dumbbell shape (30–55 nm), triangular (30–100 nm) and hexagonal shape (15-35 nm). The AuNP-OX nanobioconjugactes interact with different cancer cell lines such as lung (A549), breast (MCF-7) and colon (COLO205) show the significant inhibition of cancer cell proliferation in comparison with pristine Olax scandens leaf extract. The lung cancer (A549) incubated with AuNP-OX, shows significant brighter red fluorescence, when it is compared with cells incubated Olax extract leaf. Their results suggest that, the green synthesized AuNP-OX might be usefull in "2-in-1 system" for potential cancer diagnostics and therapeutic applications [71]. The production of biosynthesized gold nanoparticle and their applications has been rapidly growing interest from last decade. However, for stable nanoparticles synthesis the efficiency, controllable particles size and morphology need to improve in near future. Thus, using biological method, there is till lack of technological improvement [72-74] and we believe that, after a decade, biosynthesized gold nanoparticles (AuNP) will be widely applied in biomedical research, disease diagnosis and treatment.

4. Conclusions

In green biosynthesis mechanism, gold nanoparticles (AuNP) can be efficiently extracted by using different plants, bacteria and fungi. Due to high biocompatibily with chemical stability, surface bioconjugation, high surface plasmon resonance, higher surface to volume ratio, lower toxicity, gold nanoparticles (AuNP) can be used in various biomedical applications related to cancer diagnostics and therapeutics for the benefit of human civilization. This biocompatible nature of the gold nanoparticles is safe and efficacy for consumer health and environment. The gold nanoparticles can act as catalyst resulting to improve drug delivery efficiency, especially for the interaction between anticancer drug and DNA. However, till to date, this green biosynthesis technique is in under developed stage. This article provides some idea about biosynthesized gold nanoparticles and their applications for creating interest of the readers in this important research field. Researchers must need to give more attention to develop stable gold nanoparticles (AuNP) from different biological systems, which might be beneficial for future clinical trials.

Acknowledgement

We thank to Dr. Mukesh Singh, Assistant Professor, Department of Biotechnology, Haldia Institute of Technology, Haldia, for his valuable discussions with us to prepare this manuscript.

References

[1]   R. Raghavendra, K. Arunachalam, S. K. Annamalai and A. M. Arunachalam, "Diagonistics and therapeutic application of gold nanoparticles," International Journal of Pharmacy and Pharmaceutical Sciences, Vol. 6, pp. 74-87, 2014.

[2]   J. Siemieniec and P. Kruk, "Synthesis of silver and gold nanoparticles using method of green chemistry," CHEMIK, Vol. 67, pp. 842-847, 2013.

[3]   W. Cai, T. Gao, H. Hong and J. Sun, "Applications of gold nanoparticles in cancer nanotechnology," Nanotechnology, Science and Applications, Vol. 1, pp. 17–32, 2008.

[4]   H. Liao, C. L. Nehl and J. H. Hafner, "Biomedical applications of plasmon resonant metal nanoparticles," Nanomedicine, Vol. 1(2), pp. 201-208, 2006.

[5]   J. J. Diao and Q. Cao, "Gold nanoparticle wire and integrated wire array for electronic detection of chemical and biological molecules," AIP Advances, Vol. 1, pp. 012115-1-012115-5, 2011.

[6]   E. Hutter and D. Maysinger, "Gold nanoparticles and quantum dots for bioimaging," Microscopy Research and Technique, Vol. 74, pp. 592-604, 2011.

[7]   G. Schider, J. R. Krenn, A. Hohenau, H. Ditlbacher, A. Leitner and F. R. Aussenegg, "Plasmon dispersion relation of Au and Ag nanowires," Physical Review B, Vol. 68, pp. 155427-1-155427-4, 2003.

[8]   X. Lou, Z. Yi, J. Qin and Z. Li, "A highly sensitive and selective fluorescent probe for cyanide based on the dissolution of gold nanoparticles and its application in real samples," Chemistry-A European Journal, Vol. 17, pp. 9691-9696, 2011.

[9]   P. Yanez-Sedenoand J. M. Pingarron, "Gold nanoparticle-based electrochemical biosensors," Analytical and Bioanalytical Chemistry, Vol. 382, pp. 884-886, 2005.

[10] C. D. Gaddes, A. Perfenov, I. Gryczynski and J. R. Lakowicz, "Luminescent blinking of gold nanoparticles," Chemical Physics Letters, Vol. 380, pp. 269-272, 2003.

[11] M. A. Hayat, (Ed.), "Colloidal Gold: Principles, Methods and Applications," San Diego, CA: Academic Press, Vols. 1 and 2, 1989.

[12] J. Turkevich and P. H. J. Stevenson, A study of nucleation and growth process in the synthesis of colloidal gold," Discuss. Faraday Soc., Vol. 11, pp. 55-75, 1951.

[13] M. Brust, M. Walker, D. Bethell, D. J. Schiffrin and R. Whyman, "Synthesis of thiol-derivatised gold nanoparticles in a two-phase liquid–liquid system," J. Chem. Soc. Chem. Commun., issue-7 pp. 801-808, 1994. DOI: 10.1039/C39940000801.

[14] L. O. Brown, and J. E. Hutchison, "Convenient preparation of stable, narrow-dispersity, gold nanocrystals by ligand exchange reactions," J. Am. Chem. Soc., Vol. 119, pp. 12384-12385, 1997.

[15] M. Brust, J. Fink, D. Bethell, D. J. Schiffrin and C. J. Kiely, "Synthesis and reactions of functionalised gold nanoparticles," J. Chem. Soc. Chem. Commun., pp. 1655–1656, 1995. DOI: 10.1039/C39950001655.

[16] M. J. Hostetler, S. J. Green, J. J. Stokes, and R. W. Murray, "Monolayers in Three Dimensions: Synthesis and Electrochemistry of ω-Functionalized Alkanethiolate-Stabilized Gold Cluster Compounds," Am. Chem. Soc., Vol. 118, pp. 4212-4213, 1996.

[17] R. S. Ingram, M. J. Hostetler and R. W. J. Murray, "Poly-hetero-ω-functionalized Alkanethiolate-Stabilized Gold Cluster Compounds," Am. Chem. Soc., Vol. 119, pp. 9175-1978, 1997.

[18] X. Li, H. Xu, Z-S. Chen and G. Chen, "Biosynthesis of nanoparticles by microorganism and their application," Journal of nanomaterials, Vol. 2011, 2011. Doi:10.1155/2011/270974.

[19] D. S. Goodsell, Editor, "Bionanotechnology: Lessons from Nature," John Wiley & Sons Inc. Publication, 2004.

[20] S. Guo and E. Wang, "Synthesis and electrochemical applications of gold nanoparticles," Analytica Chimica Acta, Vol. 598, pp. 181-192, 2007.

[21] A. R. Sperling, R. P. Gil, F. Zhang, M. Zanella and J. W. Parak, "Biological applications of gold nanoparticles," Chemical Society Reviews, Vol. 37, pp. 1896-1908, 2008.

[22] J. A. Ho, H. C. Chang, N. Y. Shih, L-C. Wu, Y-F. Chang, C-C. Chen and C. Chou, "Diagnostic detection of human lung cancer-associated antigen using a gold nanoparticle-based electrochemical immunosensor," Anal. Chem., Vol. 82(14), pp. 5944–5950, 2010.

[23] E. Boisselier and D. Astruc, "Gold nanoparticles in nanomedicine: preparations, imaging, diagnostics, therapies and toxicity," Chem. Rev., Vol. 38, pp. 1759–1782, 2009.

[24] Y. Konishi, T. Tsukiyama, K. Ohno, N. Saitoh, T. Nomura and S. Nagamine, "Intracellular recovery of gold by microbial reduction of $AuCl_4$ ions using the anaerobic bacterium Shewanella algae," Hydrometallurgy, Vol. 81(1), pp. 24–29, 2006.

[25] E. Castro-Longoria, A. R. Vilchis-Nestor and M. Avalos-Borja, "Biosynthesis of silver, gold and bimetallic nanoparticles using the filamentous fungus Neurosporacrassa," Colloids and Surfaces B, Vol. 83(1), pp. 42–48, 2011.

[26] S. S. Shankar, A. Rai, A. Ahmad and M.S astry, "Rapid synthesis of Au, Ag, and bimetallic Au core-Ag shell nanoparticles using Neem (Azadirachtaindica) leaf broth," Journal of Colloid and Interface Science, Vol. 275, pp. 496-502, 2004.

[27] J. Huang, Q. Li, D. Sun, Y. Lu, Y. Su, X. Yang, H. Wang, Y. Wang, W. Shao, N. He, J. Hong, and C. Chen, "Biosynthesis of silver and gold nanoparticles by novel sun dried Cinnamomumcamphora leaf," Nanotechnology, Vol. 18, p. 105104, 2007.

[28] J. L. Gardea-Torresdey, J. G. Parsons, E. Gomez, J. Peralta-Videa, H. E. Troiani, P. Santiago, and M. Jose Yacaman, "Formation and growth of Au nanoparticles inside live Alfalfa plants," Nano Letters, Vol. 2, pp. 397-401, 2002.

[29] A. D. Dwivedi and K. Gopal, "Biosynthesis of silver and gold nanoparticles using Chenopodium album leaf extract," Colloids and Surfaces A: Physicochem. Eng. Aspects, Vol. 369, pp. 27-33, 2010.

[30] S. S. Shankar, A. Ahmad, R. Pasrichaa and M. Sastry, "Bioreduction of chloroaurate ions by geranium leaves and its endophytic fungus yields gold nanoparticles of different shapes," J. Mater. Chem., Vol. 13, pp. 1822–1826, 2003.

[31] S. S. Shankar, A. Rai, B. Ankamwar, A. Singh, A. Ahmed and M. Sastry, "Biological synthesis of triangular gold nanoprisms," Nat. Mater., Vol. 3, pp. 482-488, 2004.

[32] S. S. Shankar, A. Rai, A. Ahmed and M. Sastry, "Controlling the optical properties of lemongrass extract synthesized gold nanotriangles and potential application in infrared-absorbing optical coatings," Chem. Mater., Vol. 17, pp. 566-572, 2005.

[33] V. Armendariz, I. Herrera, R. Jose, P. Videa, M. J. Yacaman, H. Troiani, P. Santiago, L. Jorge and L. Gardea-Torresdey, "Size controlled gold nanoparticle formation by Avena sativa biomass: use of plants in nanobiotechnology," J. Nanopart. Res., Vol. 6, pp. 377-382, 2004.

[34] K. Ghule, A. V. Ghule, J. Y. Liu and Y. C. Ling, "Microscale size triangular gold prisms synthesized using Bengal gram beans (Cicerarietinum L.) extract and $HAuCl_4 x3H_2 0$: a green biogenic approach," J. Nanosci. Nanotechnol., Vol. 6, pp. 3746–3751, 2006.

[35] K. Badrinarayanan and N. Sakthivel, "Coriander leaf mediated biosynthesis of gold nanoparticles," Mater. Lett., Vol. 62, pp. 4588-4590, 2008.

[36] B. Nair and T. Pradeep, "Coalescence of nanoclusters and formation of submicron crystallites Assisted by Lactobacillus Strains," Crystal Growth and Design, Vol. 2(4), pp. 293-298, 2002.

[37] T. K-Joerger, R. Joerjer, E. Olsson and Cl-G. Granqvist, "Bacteria as workers in the living factory: metal-accumulating bacteria and their potential for materials science," Trends in Biotechnology, Vol. 19, pp. 15-20, 2001.

[38] V. Armendariz, J. G. Parsons, M. L. Lopez, J. R. Peralta-Videa, M. J. Yacaman, and J. L. Gardea-Torresdey, "The extraction of gold nanoparticles from oat and wheat biomasses using sodium citrate and cetyltrimethylammonium bromide, studied by X-ray absorption spectroscopy, high-resolution transmission electron microscopy, and UV-visible spectroscopy," Nanotechnology, Vol. 20(10), pp. 105607, 2009.

[39] D. Inbakandan, R. Venkatesan and S. Ajmal Khan, "Biosynthesis of gold nanoparticles utilizing marine sponge Acanthella elongate (Dendy, 1905)," Colloids. Surf. B, Vol. 81, pp. 634-639, 2010.

[40] T. Elavazhagan and K. D. Arunachalam, "Memecylonedule leaf extract mediated green synthesis of silver and gold nanoparticles," International Journal of Nanomedicine, Vol. 6, pp. 1265-1278, 2011.

[41] K. D. Arunachalam, S. K. Annamalai and S. Hari, "One-step green synthesis and characterization of leaf extract-mediated biocompatible silver and gold nanoparticles from Memecylonumbellatum," International Journal of Nanomedicine, Vol. 8, pp. 1307-1315, 2013.

[42] A. Yasmin, K. Ramesh and S. Rajeshkumar, "Optimization and stabilization of gold nanoparticles by using herbal plant extract with microwave heating," Nano Convergence, Vol. 1, p. 12, 2014.

[43] M. F. Lengke, M. E. Fleet and G. Southam, "Morphology of gold nanoparticles synthesized by filamentous cyanobacteria from gold(I)–thiosulfate and gold (III)–chloride complexes," Langmuir, Vol. 22(6), pp. 2780–2787, 2006.

[44] G. Singaravelu, J. S. Arockiamary, V. G. Kumar and K. Govindaraju, "A novel extracellular synthesis of monodisperse gold nanoparticles using marine alga, SargassumwightiiGreville," Colloids and Surfaces B, Vol. 57(1), pp. 97–101, 2007.

[45] M. Agnihotri, S. Joshi, A. R. Kumar, S. Zinjarde and S. Kulkarni, "Biosynthesis of gold nanoparticles by the tropical marine yeast Yarrowialipolytica NCIM 3589," Materials Letters, Vol. 63 (15), pp.1231–1234, 2009.

[46] A. K. Suresh, D. A. Pelletier, W. Wang, M. L. Broich, J. W. Moon, B. Gu, D. P. Allison, D. C. Joy, T. J. Phelps and M. J. Doktycz, "Biofabrication of discrete spherical gold nanoparticles using the metal-reducing bacterium Shewanellaoneidensis," Acta Biomaterialia, Vol. 7(5), pp. 2148–2152, 2011.

[47] M. M. Juibari, S. Abbasalizadeh, G. S. Jouzani and M. Noruzi, "Intensified biosynthesis of silver nanoparticles using a native extremophilic Ureibacillus thermosphaericus strain," Materials Letters, Vol. 65(6), pp. 1014–1017, 2011.

[48] N. Sharma, A. K. Pinnaka, M. Raje, F. N. U. Ashis, M. S. Bhattacharyya and A. R. Choudhury, "Exploitation of marine bacteria for production of gold nanoparticles," Microbial Cell Factories, Vol. 11, p. 86, 2012.

[49] S. R. Radhika Rajasree and T. Y. Suman, "Extracellular biosynthesis of gold nanoparticles using a gram negative bacterium Pseudomonas fluorescence," Asian Pacific Journal of Tropical Disease, pp. S795-S799, 2012.

[50] Z. Sadowski, "Biosynthesis and application of silver and gold nanoparticles," Edited Book "Silver Nanoparticles", Editor – D. P. Perez, Chapter-13, InTech Open Access Publisher, pp. 257-276 (2010).

[51] V. Armendariz, J. L.Gardea-Torresdey, M. Jose-Yacaman, J. Gonzalez, I. Herrera and J. G. Parsons, "Gold nanoparticles formation by oat and wheat biomasses," in Proceedings –Waste Research Technology Conference at the Kansas City, Mariott-Country Club Plaza, July 30–Aug 1, (2002).

[52] A. Singh, M. Chaudhary and M. Sastry, "Construction of conductivemultilayer films of biogenic triangular gold nanoparticles and their application in chemical vapour sensing," Nanotechnology, Vol. 17, pp. 2399–2405, 2006.

[53] J. Liuand Y. Lu, "Colorimetric biosensors based on DNA zyme-assembled gold nanoparticles," J. Fluoresc., Vol. 14, pp. 343–354, 2004.

[54] D. Andreeva, "Low temperaturewater gas shift over gold catalysts," Gold Bull., Vol. 35, pp. 82–88, 2002.

[55] R. Grisel, K. J. Weststrate, A. Gluhoi and B. E. Nieuwenhuys, "Catalysis by gold nanoparticles," Gold Bull., Vol. 35, pp. 39–45, 2002.

[56] G. J. Hutchings and M. Haruta, "A golden age of catalysis: a perspective," Appl. Catal. A, Vol. 291, pp. 2–5, 2005.

[57] V. Kuamr and S. K. Yadav, "Plant-mediated synthesis of silver and gold nanoparticles and their applications," J. Chem. Technol. Biotechnol., Vol. 84, pp. 151-157, 2009.

[58] R. Groning, J. Breitkreutz, V. Baroth and R. S. Muller, "Nanoparticles in plant extracts: factors which influence the formation of nanoparticles in black tea infusions," Pharmazie, Vol. 56, pp. 790–792, 2001.

[59] D. Tang, R. Yuan and Y. Chai, "Ligand-functionalized core-shell Ag–Au nanoparticles label-free amperometricimmun-biosensor," Biotechnol. Bioeng.,Vol. 94, pp. 996–1004, 2006.

[60] G. F. Paciotti, L. Myer, D. Weinreich, D. Goia, N. Pavel, R. E. McLaughlin and L. Tamarkin, "Colloidal gold: a novel nanoparticle vector for tumor directed drug delivery," Drug Deliv., Vol. 11, pp. 169–183, 2004.

[61] D. Zheng, C. Hu, T. Gan, X. Dang and S. Hu, "Preparation and application of a novel vanillin sensor based on biosynthesis of Au–Ag alloy nanoparticles," Sensors and Actuators B: Chemical, Vol. 148 (1), pp. 247-252, 2010.

[62] P. Zhang, X. Zhang, S. Zhang, X. Lu, Q. Li, Z. Su and G. Wei, "One-pot green synthesis, characterizations, and biosensor application of self-assembled reduced graphene oxide–gold nanoparticle hybrid," Journal of Materials Chemistry B, Vol. 1, pp. 6525-6531, 2013.

[63] Y. Hua, S. Huab, F. Lia, Y. Jianga, X. Baib, D. Lib and L. Niua, "Green-synthesized gold nanoparticles decorated graphene sheets for label-free electrochemical impedance DNA hybridization biosensing," Biosensors and Bioelectronics, Vol. 26(11), pp. 4355-4361, 2011.

[64] P. Kuppusamy, M. M. Yusoff, G. P. Maniam and N. Govindan, "Biosynthesized gold nanoparticle developed as a tool for detection of HCG hormone in pregnant women urine sample" 1st International Conference on Molecular Diagnostic and Biomarker Discovery/Asian Pac. J. Trop. Dis., Vol. 4(3), pp. 223-252, 2014.

[65] A. Syed, R. Raja, G. C. Kundu, S. Gambhir and A. Ahmad, "Extracellular biosynthesis of monodispersed gold nanoparticles, their characterization, cytotoxicity assay, biodistribution and conjugation with the anticancer drug doxorubicin," Nanomedicine & Nanotechnology, Vol. 4(1), p. 156, 2013. http://dx.doi.org/10.4172/2157-7439.1000155.

[66] S. Malathi, M. D. Balakumaran, P. T. Kalaichelvan and S. Balasubramanian, "Green synthesis of gold nanoparticles for controlled delivery," Advanced Materials Letters, Vol. 4(12), pp. 933-940, 2013.

[67] S. Fazal, A. Jayasree, S. Sasidharan, M. Koyakutty, S. V. Nair and D. Menon, "Green synthesis of anisotropic gold nanoparticles for photothermal therapy of cancer," ACS Appl. Mater. & Interfaces, Vol. 6(11), pp. 8080-8089, 2014.

[68] C. Krishnaraj, P. Muthukumaran, R. Ramachandran, M. D. Balakumaran and P. T. Kalaichelvan, "Acalyphaindica Linn: Biogenic synthesis of silver and gold nanoparticles and their cytotoxic effects against MDA-MB-231, human breast cancer cells," Biotechnology Reports, Vol. 4, pp. 42-49, 2014.

[69] E. Hamppa, R. Botaha, O. S. Odusanyaa, N. Anukua, K. A. Malatestaa and W. O. Soboyejo, "Biosynthesis and adhesion of gold nanoparticles for breast cancer detection and treatment," Journal of Materials Research, Vol. 27(22), pp. 2891-2901, 2012.

[70] G. A. Craig, P. J. Allen and M. D. Mason, "Synthesis, characterization, and functionalization of gold nanoparticles for cancer imaging," Methods Mol. Biol., Vol. 624, pp. 177-193, 2010.

[71] S. Mukherjee, B. Vinothkumar, S. Prashanthi, P. R. Bangal, B. Sreedharb and C. R. Patra, "Potential therapeutic and diagnostic applications of one-step in situ biosynthesized gold nanoconjugates (2-in-1 system) in cancer treatment," RSC Advances, Vol. 3, pp. 2318-2329, 2013.

[72] L. Xiang, W. Bin, J. Huali, J. Wei, T. Jiesheng, G. Feng and L. Ying, "Bacterial magnetic particles (BMPs)-PEI as a novel and efficient non-viral gene delivery system," J. Gene Med., Vol. 9(8), pp. 679-90, 2007.

[73] R. Hergta, R. Hiergeista, M. Zeisbergera, D. Schülerb, U. Heyenb, I. Hilgerc and W. A. Kaiserc, "Magnetic properties of bacterial magnetosomes as potential diagnostic and therapeutic tools," Journal of Magnetism and Magnetic Materials, Vol. 293, pp. 80–86, 2005.

[74] R. Hergt and S. Dutz, "Magnetic particle hyperthermia—biophysical limitations of a visionary tumour therapy," Journal of Magnetism and Magnetic Materials, Vol. 311, pp. 187–192, 2007.

24

A flexible research reactor for atomic layer deposition with a sample-transport chamber for in Vacuo analytics

Axel Sobottka[1], Lutz Drößler[1], C. Hossbach[2], Bernd Abel[1], Ulrike Helmstedt[1]

[1]Leibniz-Institute of Surface Modification, Permoserstraße 15, 04318 Leipzig, Germany
[2]Technische Universität Dresden, Institute of Semiconductors and Microsystems, Nöthnitzer Straße 64, 01187 Dresden, Germany

Email address:

ulrike.helmstedt@iom-leipzig.de (U. Helmstedt)

Abstract: A modular reactor for thermal atomic layer deposition (ALD) was designed, which allows changes of all reactor components in order to obtain a flexible set-up for research purpose. A sample transport chamber is included for dual purpose. It allows for *in vacuo* transport of samples to analytical devices such as an XPS instrument. Surface activation of the samples is possible in the same chamber via an irradiation-induced approach.

Keywords: Atomic Layer Deposition, Reactor Design, *in Vacuo* Sample Transport, UV Irradiation

1. Introduction

Atomic layer deposition (ALD) is a method for gas phase deposition of conformal films with control over thickness and chemical composition at atomic level. It is known to produce nearly defect free thin films making it a valuable tool for production of e.g. high-quality dielectrics, luminescent films or gas permeation barriers [1]. The latter ones are in the focus of our research, which deals with technically feasible methods for producing gas barrier laminates. For protection of electronic devices like thin film solar cells single barrier SiO_2 films with water vapor transmission rates (WVTR) below 10^{-2} g m^{-2} d^{-1} were prepared with an irradiation-induced approach [2]. Al_2O_3 thin films produced by different ALD processes exhibit WVTRs below 10^{-3} g m^{-2} d^{-1} [3] and are thus extremely interesting for gas barrier research purpose [4]. In addition to that it comprises an interesting tool for preparing 2D or 3D nanostructures [1d], which are of interest for our institution with respect to areas like the surface modification of fuel cell electrodes, of membranes for waste water treatment or of medical implants.

We evaluated different reactor concepts and construction designs to design an ALD reactor, which is modular and thus adaptable for those various research interests on a cost efficient basis. Herein we describe the design of a reactor for thin film deposition from the gas phase which complies with the flexibility demand of a research institution. It is based on the hot wall viscous flow tubular reactor already described by S. M. George in 2002 [5]. It is extended to a modular setup for higher variability in process and precursor choice. In addition, it enables *in vacuo* transport to analytical devices that allow attachment of CF flanges to their own sample transport system. Radiation-induced substrate preparation is possible in addition to standard preparation procedures.

Designs of reactors for atomic layer deposition have been described in the literature for various special and highly defined foci: concepts for *in situ* analytical studies with XPS [6], spectroscopic ellipsometry [7], mass spectrometry [8], etc. and a transportable reactor for *in situ* synchrotron photoemission studies [9]. Especially within the last decade gas flow characteristics of different reactors have been numerically simulated [10]. Roll-to-roll variants of the process are being developed [3]. Various commercial designs are offered, which are highly adapted to industrial needs of deposition speed and reproducibility.

2. Design and Construction

2.1. Overview

The basic idea of the ALD process is to introduce reactant gases in a timely or spatially separated regime into the reaction chamber. This separation allows sequential self-limiting chemical surface reactions and prevents gas phase reactions between the precursors as well as transport limitations of film growth.

Due to self-limiting surface reactions in each precursor/purge sequence, also called a half-cycle of the ALD-process, a maximum of one monolayer can be deposited on the surface. Sequentially pulsing of appropriate precursors of complementary chemical reactivity will lead to the homogeneous growth of the desired material on all surfaces of appropriate temperature and chemical functionality. Even on high aspect ratio substrates thin conformal layers, which are extremely poor in defects, are thus formed in an ALD process.

A scheme of the herein described viscous flow reactor is shown in Figure 1. It principally consists of the modules gas supply, flow tube, handling cross, heating devices, and vacuum system. All of them are discussed in detail in the following sections.

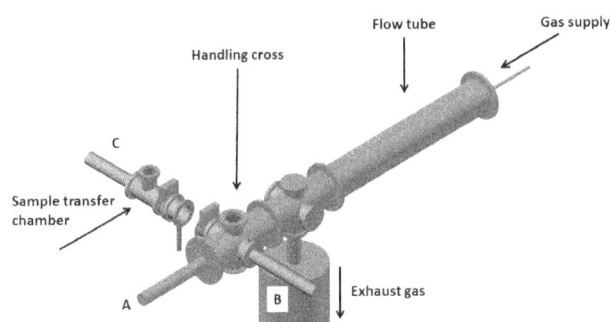

Figure 1. Modular setup of the hot wall ALD flow tube reactor; A, B, C denominate the three transfer arms for explanation in chapter 2.5.

Where not stated differently, the whole reactor has been constructed of standard vacuum stainless steel (AISI 304L) components connected with CF flanges and silver coated copper gaskets. Connections that are opened on a regular basis are set using ISO-K clamping flanges. For sample sizes up to $20 \cdot 20 \cdot 5$ mm^3 a system with inner diameter of 40 mm is sufficient in order to minimize cost (flanges of diverging size are stated within the text). If bigger samples or sophisticated three-dimensional samples are to be considered, flange size has to be adjusted accordingly.

2.2. Gas Supply

The most common setup to introduce precursor pulses into a reactor chamber is to use a gas piping system combined with special ALD diaphragm valves with high speed actuation (opening or closing time of max. 5 ms). Since for research purpose we did not aim at optimized cycle times, we chose standard pneumatically actuated diaphragm valves for ultra-high purity applications. We determined the minimum possible pulse duration to be 25 ms, if time delay caused by pneumatic actuation is considered. Two precursor lines were equipped with special high-temperature formatted valves, which allow heating up to 300 °C (elgiloy® diaphragms).

Gas pipes for carrier and reactant gases have been constructed from ¼ in. x 0,035in. AISI 316/316L stainless steel piping using VCR® metal gasket face seal fittings for

connections, which allow easy maintenance and flexible experimental setups.

For the carrier and purge gas we use nitrogen of purity 5.7 which at first flows through an activated charcoal trap to remove hydrocarbon residues and then through an oxygen/moisture trap consisting of catalyst coated molecular sieve. The gas flow is then split into six lines, four for liquid/solid precursor sources, one for reactant gas line and one for purge gas line. Each of them runs through a mass flow controller surrounded by two diaphragm valves actuated manually for easy cut-off during maintenance.

The setup for introducing precursor gas pulses into the carrier gas stream follows downstream of mass flow control for each of the liquid/solid precursor lines. We extended the principle valve setup by S. M. George et al. [5] by one additional valve, valve 3 in figure 2, in front of the precursor container in order to be able to use bubbler systems for lower vapor pressure precursors. We propose a gas switching setup shown in Figure 2.

Valves are normally off and opened by the corresponding signals from the valve controlling unit. In the off mode for the respective precursor only valve 1 (three-way valve) is open. The carrier gas is lead through the valve to a mechanical pump keeping the pressure of the line constant and avoiding back-pressure.

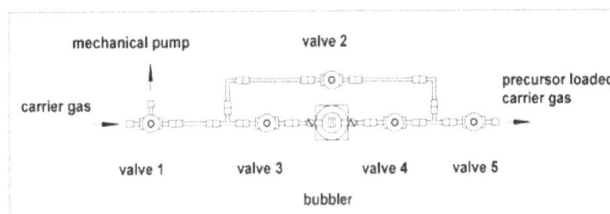

Figure 2. Gas switching setup viable for bubbler use

In the purge mode of the pulse sequence, only valves 2 and 5 are open purging the line/reactor with carrier gas, the pump outlet for valve 1 is closed, the gas flow is directed towards the gas switching system. This step is important to avoid pipe plugging, which would occur in the piping downstream of valve 5, where all precursor lines are reconnected to form one inlet into the reactor tube. If residual precursor is not purged from this reconnecting tube, it might react with the next precursor pulse in the gas phase. For injection of a precursor pulse, valve 2 closes instantaneously with the opening of valves 3 and 4 for the duration of the pulse and falls back into the purge mode after desired pulse duration.

2.3. ALD Reactor – Flow Tube

All parts within the flow tube are made of AISI 304 stainless steel. The flow tube itself consists of an 800 mm long tubular reactor of an inner diameter of 76 mm. The surface of the flow tube was electro-polished. At both ends, the reactor tube is terminated by 100 mm standard CF flanges.

At 400 mm, it contains a stainless steel sample stage, on which the sample holder can be placed. The principle of sample placement into the reactor is shown in Figure 3.

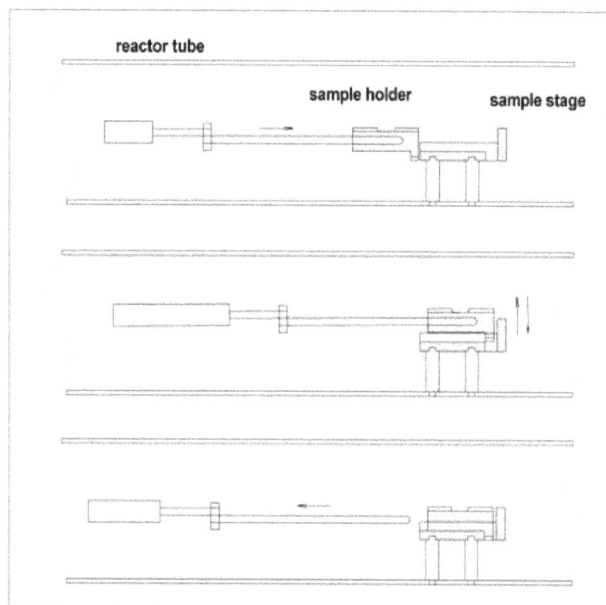

Figure 3. Scheme for sample placement and removal into/from the reactor tube

2.4. Heating Devices

If reaction temperatures up to a maximum of 300 °C are sufficient for the ALD process and highest possible flexibility is desired, a simple heating system built from glass yarn fabric heating mats and tapes has been shown to be sufficient. Both are commercially available in various dimensions with integrated NiCrNi temperature sensors and can simply be wrapped around reactor walls, gas piping and valves.

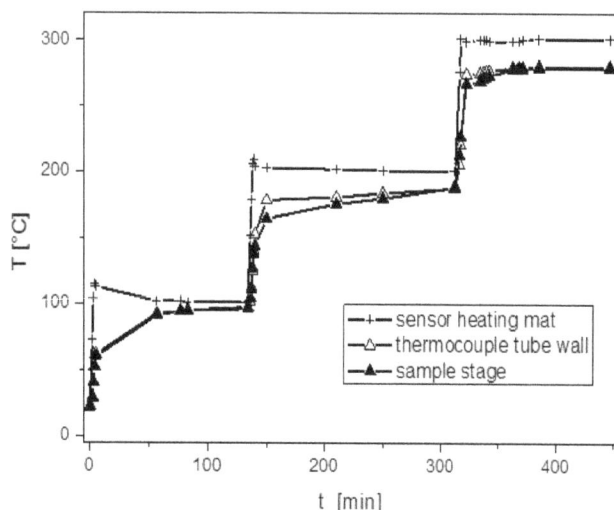

Figure 4. Temperatures measured in the heating mat, the tube wall and on the sample stage after sequentially increasing the desired temperature to 100, 200 and 300 °C, p = 2 mbar, N_2 atmosphere

As close as possible to the sample stage, a holding fixture is introduced into the reactor wall to accommodate a Pt100 resistance temperature detector for monitoring the temperature of the sample. The sample temperature is measured in the holding fixture described in chapter 2.3. During experiments for determination of the temperature

regime of the flow tube, an additional thermocouple was placed on the sample holder. Results of these experiments are shown in Figure 4. The flow tube needs on average 70 minutes to reach a stable temperature within ±1 °C precision. After that, the average temperature difference between the temperature on the sample stage and the Pt100 detector is 1 ± 1 °C which allows sufficient temperature control of the ALD process.

In optimal reactor conditions, the temperature in the reactor is advised to be 10 °C higher than in the precursor piping and 20 °C higher than in the bubbler from experience.

If higher temperatures than 300 °C are needed for special processes and/or precursors, we highly recommend commercial heating mantle systems, which are individually tailor-made by manufacturers for the very precise dimensions of reactor and gas pipes. For high temperature processes, a cooling system for flanges has to be considered in order to keep connections gas-tight over many temperature cycles.

2.5. Handling Cross / Sample Transport Chamber

In order to avoid reactions of the sample with ambient air during transport to XPS analysis, we designed a sample transport chamber. It allows simple manual transport of the sample at ALD process conditions (≈ 1 mbar). Once detached from the vacuum pump, the chamber stays at that pressure over a period of up to three hours. It can then be attached to analytical devices, which allow attachment of CF flanges to their very own sample loading setup.

A handling cross was attached behind the exit of the reactor to direct the prepared sample. Using the process sample holder and transfer arm A, the sample can be de-loaded directly to the atmosphere. If in vacuo transport to analytical devices is wanted, transfer to the transport chamber is achieved via transfer arm B directly onto the XPS-sample holder attached to transfer arm C (see Figure 1), the chamber would be attached to the sample cross, like in setup I in fig. 5.

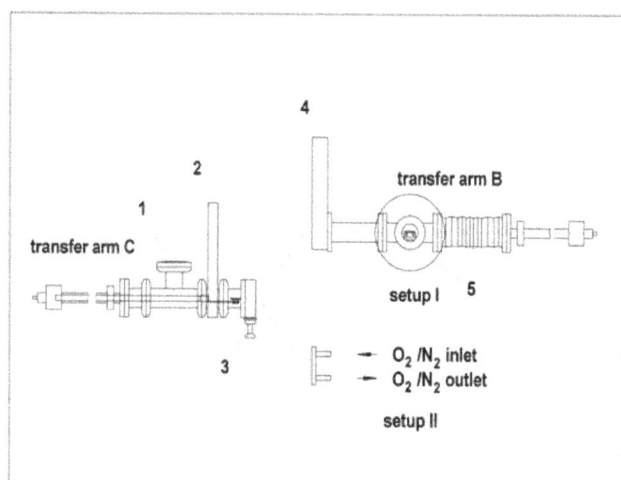

Figure 5. Side view of possible setups of the sample transport chamber: setup I: sample transport for XPS analysis, setup II: sample surface activation; 1: quartz window, 2 and 4: gate valves, 3: connection to vacuum pump, 5: flow tube.

An additional feature of the transfer chamber is a quartz window, situated directly above the sample holder. After sample loading and directly before the ALD process, the surface of the sample can thus be activated for chemical reactions e.g. by treatment with UV light in O_2/N_2 gas mixtures. For that purpose, a blank flange was equipped with two connections for in- and outlet of the gas mixture and can be attached to the transfer chamber during sample preparation. Then the sample can be loaded into the reactor without contact to air (Figure 5, setup II).

2.6. Vacuum System and Exhaust Gas Management

Inherent to the ALD process, it is necessary to keep up a constant flow distribution of precursor and purge gas through the reactor in order to obtain best possible separation of the precursors. Commonly used pressure regimes range between one to several mbar achieved with vacuum pumps. In our case a ceramic pump is used to avoid back-diffusion of pump oil into the reactor chamber and incorporation into deposit. To protect the pump from accidentally formed or abraded particles, we use particle filters with a pore size of 2 μm directly in front of it.

To prevent precursor mixing within the exhaust line which in our case is of ten meters length in total, we usually run the process with a purging gas flow of 600 ml · min⁻¹ (T = 23 °C, p = 2 mbar). Using this construction during the ALD of Al_2O_3 from Me_3Al and water, we did not observe precursor condensation at the exit of the reactor flow tube. For precursors with a lower vapor pressure, this problem might arise and might make heating of the exhaust tubing necessary.

In ALD processes only a small percentage of the dosed precursor is actually used for layer deposition, the majority is going through the pump and has to be removed from the exhaust gas before release to the atmosphere. Therefore, behind the pump the exhaust gas runs through a system of gas washing bottles that can be filled with water, bases, or acids according to the removal process necessary for the desired precursor substance. We use a safety bottle to prevent backflow of water into the exhaust tubing, followed by a gas wash bottle and a second one equipped with a glass frit at the outlet in order to enable optimal distribution of the gas throughout the solution.

2.7. Software

The designed software operates all mass flow controllers as well as the valves individually, the latter ones via pneumatic impulses supplied by a valve controlling unit. It allows individual setting of the gas flow of each gas line, of the pulse lengths for each valve, of the pulse sequence of the ALD cycle, of a purge gas pulse for reproducible starting conditions, as well as of the cycle number. Thus, all needed parameters are adjustable individually to fully control the ALD processes. An exemplary experimental setup for the standard Al_2O_3 ALD process and a XPS depth profile of the resulting Al_2O_3 film are shown in Figure 6.

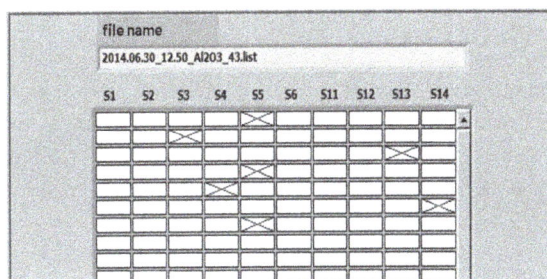

a

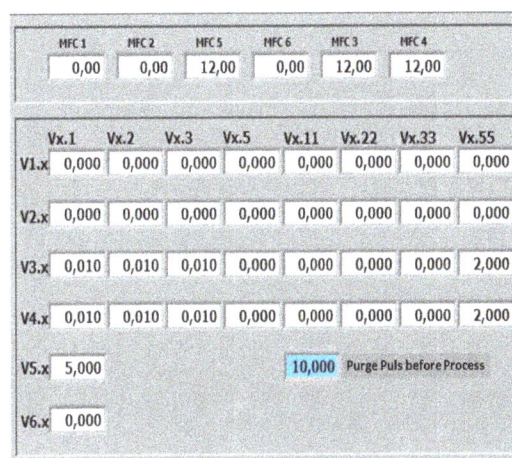

b

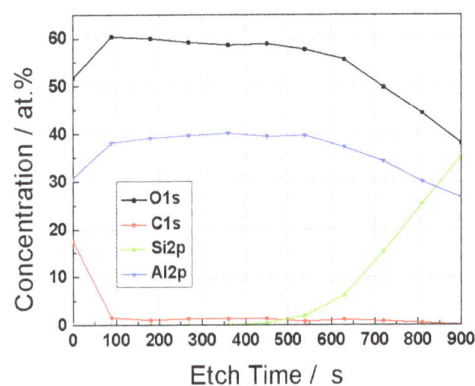

c

Figure 6. a) Screenshots of the input mask for set up of ALD experiments, the upper part defines the order, in which the valves defined in the lower window are opened. The length of the pulses is defined in the lower window (unity: s); b) XPS depth profile of a 40 nm thick Al_2O_3 thin film, for experimental details see [11]

3. Conclusion

A description of a modular reactor for thermal atomic layer deposition has been given. The design allows for deposition temperatures of up to 300 °C with up to five precursors, one of them to be gaseous. A handling cross allows for *in vacuo* transfer of the sample to sample holders for external analytics which allow attachment of standard flanges to the sample transport system. The very same transport chamber allows for irradiation induced sample preparation *in vacuo*.

Acknowledgements

The authors gratefully acknowledge helpful discussions with Martin Knaut, and Dr. Lutz Prager as well as the work of the IOMs workshop employees. We acknowledge Fa. Steffen Quellmalz for software support.

References

[1]   a) T. Kääriäinen, D. Cameron, M.-L. Kääriäinen, A. Sherman, "Atomic Layer Deposition: Principles, Characteristics, and Nanotechnology Applications," John Wiley & Sons, Inc. Hoboken, New Jersey, 2013; b) R. W. Johnson, A. Hultqvist, S. F. Bent "A brief review of atomic layer deposition: from fundamentals to applications," Materials Today, vol. 17, 2014, pp. 236-246; c) M. Leskelä, M. Ritala, "Atomic Layer Deposition Chemistry: Recent Developments and Future Challenges," Angew. Chem. Int. Ed., vol. 42, 2003, pp. 5548 – 5554; d) N. Pinna, M. Knez, "Atomic Layer Deposition of Nanostructured Materials," Wiley VCH, Weinheim, 2011

[2]   L. Prager, U. Helmstedt, H. Herrnberger, O. Kahle, F. Kita, M. Münch, A. Pender, A. Prager, J.W. Gerlach, M. Stasiak, "Photochemical approach to high-barrier films for the encapsulation of flexible laminary electronic devices," Thin Sol. Films, vol. 570, 2014, pp. 87-95

[3]   A. Singh, H. Klumbies, U. Schröder, L. Müller-Meskamp, M. Geidel, M. Knaut, C. Hoßbach, M. Albert, K. Leo and T. Mikolajick, "Barrier performance optimization of atomic layer deposited diffusion barriers for organic light emitting diodes using x-ray reflectivity investigations," Appl. Phys. Lett, vol. , 2013, 233302,

[4]   a) E. Langereis, M. Creatore, S.B.S. Heil, M.C.M. Van de Sanden,W.M.M. Kessels, "Plasmaassisted atomic layer deposition of Al$_2$O$_3$ moisture permeation barriers on polymers," Appl. Phys. Lett, vol. 89, 2006, 081915; b) T. Kääriäinen, D. Cameron, M.-L. Kääriäinen, A. Sherman, Atomic Layer Deposition — Principles, Characteristics, and Nanotechnology Applications, John Wiley & Sons, and Scrivener Publishing, Hoboken (NJ) and Salem (Ma), 2013

[5]   J. W. Elam, M. D. Groner, and S. M. George, "Viscous Flow Reactor with Quartz Crystal Microbalance for Thin Film Growth by Atomic Layer Deposition," Rev. Sci. Instr., vol. 73, 2002, pp. 2981-2987.

[6]   M. Geidel, M. Junige, M. Albert, J. W. Bartha, „In-situ Analysis on the initial Growth of ultra-thin Ruthenium Films with Atomic Layer Deposition," Mircoelectr. Engin., vol. 107, 2013, pp. 151-155.

[7]   a) W. A. Kimes, E. F. Moore, J. E. Maslar, "Perpendicular-Flow, single-Wafer Atomic Layer Deposition Reactor Chamber Design for Use with in situ Diagnostics," Rev. Sci. Instr., vol. 83, 2012, pp. 083106; b) J. Dendooven, K. Devloo-Casier, E. Levrau, R. Van Hove, S. P. Sree, M. R. Baklanov, J. A. Martens, C. Detavernier, "In Situ Monitoring of Atomic Layer Deposition in Nanoporous Thin Films Using Ellipsometric Porosimetry," Langmuir, vol. 28, 2012, pp. 3852 - 3859

[8]   M. Ritala, M. Juppo, K. Kukli, A. Rahtu, M. Leskela, "In situ characterization of atomic layer deposition processes by a mass spectrometer," J. Physique IV, vol. 9, 1999, pp. 1021 – 1028.

[9]   R. Methaapanon, S. M. Geyer, C. Hagglund, P. A. Pianetta, and S. F. Bent, "Portable Atomic Layer Deposition Reactor for in situ Synchrotron Photoemission Studies," Rev. Sci. Instr., vol. 84, 2013, pp. 015104.

[10] a) A.-A. D. Jones, A. D. Jones, "Numerical Simulation and Verification of Gas Transport during an Atomic Layer Deposition Process," Mater. Sci. Semicond. Proc., vol. 21, 2014, pp. 82-90; b) D. Q. Pan, T. Li, T.C. Jen, C. Yuan, "Numerical Modelling of Carrier Gas Flow in Atomic Layer Deposition Vacuum Reactor: A comparative Study of Latice Boltzmann Models," J. Vac. Sci. Tech. A, vol. 32, 2013, pp. 01A110; c) A.M. Lankhorst, B.D. Paarhuis, H.J.C.M. Terhorst, P.J.P.M. Simons, C.R. Klein, "Transient ALD Simulations for a multi-Wafer Reactor with trenched Wafers," Surf. Coat. Tech., vol. 201, 2007, pp. 22-23.

[11] To verify the reactor Al$_2$O$_3$ films were deposited with the instrument described above. AlMe$_3$ (pur. ≥ 98 %) was used as obtained from Strem Chemicals Inc., Millipore ® grade water was degassed before filling of the precursor containers. Containers were kept at room temperature during deposition. Nitrogen was used as a carrier gas at a flow rate of 600 ml/min. Pulsing times were 1 s for the precursors and 4 s for purge gas. The deposition temperature was 200 °C. X-ray photoelectron spectra (XPS) were measured using an AXIS ULTRA Probe instrument from KRATOS Analytical Ltd., Manchester, UK, equipped with a monochromatic Al Kα X-ray source (15 kV, 10 mA) and a magnetic immersion lens. Depth profiles were determined by alternating XPS measurements and stepwise depth sputtering with an Ar$^+$ beam (1 kV, area 2 2 mm^2).

Recent advances in self-assembled DNA nanosensors

Karina M. M. Carneiro[1], Andrea A. Greschner[2], [*]

[1]School of Dentistry, Department of Preventive and Restorative Dental Science, UCSF, San Francisco, USA
[2]Institut National de la Recherche Scientifique, Centre d'Énergie, Matériaux et Télécommunications, Varennes, Canada

Email address:
andrea.greschner@emt.inrs.ca (A. A. Greschner)

Abstract: Over the past 30 years DNA has been assembled into a plethora of structures by design, based on its reliable base pairing properties. As a result, many applications of DNA nanotechnology are emerging. Here, we review recent advances in the use of self-assembled DNA nanostructures as sensors. In particular, we focus on how defined nanostructures, such as rigid DNA tetrahedra, provide an advantage over traditional nanosensors consisting of arrays of single-stranded DNA. We also explore advances in DNA origami that have resulted in consistent detection of single molecules.

Keywords: Self-Assembly, DNA, Nanosensors, Tetrahedron, DNA Origami

1. Introduction

The simple four-letter alphabet and predictable hydrogen bonding assembly patterns associated with the genetic code – adenine (A) binds to thymine (T) and cytosine (C) binds to guanine (G) – make DNA a reliable material for the construction of both nanostructures and nanomaterials.[1]This same high specificity also makes nucleic acids ideal candidates for use as sensors, and for binding and catalysis applications, coining the term 'functional nucleic acids'.[2, 3] Being on the same length scale as many biological molecules - such as proteins, enzymes, and antibodies - rationally controlling the organization of molecules in the nanometer regime is particularly important in medical diagnostics.

Sensors detect the presence of specific analytes, ideally with high affinity and specificity, a fast response time, a long shelf life and reusability. In broad terms, a sensor consists of two parts: target recognition and signaling moieties. Nucleic acid aptamers (single-stranded nucleic acid molecules that have a well-defined three-dimensional structure and a high affinity to their target molecule) have been extensively used for target recognition within sensors due to their sequence-specific properties. A noteworthy combinatorial method called systematic evolution of ligands by exponential enrichment (SELEX) can be used to identify nucleic acid sequences that bind to a desired target with high affinity.[4]This step-wise protocol selects for the best aptamer sequences from a large pool of DNA strands, thereby streamlining aptamer development substantially. It has been reported that aptamers can rival the binding performance of antibodies in certain aspects, and are thus valid target recognition alternatives to antibodies, the development of which presents its own challenges and limitations.[5]Hundreds of aptamers have been designed by scientists, and the widespread applicability of these aptamers has even motivated the creation of a searchable online database.[6] The range of analytes that have been targeted by aptamers is exceptionally wide; two noteworthy model aptamer targets are thrombin and adenosine triphosphate (ATP), although the full range also encompasses metal ions, organic small molecules, macromolecular examples of both natural and synthetic origin, and live organisms such as bacteria and eukaryotic cells. For an extensive review on functional nucleic acid sensors, please refer to Lu *et al.*[2]A second major class of nucleic acid sensors is referred to as molecular beacons (single-stranded DNA hairpins functionalized with a FRET pair). These functional nucleic acid sensors are typically employed to probe for the presence of a complimentary target such as DNA or RNA, which may provide useful genetic or diagnostic information.[7]They have also been used to probe for the presence of the enzymatic target ligase, showcasing their potential beyond the area of nucleic acid sequence detection. The sensitivity of molecular beacons can be increased, for example, through conjugation with superquenchers and conjugated polymers. For more information on molecular beacons, please refer to Tan *et al.*[3]

Nucleic acids are already playing a large role in sensing applications. While the traditional sensors employ relatively simple or locally structured nucleic acid moieties, multiple new developments in the hierarchical self-assembly of DNA have introduced DNA nanostructures to the field of molecular sensing. Here, we will present the most recent advances in using self-assembled DNA nanostructures to construct nanosensors. We will explore how the rational design of individual DNA nanostructures can increase sensitivity, amplify a signal, and allow detection of a variety of targets. In particular, we will review developments favoring the use of a DNA tetrahedron over the more traditional single duplex and hairpin sensors, as well as both static and dynamic DNA origami sensors.

2. 3D DNA Nanostructures as Nanosensors

2.1. The Tetrahedron

In 2004, the emergence of a simple and reliable assembly strategy for small DNA nanostructures[8] led to the development of a new class of tetrahedron detectors. The tetrahedral shape of this molecule, obtained through the one-pot assembly of four DNA strands, had several characteristics that were ideal for nanosensing applications. Initial studies determined that the shape, size, and double-stranded nature of the tetrahedron makes it resistant to many types of enzyme digestion,[9] an important feature in developing nanosensors that may be exposed to serum samples during routine testing (Scheme 1ai).In addition, the four-strand assembly pattern, with each strand terminating at a vertex of the tetrahedron, made it amenable to functionalization. Howorka et al., were the first to take advantage of this feature. Through chemical modification of the 5' end of each DNA strand, they were able to append a disulfide molecule on each of three vertices, with either a biotin or Cy3 dye on the fourth (Scheme 1a ii).The three disulfide tags permitted strong adhesion to gold surfaces (Scheme 1a iii).A kinetic dissociation experiment determined that only 5% of triple-tagged tetrahedra were displaced after 2 hours, whereas 50% of doubly-tagged and 90% of singly-tagged tetrahedra were lost.AFM experiments with the biotin functionalized molecule confirmed that the fourth vertex was vertically oriented and remained accessible for complexation with streptavidin.[10, 11].

In sensor applications, the spacing of nanostructures on surfaces plays a large role in determining sensitivity. Common problems with single-stranded (ss) or hairpin-based nanosensors include molecular crowding at high concentration or long strand length, and probe strands adhering along the surface at low concentrations and/or short strand length. In both cases, access of the target to the sensing strand is diminished.[12, 13]In contrast, the three-dimensional shape and 3-point, 4-5.8 nm footprint[8, 14] of a tetrahedron ensures consistent spacing and proper orientation of the sensor, allowing easy access for targets (figure 1a iii).

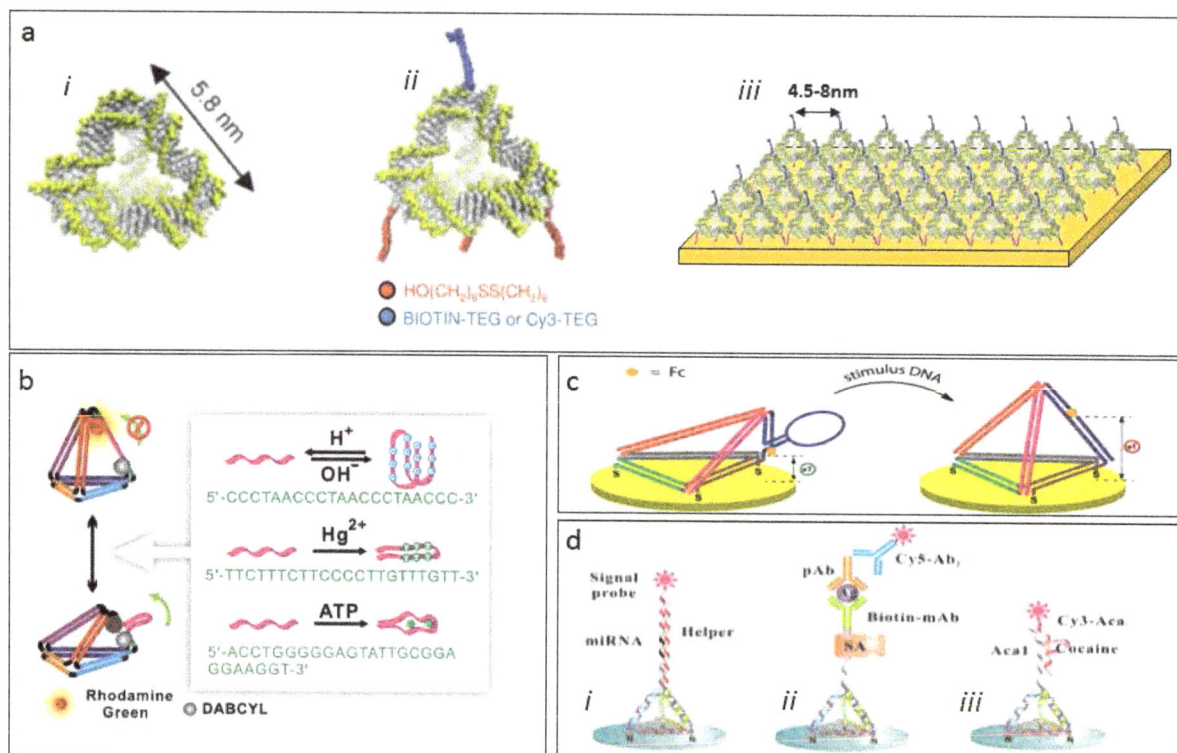

Scheme 1. a) The DNA tetrahedron is constructed by combining four oligonucleotide strands and allowing them to self assemble (i). Adding a functional group to the end of each strand allows for specific placement at each vertex after assembly (ii). In this case, three thiol modifications and one sensor group were attached. The three thiols anchor the tetrahedra to the gold substrate, resulting in a 4.5-8 nm separation between sensor groups. b) Dynamic structural changes can be achieved by introducing specific single-stranded DNA sequences to one side of the tetrahedron. Upon analyte recognition the sequence folds, quenching the fluorescence. c) Electrochemical sensing of a target strand fully complementary to the built-in hairpin. d) By functionalizing with amines in the place of thiols, glass substrates can be used. Three targets were tested: i) miRNA, ii) prostate-specific antigen, and iii) cocaine. Adapted from references 10, 15, 16, and 19 with permission from Wiley-VCH Verlag and the American Chemical Society.

2.2. Signaling Schemes

The signaling action of nanosensors can take many forms and has been a forum for recent advances.Similar to the classical hairpin probes, tetrahedral probes can signal a change in conformation.An elegant example of this is a system designed by Fan et al.In their study, they modified the traditional tetrahedron such that one side remained single-stranded.A dabcyl quencher and rhodamine green fluorophore were added to either end of the ssDNA portion.The ssDNA sequence was specifically chosen to respond to the presence of certain molecules or conditions by folding, thus bringing the FRET pair together and creating a detectable quench in fluorescence.By varying only a single strand, they were able to sense changes in pH (using an i-motif), the presence of mercury (T-rich mercury-specific oligonucleotide), and ATP (anti-ATP aptamer) (Scheme 1b).In the case of the pH sensor, the process was completely reversible.Sensor capabilities were expanded by creating two ssDNA sides with orthogonal targets.Sensitivity varied depending on the target, with detection limits of 2 μM (~1 μg/mL) for ATP and 20 nM (~4 ng/mL) for mercury.Regardless, selectivity was high, with similar molecules such as CTP, GTP, and UTP and other metal ions producing very little change in signal compared to the real targets.[15]

Electrochemical assays are gaining in popularity due to their low detection limits compared to fluorescence-based assays.The gold substrate commonto many tetrahedron-based assays is appealing, as it can act as an electrode.While maintaining the substrate and tetrahedral design, many other adjustments have been pursued to obtain the lowest possible detection limits. Dynamic, surface-tethered DNA structures were combined with ferrocene to produce a sensor that modulates energy transfer based on surface-to-ferrocene distance.The ferrocene label was introduced above the hairpin near the peak vertex of the tetrahedron.A fully complementary target binding to the hairpin causes the tetrahedron to open, increasing the distance between ferrocene and the electrode and decreasing the signal substantially. The rigidity lent by the tetrahedral shape ensures that the ferrocene remains at the intended distance from the surface (Scheme 1c).[16]

Leong et al., used an electrochemical antibody sandwich technique to detect IgG antigens (Scheme 2a-c).The first IgG antibody was coupled to the free vertex of the tetrahedron.Carbodiimide coupling was used to attach the second antibody to a ferrocene.When exposed to target, the IgG antigen binds to the antibody on the tethered tetrahedron and is sandwiched by the antibody-labeled ferrocene.The resulting electrochemical signal had a detection limit of 2.8 pg/mL.[17]

The sandwich motif can also be used for signal amplification.Zuo et al. constructed a sandwich-based tetrahedral nanosensor to detect prostate-specific antigen.They modified the detection method by using a gold nanoparticle multiply-labelled with antibody and horseradish

peroxidase (HRP) instead of ferrocene.Because multiple HRPs were associated with each antibody binding action, the resulting signal was greatly augmented.The combination of ideal spacing (provided by the tetrahedron) and signal amplification (via the gold nanoparticle) resulted in a detection limit of 1 pg/mL, a full order of magnitude greater than singly-labeled sensors (Scheme 2d).[18]

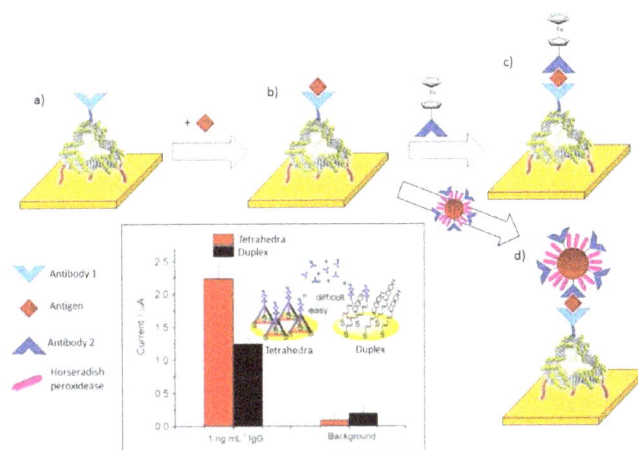

Scheme 2. a) A gold-tethered tetrahedron functionalized with an antibody. b) When a suitable antigen is added to the solution, it binds to the antibody. c) Addition of a second, ferrocene-labeled antibody creates a sandwich and results in electron transfer. d) Alternatively the sandwich can be completed with a gold nanoparticle multiply labeled with antibodies and HRPs. The many HRPs result in signal amplification. Inset: Compared to single-duplex sensors, tetrahedra provide twice the signal.Adapted from references 17 and 18 with permission from Nature Publishing Group and the American Chemical Society.

2.3. Accessibility

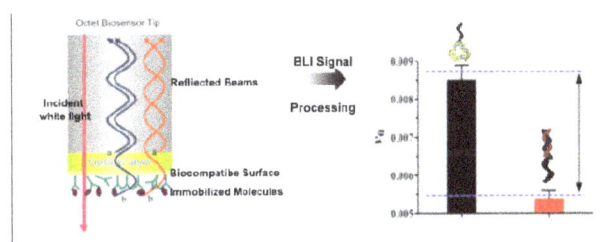

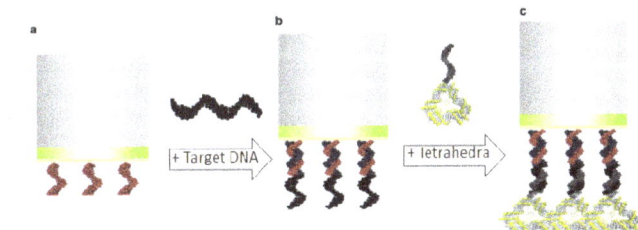

Scheme 3. Inset: General BLI sensor consists of a fibre optic (grey), with an optical biolayer (green) and a biocompatible surface (yellow).Light shines through the fibre optic and reflects off both a reference layer and biolayer.The signal from the biolayer depends on thickness and density, and is greatly increased when a pendant tetrahedron is used. a) Fibre optic functionalized with half an aptamer (red). b) Target molecule (black) is partially recognized by the immobilized half-aptamer. c) A tetrahedron, conjugated to the remaining half of the aptamer, also recognizes the target, amplifying the signal.Adapted from references 22, 23 with the permission of Elsevier and the American Chemical Society.

While the above methods have excellent detection limits, their cost can be prohibitive, with both gold substrate and chemical labelling required. Several steps are being taken towards making these techniques more accessible. One approach is to use a different substrate. Fan *et al.* modified the DNA tetrahedron by replacing the thiol functional groups used previously with amines, allowing for immobilization on glass substrates. The group surveyed several possible targets using a variety of sandwich assays. Even though unamplified fluorescence detection methods were used in lieu of electrochemical methods, they were able to achieve respectable detection limits of 40 pg/mL for prostate-specific antigen and 100 nM (~30 ng/mL) for cocaine and demonstrate good correlation with clinical tests, while using a readily available, inexpensive substrate (Scheme 1d).[19]

Still others are moving towards label-free analyses. These nanosensors are easier to work with, less expensive, and often quicker to prepare, although they lack somewhat in sensitivity.[20, 21]One label-free method is biolayer interferometry (BLI). In general, the tip of a fibre optic sensor is coated with a layer of molecules for trapping the designated targets. White light is shone through the fibre optic, and partially reflected by the two layers at the tip (the reference layer and biolayer).Reflected light is captured by a spectrometer. When the target is bound to the biolayer, it alters the reflectance and can be detected (Scheme 3, inset).[22]Ye *et al.* used tetrahedra to enhance this signal by splitting the sequence of an aptamer. Half of the sequence was immobilized on the fibre optic tip (Scheme 3a), the other half was pendant on a vertex of a tetrahedron. Dipping the fibre optic into a solution containing target allowed for initial binding to the sensor (Scheme 3b).When tetrahedra are added, the remaining half of the aptamer was also recognized by the target, creating a tetrahedron-target-sensor sandwich (Scheme 3c).The added bulk of the tetrahedra created a thicker and denser biolayer, enhancing the signal by two orders of magnitude and decreasing the detection limit to 200 pM (~1.4 ng/mL).[23]

2.4. Summary

Moving from ssDNA and hairpin sensors to discrete self-assembled DNA nanostructures has many benefits. Easily addressed vertices allow for a variety of functionalities, including surface tethers (thiol for gold substrates, and amines for glass substrates), fluorophores, and aptamers. The size and shape of the tetrahedron provides ideal spacing between sensors, and ensures that they are oriented in an upright position for easy target access. The programmability of the DNA alphabet allows for target-triggered dynamic movement of the tetrahedron itself, or it can be used as a scaffold for electrochemical, fluorescence, sandwich, and label-free assays.

3. DNA Origami in Nanosensing

The above-presented techniques are excellent for detecting target analytes. However, the signal obtained in response to

target detection is inevitably an average of all the sensor-target interactions.

DNA origami is a system for making self-assembled nanoscale DNA structures. Each shape is predesigned and based on a single long DNA strand folded by a series of added staple strands.[24] The resulting structures can be static or dynamic,[25, 26] and are characterized via atomic force microscopy (AFM), precluding the need for fluorescent tags or electrochemical sensing techniques.DNA origami offers an opportunity to isolate and identify specific interactions. These structures can be functionalized in a variety of ways, based on their shape and design.

3.1. Static DNA Origami Sensors

Static origami nanosensors rely on a change of surface characteristics for detection. Yan and co-workers designed a DNA origami tile (60 x 90 nm) with distinct bar codes or labels (specifically positioned DNA dumbbell structures on the surface of the tiles) to differentiate between tiles targeting individual RNA sequences.[27]Each tile was functionalized with adjacent 20-nucleotide single-stranded regions for hybridization with a 40-nucleotide RNA target (namely *Rag*-1, c-*myc* and ß-*actin*) as seen in Scheme 4a.Once the target hybridized to the complementary sequences on the tile, a V-shaped structure formed, visible via AFM. The authors found that the edge of the tile provided the best position for target hybridization, presumably due to steric interactions and charge repulsion being greater at the center of the tile. These tiles could be used at a 5:1 ratio with the target, with 30 minutes incubation time, no stirring and at room temperature, for visualization by AFM. The targets were also able to successfully detect an RNA sequence in the presence of a large excess of cellular RNA, with no cross-hybridization events reported. This system is a proof-of-concept for nucleic acid sensing, but is limited by tile concentration (minimum reported 200 pM) and AFM as a characterization technique.

An asymmetric DNA origami (with the shape of the map of China) was also used for the detection of specific DNA strands in solution.[28]This structure differs from the one proposed by Yan *et al.*[27] due to its asymmetric nature, and therefore relinquishes the need of a label within the assembly. It is also different with regards to the DNA probe used: linear versus V-shaped. The group used this asymmetric tile to detect DNA strands with a specific sequence in solution through a sandwich approach, whereby a short overhang strand hybridized to half of the target strand, and the other half of the strand hybridized to a biotin-functionalized DNA strand which is subsequently conjugated to a streptavidin. The increased height due to the presence of streptavidin on the surface of the map can be visualized by AFM.A similar streptavidin-based approach was used for the detection of DNA strands containing a single nucleotide polymorphism (SNP). In this case, the tile was labelled with a toehold strand pre-labelled with a streptavidin. The term 'toehold' indicates that the strand is slightly longer than its complement, allowing for displacement of the shorter (streptavidin-labelled) strand if the fully complementary strand is introduced. The group was able to use the toehold-displacement strategy to distinguish between fully complementary targets and those with a single mismatch (Scheme 4b).[29]

Seeman and co-workers designed a DNA origami tile decorated with hairpins in the shape of the DNA base letters: A, C, T and G, which were visualized by AFM.[30]These hairpins had a toehold, such that they could be displaced in the presence of a DNA strand fully complementary to the hairpin strand (via the strand displacement method).In the presence of an analyte fully complementary to the strand, the hairpin structure is destabilized and removed as a duplex from the surface of the origami tile, therefore erasing the letter from the surface of the tile. This method provides a visual readout of the SNP type by AFM. The authors also created a computer program that calculates an average of 25 images and subtracts the background for a direct readout of the mismatched base (Scheme 4c).

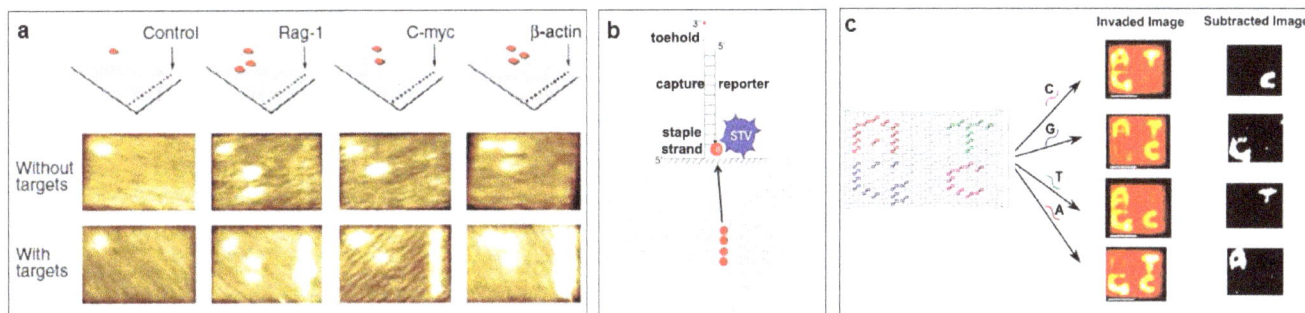

Scheme 4. a) Rectangular DNA origami tile functionalized for label-free detection of specific RNA targets. The 'barcode' of each tile is visible as dots on the main tile surface, whereas the strip along the right edge indicates target detection. b) DNA origami in the shape of the map of China for the detection of single nucleotide polymorphisms (SNPs). c) Rectangular DNA origami tile used for a visual readout of SNPs based on AFM images. Reproduced and adapted with permission from references 27, 29, 30 with permission from the American Association for the Advancement of Science, John Wiley & Sons, and the American Chemical Society.

3.2. Dynamic Origami Sensors

Dynamic origami nanosensors are those that change their overall shape in response to the presence of an analyte. The changes are distinctly visible through AFM.

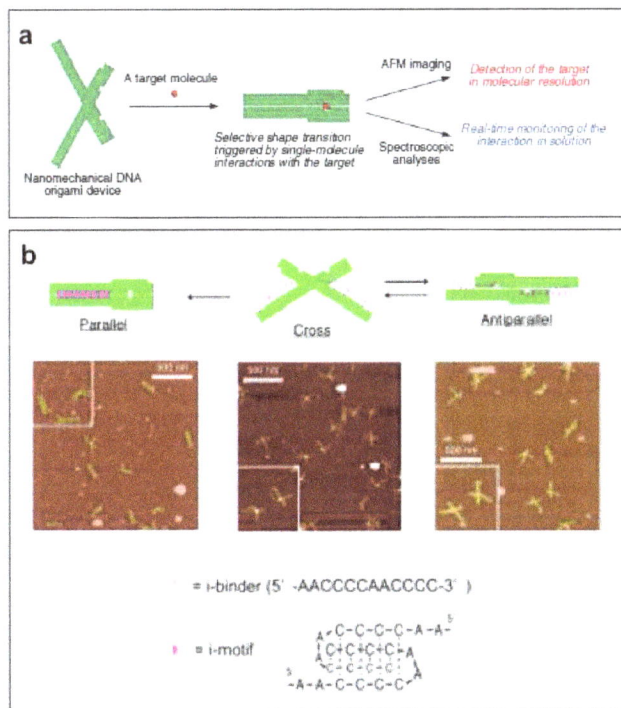

Scheme 5. a) Dynamic DNA origami assembly changes shape in the presence of target molecules. b) pH responsive shape change of a nanomechanical DNA origami sensor. Adapted from references 31 and 32 with permission from Nature Publishing Group and MDPI.

Komiyama *et al.*, designed what they referred to as DNA origami 'pliers' and 'forceps'. These designs consisted of two origami arms (170 x 20 nm) connected by a type of four-way junction referred to as a Holliday junction. The natural form of the Holliday junction is x-shaped. As such, when the entire origami structure settles on mica prior to imaging, the preferred orientation of the arms is a cross shape. In order to detect individual molecules, ligands were placed inside a small notch on each arm. When the target is recognized by the ligands, the two notches are aligned, resulting in a parallel arm orientation (Scheme 5a).This switch from cross to parallel is easily visible via AFM, and indicates that a single molecule of analyte has been detected and immobilized. The proof-of-concept was performed using biotin/streptavidin and FAM/antifluorescein IgG as ligand/target pairs.[31]

The same team also investigated 'zipper' sensors. The tweezer shape and cross versus parallel detection method remained the same, but instead of ligands, they attached pendant DNA strands to each lever. The pendant strands were chosen for their ability to self-assemble under target conditions. For example, by using a 12-base telomeric excerpt, the pendant arms will join together to form a G-quadruplex in the presence of K^+ or Na^+ (depending on the sequence).[31]The system has since been adapted as a pH sensor, using the acid-sensitive i-motif sequence as pendant strands (Scheme 5b).[32]

Using origami assemblies as sensors for nucleic acids and RNA allows for aqueous quantification and very small sample amounts are required. However, the sensor results can only be interpreted by AFM imaging. This would cause a technical challenge in clinical settings, as personnel would need to be trained on the instrument, and imaging can be time-consuming and challenging. This difficulty may be overcome by the continuing efforts in simplifying AFM characterization, and with fast scan AFM.

4. Conclusion

While DNA aptamers, hairpins, and molecular beacons have become quite common in sensor applications, the use of self-assembled DNA nanostructures remains fairly new.Replacing molecular beacons and single-stranded sensors with rigid, three-dimensional tetrahedra promotes proper probe spacing, thereby reducing steric hindrance and improving target access.The stiffness of the structure also keeps the probes properly oriented.Because of the innate programmability of DNA, dynamic sensors can be designed to utilize changes in tetrahedron side lengths as signaling devices.A shift has also been seen towards more accessible DNA nanosensors, with alternative substrates and label-free detection methods being explored.

Functionalization of DNA origami has also led to advances in molecular-level sensing.Based on AFM characterization, target recognition can be detected either through changes in the surface of a DNA origami tile, or through dynamic shape changes in the overall origami structure.Clever design techniques have enabled static DNA origami to be programmed to not only indicate the presence of a target, but to also identify the target. Dynamic DNA origami, on the other hand, is capable of signaling the detection of a single analyte molecule.

Overall, research into self-assembled DNA nanosensors is yielding a variety of benefits.Work towards accessible and affordable substrates, label-free sensors, and fast-scan microscopy techniques will continue to drive this field.

References

[1] K.M. Carneiro, N. Avakyan, H.F. Sleiman, "Long-range assembly of DNA into nanofibers and highly ordered networks", Wiley interdisciplinary reviews. Nanomedicine and nanobiotechnology, 2013, pp.

[2] J. Liu, Z. Cao, Y. Lu, "Functional nucleic acid sensors", Chemical Reviews, 109, 2009, pp. 1948-1998.

[3] K. Wang, Z. Tang, C.J. Yang, Y. Kim, X. Fang, W. Li, Y. Wu, C.D. Medley, Z. Cao, J. Li, P. Colon, H. Lin, W. Tan, "Molecular engineering of DNA: Molecular beacons", Angewandte Chemie International Edition, 48, 2009, pp. 856-870.

[4] C. Tuerk, L. Gold, "Systematic evolution of ligands by exponential enrichment: Rna ligands to bacteriophage t4 DNA polymerase", Science, 249, 1990, pp. 505-510.

[5] S.D. Jayasena, "Aptamers: An emerging class of molecules that rival antibodies in diagnostics", Clinical Chemistry, 45, 1999, pp. 1628-1650.

[6] K. Robison, A.M. McGuire, G.M. Church, "A comprehensive library of DNA-binding site matrices for 55 proteins applied to the complete escherichia coli k-12 genome", Journal of Molecular Biology, 284, 1998, pp. 241-254.

[7] S. Tyagi, F.R. Kramer, "Molecular beacons: Probes that fluoresce upon hybridization", Nat Biotech, 14, 1996, pp. 303-308.

[8] R.P. Goodman, R.M. Berry, A.J. Turberfield, "The single-step synthesis of a DNA tetrahedron", Chemical Communications, 2004, pp. 1372-1373.

[9] J.-W. Keum, H. Bermudez, "Enhanced resistance of dnananostructures to enzymatic digestion", Chemical Communications, 2009, pp. 7036-7038.

[10] N. Mitchell, R. Schlapak, M. Kastner, D. Armitage, W. Chrzanowski, J. Riener, P. Hinterdorfer, A. Ebner, S. Howorka, "A DNA nanostructure for the functional assembly of chemical groups with tunable stoichiometry and defined nanoscale geometry", Angewandte Chemie International Edition, 48, 2009, pp. 525-527.

[11] M. Leitner, N. Mitchell, M. Kastner, R. Schlapak, H.J. Gruber, P. Hinterdorfer, S. Howorka, A. Ebner, "Single-molecule afm characterization of individual chemically tagged DNA tetrahedra", ACS Nano, 5, 2011, pp. 7048-7054.

[12] D.Y. Petrovykh, V. Pérez-Dieste, A. Opdahl, H. Kimura-Suda, J.M. Sullivan, M.J. Tarlov, F.J. Himpsel, L.J. Whitman, "Nucleobase orientation and ordering in films of single-stranded DNA on gold", Journal of the American Chemical Society, 128, 2005, pp. 2-3.

[13] H. Pei, X. Zuo, D. Pan, J. Shi, Q. Huang, C. Fan, "Scaffolded biosensors with designed DNA nanostructures", NPG Asia Mater, 5, 2013, pp. e51.

[14] H. Pei, N. Lu, Y. Wen, S. Song, Y. Liu, H. Yan, C. Fan, "A DNA nanostructure-based biomolecular probe carrier platform for electrochemical biosensing", Advanced Materials, 22, 2010, pp. 4754-4758.

[15] H. Pei, L. Liang, G. Yao, J. Li, Q. Huang, C. Fan, "Reconfigurable three-dimensional DNA nanostructures for the construction of intracellular logic sensors", Angewandte Chemie International Edition, 51, 2012, pp. 9020-9024.

[16] A. Abi, M. Lin, H. Pei, C. Fan, E.E. Ferapontova, X. Zuo, "Electrochemical switching with 3d DNA tetrahedral nanostructures self-assembled at gold electrodes", ACS Applied Materials & Interfaces, 6, 2014, pp. 8928-8931.

[17] L. Yuan, M. Giovanni, J. Xie, C. Fan, D.T. Leong, "Ultrasensitive igg quantification using DNA nano-pyramids", NPG Asia Mater, 6, 2014, pp. e112.

[18] X. Chen, G. Zhou, P. Song, J. Wang, J. Gao, J. Lu, C. Fan, X. Zuo, "Ultrasensitive electrochemical detection of prostate-specific antigen by using antibodies anchored on a DNA nanostructural scaffold", Analytical Chemistry, 86, 2014, pp. 7337-7342.

[19] Z. Li, B. Zhao, D. Wang, Y. Wen, G. Liu, H. Dong, S. Song, C. Fan, "DNA nanostructure-based universal microarray platform for high-efficiency multiplex bioanalysis in biofluids", ACS Applied Materials & Interfaces, 6, 2014, pp. 17944-17953.

[20] M. Zhang, B.-C. Ye, "Label-free fluorescent detection of copper (ii) using DNA-templated highly luminescent silver nanoclusters", Analyst, 136, 2011, pp. 5139-5142.

[21] M. Zhang, S.-M. Guo, Y.-R. Li, P. Zuo, B.-C. Ye, "A label-free fluorescent molecular beacon based on DNA-templated silver nanoclusters for detection of adenosine and adenosine deaminase", Chemical Communications, 48, 2012, pp. 5488-5490.

[22] T. Do, F. Ho, B. Heidecker, K. Witte, L. Chang, L. Lerner, "A rapid method for determining dynamic binding capacity of resins for the purification of proteins", Protein Expression and Purification, 60, 2008, pp. 147-150.

[23] M. Zhang, X.-Q. Jiang, H.-N. Le, P. Wang, B.-C. Ye, "Dip-and-read method for label-free renewable sensing enhanced using complex DNA structures", ACS Applied Materials & Interfaces, 5, 2013, pp. 473-478.

[24] P.W.K. Rothemund, "Folding DNA to create nanoscale shapes and patterns", Nature, 440, 2006, pp. 297-302.

[25] E.S. Andersen, M. Dong, M.M. Nielsen, K. Jahn, A. Lind-Thomsen, W. Mamdouh, K.V. Gothelf, F. Besenbacher, J. Kjems, "DNA origami design of dolphin-shaped structures with flexible tails", ACS nano, 2, 2008, pp. 1213-1218.

[26] E.S. Andersen, M. Dong, M.M. Nielsen, K. Jahn, R. Subramani, W. Mamdouh, M.M. Golas, B. Sander, H. Stark, C.L.P. Oliveira, J.S. Pedersen, V. Birkedal, F. Besenbacher, K.V. Gothelf, J. Kjems, "Self-assembly of a nanoscale DNA box with a controllable lid", Nature, 459, 2009, pp. 73-U75.

[27] Y. Ke, S. Lindsay, Y. Chang, Y. Liu, H. Yan, "Self-assembled water-soluble nucleic acid probe tiles for label-free rna hybridization assays", Science, 319, 2008, pp. 180-183.

[28] Z. Zhang, Y. Wang, C. Fan, C. Li, Y. Li, L. Qian, Y. Fu, Y. Shi, J. Hu, L. He, "Asymmetric DNA origami for spatially addressable and index-free solution-phase DNA chips", Advanced Materials, 22, 2010, pp. 2672-2675.

[29] Z. Zhang, D. Zeng, H. Ma, G. Feng, J. Hu, L. He, C. Li, C. Fan, "A DNA-origami chip platform for label-free snp genotyping using toehold-mediated strand displacement", Small, 6, 2010, pp. 1854-1858.

[30] H.K.K. Subramanian, B. Chakraborty, R. Sha, N.C. Seeman, "The label-free unambiguous detection and symbolic display of single nucleotide polymorphisms on DNA origami", Nano Letters, 11, 2011, pp. 910-913.

[31] A. Kuzuya, Y. Sakai, T. Yamazaki, Y. Xu, M. Komiyama, "Nanomechanical DNA origami 'single-molecule beacons' directly imaged by atomic force microscopy", Nat Commun, 2, 2011, pp. 449.

[32] A. Kuzuya, R. Watanabe, Y. Yamanaka, T. Tamaki, M. Kaino, Y. Ohya, "Nanomechanical DNA origami ph sensors", Sensors, 14, 2014, pp. 19329-19335.

26

Efficient route to high-quality graphene materials: Kinetically controlled electron beam induced reduction of graphene oxide in aqueous dispersion

Roman Flyunt[1, *], Wolfgang Knolle[1], Axel Kahnt[2], Siegfried Eigler[3], Andriy Lotnyk[1], Tilmann Häupl[1], Andrea Prager[1], Dirk Guldi[2], Bernd Abel[1]

[1]Leibniz-Institut für Oberflächenmodifizierung (IOM), Permoserstr. 15, 04303 Leipzig, Germany
[2]Department of Chemistry and Pharmacy & Interdisciplinary Center for Molecular Materials (ICMM), Friedrich-Alexander-Universität Erlangen-Nürnberg (FAU), Egerlandstraße 3, 91058 Erlangen, Germany
[3]Department of Chemistry and Pharmacy & Central Institute for New Materials and Processing Technology, Friedrich-Alexander-Universität Erlangen-Nürnberg (FAU), Dr.-Mack Str. 81, 90762 Fürth, Germany

Email address:

roman.flyunt@iom-leipzig.de (R. Flyunt)

Abstract: This work is presenting a highly efficient, cost-efficient and environmentally friendly method for the production of graphene materials (reduced graphene oxide, RedGO) via electron beam (EB) irradiation of aqueous dispersions of graphene oxide (GO). Our strategy here is based on a reduction of GO via EB irradiation under optimally controlled conditions, i.e. dose and dose rate, reducing species, and taking the environmental impact of educt and product into account. The preparation of highly conductive RedGO under these conditions takes only 10-20 minutes at ambient temperature. After our first approach [1], a somewhat similar study was reported by Jung et al. [2] for GO dispersions in H₂O/EtOH (50:50). However, the latter route [2], although being similar in spirit, has serious drawbacks for large-scale production because of the formation of acetaldehyde, a very toxic compound, derived from the ethanol in the solvent. The advantages of the present approach compared to [2] are: (i) the use of water as a solvent with only a small content (0.03 - 2 wt.-%) of 2-PrOH allows the scaling-up, since neither 2-PrOH nor its final product acetone are of high technological or environmental concerns; (ii) a much lower dose is required for GO reduction (about 20 vs. 200 kGy, corresponding to only 1/10 of energy consumed); (iii) the conductivity of RedGO is over 60 times higher. Based on the XPS and conductivity measurements, it was established that the EB treatment is leading also to a more efficient reduction of GO compared to the hydrazine method. The highest conductivity in our systems is identical to the best known value of 3×10^4 S/m for RedGO obtained via HI / acetic acid treatment which takes, however, 40 h at 40 °C.

Keywords: Electron Beam, Reducing Free Radicals, Reduced Graphene Oxide, Highly Conductive Carbon Nanomaterials, Graphene Oxide

1. Introduction

The discovery of graphene [3] is undoubtedly one of the most important events in the chemistry and physics of carbon materials within the past two decades. The chemistry and numerous applications of graphene, GO, and their derivatives are highlighted in recent reviews [4-10]. The most effective route to obtain low cost, good quality graphene in the form of highly reduced GO is through the reduction of GO (see eq.1) in its colloidal suspensions.

A number of conceptually different methods to reduce GO have been published, employing chemical reductants such as hydrazine, dimethylhydrazine, hydroquinone, NaBH₄, vitamine C, hydriodic/acetic acids, thermal means, and photocatalysis based on TiO₂ or ZnO etc. [8]. A detailed overview covering the most frequently used methods including a short discussion is given as Table 1. In chemical terms, reduction of GO is sketched in equation 1 for GO lacking major structural defects:

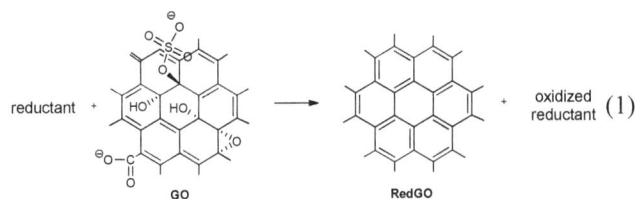

$$\text{reductant} + \text{GO} \longrightarrow \text{RedGO} + \text{oxidized reductant} \quad (1)$$

When comparing the different reduction processes a number of factors should be taken into account. The majority of recent investigations focuses on the quality of received RedGO as reflected in conductivity characteristics. Aspects like environmental hazard stemming from the reductant and/or its final product(s), on one hand, and process scalability, on the other hand, are, at least, of equal importance but are often neglected. The most important drawbacks of the well-known and favored reduction methods can be shortly summarized as follows. Firstly, a particular low conductivity is found for RedGO when hydroquinone, pyrogallol, KOH, or $NaBH_4$ are used. Secondly, reaction times reach up to 40 hours in the case of gamma irradiating GO dispersions, which is hardly acceptable. Thirdly, the toxicity of reducing agents such as hydrazine and hydroxylamine and their products is a problem. Fourthly, the highly explosive nature of hydrazine creates technological problems with respect to handling. Finally, the separation of the resulting RedGO from, for example, other components is also a big technological challenge, since vacuum filtration takes too long and waste disposal is too expensive. Zhang et al. [33] have investigated the γ-irradiation induced reduction of GO in dispersion. During reduction, the starting GO is subject to a significant deoxygenation, that is, transforming from a C/O ratio of 1.1 to 10.1, in N_2-purged H_2O/EtOH (1:1 v/v) solution. A distinct disadvantage of this approach is the

Table 1. Comparison of the methods for the reduction of GO.

Reducing compound or Method	Advantage	Disadvantage
Hydrazine [11-13]	very efficient	high temp. (95 °C), hydrazine is explosive and highly toxic
Dimethylhydrazine [14]	very efficient	as for hydrazine
Hydroxylamine [15]	low price and lower toxicity compared to hydrazine	Explosive and toxic compound; slightly less efficient than hydrazine
Hydroquinone [16]	less toxic than hydrazine	much less efficient than hydrazine; 95 °C
$NaBH_4$ [13, 17]	unknown	10^4 fold lower conductivity compared to hydrazine method, too expensive, 95 °C
Pyrogallol [13]	less toxic than hydrazine	too expensive, 95 °C
Hydriodic/acetic acid [18]	conductivity higher than for hydrazine method, treatment at 40 °C	too expensive, too long process (40 h), waste management
$NaHSO_3$ and other sulfur-containing compounds [19]	cheap, low toxicity, conductivity is comparable with hydrazine method, duration 3 h	95 °C; utilization of oxidation products derived from sulfur-containing compounds
vitamin C at 95 °C [13]	short time (30 min), efficient almost as hydrazine, no toxicity	too expensive; 95 °C; utilization of products from vitamin C;
vitamin C at room temperature [20]	no heating, no further oxidation processes	too long treatment (48 h), ~ 550 times lower conductivity compared to vitamin C at 95°C;
vitamin C + l-tryptophan [21]	gives a stable aqueous RedGO dispersion	even more expensive
vitamin C + Triton-X100 [22]	gives a stable aqueous RedGO dispersion	5 times lower conductivity than with vitamin C at 95 °C
Aluminium powder [23]	Simple and fast reaction (30 min) without external heating; relatively low cost	5 times lower conductivity than that for the hydrazine method, product is in a form of precipitate; utilization of 0.5M HCl, Al and $AlCl_3$
Reduction at high pH [24]	Very fast (a few minutes at 50-80 °C)	use of highly concentrated hydroxide; 5×10^4 fold lower conductivity compared to hydrazine method [13]
Thermal reduction [25, 26]	no need for dispersion in a solvent	1000° C; 30 % weight loss, graphene sheets highly wrinkled
Hydrothermal [27]	leads to a stable aqueous RedGO dispersion	180 °C, 6 h, high content of oxygen in final product; no data on conductivity
Solvothermal [28]	good degree of deoxygenation, stable dispersion in propylene carbonate	150 °C, conductivity is still 4 times lower compared to hydrazine method
Electrochemical reduction [29, 30]	very efficient, ambient temperature, no toxicity	high energy consumption; deposition of the product onto electrodes
TiO_2- or ZnO-assisted photocatalytical reduction [31, 32]	no toxicity, ambient temperature	only partial reduction; separation of graphene and semiconductor difficult due to deposition of the product on the photocatalyst
Gamma-irradiation of water-alcohol dispersions of GO [33]	Ambient temperature, low toxicity, high degree of deoxygenation, low costs	too long process (40 h), conductivity improvement (4-5 orders of magnitude) is still considerably lower compared to hydrazine method (6-7 orders)
EB treatment of water-alcohol dispersions of GO [2]	fast (20 minutes), ambient temperature	high dose of 200 kGy, moderate conductivity of RedGO, high toxicity of formed acetaldehyde

long irradiation time in the range of 40 h to realize a total absorbed energy of 35 kGy. In contrast to the latter, EB treatment generates free radicals in very high concentrations. Thus, short processing times and easy scalability become feasible.

It was recently demonstrated that the irradiation time can be significantly reduced to 5-20 minutes (corresponding to 50-200 kGy) by EB irradiation [2]. But, even with absorbed doses of 200 kGy the electrical characteristics of the correspondingly formed RedGO are still far from optimum when compared to benchmarks based on hydrazine or HI/AcOH treatment – vide infra. Moreover, the employed solvent was an ethanol-water mixture (1:1 v/v) [2, 33]. To this end, acetaldehyde is produced from ethanol, which poses high environment risks.

Very recently, we have reported a basic study on the radiation-induced reduction of GO, using water as solvent, where only small content (at most 2 wt.-%) of simple organic compounds are required to reduce GO [34]. In the current contribution, we demonstrate the highly efficient EB-induced reduction of GO with focus on its environmental compatibility, low cost, and scalability, and on overcoming the drawbacks described in any of the aforementioned approaches [2, 33]. It will be shown, that high quality RedGO with superior conductivity is obtained at much lower absorbed dose, using water containing $\leq$ 2 wt.-% of 2-PrOH. Importantly, acetone as the main product of the water / 2-PrOH radiolysis is of low environmental and technological concerns.

Three different samples of GO (from different suppliers) have been investigated to prove the general concept and to find out the peculiarities of each kind of GO. The study shows that the results strongly depend upon and vary with the initial material.

2. Experimental

All chemicals (Sigma-Aldrich) were of analytical grade and used without further purification.

Three different GO samples have been investigated in this work: a) single-layered GO purchased from Cheaptubes.com (USA; further called CT-GO), b) GO from Nanoinnova Technologies (Madrid, Spain; further called NT-GO) and c) home-made almost intact GO (further called ai-GO) synthesized according to Eigler et al. [35].

The aqueous 0.5 g/L GO solution was prepared by agitation in ultrasonic bath for 1 h, followed by vacuum filtration through *Millipore* HVLP *0.45 µm* membrane filter. The GO settled on the filter was then washed with copious amount of water and redispersed in Millipore water in ultrasonic bath for 15 min. The obtained GO solution with a natural pH of ca. 4.0 was then centrifuged at 4000 rpm for 4 hours. The decanted GO dispersion was used for described experiments. This procedure gives stable (> 1 month) dispersions of a few-layers GO as confirmed by TEM measurements with final concentration of 0.4 g/L. If needed, pH was adjusted with KOH solution. The preparation of

stock solution of Nanoinnova Technologies GO (further NT-GO) was done in a similar manner, excepting much shorter ultracentrifugation time (a few times repeated ultracentrifugation for 15 min at 4000 rpm). This was due to much lower solubility (ca. 7-8 times) of NT-GO compared to CT-GO (final concentration is 0.05 g/L). The desired GO concentration was obtained by dilution of stock solution with Millipore water. Solubility of ai-GO is much higher due to the presence of organosulfate groups [36].

A stock solution (5 g/L of ai-GO) was diluted with Millipore water to a final GO concentration of 6.9 mg/L. This solution was divided into two equal aliquots. One was used at its original pH of 4, the pH of the second aliquot was adjusted to pH 10.

All GO solutions were purged with nitrogen (excepting the experiment with air-saturated solution) and irradiated with EB at 1 kGy/step. Irradiated solutions were vacuum filtrated (from Whatman, $\varnothing$ 47 mm, 0.2 µm pore size), washed with copious amount of Millipore water and dried in vacuum desiccator.

For the preparative irradiation of GO systems a 10 MeV linear accelerator ELEKTRONIKA (Toriy Company, Moscow) was employed, which is operating at 50 Hz repetition rate and 4µs pulse length.

Spectral changes in the EB-irradiated GO solutions were followed using UV-VIS spectrometer TIDAS-II (Spectralytics GmbH, Essingen, Germany).

TEM observations were performed with a probe Cs-corrected Titan^3 G2 60-300 microscope equipped with HAADF, BF, DF, ABF and Super-X EDX detectors as well as with GIF Quantum Gatan imaging filter. The microscope was operated at 80 kV to minimise sample damages.

Raman spectra were recorded from 1050 to 3410 cm^{-1} under 532 nm (2.33eV) excitation with two confocal microscope setups, a LabRAM ARAMIS (HORIBA Jobin Yvon) and an inverted microscope IX71 (Olympus) fiber coupled to a spectrometer (iHR320, synapse CCD, HORIBA Jobin Yvon). GO samples from Cheaptubes and Nanoinnova Technologies were measured dip coated on a silicon-oxide wafer (300 nm) with the former, vacuum filtrated ai-GO or ai-RedGO samples were performed with the latter setup.

XPS spectra were recorded on Axis Ultra (Kratos) using monochromatized Al-K$_\alpha$ radiation. Conductivity measurements of graphene-like materials were done on a MDC four-point probe system (MDC S.A, Geneve, Switzerland).

3. Results and Discussion

Irradiation of aqueous systems by EB generates three highly reactive radical species, namely hydrated electrons (e_{aq}^-), hydrogen atoms (H$^\bullet$), and hydroxyl radicals ($^\bullet$OH), as summarized in reaction 2 [37]:

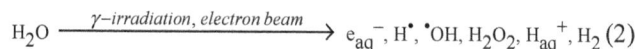

$$H_2O \xrightarrow{\gamma-irradiation,\ electron\ beam} e_{aq}^-,\ H^\bullet,\ {}^\bullet OH,\ H_2O_2,\ H_{aq}^+,\ H_2 \quad (2)$$

The radiation chemical yields of the primary species amount to 0.6×10^{-7} mol J^{-1} for H$^\bullet$ and 2.9×10^{-7} mol J^{-1} for

e_{aq}^- and ${}^\bullet OH$. Both, e_{aq}^- and $H^\bullet$, are strong reductants with reduction potentials of -2.9 and -2.4 V, respectively [38] that are directly employable to reduce GO. In contrast, ${}^\bullet OH$ radicals are known as one of the strongest oxidants ($E({}^\bullet OH,$ $H^+/H_2O) = +2.73$ V) [38] and, therefore, have to be converted into reducing radicals by a reaction with, for example, alcohols :

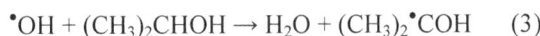

$${}^\bullet OH + (CH_3)_2CHOH \rightarrow H_2O + (CH_3)_2{}^\bullet COH \quad (3)$$

Hydrogen atoms react with most alcohols in a similar manner to ${}^\bullet OH$ (reaction 3), giving the same reducing species. The reductants derived from simple alcohols such as 2-PrOH, EtOH, or MeOH feature reduction potentials of -1.9 V for $(CH_3)_2{}^\bullet C(OH)$, -1.25 V for $CH_3{}^\bullet CHOH$, and -1.18 V for ${}^\bullet CH_2OH$ [38]. As such, these values are comparable or even higher than those established for hydrazine (-1.16 V) or borohydride (-1.24 V) [23], known to efficiently reduce GO. Thus, the chosen free radical species are expected to render good reducing agents for GO as well.

The primary reductants e_{aq}^-, $H^\bullet$ and the secondary reductants – radicals formed via reaction 3 – react fast with oxygen with rate constants in the range of 10^9 - 10^{10} dm^3 $mol^{-1}s^{-1}$) [37] to afford the corresponding peroxyl radicals. The latter are, however, unable to reduce GO. To avoid this undesired side reaction, it is favorable to conduct the irradiation of GO dispersions in the absence of oxygen. Thus, in N_2-saturated aqueous dispersions of GO in the presence of any of the aforementioned ${}^\bullet OH$-radical scavengers two major reductants, namely e_{aq}^- and radicals derived from the scavengers, are generated.

Upon EB irradiation in the presence of small amounts of MeOH, EtOH, or 2-PrOH, the initially yellow-brown aqueous GO dispersions turned black within a few minutes. Reduction of GO is easily followed employing UV-Vis spectroscopy. An example is given in Fig. 1 for the case of CT-GO and 2-PrOH as ${}^\bullet OH$-scavenger. The absorption maximum is shifting from ca. 227 nm for the starting GO to ca. 266 nm for the final RedGO. Similar trends evolved for NT-GO and ai-GO – Fig. 2 and 3. Shifts to 268 ± 2 nm after the reduction of GO with hydrazine and vitamin C (pH ~ 9-10) were reported in the literature [39, 40] and the resulting spectra were assigned to RedGO. Considering that the long-wavelength absorbance at $\lambda > 600$ nm is almost exclusively due to the absorption of RedGO – Fig. 1 – a rough estimate of the RedGO yield at a defined dose of EB irradiation is made with the maximum absorbance at, for example, 700 nm at hands. When comparing solutions containing 2 wt.-% of different OH scavengers, the highest absorbance was obtained in the EtOH system and taken as 100%. The corresponding values for aqueous dispersion with either 2-PrOH or MeOH at the same dose are 96 and 92%, respectively. Much lower values were derived for t-BuOH with 43%. In the latter, ${}^\bullet OH$ radicals are transformed into non-reducing species leaving electrons as the only reductants. In addition, we studied the reduction of GO with a number of other free radicals coming from the reaction of ${}^\bullet OH$ radicals

with simple organics. Among those, formate is giving the same high yield of RedGO as EtOH, while ethylene glycol and glycerine afford somewhat smaller yields of 87 and 81%, respectively.

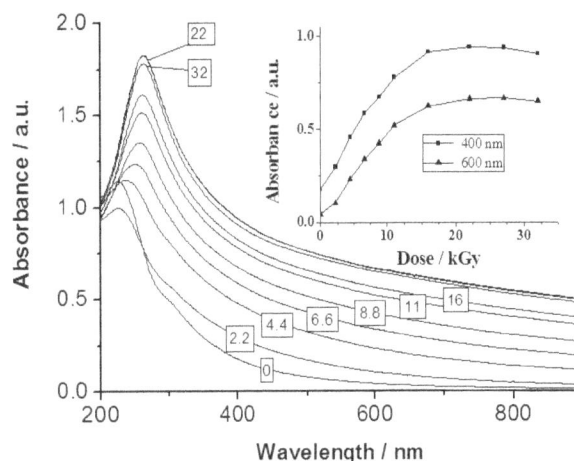

Figure 1. *UV-Vis spectra of EB-irradiated N_2-saturated solution containing 0.4 g/L CT-GO, 2 wt.-% 2-PrOH at pH 4. Absorbed doses are indicated in inset, irradiation followed in 1.1-1.2 kGy/step. Probes were diluted with water (1:9) before recording of spectra; Inset: absorbance bild-up at different wavelengths.*

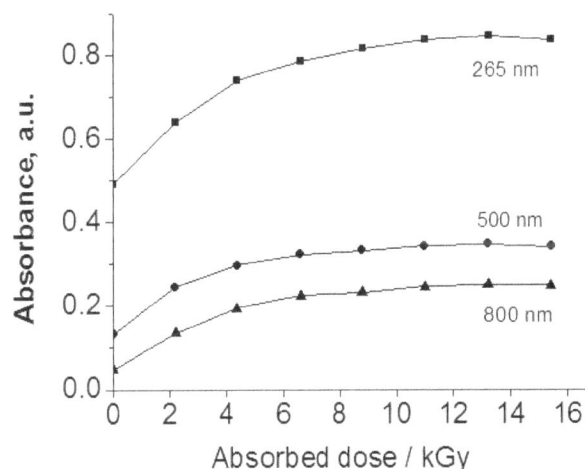

Figure 2. *Absorbance bild-up at different wavelengths during EB treatment of N_2-saturated solution containing NT-GO and 2 wt.-% iso-PrOH at pH 10.*

Other free radicals, which were formed from sucrose, glyoxylate and lactate are also able to reduce GO. But, the corresponding RedGO yields are significantly lower with 57% for sucrose, 42% for glyoxylate, and 40% for lactate. In conclusion, reducing radicals derived from MeOH, EtOH and 2-PrOH are most efficient in terms of GO reduction. To select any of them for industrial up-scaling, additional criteria are considered: on one hand, MeOH possesses incomparably higher toxicity than EtOH and 2-PrOH. On the other hand, formaldehyde and acetaldehyde, the final products of the radiolysis / reduction with MeOH and EtOH, respectively [37], are known to be very toxic, whereas acetone as it evolves from 2-PrOH (eq. 4) is neither a major environmental nor technological concern. Therefore, to up-scale the GO reduction, *the preference should be given to 2-*

PrOH. In this system, reduction of GO with $(CH_3)_2$·C(OH) radicals follows:

$$n\,(CH_3)_2\text{·C(OH)} + GO \rightarrow n\,(CH_3)_2CO + RedGO + n\,H_2O \quad (4)$$

In competition to the desired reaction 4, $(CH_3)_2$·C(OH) may undergo bimolecular termination reactions

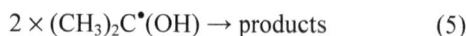

$$2 \times (CH_3)_2C\text{·}(OH) \rightarrow \text{products} \quad (5)$$

Although the contribution of such unwanted reactions can be dramatically reduced by lowering the dose rate, the very low dose rates of, for example, gamma-irradiation results in an unacceptable long irradiation time [33]. High dose rates of 10 kGy/min as used in [2] will facilitate the termination reaction. In light of the latter and to realize optimum conditions, our experiments were performed at a dose rate of about 1 kGy/step.

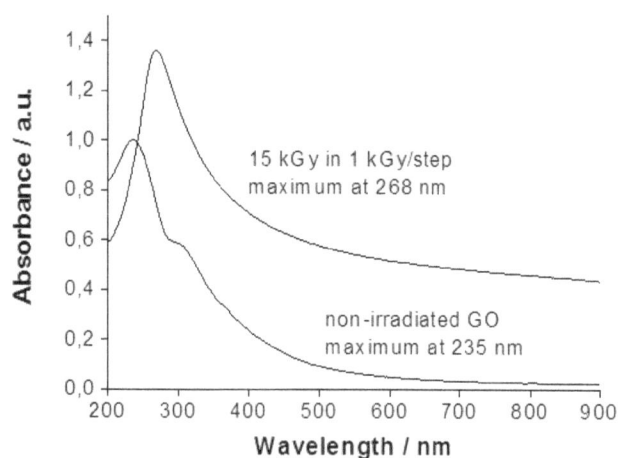

Figure 3. *UV-Vis spectra of aqueous dispersions of ai-GO at pH 10 (2 wt.-% 2-PrOH, N_2-sat.) before and after EB-treatment.*

It was established, that an efficient RedGO formation was also observed with much smaller concentration of ·OH scavengers. For example, decreasing EtOH from 2 to 0.03 wt.-% led to only a 15% shrinking of the RedGO yield at the same dose [34]. On a final note, EB-induced GO reduction is possible with an acceptable efficiency even in air-saturated solutions. We have found, that absorbance of the EB irradiated air-saturated solution of GO is only slightly lower (ca. 15 ± 5 %) compared to one for N_2-saturated solution in the whole UV-Vis spectrum. This is in a sharp contrast to gamma-irradiated GO [33], where almost no sign of GO reduction was seen in air-saturated solution. Dissolved oxygen (0.28 mM) is consumed via fast reactions with radical species after absorbance of ca. 1 kGy, i.e. in about 0.5 minute in our experiment and in appr. 1 h in the case of gamma-irradiation. Further diffusion of oxygen from the air into solution seems to be inefficient compared to GO reduction by means of EB, which is presumably the main reason for drastic difference observed for two methods. These aspects are advantageous for scaling-up.

The optimal dose for EB reduction of GO is determined by means of UV-Vis spectroscopy. As illustrated in Fig. 1, the absorbance from 200 to 900 nm maximizes at 22 kGy. Prolonged irradiation leads to an incremental decrease in absorbance. Considering, for example, spectra taken at 32 kGy, the effects are rationalized on grounds of RedGO agglomeration as described in [39].

In considering the ratio of absorbancies for RedGO and GO, namely A_{RedGO}/A_{GO}, we established a simple UV spectroscopic criterion to relate to the degree of GO reduction in aqueous solutions. For EB treated CT-GO it is 1.6. Control experiments, in which the GO solutions were reduced with hydrazine gave identical values for λ_{max}, A_{RedGO} and A_{RedGO}/A_{GO}. Similar A_{RedGO}/A_{GO} ratios in the range from 1.5 to 1.6 were calculated from the data reported in [39] and [40]. The values obtained for NT-GO and ai-GO are slightly lower with 1.45 and 1.35, respectively, in the latter case precipitation sets in.

To this date, Raman spectra of RedGO are interpreted in different, sometimes controversial ways. Based on a recent work [41], we applied Lorentz functions to fit the obtained Raman spectra. Overall, they reveal the characteristics of few-layer-graphene with broadened D and G bands. They were deconvoluted into D, D** and G bands – Fig. 4. Contributions stemming from D* and D′ bands were, however, negligible. It should be noted that the Raman spectra of ai-GO samples were fit satisfactory without the needs of adding D** bands to the deconvolution routine. Similarly, the so-called 2 D regions in the Raman spectra were fit with three peaks located at around 2690 ± 30, 2940 ± 30, and 3200 ± 30 cm^{-1} – Fig. 4 and Table 2 – which are assigned to 2 D (or G′), D + D′ and 2 D′ modes, respectively. All bands in the 2 D region are markedly broadened. As a matter of fact this is typical for batches featuring sufficient numbers of defects.

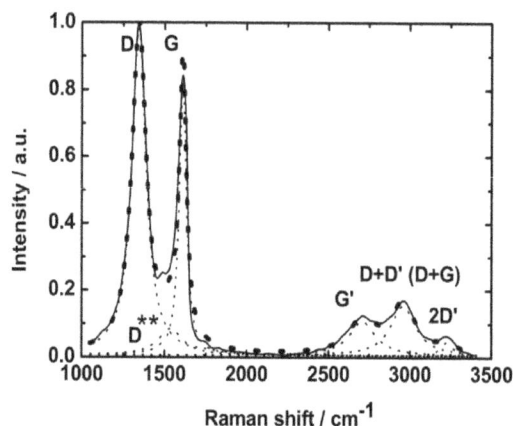

Figure 4. *Raman spectrum of RedGO obtained after EB treatment of CT-GO dispersion at pH 10 (solid line) and the deconvoluted (thin dashed line) and fitted (thick dashed line) spectra. To note: all presented Raman spectra (Fig. 4 - Fig. 6) are normalised taking the intensity of D band as unity.*

From our Raman measurements we conclude that the D and G bands sharpen considerably upon reduction – see Fig. 5, 6 and Table 2). This effect is quantified by the full width at half maximum (FWHM) values denoted as Γ. Especially pronounced is a drop of Γ_D for ai-GO from 134 to about 75 cm^{-1} prior and after its reduction, respectively – Table 2. In contrast, the G band sharpening is rather moderate with an initial Γ_G of ca. 85 cm^{-1} and a final Γ_G of ca. 65 ± 4 cm^{-1}. A sharpening of the Raman spectra of RedGO correlates with an increased quality. Earlier investigations on ai-GO were performed exclusively with films of single flakes, while Raman spectra of few-layer and multi-layer graphene have been filtered out [35, 42, 43]. In the present study, we analyze, however, the complete sample reduced in dispersion followed by filtration.

D bands of treated GO are usually downshifted compared to the non-treated one (see Table 2). For example, in the case of CT-GO, the D band has maxima at 1344 cm^{-1} for EB and hydrazine treated solution and at 1353 cm^{-1} for the starting GO. The maxima for D bands for all irradiated ai-GO probes are situated at 1332-33 cm^{-1}, which is about 15 cm^{-1} downshifted compared to non-treated sample. A different picture was observed for hydrazine treated ai-GO, namely there is almost no shift of the D band (1346 vs 1348 cm^{-1}). Non-treated and irradiated samples NT-GO have the same maxima positions for D band as well.

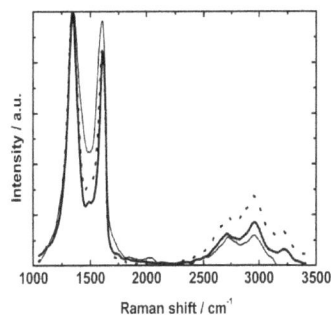

Figure 5. *Raman spectra of CT-GO (thin grey line curve) and RedGO obtained after hydrazine (dashed line curve) and EB-treatment (thick line curve).*

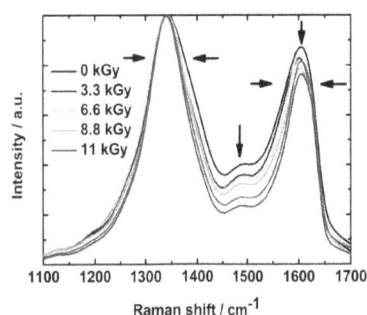

Figure 6. *Raman spectra of NT-GO before and after EB-irradiation.*

Table 2. Results of Raman analysis for GO and RedGO

Band assignment/Ratios of intensities	D	D**	G	2D	D + D′	2 D′	A_D/A_G
CT-GO *non-treated*							
Peak maxima / cm^{-1}	1353	1490	1598	2725	2966	3230	1.77
Γ, cm^{-1}	128	155	75	222	177	208	
EB treated							
Peak maxima / cm^{-1}	1344	1541	1610	2708	2958	3225	2.36
Γ, cm^{-1}	105	134	51	196	194	84	
hydrazine treated							
Peak maxima / cm^{-1}	1344	1525	1604	2708	2954	3222	2.48
Γ, cm^{-1}	113	109	55	225	224	193	
NT-GO *non-treated*							
Peak maxima / cm^{-1}	1344	1535	1606	2719	2950	3207	2.2
Γ, cm^{-1}	118	120	63	339	218	253	
EB treated							
Peak maxima / cm^{-1}	1344	1520	1603	2713	2956	3227	2.4
Γ, cm^{-1}	104	120	54	256	219	246	
ai-GO *non-treated*							
Peak maxima / cm^{-1}	1348	n.u.*	1588	2681	2926	3171	1.68
Γ, cm^{-1}	135	-	85	262	182	51	
ai-GO *hydrazine treated*							
Peak maxima / cm^{-1}	1346	n.u.*	1594	2681	2939	3189	1.96
Γ, cm^{-1}	77	-	68	216	148	148	
ai-GO *pH 10, EB treated*							
Peak maxima / cm^{-1}	1333	n.u.*	1585	2676	2923	3187	1.94
Γ, cm^{-1}	74	-	65	148	127	92	

* - the Raman spectra of ai-GO samples were fitted satisfactory without the use of D** band.

Usually, a downshift for G band of RedGO is associated with restoration of the hexagonal network of conjugated sp2 C-C bonds. In contrast, an upshift for G band is attributed to the formation of isolated C=C bonds [44]. It should be, however, noted that a typical error for the Raman measurements is equal to ± 4 cm-1, so the observed small shifts have to be taken with precaution.

With the measured D and G peak areas in hand the values of AD/AG as a measure of the lattice disorder [45] have been calculated. In the case of CT-GO samples, they were equal to 1.77, 2.36 and 2.48 for starting GO, EB treated GO, and hydrazine treated GO, respectively. Weaker were the increases of the A_D/A_G ratio for NT-GO (ca. 8%) and ai-GO samples (at most 15%). From the fact that EB and hydrazine treated samples give similar A_D/A_G ratios we conclude that both methods are comparable. An increase in the I_D/I_G ratio upon treatment with a variety of reductants is well documented in the literature [11, 33, 46-48]. In the current study we used the A_D/A_G ratio rather than the I_D/I_G ratio, owing to the fact that peaks with different line width are compared. Overall, this effect is rationalized by decreasing sizes of in-plane sp^2-domains of graphene as well as partially ordered crystal structures of RedGO [44]. The lower A_D/A_G ratios as they were observed for treated ai-GO relative to CT-GO and NT-GO samples, infers that the former is of better quality with a lower defect density.

Fig. 7a illustrates a bright-field TEM image of CT-GO sample. As such, the sheet contains defects like holes – Fig. 7a, bottom left. A selected area electron diffraction (SAED) pattern of GO is shown in Fig. 7b. In the latter, the measured d-spacing indicates graphitic crystalline structures of GO, which was further corroborated by electron energy loss spectroscopy (EELS) investigations – Fig. 8.

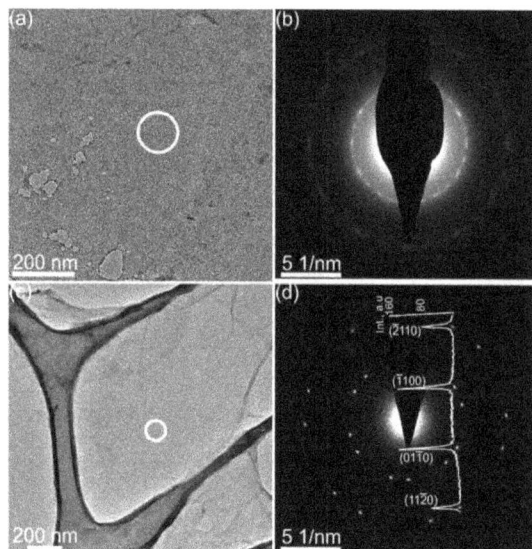

Figure 7. TEM images of (a) GO and (c) RedGO. SAED patterns of (b) GO and (d) RedGO. The patterns were taken from the areas marked by the white circles in images (a) and (c). All measurements were done with CT-GO.

However, the ring-shaped pattern consisting of many diffraction spots suggests multilayer structure of starting GO. The average thickness of the GO specimen based on EELS measurements ranged from 2.3 to 9 nm and is approximated as 2 to 8 layers. In stark contrast, a bright-field TEM image of a RedGO specimen – Fig 7c –, reveals wrinkling and folding of the RedGO sheet. In a typical SAED pattern of the RedGO sample – Fig. 7d – the slight spot broadening in the pattern is due to the fact that the RedGO sheets are not exactly flat [49]. Additionally, analysis of diffracted intensities of the SAED pattern in Fig. 7d confirmed the bilayer structure of the RedGO sheet [49, 50], a representative one for this sample. The profile along the (-2110) to the (-1120) reflections gives ratios of $I_{\{1110\}}/I_{\{2110\}}$ close to 1.8. Notably, ratios larger than 1 indicate single layer graphene structures [49, 50]. A quantitative similar behavior of $I_{\{1110\}}/I_{\{2110\}}$ was observed for all others diffracted

intensities in Fig. 7d. Thus, the RedGO sheet is two layers thick. Measurements of thickness on different sample areas using EELS resulted in thicknesses of RedGO sheets in the range from 2 to 4 nm, that is approximately 2 to 4 layers. At this point we hypothesize that further exfoliation took place during the EB induced GO reduction.

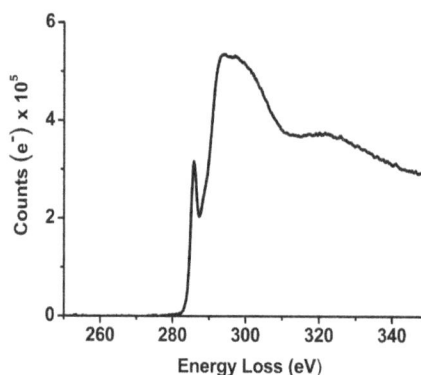

Figure 8. EELS spectra of C K-edge taken from CT-GO.

Reduction of GO under EB irradiation is clearly confirmed by XPS. The two main peaks in the XPS spectrum of the starting ai-GO – Fig. 9 – are situated at 284.5 and 286.5 eV with a relative distribution of 48 and 52%, respectively. Upon GO reduction, the relative intensity of the first peak strongly increased, whereas that of the second peak dramatically decreased. In the deconvoluted XPS spectra of ai-RedGO obtained via EB treatment – inset to Fig. 9 – the contributions of C-C sp^2 at 284.6 eV, C-OH at 285.1 eV, epoxy/ether at 286.1 eV, C=O at 287.2 eV, and carboxyl at 289.1 eV are discernable.

XPS analyses shed light onto the C/O ratios, which increased from 2.60 for ai-GO prior to EB treatment to 10.9 (pH 4) and 9.6 (pH 10), respectively, after EB treatment. Any of these values are higher than the 7.1 obtained for the hydrazine treated ai-GO - Table 3. Additionally, the intensity of the C1 peak due to C-C sp^2 is higher and the intensities of the other peaks are lower for the EB method – Fig. 9. Based on these findings, one can state that the EB treatment is leading to a more efficient reduction of GO compared to the hydrazine method.

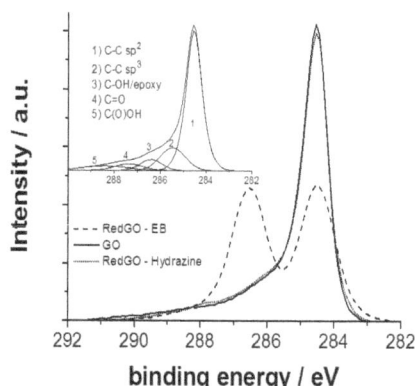

Figure 9. Core level C1s XPS spectra of ai-GO (dashed line) and ai-RedGO obtained by its treatment with hydrazine (dotted line) and EB (full line). Inset: curve fit of XPS spectrum of ai-RedGO (EB method).

The highest conductivity of 29000 S/m was seen for the ai-GO with pH 10. The RedGO film obtained from the same GO, but reduced at pH 4, gives rise to a 5 times lower conductivity of 5400 S/m. This indicates that GO at pH 10, where the carboxylic groups (pK_a = 6.1) are fully deprotonated [39], is more susceptible for reduction. A likely factor is a higher degree of exfoliation of ai-GO at pH 10 due to a strong electrostatic repulsion of the GO sheets with completely dissociated carboxylic groups.

These results underline that deoxygenation of GO is necessary but not decisive to obtain highly conductive RedGO. The conductivity of RedGO depends not only on the reduction conditions and more efforts are needed to map out this aspect with sufficient care. The nature of a GO sample, that is, the kind and density of defects, the lateral sizes of sheets, the presence of impurities, etc. plays a large role. As it is illustrated in Table 3, the conductivity of RedGO from CT-GO is low compared to ai-RedGO. It should be noted that the highest conductivity measured in this study is 2.9×10^4 S/m for ai-GO at pH 10 which, in turn, is more than 60 times higher than the reported 450 S/m for RedGO obtained by EB treatment of GO dispersions in H_2O/EtOH (1:1) [2]. It should also be recognized that the dose applied in our experiments is nearly 10 times lower – 22 versus 200 kGy – which infers the needs for much lower energy input. Another important advantage is the use of H_2O/2-PrOH (98:2 v/v) as solvent instead of H_2O/EtOH (50:50 v/v) enabling the scale-up as discussed above. Importantly, the highest conductivity in our study is identical to the best known value for RedGO obtained with HI-acetic acid method [48]. However, the latter requires 40 h treatment at 40 °C. It is interesting to note that the conductivity of films made of graphene nanoplatelets is only two times higher (6×10^4 S/m) [51].

Table 3. Conductivity and C/O ratio measurements for different RedGO probes.

Probe and conditions	Conductivity, S/m	Measured C/O ratio*
ai-GO, pH 4, 15 kGy	5400	10.9
ai-GO, pH 10, 15 kGy	29000	9.6
hydrazine treated ai-GO, pH 10	14000	7.1
CT-GO pH 10, 22 kGy	480	7.3
EB treated GO in H_2O:EtOH (1:1), at 50 kGy/200 kGy [2]	1.5 / 450	not reported
HI/acetic acid treated GO (40 h at 40 °C) [48]	30000	6.7

* - the calculated C/O ratios represent a lower limit values. A certain amount of oxygen originates from of Si and S compounds which were detected in the range of a few percents. However, as their stoichiometry with oxygen is unknown, a correction for the C/O ratio is unreliable.

Our present study reveals the great potential of EB reduction of GO for the preparation of RedGO, with a conductivity well comparable with the best known wet-chemical methods.

In conclusion, we highlighted a rational synthesis strategy, appreciating the advantages of high energy radiation to provide optimal reduction conditions, which appears to be superior to other approaches lacking options in terms of environmentally friendliness, cost-effectiveness, and up-scaling for a high quality graphene material.

Acknowledgments

The financial support of the European Union and the Free State of Saxony (LenA project) is greatly acknowledged. AK, BA and DG are gratefully acknowledging funding from the Deutsche Forschungsgemeinschaft (DFG) via grants KA 3491/2-1/AB 63/14-1 and EI 938/3-1. We like to thank Dr. Jenny Malig for her support during the Raman measurements. We thank Prof. Dr. Andreas Hirsch for his support at FAU Erlangen-Nürnberg. This work is also supported by the Cluster of Excellence 'Engineering of Advanced Materials (EAM)' and SFB 953 funded by the DFG.

References

[1]   R. Flyunt, W. Knolle, B. Abel, B. Rauschenbach, Verfahren zur Herstellung von reduziertem Graphenoxid sowie damit hergestelltes reduziertes Graphenoxid und dessen Verwendung 2012.

[2]   J.-M. Jung, C.-H. Jung, M.-S. Oh, I.-T. Hwang, C.-H. Jung, K. Shin, J. Hwang, S.-H. Park, J.-H. Choi, Rapid, facile, and eco-friendly reduction of graphene oxide by electron beam irradiation in an alcohol–water solution, Materials Letters, 126 (2014) 151-153.

[3]   K.S. Novoselov, A.K. Geim, S.V. Morozov, D. Jiang, Y. Zhang, S.V. Dubonos, I.V. Grigorieva, A.A. Firsov, Electric field effect in atomically thin carbon films, Science, 306 (2004) 666-669.

[4]   S. Park, R.S. Ruoff, Chemical methods for the production of graphenes, Nature Nanotechnology, 4 (2009) 217-224.

[5]   O.C. Compton, S.T. Nguyen, Graphene Oxide, Highly Reduced Graphene Oxide, and Graphene: Versatile Building Blocks for Carbon-Based Materials, Small, 6 (2010) 711-723.

[6]   D.R. Dreyer, S. Park, C.W. Bielawski, R.S. Ruoff, The chemistry of graphene oxide, Chemical Society Reviews, 39 (2010) 228-240.

[7]   H. Kim, A.A. Abdala, C.W. Macosko, Graphene/Polymer Nanocomposites, Macromolecules, 43 (2010) 6515-6530.

[8]   C.K. Chua, M. Pumera, Chemical reduction of graphene oxide: a synthetic chemistry viewpoint, Chemical Society Reviews, 43 (2014) 291-312.

[9]   S.F. Pei, H.M. Cheng, The reduction of graphene oxide, Carbon, 50 (2012) 3210-3228.

[10] S. Eigler, A. Hirsch, Chemistry with Graphene and graphene oxide - challenges for synthetic chemists, Angewandte Chemie International Edition, (2014).

[11] S. Stankovich, D.A. Dikin, R.D. Piner, K.A. Kohlhaas, A. Kleinhammes, Y. Jia, Y. Wu, S.T. Nguyen, R.S. Ruoff, Synthesis of graphene-based nanosheets via chemical reduction of exfoliated graphite oxide, Carbon, 45 (2007) 1558-1565.

[12] C. Gomez-Navarro, R.T. Weitz, A.M. Bittner, M. Scolari, A. Mews, M. Burghard, K. Kern, Electronic transport properties of individual chemically reduced graphene oxide sheets, Nano Letters, 7 (2007) 3499-3503.

[13] M.J. Fernandez-Merino, L. Guardia, J.I. Paredes, S. Villar-Rodil, P. Solis-Fernandez, A. Martinez-Alonso, J.M.D. Tascon, Vitamin C Is an Ideal Substitute for Hydrazine in the Reduction of Graphene Oxide Suspensions, Journal of Physical Chemistry C, 114 (2010) 6426-6432.

[14] S. Stankovich, D.A. Dikin, G.H.B. Dommett, K.M. Kohlhaas, E.J. Zimney, E.A. Stach, R.D. Piner, S.T. Nguyen, R.S. Ruoff, Graphene-based composite materials, Nature, 442 (2006) 282-286.

[15] X.J. Zhou, J.L. Zhang, H.X. Wu, H.J. Yang, J.Y. Zhang, S.W. Guo, Reducing Graphene Oxide via Hydroxylamine: A Simple and Efficient Route to Graphene, Journal of Physical Chemistry C, 115 (2011) 11957-11961.

[16] G.X. Wang, J. Yang, J. Park, X.L. Gou, B. Wang, H. Liu, J. Yao, Facile synthesis and characterization of graphene nanosheets, Journal of Physical Chemistry C, 112 (2008) 8192-8195.

[17] Y. Si, E.T. Samulski, Synthesis of water soluble graphene, Nano Letters, 8 (2008) 1679-1682.

[18] I.K. Moon, J. Lee, R.S. Ruoff, H. Lee, Reduced graphene oxide by chemical graphitization, Nat Commun, 1 (2010) 73.

[19] W.F. Chen, L.F. Yan, P.R. Bangal, Chemical Reduction of Graphene Oxide to Graphene by Sulfur-Containing Compounds, Journal of Physical Chemistry C, 114 (2010) 19885-19890.

[20] J. Zhang, H. Yang, G. Shen, P. Cheng, J. Zhang, S. Guo, Reduction of graphene oxide via L-ascorbic acid, Chem Commun, 46 (2010) 1112-1114.

[21] J. Gao, F. Liu, Y.L. Liu, N. Ma, Z.Q. Wang, X. Zhang, Environment-Friendly Method To Produce Graphene That Employs Vitamin C and Amino Acid, Chemistry of Materials, 22 (2010) 2213-2218.

[22] V. Dua, S.P. Surwade, S. Ammu, S.R. Agnihotra, S. Jain, K.E. Roberts, S. Park, R.S. Ruoff, S.K. Manohar, All-Organic Vapor Sensor Using Inkjet-Printed Reduced Graphene Oxide, Angewandte Chemie International Edition, 49 (2010) 2154-2157.

[23] Z.J. Fan, K. Wang, T. Wei, J. Yan, L.P. Song, B. Shao, An environmentally friendly and efficient route for the reduction of graphene oxide by aluminum powder, Carbon, 48 (2010) 1686-1689.

[24] X. Fan, W. Peng, Y. Li, X. Li, S. Wang, G. Zhang, F. Zhang, Deoxygenation of Exfoliated Graphite Oxide under Alkaline Conditions: A Green Route to Graphene Preparation, Adv Mater, 20 (2008) 4490-4493.

[25] M.J. McAllister, J.L. Li, D.H. Adamson, H.C. Schniepp, A.A. Abdala, J. Liu, M. Herrera-Alonso, D.L. Milius, R. Car, R.K. Prud'homme, I.A. Aksay, Single sheet functionalized graphene by oxidation and thermal expansion of graphite, Chemistry of Materials, 19 (2007) 4396-4404.

[26] P. Steurer, R. Wissert, R. Thomann, R. Mulhaupt, Functionalized Graphenes and Thermoplastic Nanocomposites Based upon Expanded Graphite Oxide, Macromolecular Rapid Communications, 30 (2009) 316-327.

[27] Y. Zhou, Q. Bao, L.A.L. Tang, Y. Zhong, K.P. Loh, Hydrothermal Dehydration for the "Green" Reduction of Exfoliated Graphene Oxide to Graphene and Demonstration of Tunable Optical Limiting Properties, Chemistry of Materials, 21 (2009) 2950-2956.

[28] Y.W. Zhu, M.D. Stoller, W.W. Cai, A. Velamakanni, R.D. Piner, D. Chen, R.S. Ruoff, Exfoliation of Graphite Oxide in Propylene Carbonate and Thermal Reduction of the Resulting Graphene Oxide Platelets, Acs Nano, 4 (2010) 1227-1233.

[29] M. Zhou, Y.L. Wang, Y.M. Zhai, J.F. Zhai, W. Ren, F.A. Wang, S.J. Dong, Controlled Synthesis of Large-Area and Patterned Electrochemically Reduced Graphene Oxide Films, Chem-Eur J, 15 (2009) 6116-6120.

[30] R.S. Sundaram, C. Gomez-Navarro, K. Balasubramanian, M. Burghard, K. Kern, Electrochemical modification of graphene, Adv Mater, 20 (2008) 3050-3053.

[31] G. Williams, P.V. Kamat, Graphene-Semiconductor Nanocomposites: Excited-State Interactions between ZnO Nanoparticles and Graphene Oxide, Langmuir, 25 (2009) 13869-13873.

[32] G. Williams, B. Seger, P.V. Kamat, TiO_2-graphene nanocomposites. UV-assisted photocatalytic reduction of graphene oxide, Acs Nano, 2 (2008) 1487-1491.

[33] B. Zhang, L. Li, Z. Wang, S. Xie, Y. Zhang, Y. Shen, M. Yu, B. Deng, Q. Huang, C. Fan, J. Li, Radiation induced reduction: an effective and clean route to synthesize functionalized graphene, Journal of Materials Chemistry, 22 (2012) 7775-7781.

[34] R. Flyunt, W. Knolle, A. Kahnt, A. Prager, A. Lotnyk, J. Malig, D. Guldi, A. Abel, Mechanistic Aspects of the Radiation-Chemical Reduction of Graphene Oxide to Graphene-Like Materials International Journal of Radiation Biology, 90 (2014) 486-494.

[35] S. Eigler, M. Enzelberger-Heim, S. Grimm, P. Hofmann, W. Kroener, A. Geworski, C. Dotzer, M. Rockert, J. Xiao, C. Papp, O. Lytken, H.P. Steinruck, P. Muller, A. Hirsch, Wet Chemical Synthesis of Graphene, Adv Mater, 25 (2013) 3583-3587.

[36] S. Eigler, C. Dotzer, F. Hof, W. Bauer, A. Hirsch, Sulfur Species in Graphene Oxide, Chem-Eur J, 19 (2013) 9490-9496.

[37] G.V. Buxton, C.L. Greenstock, W.P. Helman, A.B. Ross, Critical-Review of Rate Constants for Reactions of Hydrated Electrons, Hydrogen-Atoms and Hydroxyl Radicals ($°OH/°O^-$) in Aqueous-Solution, Journal of Physical and Chemical Reference Data, 17 (1988) 513-886.

[38] P. Wardman, Reduction Potentials of One-Electron Couples Involving Free-Radicals in Aqueous-Solution, Journal of Physical and Chemical Reference Data, 18 (1989) 1637-1755.

[39] D. Li, M.B. Muller, S. Gilje, R.B. Kaner, G.G. Wallace, Processable aqueous dispersions of graphene nanosheets, Nature Nanotechnology, 3 (2008) 101-105.

[40] J.I. Paredes, S. Villar-Rodil, P. Solis-Fernandez, A. Martinez-Alonso, J.M.D. Tascon, Atomic Force and Scanning Tunneling Microscopy Imaging of Graphene Nanosheets Derived from Graphite Oxide, Langmuir, 25 (2009) 5957-5968.

[41] A. Kaniyoor, S. Ramaprabhu, A Raman spectroscopic investigation of graphite oxide derived graphene, Aip Adv, 2 (2012).

[42] S. Eigler, S. Grimm, M. Enzelberger-Heim, P. Muller, A. Hirsch, Graphene oxide: efficiency of reducing agents, Chem Commun, 49 (2013) 7391-7393.

[43] S. Eigler, F. Hof, M. Enzelberger-Heim, S. Grimm, P. Müller, A. Hirsch, Statistical Raman Microscopy and Atomic Force Microscopy on Heterogeneous Graphene Obtained after Reduction of Graphene Oxide, The Journal of Physical Chemistry C, 118 (2014) 7698-7704.

[44] A.C. Ferrari, J. Robertson, Interpretation of Raman spectra of disordered and amorphous carbon, Phys Rev B, 61 (2000) 14095-14107.

[45] M.M. Lucchese, F. Stavale, E.H.M. Ferreira, C. Vilani, M.V.O. Moutinho, R.B. Capaz, C.A. Achete, A. Jorio, Quantifying ion-induced defects and Raman relaxation length in graphene, Carbon, 48 (2010) 1592-1597.

[46] V.H. Pham, H.D. Pham, T.T. Dang, S.H. Hur, E.J. Kim, B.S. Kong, S. Kim, J.S. Chung, Chemical reduction of an aqueous suspension of graphene oxide by nascent hydrogen, Journal of Materials Chemistry, 22 (2012) 10530-10536.

[47] Y.X. Xu, K.X. Sheng, C. Li, G.Q. Shi, Highly conductive chemically converted graphene prepared from mildly oxidized graphene oxide, Journal of Materials Chemistry, 21 (2011) 7376-7380.

[48] I.K. Moon, J. Lee, R.S. Ruoff, H. Lee, Reduced graphene oxide by chemical graphitization, Nat Commun, 1 (2010).

[49] J.C. Meyer, A.K. Geim, M.I. Katsnelson, K.S. Novoselov, D. Obergfell, S. Roth, C. Girit, A. Zettl, On the roughness of single- and bi-layer graphene membranes, Solid State Commun, 143 (2007) 101-109.

[50] Y. Hernandez, V. Nicolosi, M. Lotya, F.M. Blighe, Z.Y. Sun, S. De, I.T. McGovern, B. Holland, M. Byrne, Y.K. Gun'ko, J.J. Boland, P. Niraj, G. Duesberg, S. Krishnamurthy, R. Goodhue, J. Hutchison, V. Scardaci, A.C. Ferrari, J.N. Coleman, High-yield production of graphene by liquid-phase exfoliation of graphite, Nature Nanotechnology, 3 (2008) 563-568.

[51] Y. Geng, S.J. Wang, J.-K. Kim, Preparation of graphite nanoplatelets and graphene sheets, Journal of Colloid and Interface Science, 336 (2009) 592-598.

27

Bactericidal Evaluation of Nano-coated Cotton Fabrics

Hanan Basioni Ahmed[1, *], Mohammed Hussein El-Rafie[2], Magdy Kandil Zahran[1]

[1]Chemistry Department, Faculty of Science, Helwan University, Ain-Helwan, Cairo, Egypt
[2]Textile Research Division, National Research Centre, Dokki, Cairo, Egypt

Email address:
hananbasiony@gmail.com (H. B. Ahmed)

Abstract: Nano-sized silver particles (AgNPs) were synthesized by easy and quite simple method, using pectin as both reducing and stabilizing agent. Solutions of AgNPs were applied to cotton fabrics in presence/absence of binder. The finished fabrics were examined for morphological and topographical features by using scanning electron microscopy which reveals that AgNPs- pectin composite are deposited on the surface of coated fabrics. Also, color coordinates were measured for the uncoated and coated fabrics to show the effect of nanosilver loading on the color of coated fabrics. The antibacterial activity of the treated fabrics loaded with AgNPs was evaluated against *Escherichia coli*, *Pseudomonas aeruginosa* and *Staphylococcus aureus*.

Keywords: AgNPs- Pectin Composite, SEM, Color Coordinates, Bactericidal Activities

1. Introduction

In the last two decades, the study and preparation of inorganic crystalline particles in the order of nanometer range has attracted considerable attention of scientists from both fundamental and applied research field [1]. Metal nanoparticles (MNPs), such as silver[2, 3, 4, 6, 7, 8, 9], gold and copper have received special attraction because of their catalytic[12], electronic[13] and unique optical properties [14] making them very attractive in the fields of particularly sensing, bio-conjugation, and surface enhancement Raman spectroscopy (SERS) [15, 16].

Among the noble MNP's, silver has wide recognition for its application in semiconductors [17, 18, 19], superconductors [20, 21], super magnets [22], micro-electronics [23, 24], lithography [16, 25], etc. Recently, researchers have shown that the silver nanoparticles interact with a human immunodeficiency virus and prevent virus from binding to the host cells [26]. The antimicrobial activity of silver nanoparticles is comparatively better than the broad-spectrum most prominent antibiotics used worldwide [27].

Lenard reported that colloidal silver atoms can kill almost any germ which comes in contact with it. Most antibacterial agents available today inactivate or kill only a limited spectrum of bacteria, viruses, or fungi and also these agents often develop the resistance species but silver formulas are exception to this rule [28]. Recently, many studies [7, 39, and 41] reported that, treated cotton fabrics with nanocrystalline silver particles using pad-dry-cure method. Samples were tested against gram positive and gram negative bacteria for evaluation of the biocidal activities, and they observed that treated fabrics exhibited high levels of bacterial inhibition, while the untreated fabrics did not show any antibacterial activities. More recent researches were interested in manufacturing of multifunctional fabrics by uploading of AgNPs through exploiting fabric backbones with its reducing end groups to play the dual roles of reducer and capping agents for metal nanoparticles [38, 40, and 42].

It is known that the reactivity of polysaccharide molecule is due to terminal sugars, mainly localized in side chains. In addition, conformation features of the macromolecule, caused by intramolecular stabilization bonds between functional groups in side chains, are responsible for biological activity of a polysaccharide.

One of the widespread approaches to the synthesis of metal nanoparticles involves the reduction reaction of metal ions in a polymeric solution [5, 10 and 11]. As a rule, the high-molecular compound (polysaccharide) employed in this case acts as a protective polymeric screen ensuring both the size of metal nanoparticles and stabilization of the nanobiocomposite formed [29, 43]. Borohydrides, aluminium hydrides, aminoboranes, hypophosphites, hydroquinone, formalin, light, and radiation are used in literatures as reducing agents [30].

Pectin is a natural, non-toxic, and amorphous carbohydrate present in cell walls of all plant tissues, which functions as an intercellular and intracellular cementing material. As a secondary product of fruit juice, sunflower oil, and sugar manufacture industries, pectin is both inexpensive and abundantly available. Therefore, pectin is an excellent candidate for eco-friendly biodegradable applications. Pectin is commonly used in the food industry as a gelling and stabilizing agent. Pectin macromolecules are able to bind with some organic or inorganic substances via molecular interactions. So, pectin can be used to construct matrices to absorb desired materials and deliver them in a controlled manner [31]. Indeed, pectin has been used to fabricate delivery systems for controlled drug release [32], implantable cell carriers [33], and so on.

Currently, hybrid inorganic–organic nanocomposite materials are of great interest because of their multi-functionality owing to a combination of different compounds incorporated [34]. The incorporation of nanocrystalline silver into pectin to form nanocomposite may impart unique functionalities to the nanocomposite prepared, which could be uploaded on the surface of fabrics to be applied in medical purposes.

Herein we report the preparation of pectin- silver nanocomposites [9] to be used for coating cotton fabrics, in order to obtain new product with antimicrobial activities with a facile solution approach. This approach may find potential application in the medical field.

2. Experimental

2.1. Materials

Silver nitrate (AgNO$_3$, 99.5%), pectin (M.W= 300000 – 1000000) supplied from (Alpha Chemika Company, Mumbai, India), Sodium hydroxide (NaOH), Sodium carbonate monohydrate (Na$_2$CO$_3$.H$_2$O) and Nitric acid (HNO$_3$, 55%) were all used without further purification. Desized, scoured, and bleached 100% cotton fabrics, were kindly supplied from El-Mahalla Company for Spinning and Weaving, El- Mahalla El-Kubra, Egypt.

2.2. Coating Process

Incorporation of silver Nanoparticles in cotton fabrics was performed by pad – dry – cure method. Ag- Cellulosic fabrics was prepared by immersing of fabrics (20 cm × 20 cm) in a colloidal solution bath of silver nanoparticles (50 ppm and 100 ppm) for 30 seconds and squeezed to 100% wet pick up using laboratory pad at constant pressure. The samples were dried at 75°C for 15 minutes and cured at 120°C for 3 minutes for thermal fixation of nanoparticles on fabrics surface.

Table 1 shows synthesis process and characterization of AgNPs [9]. 50 and 100 ppm AgNPs colloidal solutions are prepared as follows: 3 g/l pectin was hydrolyzed by 20 g/l NaOH at 70°C then AgNO$_3$ (0.5 mmole/l in case of preparing 50ppm, and 1mmole/l in case of preparing 100ppm AgNPs) was gradually added, and then left for 30 minutes to complete the preparation process. In case of presence of binder (printo® FX based on acrylate), the binder (10g/l),

was added at the end of preparation process.

3. Measurements

3.1. UV-Visible Spectroscopy

Silver nanoparticles solutions exhibit an intense absorption peak due to the Surface Plasmon Resonance (SPR). Thus the UV-visible absorption spectra were used to prove the formation of AgNPs colloidal solutions. The UV-visible absorption spectra of AgNPs colloidal solutions were measured using a multi channel spectrophotometer (T80 UV/VIS, d= 10 mm, PG Instruments Ltd, Japan) at wavelengths 250 - 600 nm.

3.2. Transmission Electron Microscope (TEM)

For more characterization of the prepared silver nanoparticles, two drops of the silver nanoparticles colloidal solutions were placed on a 400 mesh copper grid coated by carbon film. The morphology and the distribution of AgNPs were characterized by means of a JEOL-JEM-1200 Transmission Electron Microscope.

3.3. Particles Size Distribution

The diameter and distribution of silver nanoparticles were calculated by 4 pi analysis software using TEM photos. The average diameter of the silver nanoparticles was determined from the diameter of at least 20 – 100 nanoparticles.

3.4. Scanning Electron Microscopy (SEM)

Scanning electron microscopy (SEM) was used to study the surface topographic features of fabrics treated with AgNPs in comparison with the untreated fabrics. Fabric samples were located on copper coated carbon tap double face, and then coated by the gold layer by evaporization of gold in argon atmosphere using sputter coater (S150 A – Edwards, UK). The surfaces of samples were scanned using JEOL X840A – Japan/ JXA – 840A electroprobe Micro Analayzer – Japan.

3.5. Moisture Content

Moisture content was measured as follows: 1 g of sample was weighed accurately and then dried at 105°C for 4 h. The dried sample was reweighed; then, moisture content was calculated according to Eq.1, which was found to be 4.54%.

$$MC = [(A – B)/A] \times 100 \qquad (1)$$

Where, MC is moisture content (%), A= initial weight (g) and B= weight of dried fabric (g).

3.6. Detection of Silver Content

Silver content in the coated cotton fabrics was measured as follows: 0.2 g of coated dried fabrics was immersed in 30 ml of 15 wt% nitric acid for 2 h at 80°C. Silver concentration was recorded by using flame atomic absorption spectroscopy

(AAS, SpectrAA, 220, Varian, Australian) equipped with silver lamp (328.1 nm). The silver content was calculated by Eq.2.

$$\textbf{Silver content } (\boldsymbol{mmole/kg}) = \frac{C_S}{W_d/(1-MC/100)} \times V \qquad (2)$$

Where C_s= silver concentration (mmole/l) in extracted solution which is detected by atomic absorption spectroscopy; V= volume of extracted solution; W_d= weight of dried coated fabric (g); MC= moisture content of coated fabrics (%)[3].

3.7. Washings and Silver Release

The washing process of AgNPs coated fabrics can be described briefly as follows(Hossam et al, 2013): treated fabric was immersed in washing solution which contained 2 g/l Na_2CO_3 and 2 g/l commercial detergent, using material to liquor ratio 1:50.Then, the samples were stirred and left for 15 minutes at 55 ± 5°C. Finally, the fabrics were gently squeezed and rinsed with tap water. This process was repeated 5, 10 and 20 times to get 5, 10 and 20 washings. The silver release was calculated with Eq.3.

$$\text{Silver release (\%)} = [(C_b - C_a)/ C_b] \times 100 \qquad (3)$$

C_b is the silver content in treated fabrics before washing, and C_a is the silver content in treated fabrics after washing. C_a and C_b were measured by extraction method which is described previously.

3.8. Color Measurements

Color measurements of AgNPs coated fabrics were recorded with a colorimeter with pulsed xenon lamps as light source (UltraScan Pro, Hunter Lab, USA). The equipment could be characterized as follows : CIE LAB color space, 10° observer with D65 illuminant, d/2 viewing geometry and measurement area of 2 mm. Color measurement parameters are lightness (L^*) from black (0) to white (100), a^* is a red (+) / green (-) ratio, b^* is yellow (+) / blue (-) ratio. Each data point was the average of two independent measurements.

3.9. Antibacterial Test

The Antimicrobial activity of AgNPs coated cotton fabrics were tested by using two different techniques; qualitative method (inhibition zone technique) and quantitative method (plate count agar method).

The qualitative method was carried out by using a modified Kirby-Bauer disc diffusion technique. Briefly, 100 µl of the tested bacteria were grown in 10 ml of fresh media until they reached a count of approximately 108 cells /ml [35]. A 100 µl of microbial suspension was spread onto agar plates corresponding to the broth in which they were maintained. Plates inoculated separately with Gram (+) bacteria as *Staphylococcus aureus* and Gram (-) bacteria as *Escherichia coli* and *Pseudomonas aeruginosa*, were incubated at 35-37°C for 24-48 hours, then the diameters of the inhibition zones were measured in millimeters .Standard

discs of Tetracycline (Antibacterial agent), served as positive controls for antimicrobial activity, however, filter discs impregnated with 10 µl of solvent (distilled water, chloroform, DMSO) were used as a negative control.

When a part of the AgNPs coated fabrics is placed on agar media, AgNPs will diffuse from fabric into the surrounding. The solubility of nanosilver and its particle size will determine the size of the area of silver infiltration around the fabric. If an organism is placed on agar it will not grow in the area around the fabric (if it is susceptible to the AgNPs). This area of no growth around the coated fabric is known as a "Zone of inhibition" or" Clear zone".

For AgNPs diffusion, the zone diameters were measured with slipping calipers of the National Committee for Clinical Laboratory Standards. The average width for zone of inhibition along a streak on either side of the tested fabric was calculated using eq. 4:

$$W= (T-D)/2 \qquad (4)$$

Where W=width of clear zone of inhibition in mm, T= total diameter of tested fabric and clear zone in mm, and D= diameter of the tested fabric in mm.

The quantitative method was performed for the washed samples against *Staphylococcus aureus* as Gram +ve bacteria according to the AATCC test method 100–1999 for Bacterial Counting. Briefly all treated fabrics were kept at 35°C prior to test. Then, 0.5 g fabrics were transferred into 100 ml of nutrient broth (*ca.* 1.5×10^8 colony forming unit per ml), and shaked vigorously for 1min. A normal saline solution was prepared with 0.9% (w/v), was exposed to serial dilution and then plated onto Mannitol salt agar plates. Plates were incubated at 37°C for 24 hours and then the colonies were counted. The reduction percentage of bacterial colonies was calculated using equation 5.

$$R\% = [(B - A) / B] \times 100 \qquad (5)$$

Where R% is the reduction percentage of bacterial colonies, A is the number of bacterial colonies on the agar plate with coated fabric, and B is the number of bacterial colonies on the agar plate for control.

4. Results and Discussion

Nanocomposites synthesis is based on the nanocomposites self-organization, where the polymers play a role of reducing agent and nano stabilizing medium [2, 3, 4, 5, 6, 7, 8, 9, 36]. In this case, synergism of properties of the polymeric matrix (biological activity, hydrodynamical characteristics) and those of the metal core (optical, biological, thermophysical, electric) takes place which provides for promising performance characteristics of the nanocomposites formed. According to this approach, nanobiocomposites have been prepared using natural polysaccharides: arabinogalactan, galactomannan, carboxymethylcellulose, heparin [36], sea seaweed polysaccharides [37], and so forth.

4.1. UV-vis Absorption Spectroscopy

A well-stabilized AgNPs solution with a concentration of 100 ppm was prepared using pectin as reducing agent for silver ions as well as stabilizer for the formed AgNPs in the optimum conditions as follows: pectin, 3g/l g; silver nitrate, 1 mmole/l; pH, 12; temperature, 70°C. Fig.1 shows the UV–vis absorption spectroscopy for AgNPs colloidal solution (concentration of 100 ppm) prepared in the above optimum conditions. It is clear that, the band becomes stronger and more symmetrical with a pronounced bell shape at λ max 409 nm. The band can be assigned to the plasmon resonance of AgNPs.

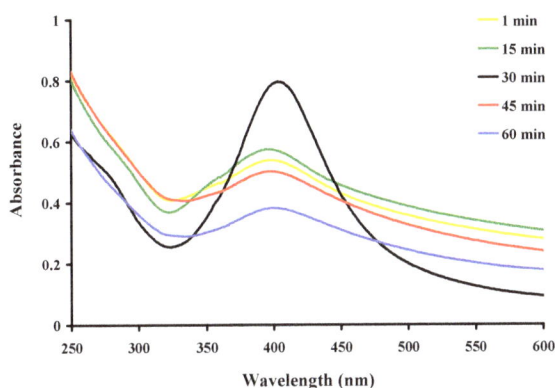

Figure 1. *UV-Vis absorption spectroscopy of silver nanoparticles (AgNPs) prepared at different durations. Reaction conditions: 3g/l pectin, 1mmole/l AgNO₃, pH12, temperature 70°C.*

4.2. TEM Image and Histogram

Figs.2a&b show the TEM image and the histogram of the size and size distribution of AgNPs in the aforementioned conditions. The obtained figures depict that the resultant product contains a well-stabilized AgNPs solution with a concentration of 100 ppm and a diameter range of (5–10 nm). AgNPs solutions with such unique characteristics are unequivocally feasible for industrial applications. The antibacterial activity of untreated/treated fabrics with colloidal solution of AgNPs is studied, and the obtained data are discussed below.

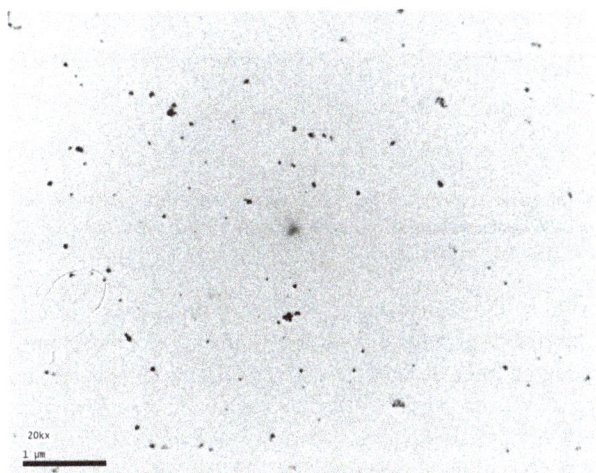

(a)

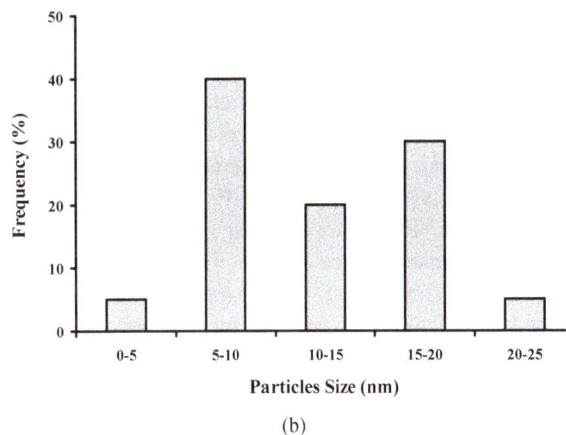

(b)

Figure 2. *(a) TEM image silver nanoparticles prepared by 1 mmole/l AgNO₃ (reaction conditions: 3 g/l pectin, pH 12 at 70°C for 30 minute). (b) Size distribution histogram for silver nanoparticles in the viewed TEM image.*

4.3. Color Measurements

The CIELAB color system is widely used in the color measurement of textiles. In this system, L* shows the lightness of the fabric and a* and b* indicate red-green (redder if positive; greener if negative) and yellow-blue colors (yellower if positive; bluer if negative), respectively.

The achieved results are proposed in Table 1 with two concentrations of silver nano particles (50 & 100 ppm), with and without washings. These data showed the effect of pectin – silver nanocomposite on the coated fabric coloration under washing process. Through increasing the amount of the silver nanoparticles on the fabrics from 50 to 100 ppm, b* values increased and the color of the fabrics tuned to creamy-yellow indicating the formation of the nanoparticles on the fabric surface. In contrast, L* values are decreased as the lightness of coated fabrics is decreased. However, by washing, b* values were decreased and L* values were increased, as the lightness is increased by washing.

By comparing the data measured for the fabric coated with 50 ppm and that were coated with 100 ppm, the difference between L* values was not significant , which may sign to that 50 ppm is sufficient concentration for the preparation of coated fabric , which could exhibiting antibacterial activities.

Table 1. *Color coordinates data (CIE lab) for silver-pectin nanocomposite treated fabrics as a function of silver content.*

Samples		L*	a*	b*
Blank (Untreated)		93.47 ± 0.15	-0.27 ± 0.08	1.59 ± 0.09
A	1	79.49 ± 1.94	2.23 ± 0.77	11.69 ± 1.55
	2	87.23 ± 0.45	0.15 ± 0.13	2.43 ± 0.18
	3	87.83 ± 0.00	0.08 ± 0.02	2.55 ± 0.06
	4	88.99 ± 0.18	0.01 ± 0.01	2.40 ± 0.16
B	1	75.58 ± 1.26	3.47 ± 0.50	14.21 ± 1.06
	2	84.29 ± 0.08	1.07 ± 0.05	8.38 ± 0.16
	3	85.24 ± 1.56	0.50 ± 0.32	7.09 ± 1.15
	4	85.63 ± 0.23	0.73 ± 0.01	6.30 ± 1.00

A: Cotton treated with 50 ppm AgNPs solution
B: Cotton treated with 100 ppm AgNPs solution
1= before washing, 2= after 5 washing cycles
3= after 10 washing cycles and 4= after 20 washing cycles.

4.4. Silver Particle-Coated Fabrics and Surface Characterization (SEM Images)

The surfaces of the nanosilver treated fabrics were examined using SEM images as depicted in Figure 3. Figure 3a showed the surface image of the untreated cotton fabric, while treated fabric with 100 ppm AgNPs colloidal solution in the presence of binder is shown in Figures 3b. It could be observed that, treated fabrics exhibited AgNPs-pectin composite on some concentrated areas of the fabric. It could be supposed that the AgNPs coated on the surface of fabrics could not be seen with SEM, due to their small size and the fact that they are embedded in the polymer matrix of the alkali hydrolyzed pectin, so it could be decided that SEM images reveal aggregates of somewhat larger units of the nanosilver – polysaccharide composite.

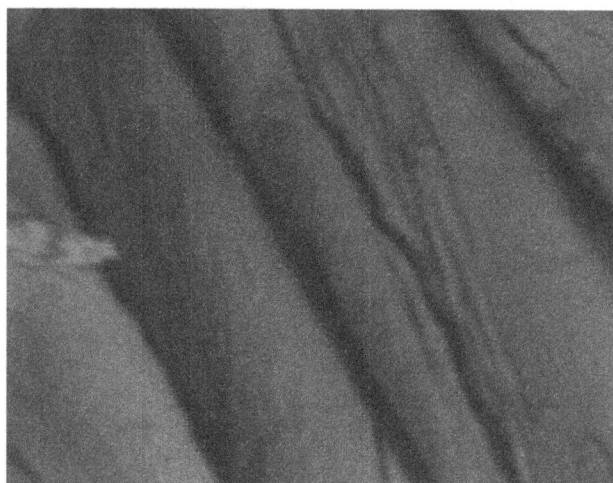

10µm   X3500

(a)

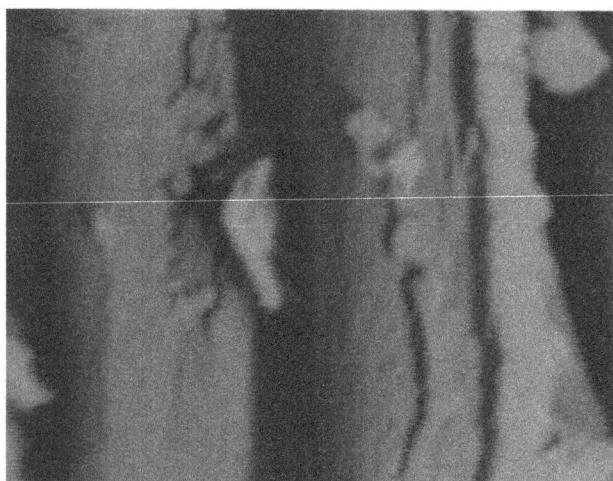

10µm   X3500

(b)

Figure 3. SEM image for (a) untreated cotton fabric and (b) treated fabric.

4.5. Silver Content in AgNPs Coated Cotton Fabrics and Silver Release

Total silver content in coated fabrics was determined by atomic absorption spectrophotometer. From Figure 4a, significantly, it could be observed that the actual amount of silver content on the surface of fabrics increased by increasing the concentration of the AgNPs colloidal solution. It could be associated with higher deposition of AgNPs, which is suggested to take a place with fixation therein through physical bonding.

The data also highlighted that, higher concentrations of AgNPs with binder improved nanosilver loading. As by coating fabrics with 100 ppm AgNPs colloids using binder, results in giving the best value of nanosilver loading (121.21mg/kg). This confirms the role of binder in fixation of the deposits of silver nanoparticles within the molecular structure of cotton.

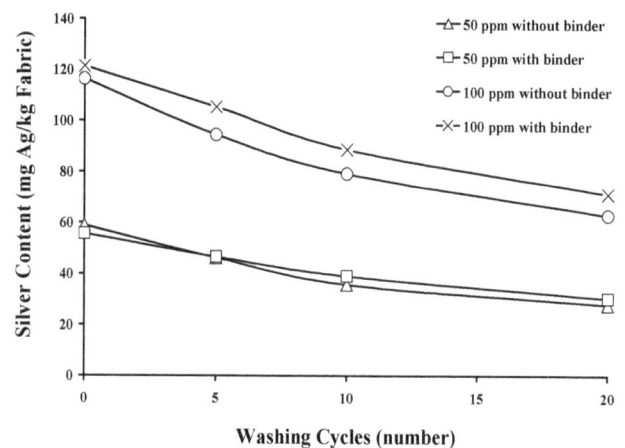

(a)

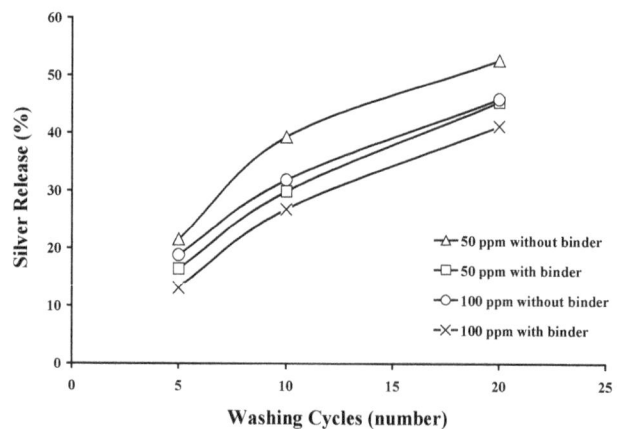

(b)

Figure 4. (a) Silver contents on the coated fabrics with different treatment conditions with number of washing cycles. (b) Silver release from the coated cotton fabrics in the washing liquor.

To evaluate the laundering durability of nanosilver coated cotton fabrics, Ag content and silver release percent were both measured after different washings. Figure 4b shows the silver release from fabrics after repeated washings. It could be observed that, silver release percent was higher in case of using AgNPs colloidal solution without binder rather than that with binder. As in case of coated fabrics treated with 100 ppm AgNPs in absence of binder, silver release percent was 18.8% after 5 washings, and became 46.0% after 20 washings. However, in the presence of binder, silver release percent was 13.1% after 5 washings, and reached 41.1% after 20 washings. So, it could be summarized that, washing the coated cotton fabrics resulting in removal of nanosilver deposits and this created the idea of incorporating fixing agent in the finishing bath.

4.6. Antimicrobial Assessment of the Silver Coated Fabrics

An antibacterial activity assessment of the fabric samples treated with nanosilver particles was conducted using the method described previously in the experimental part and was compared against untreated fabric. The growth of different bacterial strains directly under the specimens is given in Tables 2 & 3.

When the fabric specimen is placed on top of the bacterial lawn, antimicrobial agents (silver particles) from the fabric diffuse into the media that has bacterial lawn (growth). Untreated cotton specimens showed no antibacterial effect under the sample contact area, where as the unwashed treated fabrics with 50 ppm nanosilver in absence of binder, exhibited considerable antibacterial effect, as the antibacterial activity against *S. aureus, E. coli, and P. aeruginosa* which was represented in the form of inhibition zone values (mm/1cm fabric) were 2, 2 and 3 mm respectively. However, in the presence of binder, inhibition zone values were 3, 3 and 3 mm respectively.

For treated fabrics with 50 ppm in the presence of binder, washed in 20 cycles, inhibition zone values were 1, 1 and 1.5 mm respectively. For fabric treated with 100ppm nanosilver in the presence of binder and washed by 20 cycles, inhibition zone values for the plates contained the three strains mentioned before were 2, 1 and 2mm respectively.

The bacterial counting method was used as quantitative method for the detection of antibacterial efficacy against *Staphylococcus aureus* as an example of gram positive bacterial species, and data was reported in table 3. The experiment was done only for the coated fabrics subjected to 20 washings. Results of Table 3 also showed that, regardless of the concentration of AgNPs used or silver content of fabrics, the reduction of bacterial colonies was always higher than 80%. An excellent antibacterial property was acquired by the fabrics coated with 100 ppm AgNPs solution in the presence of binder, and subjecting the treated fabrics to washings leads to non sense decrement in the reduction of bacterial colonies, as after 20 washings, fabrics coated with 100 ppm AgNPs colloidal solution (with binder) caused a bacterial reduction reached to 98%. Thus, it still exhibited excellent antibacterial properties.

Thus, it could be concluded that, coating of cotton fabric with 100 ppm AgNPs colloidal solution, in the presence of binder is preferable to obtain excellent antibacterial activity against *S. aureus* as Gram positive bacterial strain and both of *E. coli* and *P. aeruginosa* as Gram negative bacterial strain.

Table 2. *Effect of washing cycles and silver content on the antibacterial activities of nanosilver coated cotton fabrics against three different bacterial strains.*

Samples		Silver content (mg/kg)	Inhibition zone diameter (mm)		
			Staphylococcus aureus (gram +ve)	Escherichia coli (gram – ve)	Pseudomonas aeruginosa (gram -ve)
Blank (Untreated)		0	0	0	0
A	1	58.90	2	2	3
	2	46.23	1	3	2
	3	35.73	1.5	2	1
	4	27.84	1	0.5	1
B	1	55.64	3	3	3
	2	46.53	2.5	3	2.5
	3	39.00	2	2	2
	4	30.36	1	1	1.5
C	1	116.30	3	3.5	3
	2	94.45	3	3	2
	3	79.22	2.5	2	2
	4	62.85	1.5	1	1.5
D	1	121.21	4	4	5
	2	105.33	3	2	3.5
	3	88.70	2	1.5	2.5
	4	71.19	2	1	2

A: Cotton treated with 50 ppm AgNPs solution
B: Cotton treated with 50 ppm AgNPs solution in presence of binder
C: Cotton treated with 100 ppm AgNPs solution
D: Cotton treated with 100 ppm AgNPs solution in presence of binder
1= before washing, 2= after 5 washing cycles
3= after 10 washing cycles and 4= after 20 washing cycles.

Table 3. *Quantitative analysis of antibacterial activities for AgNPs treated cotton fabrics after 20 washing cycles.*

Samples	Silver content (mg/kg)	Bacterial reduction (%)
		Staphylococcus aureus (gram +ve)
Blank (Untreated)	0	0
A	27.84	82
B	30.36	84
C	62.85	94
D	71.19	98

A: Cotton treated with 50 ppm AgNPs solution
B: Cotton treated with 50 ppm AgNPs solution in presence of binder
C: Cotton treated with 100 ppm AgNPs solution
D: Cotton treated with 100 ppm AgNPs solution in presence of binder

5. Conclusion

We developed a simple approach to prepare pectin–silver nanocomposite in aqueous alkaline solution. The experimental results confirm the true pectin–nanosilver composite structure and the existence of strong interaction between pectin molecules and silver, and its stabilized physical deposition on the surface of coated fabrics. As, the obtained AgNPs were successfully applied to cotton fabrics, and it is found that, using 100 ppm of AgNPs with binder showed excellent antibacterial activity against *E. coli, P.*

aeruginosa and S. aureus. The SEM analysis indicates that the AgNPs are well dispersed on the cotton fabrics. Binder was used successfully in this work to diminish the bacterial growth on the coated cotton fabrics. The result of durability to wash of the coated fabric also showed long-lasting bactericidal effect even after 20 washing cycles. Therefore, this kind of treatment could be expressed as a safe, cost effective and environmental friendly process, easily applicable for antibacterial finishing, and may be extended to prepare other hybrid inorganic–organic nanocomposite materials.

References

[1]   Elghanian, R., Storhoff, J. J., Mucic, R. C., Letsinger, R. L., & Mirkin, C. A. (1997). Selective colorimetric detection of polynucleotides based on the distance-dependent optical properties of gold nanoparticles. Science, 277, 1078–1081.

[2]   Emam, H. E., Mowafi, S., Mashaly, H. M., Rehan, M. (2014). Production of Antibacterial Colored Viscose Fibers Using In-Situ Prepared Spherical Ag nanoparticles. Carbohydrate Polymers, doi.10.1016/j.carbpol.2014.03.082.

[3]   Hossam E. E., Avinash P. M., Barbora S., Heinz D., Bernhard R., Alexandra P., Thomas B. (2013). Treatments to impart antimicrobial activity to clothing and household cellulosic-textiles e why "Nano"-silver? Journal of Cleaner Production, 39, 17-23.

[4]   Hebeish, A. A., El-Rafie, M. H., Abdel-Mohdy, F. A., Abdel-Halim, E. S., Emam, H. E., (2010). Carboxymethyl cellulose for green synthesis and stabilization of silver nanoparticles. Carbohydrate Polymers, 82, 933–941.

[5]   7. M. H. El-Rafie, Hanan B. Ahmed, M. K. Zahran. (2014).Facile precursor for synthesis of silver nanoparticles using alkali treated maize starch. International Scholarly Research Notices, Volume 2014, Article ID 702396, 12 pages.

[6]   El-Rafie, M. H., Hanan B. Ahmed, Zahran, M. K. (2014). Characterization of nanosilver coated cotton fabrics and evaluation of its antibacterial efficacy. Carbohydrate polymers, 107, 174-181.

[7]   Zahran, M. K., Hanan B. Ahmed, El-Rafie, M. H. (2014a). Surface modification of cotton fabrics for antibacterial application by coating with AgNPs-Alginate composite. Carbohydrate polymers, 108, 145-152.

[8]   Zahran, M. K., Hanan B. Ahmed, El-Rafie, M.H. (2014b). Alginate mediate for synthesis controllable sized AgNPs. Carbohydrate polymers, IN PRESS.

[9]   Zahran, M. K., Hanan B. Ahmed, El-Rafie, M. H. (2014c). Facile size regulated synthesis of silver nanoparticles using pectin. Carbohydrate polymers, 111, 971–978.

[10] Hossam E. Emam, Avinash P. Manian, Barbora Široká, Heinz Duelli, Petra Merschak, Bernhard Redl, Thomas Bechtold. 2014. Copper (I) oxide surface modified cellulose fibers— Synthesis, characterization and antimicrobial properties", *Surface and Coatings Technology, 254,* 344–351.

[11] Hossam E. Emam, Manal K. El-Bisi. (2014). Merely Ag nanoparticles using different cellulose fibers as removable reductant. *Cellulose,* 21, 4219–4230.

[12] Mallik, K., Witcomb, M. J., & Scurell, M. S. (2005). Redox catalytic property of gold nanoclusters: Evidence of an electron-relay effect. Applied Physics A, 80, 4, 797–801.

[13] Kamat, P. V. (2002). Photophysical, photochemical and photocatalytical aspects of metal nanoparticles. Journal of Physical Chemistry B, 106, 7729–7744.

[14] Liz-Marzan, L. (2006). Tailoring surface plasmons through the morphology and assembly of metal nanoparticles. Langmuir, 22, 32–41.

[15] Cao, Y. C., Jin, R., Nam, J., Thaxton, C. S., & Mirkin, C. A. (2003). Raman-dye-labeled nanoparticle probes for proteins. Journal of American Chemical Society, 125, 14676–14677.

[16] Shipway, A. N., Lahav, M., & Willner, I. (2000). Nanostructured gold colloid electrodes. Advanced Materials, 12, 13, 993–998.

[17] Henglein, A. (1989). Small-particle research: Physicochemical properties of extremely small colloidal metal and semiconductor particles. Chemical Review, 89, 1861–1873.

[18] Kamat, P. V. (1993). Photochemistry on nonreactive and reactive (semiconductor) surfaces. Chemical Reviews, 93, 267–300.

[19] Schimd, G. (1992). Large clusters and colloids metals in the embryonic state. Chemical Reviews, 92, 1709–1727.

[20] Henglein, A. (1993). Physicochemical properties of small metal particles in solution: Microelectrode reactions, chemisorption, composite metal particles, and the atom-to-metal transition. Journal of Physical Chemistry, 97, 5457–5471.

[21] Pileni, M. P. (1993). Reverse micelles as microreactors. Journal of Physical Chemistry, 97, 6961–6973.

[22] Lee, A. F., Baddeley, C. J., Hardacre, C., Ormerod, R. M., Lambert, R. M., Schmid, G., et al. (1995). Structural and catalytic properties of novel Au/Pd bimetallic colloid particles: EXAFS, XRD and acetylene coupling. Journal of Physical Chemistry, 99, 6096–6102.

[23] Deheer, W. A. (1993). The physics of simple metal clusters; experimental aspects and simple models. Reviews of Modern Physics, 65, 611–676.

[24] Littau, K. A., Szajowski, P. J., Muller, A. J., Kortan, A. R., & Brus, L. E. (1993). A luminescent silicon nanocrystal colloid via a high-temperature aerosol reaction. Journal of Physical Chemistry, 97, 1224–1230.

[25] Xia, A., Rogers, J. A., Paul, K. E., & Whitesides, G. M. (1999). Unconventional methods for fabricating and pattering nanostructures. Chemical Reviews, 99(7), 1823–1848.

[26] Elechiguerra, J. L., Burt, J. L., Morons, J. R., Camacho-bragado, A., Gao, X., Lara, H. H., et al. (2005). Interaction of nanoparticles with HIV-1. Nanobiotechnology, 3, 6–16.

[27] Roy, R., Hoover, M. R., Bhalla, A. S., Slaweekl, T., Dey, S., Cao, W. (2008). Ultradilute Ag-aquasols with extraordinary bactericidal properties; role of the system Ag–O–H$_2$O. Materials Research Innovations, 11, 3–18.

[28] Lenard, L. (2009). Silver-protein: Gold standard among antimicrobial agents, http://intelegen.com/ImmuneSystem/silver_protein.htm.

[29] Litmanovich, O. E. (2008) "Psevdomatrix sintez of polymer-metal nanocomposite sols: interacion of macromolecules with metal nanoparticles. Chinese Journal of Polymer Science C, 58, 7, 1370–1396.

[30] Pomogailo, A.D. (1997). Polymer-immobilised nanoscale and cluster metal particles, Uspekhi Khimii, 66, 8, 785–791.

[31] Liu L. S., Cooke P. H., Coffin D. R., Fishman M. L., Hicks K. B. (2004). Pectin and polyacrylamide composite hydrogels: effect of pectin on structural and dynamic mechanical properties. Journal of Applied Polymer Science, 92, 1893–1901.

[32] Vandamme, T.F., Lenourry, A., Charrueau, C., Chaumeil, J.C. (2001). The use of polysaccharides to target drugs to the colon. Carbohydrate Polymers. 48, 219.

[33] Liu, L., Fishman, M., Kost, J., Hicks, K.B. (2003). Pectin-based systems for colon-specific drug delivery via oral route. Biomaterials, 24, 3333–3343.

[34] Yao, K. X.; Zeng, H. C. (2007). ZnO/PVP nanocomposite spheres with two hemispheres. Journal of physical chemistry. C, 111, (36), 13301-13308.

[35] Pfaller, M., Burmeister, L., Bartlett, M., Rinaldi, M., (1988). Multicenter evaluation of four methods of yeast inoculum preparation. Journal of Clinical Microbiology, 26, 1437–1441.

[36] Trofimov, B. A., Sukhov, B. G., Aleksandrova G. P. (2003). Nanocomposites with magnetic, optical, catalytic, and biologically active properties based on arabinogalactan. Doklady Chemistry, 393, 287–288.

[37] Yurkova, I. N. , Panov, D. A., Ryabushko, V. A. (2009). Study of optical properties of nanobiocomposites on silver and marine alga polysaccharides base. Uchenye Zapiski Tavricheskogo Universiteta Series Biologiya and Khimiya, 22, 1, 203– 207.

[38] Mohamed Rehan, Hamada M. Mashaly, Salwa Mowafi, A. Abou El-Kheir, Hossam E. Emam. (2015). Multi-functional textile design using in-situ Ag NPs incorporation into natural fabric matrix. Dyes and Pigments 118, 9-17.

[39] Hossam E. Emam, M. K. Zahran. (2015). Ag^0 nanoparticles containing cotton fabric: synthesis, characterization, color data and antibacterial action. *International Journal of Biological Macromolecules*, 75, 106 – 114.

[40] Hossam E. Emam, N.H. Saleh, Khaled S. Nagy, M.K. Zahran. (2015) Functionalization of medical cotton by direct incorporation of silver nanoparticles", *International Journal of Biological Macromolecules,* 78, 249–256.

[41] Hossam E. Emam, M. H. El-Rafie, Hanan B. Ahmed, M. K. Zahran. (2015). Room Temperature Synthesis of Metallic Nanosilver Using Acacia to Impart Durable Biocidal Effect on Cotton Fabrics", *Fibers and Polymers*, 16(8), 1676 – 1687.

[42] Hossam E. Emam, Thomas Bechtold. (2015). Cotton fabrics with UV blocking properties through metal salts deposition. *Applied Surface Science*, http://dx.doi.org/10.1016/j.apsusc.2015.09.095.

[43] Hossam E. Emam, Hanan B. Ahmed. (2016). Review: Polysaccharides templates for assembly of nanosilver", *Carbohydrate Polymers*, 135, 300 – 307.

28

Synthesis of a novel L-tartaric acid derived homochiral nanoscale framework and its application in L-proline detection and acetalization catalysis

Xiong Peng, Radoelizo S. A., Liping Liu, Yi Luan[*]

School of Materials Science and Engineering, University of Science and Technology Beijing, 30 Xueyuan Road, Haidian district, Beijing, 100083 (P. R. China)

Email address:

xiongpengvici@126.com (Xiong Peng), yiluan@ustb.edu.cn (Yi Luan)

Abstract: A novel homochiral nanoscale compound, $[Ca(L-C_4H_4O_6)(H_2O)_2] \cdot 2H_2O$ $(Ca(L-tart)(H_2O)_2)$, which is derived from calcium ions and L-tartaric acid (L-tart $=C_4H_4O_6$), was synthesized under hydrothermal condition. It has been characterized by single crystal X-ray diffraction, SEM, XRD, FTIR and TG. The calcium atoms adopt a tetrahedron geometry and each atom coordinates with eight oxygen atoms. The compound forms a two-dimensional network structure in the solid state via hydrogen bonds. Its performance of L-proline detection was tested, which attained effective result for the porous framework. Meanwhile, the high activity was also shown in acetalization catalysis.

Keywords: L-Tartaric Acid, Inorganic-Organic Framework Compound, L-Proline Detection

1. Introduction

In the past decades, researchers have made notable work in the field of molecular inorganic-organic framework compounds, especially in terms of phosphonates and carboxylates.[1],[2] They were constituted of metal centers assembled by various functional organic ligands via an infinite, regular patterns, and form 1-, 2-, and 3-dimensional structures. The great interest has been attracted for the possible applications in areas such as separation,[3] catalysis[4] and hydrogen storage.[5] One of the particularly interesting hybrid frameworks related to those that have chiral properties.[6] The unique materials have potential application in enantioselective separation,[7] sensors,[8] catalysis[9],[10] and non-linear optics[11]. As we all known that the network of the coordination polymers were controlled by both the metal ions geometry and the nature of ligands.[12] So it was necessary to choose the desired polymers and appropriate ligands to obtain the intended structures and properties. The polycarboxylate ligands coordinating with metal ions should own a feature that they can act as precursors for various polynuclear cluster-based metal compounds for both discrete entities and multi-dimensional systems.[13]

L-tartaric acid, which was hydroxyl-containing polycarboxylate, was an versatile organic carboxylic acid.[14],[15] In recent years, the acid and its derivatives have been applied in hydrogels,[16],[17] chiral extractant,[18] asymmetric epoxidation[19] for its chiral feature. Various metal-tartrate compounds have been synthesized and researched. Xu et al. reported cobalt tartrate compound synthesized via hydrothermal reaction and researched the water chains in the framework.[20] Gübitz et al. reported copper compound of L-tartaric acid through hydrothermal reaction and applied it in the chiral separation of β-blockers and sympathomimetics.[21]

Herein, a novel chiral compound derived from calcium ions and L-tartaric acid, $[Ca(L-C_4H_4O_6)(H_2O)_2] \cdot 2H_2O$, which owns 2-dimensional chiral nanoscale framework, has been successfully synthesized. The structure of the compound was analyzed through the single crystal X-ray diffraction. Meanwhile, its properties in L-prolline detection was furtherly tested, which attained effective result. Further more, the highly activity was also shown in the acetalization catalysis.

2. Experimental Section

2.1. General Procedures

All reagents, calcium nitrate tetrahydrate [Ca(NO$_3$)$_2$·4H$_2$O] (Beijing Chemical Reagent Company), L-(+)-tartaric acid [L-C$_4$H$_6$O$_6$] (99%,Alfa Aesar) and LiOH·H$_2$O (Guangdong Guanghua Sci-Tech Co., Ltd) were commercially available and used without further purification.

2.2. Synthesis of Ca(L-Tart)(H$_2$O)$_2$

Ca(NO$_3$)$_2$·4H$_2$O (0.56 mmol, 0.125 g), L-(+)tartaric acid (0.66 mmol, 0.099 g), and LiOH·H$_2$O (1.12 mmol, 0.047 g) were added to 10 mL deionized water in a 50 ml scintillation vial. The mixture was stirred until a clear solution was formed. The obtained resulting solution was sealed in a 20 ml Teflon-lined autoclave and put into an oven at 150 $^{\circ}$C for 48 hours. Upon cooling during 12 hours, the crystals were washed with EtOH for three times and dried in air at 40 $^{\circ}$C.

2.3. Detection Experiment

In this experiment, the calcium tartrate compound (3.9 mmol, 1 g) was added in 50 mL aqueous solution of 0.01 M L-proline. The mixture was filtered after stirring for 6 hours and then dried at 40 $^{\circ}$C in the oven. The performance of detection was analyzed via FTIR and TGA.

2.4. Catalytic Acetalization Reactions

Generally, the properties of the compound in catalysis were examined for the acetalization in 25 mL round-bottom vial. The aldehydes were added in the mixed solution of 15 mL methanol and 15 mL dichloromethane, together with 20 mg (0.078 mmol) catalyst. The reaction mixture was stirred for 24 h at room temperature. Then the filtered liquid was analyzed by GC-MS, and nitrobenzene was added as an internal standard.

2.5. Characterization

The single crystal X-ray diffraction analysis and data collection were demonstrated on a Rigaku RAXIS-RAPID. And the basic information with regard to crystal parameters and structures was acquired from Diamond 3.1 and summarized in table 1 and table 2. The phase and structure of the samples were collected by X-ray powder diffraction (XRD, Rigaku DMAX-RB 12 KW) with Cu Kα radiation (λ=0.15406 nm). The morphology of the samples was characterized by scanning electron microscopy (SEM, ZEISS SUPRA55). The thermal characteristic of the samples was performed by thermogravimetric analysis (TG) using Netzsch STA449F3 instrument at a heating rate of 10 $^{\circ}$C/min under a N$_2$ flow. Fourier transform Infrared spectras (FT-IR) were obtained on a Nicolet 6700 using the potassium bromide (KBr) pellet technique. The reaction products were analyzed via Gas Chromatography-Mass Spectrum (Agilent 7890A/5975C-GC/MSD), and nitrobenzene was added as an internal standard.

Table 1. *Crystal data and structure for Ca-tart*

Phase data	
Formula sum	Ca$_4$ O$_{40}$ C$_{16}$ H$_8$
Formula weight	1000.53 g/mol
Crystal system	orthorhombic
Space-group	P 21 21 21 (19)
Cell parameters	a=9.229(2) Å b=9.6087(30) Å c=10.5827(22) Å
Cell ratio	a/b=0.9605 b/c=0.9080 c/a=1.1467
Cell volume	938.46(41) Å^3
Z	
Calc. density	1.77026 g/cm^3
Meas. density	
Melting point	
RAll	
RObs	
Pearson code	oP68
Formula type	NO2P4Q10
Wyckoff sequence	a17

Table 2. *Atomic parameters for Ca-L-tart.*

Atom	Ox.	Wyck.	Site	S.O.F.	x/a	y/b	z/c	U [Å^2]
CA1		4a	1		0.18639	0.81757	1.17719	0.0174
O1		4a	1		0.01439	0.76874	1.00959	0.0290
O2		4a	1		-0.04850	0.73695	0.80926	0.0252
O3		4a	1		0.27423	0.85763	0.95462	0.0214
O4		4a	1		0.16270	0.97363	0.71502	0.0219
O5		4a	1		0.46181	0.74112	0.71549	0.0245
O6		4a	1		0.44272	0.96775	0.67231	0.0248
C1		4a	1		0.04312	0.76129	0.89315	0.0181
C2		4a	1		0.20215	0.77919	0.85836	0.0168
H2		4a	1		0.24546	0.68607	0.85860	0.0800
C3		4a	1		0.22831	0.84015	0.72773	0.0188
H3		4a	1		0.18612	0.77748	0.66461	0.0800
C4		4a	1		0.39093	0.85096	0.70389	0.0168
O1W		4a	1		0.19505	0.56438	1.16892	0.0445
O2W		4a	1		0.22555	0.83698	1.41017	0.0316
O3W		4a	1		0.57193	0.91515	0.42630	0.0421
O4W		4a	1		0.42929	0.57609	0.43705	0.0544

3. Results and Discussion

3.1. Structure Characterization

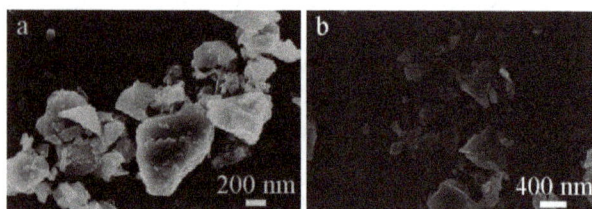

Figure 1. The SEM of (a) fresh sample; (b) the sample after catalytic reaction.

To investigate the structure of the novel compound, the scanning electron microgragh (SEM) and single crystal X-ray diffraction were analyzed. Figure 1 shows the SEM of the fresh sample, which was formed in nanoscale. And the unit of the compound, as demonstrated in Figure 2, which was obtained from Materials Studio 7.0, was consisted of four calcium atoms, four L-tartrate acid anions, eight consorted water molecules and eight uncoordinated water molecules. The calcium atoms lie in tetrahedron coordination environment and each atom coordinates with three different tartrate ligands eight oxygen atoms. Two of the independent tartrate anions complex with the metal centers via a hydroxyl oxygen atom and a carboxylate oxygen atom. Another two oxygen atoms of tartrate anions from neighboring dimer assemble through a carboxylate oxygen atom. The last two coordination sites were given to two aquo ligands. The distance between two metal ions was long enough that there was no interaction between them. What was more interesting , there exist pore structure in the 2D channeled framework which can absorb the water molecules in the channels.[22],[23]

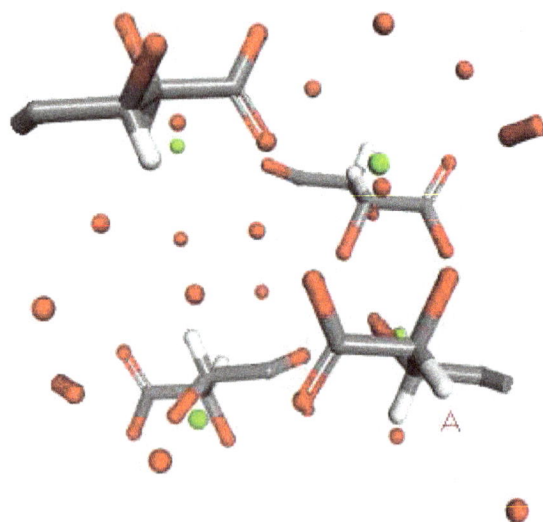

Figure 2. The molecular structure for the Ca-L-tart compound.

The crystal phases of the compounds were further confirmed by the X-ray diffraction (XRD). The simulated XRD was also achieved according to the single crystal data. As shown in figure 3, the XRD of as-prepared samples agrees

well with the simulated result. The strong unexpected peak around 25 was result from the L-tartaric acid residing in the compound.

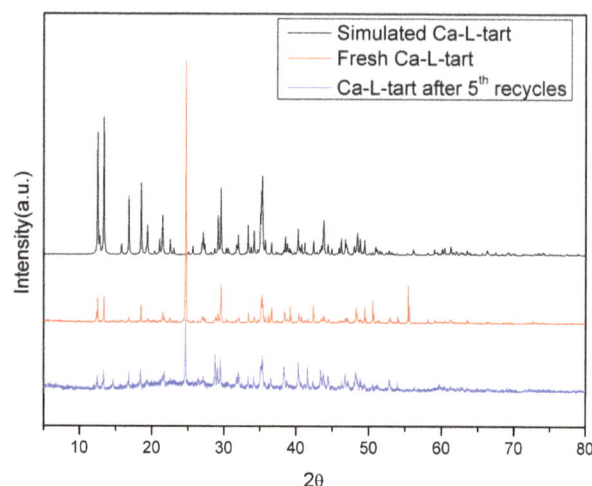

Figure 3. The powder X-ray diffraction of the compound that simulated ,before and after catalytic reaction.

3.2. The use of Ca-L-Tart for L-Proline Detection

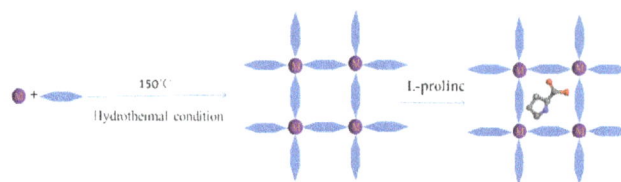

Scheme 1. Synthesis and detection of Ca-L-tart.

Materials in nanoscale have great promise for medicine and biology in view of its properties, unique size and abilities that can be functionalized with recognizable biological elements.[24] The application of the porous compound comprised of metal and organic ligand in detection have been reported.[25] Herein, as shown in the scheme 1, the utilize of Ca-L-tart for L-proline detection , which was an important component of human proteins, was analyzed.

The recorded FTIR spectras were analyzed comparing with the standard spectra character of the functional groups. The strong peak at 3321 cm^{-1} for the O-H in the carboxyl group and the weak peak at 1408 cm^{-1} due to O-H deformation and C-O streching of L-tartaric acid has disppear because of the formation of the compound.[26] The band appears at 1589 cm^{-1} ascribed to protonated carboxylate groups indicates deprotonation upon reaction with Ca^{2+} in the compound, which was consistent with the analysis of the single crystal X-ray diffraction. Especially, the special band appears around 3000 cm^{-1} was attributed to the amino group in the L-proline, which confirmed the result of detection.

The TGA result for the compound was shown in Fig 4. The gradual weight loss from 80 to 190 °C was result from the liberation of the free water molecules. And the further weight loss, observed up to 330 °C was due to the dehydration of water molecules that coordinated with Calcium ions in the compound.[27] Additional weight loss between 330 and 450 °C, followed by the final step between 620 and 750 °C, was attributed to the decomposition of the L-tartaric acid ligand and the formation of calcium oxide with amorphous character. The total weight loss of the initial calcium tartrate compound was 74.2%, which agrees well with the caculated value of 78.1%. Moreover, the weight loss of the compound after detection was 77.7%, which indicate the compound has absorb the L-proline. The decomposition of L-proline started at 200 °C, and the amount of detection was caculated to be 0.035g/g.

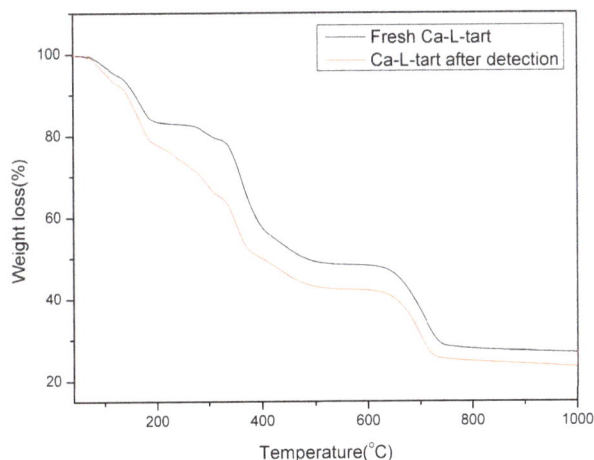

Figure 4. The TG profile of the fresh sample and the sample after detection.

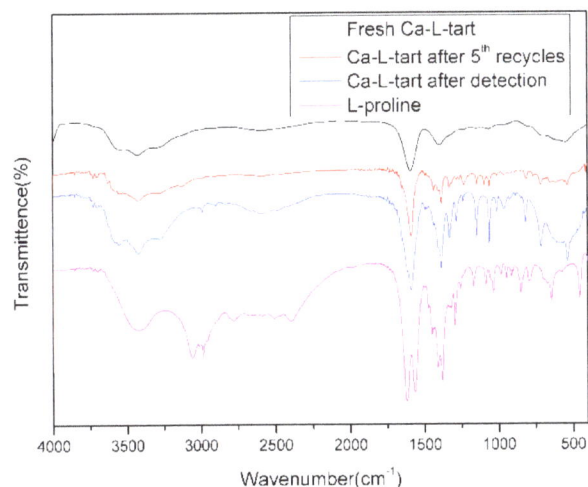

Figure 5. The FTIR spectra of the fresh sample and the sample after catalytic reaction and detection.

3.3. Catalytic Performance of the Catalyst In the Acetalization

The ability of the calcium tartrate compound was studied by catalyzing acetalization, since acetals were important intermediates in carbohydrate and synthetic chemistry as masked carbonyl derivatives.[28] The acetalization of methanol and aldehydes catalyzed by solid heterogeneous catalyst at home temperature has been reported in some literature.[29],[30] However, this compound which was inexpensive and easy to get offers effective catalysis with lowered catalyst loading. As shown in the table 3, benzaldehyde was converted to

corresponding dimethyl acetal in good yield (entry 1). Electron-rich benzaldehyde resulted in relative low conversion, which ascribe to the higher electron-density around the benzylic carbon. Furthermore, 4-fluorobenzaldehyde was evaluated as electron-deficient aldehyde, which give better results.

Table 3. Acetalization of several aldehydes with methanol using Ca-L-tart as solid heterogeneous catalyst.[a]

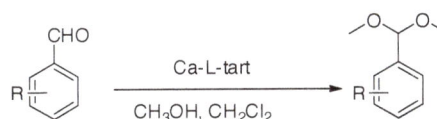

Entry	Substrate	Time(h)	Conversion(%)[b]	Selectivity(%)[b]
1		24	84	>99
2		24	55	>99
3		24	72	>99

[a]Reaction conditions: substrate (1.0 mmol), Ca-L-tart catalyst (0.078 mmol), methanol 15 mL and dichloromethane 15 mL, r.t.; [b]Determined by GC-MS.

The catalyst was recovered by centrifugation, then washed with alcohol and dried at home temperature for the next cycle reaction. The conversion of the acetal product remained >80% after 5th recycles (Figure 6). Furthermore, the SEM and crystallinity of sample after 5th recycles have no change at all compared with the fresh one, as shown in Figure 3.

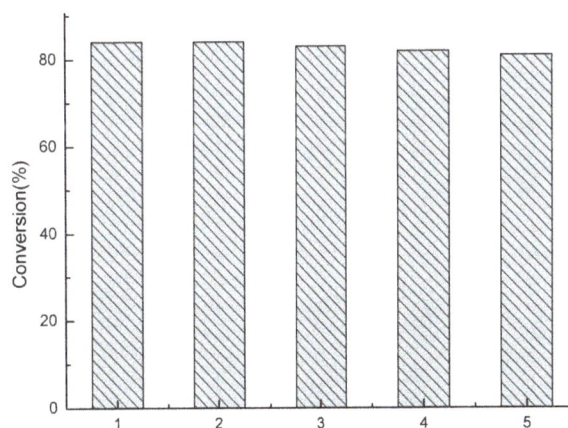

Figure 6. Recycling test of Ca-L-tart.

4. Conclusion

In summary, a novel calcium tartrate porous coordination

compound have been obtained via a simple strategy and the structure was discussed. Each calcium atom coordinated with three different tartrate ligands, which was unexpected complexity for the chiral character of the tartrate ligand. Nanoscale compound showed excellent ability in L-proline detection. Meanwhile, the catalytic activity of Ca-L-tart was also efficient in acetalization of methanol and aldehydes under mild temperature.

References

[1]   S. Kitagawa, R. Kitaura, S. I. NoroAngew. Chem. Int. Ed., 2004, 43, 2334-2375.

[2]   A. K. Cheetham, C. N. R. Rao, R. K. Feller, Chem. Commun.,2006, 46, 4780-4795.

[3]   B. L. Chen, C. D. Liang, J. Yang, D. S. Contreras, Y. L. Clancy, E. B. Lobkovsky, O. M. Yaghi, S. Dai, Angew. Chem. Int. Ed., 2006, 45, 1390-1393.

[4]   A. Clearfield, Z. K. Wang, J. Chem. Soc. Dalton Trans., 2002, 2937-2947.

[5]   X. B. Zhao, B. Xiao, A. J.Fletcher, K. M.Thomas, D. Bradshaw, M. J. Rosseinsky, Science, 2004, 306, 1012-1015.

[6]   A. S. F. Au-Yeung, H. H. Y. Sung, J. A. K. Cha, A. W. H. Siu, S. S. Y. Chui, I. D. Williams, Inorg. Chem. Commun., 2006, 9, 507-511.

[7]   J. S. Seo, D. Whang, H. Lee, S. I. Jun, J. Oh, Y. J. Jeon, K. Kim, Nature, 2000, 404, 982-986.

[8]   M. M. Wanderley , C. Wang , C. D. Wu , W. B. Lin, J. Am. Chem. Soc., 2012, 134, 9050–9053.

[9]   G. Tuci, G. Giambastiani, S. Kwon, P. C. Stair, R. Q. Snurr, A. Rossin, ACS Catal.2014, 4, 1032–1039.

[10] K. Mo, Y. H. Yang, Y. Cui, J. Am. Chem. Soc.2014, 136, 1746-1749.

[11] X. M. Jiang, M. J. Zhang, H. Y. Zeng, G. C. Guo, J. S. Huang, J. Am. Chem. Soc., 2011, 133, 3410-3418.

[12] F. F. Jian, P. S Zhao, Q. X. Wang, J. Coord. Chem., 2005, 58, 1133-1138.

[13] X. F. Wang, X. Y. Zhang, S. Black, L. P. Dang, Ho. Y. Wei, J. Chem. Eng. Data, 2012, 57, 1779-1786.

[14] H. H. M. Yeung, M. Kosa, M. Parrinello, A. K. Cheetham, Cryst. Growth Des., 2013, 13, 3705-3715.

[15] D. H. Wu, J. Z. Ge, H. L. Cai, W. Zhang, R. G. Xiong, CrystEngComm, 2011, 13, 319–324.

[16] F. J. Zhang, Z. H. Xu, S. L. Dong, L. Feng, A. X. Song, C. H. Tung, J. C. Hao, Soft Matter, 2014, 10, 4855-4862.

[17] M. Dubey, A. Kumar, R. K. Gupta, D. S. Pandey, Chem. Commun., 2014, 50, 8144-8147.

[18] Z. Q. Ren, Y. Zeng, Y. T. Hua, Y. Q. Cheng, Z. M. Guo, J. Chem. Eng. Data, 2014, 59, 2517–2522.

[19] N. N. Reed, T. J. Dickerson, G. E. Boldt, K. D. Janda, J. Org. Chem., 2005, 70, 1728-1731.

[20] Jing Lu,Jie-Hui Yu, Xiao-Yan Chen,Peng Cheng,Xiao Zhang, Ji-Qing Xu, Inorg. Chem. 2005, 44, 5978–5980.

[21] H. Hödl, A. Krainer, K. Holzmüller, J. Koidl, M. G. Schmid, G. Gübitz, Electrophoresis, 2007, 28, 2675-2682.

[22] K.C. Kam, K.L.M. Young, A. K. Cheetham, Cryst. Growth Des., 2007, 8, 1522-1532.

[23] J. A. Rood, B. C. Noll, K. W. Henderson, J. Solid State Chem., 2010, 183, 270-276.

[24] D. Zheng, D. S. Seferos, D. A. Giljohann, P. C. Patel, C. A. Mirkin, Nano Lett., 2009, 9 3258-3261.

[25] S. Dang, E. Ma, Z. M. Sun, H. J. Zhang, J. Mater. Chem., 2012, 22, 16920-16926.

[26] K. Moovendaran, S. Natarajan, J. Appl. Cryst., 2013, 46, 993–998.

[27] K. C. Kam, K. L. M. Young, A. K. Cheetham, Cryst. Growth Des., 2007, 7, 1522-1532.

[28] M. Kotke, P. R. Schreiner, Tetrahedron, 2006, 62, 434-439.

[29] B. Mallesham , P. Sudarsanam , G. Raju, B. M. Reddy, Green Chem., 2013, 15, 478-489.

[30] Y. Luan, N. N. Zheng, Y. Qi, J. Tang, Ge Wang, Catal. Sci. Technol., 2014, 4, 925-929.

Comparison of four ionic liquid force fields to an *ab initio* molecular dynamics simulation

Stefan Zahn[*], **Richard Cybik**

Wilhelm-Ostwald-Institut für Physikalische und Theoretische Chemie, University of Leipzig, Leipzig, Germany

Email address:
stefan.zahn@uni-leipzig.de (S. Zahn)

Abstract: The reliability of four force fields developed for 1-alkyl-3-methylimidazolium bis(trifluoromethylsulfonyl)imide ionic liquids are compared to an *ab initio* molecular dynamics simulation regarding structural properties. Except the hydrogen bond structure between the most acidic hydrogen atom of the imidazolium ring and the nitrogen atom of the anion as well as the intramolecular potential surface of the anion in solution, structural properties are reproduced very well by all investigated force fields. Most recommended can be the force field developed by Canongia Lopes and Pádua because it reproduces best the hydrogen bond structure between the most acidic hydrogen atom of the imidazolium ring and the nitrogen atom of the anion.

Keywords: Ionic Liquids, Classical Molecular Dynamics Simulations, Ab Initio Molecular Dynamics Simulations

1. Introduction

A large variety of cations and anions can be combined to ionic liquids (ILs), solvents consisting solely of ions. Since the number of possible ILs exceeds the number of common solvents, the discussion of general properties is complex [1,2]. Most ILs possess a high thermal and electrochemical stability as well as a low vapor pressure. Already in 1914, Paul Walden reported the first systematic study of ionic liquids [3]. However, the scope of ILs was recognized barely until the development of air and water stable imidazolium-based ILs in 1992 [4]. Today, ionic liquids are also applied in nano chemistry [5,6].

Reliable computational models are necessary to predict properties of ILs. Unfortunately, ILs are challenging for computational approaches since induction forces and dispersion forces influence significantly equilibrium distance and interaction energy of IL ions [7,8]. Thus, large systems must be investigated for which only density functional theory can be employed to consider induction forces. Several studies have shown that the well-known error of Kohn-Sham density functional theory to consider dispersion forces can be corrected if an empirical dispersion correction is employed [9-13].

A large number of *ab initio* molecular dynamics simulations of IL systems were carried out over the last decade, mainly by the group of Barbara Kirchner [14-42]. Nonetheless, *ab initio* molecular dynamics simulations of ILs are still at the limit of computational resources. Furthermore, it was shown that the limit of *ab initio* molecular dynamics simulation of about 50 ion pairs is sufficient for structural properties but at least 500 ion pairs are needed to obtain correct dynamics of ionic liquid systems [43]. Therefore, classical molecular dynamics simulations are very often the method of choice to investigate properties of ILs.

The first force field for ILs was proposed by Hanke, Price, and Lynden-Bell in 2001 [44]. Since then many all-atom force fields for ionic liquids were developed [45-54]. Most popular is the force field of Canongia Lopes and Pádua because it is available for a large number of typical IL ions [55-60]. Most parameters were taken from the OPLS-AA [61] and AMBER [62] force fields. Missing dihedral angle force constants were obtained from *ab initio* torsion energy profiles of isolated ions in the gas phase. Thus, no force field parameter was fitted to experimental values. Nonetheless, the calculated density shows only a deviation between 1 % and 5 % to experimental values. Furthermore, calculated crystal structures match very well to experimental data. Unfortunately, dynamical properties such as diffusion coefficients are too sluggish in the force field of Canongia Lopes and Pádua [63,64]. Therefore, Köddermann *et al.* fitted Lennard-Jones potentials of 1-alkyl-3-methylimidazolium bis(trifluoromethylsulfonyl) imide ILs to experimental diffusion coefficients and to NMR rotational correlation times [63]. Calculated heats of vaporization and shear viscosities match close to experimental values in the proposed force field by Köddermann et al. Zhao et al. took a different ansatz than Köddermann et al. to improve dynamical properties obtained from the force field of Canongia Lopes and Pádua for 1-butyl-3-methylimidazolium bis(trifluoromethylsulfonyl)-imide [64]. An effective dielectric constant $\varepsilon_{eff}=1.8$ was introduced into the calculation of electrostatic interactions which is equal to a charge scaling of 0.75. As shown by Youngs and Hardacre, absolute ion charges below 1 systematically fluidize ionic liquids [65]. Thus, it seems reasonable to reduce the absolute ion charge of 1 used in the force field of Canongia Lopes and Pádua. Additionally, Zhao et al. took Lennard-Jones parameters of the acidic hydrogen atoms of the imidazolium ring from the force field of Bhargava and Balasubramanian which employed downscaled ion charges, as well [66]. Finally, Lennard-Jones potentials of oxygen and fluorine were adjusted to the experimental density and ion self-diffusion coefficients. Similar as the force field of Köddermann et al., a good agreement to experimental data such as heat of vaporization was obtained for the force field refined by Zhao et al. [64]. However, since both improved force fields changed different parameters to reproduce experimental properties such as diffusion coefficients, the question arises how are structural properties affected compared to the force field of Canongia Lopes and Pádua or ab inito molecular dynamics simulations?

Morrow and Maginn already used absolute ion charges below 1 in 2002 [67]. Recently, Liu and Maginn proposed an internally consistent ansatz for a non-polarizable all atom force field of 1-alkyl-3-methylimidazolium bis(trifluoro-methylsulfonyl)imide ILs [68]. Force field constants were taken from the generalized AMBER force field [69] (GAFF) while partial charges were calculated by the restrained electrostatic potential method [70] (RESP) and scaled uniformly by 0.8. Finally, all dihedral angle force constants were fitted versus ab initio data. Densities, heat capacities, and thermal expansivities were in agreement with experimental data [68]. Unfortunately, transport properties deviate from experimental references, especially at low temperature. However, trends of dynamical properties are reproduced very well [68].

Within this work, we compare the force fields of Canongia Lopes and Pádua [55-57], Köddermann et al. [63], Zhao et al. [64], and Liu and Maginn [68] to ab initio molecular dynamic simulations of 1-ethyl-3-methylimidazolium bis(trifluoro-methylsulfonyl)imide ([C_2C_1im][NTf_2]). Main focus will be the liquid structure. While the calculation of macroscopic properties such as viscosity is challenging for ionic liquids due to the sluggish dynamics [49], well reproduced structural properties by classical molecular dynamics simulations might help to understand unique properties of ionic liquids or even allow to forecast properties of ionic liquids. For example, nanoscale segregation in polar and nonpolar domains was predicted for ionic liquids by classical molecular dynamics simulations [71,72] before it was found by X-ray diffraction

[73] and by Raman-induced Kerr effect spectroscopy [74]. Thus, a force field reproducing structural properties very well might be preferred over a force field which was fitted to match dynamical properties at the cost of accuracy in structural properties.

2. Computational Details

2.1. Ab initio Molecular Dynamics Simulations

Born-Oppenheimer molecular dynamics simulations of 27 1-ethyl-3-methylimidazolium bis(trifluoromethylsulfonyl)-imide ion pairs were carried out with CP2K [75,76]. Periodic boundary conditions were applied in the NVT simulations in which the cubic box length was set to 2295 pm to reproduce the density of 1.45 g/cm^3 at 350 K [77]. Temperature was keept constant by a Nosé-Hoover chain thermostat [78-80]. The Kohn-Sham density functional calculations employed the BLYP-D2 [81-83] functional which includes an empirical dispersion correction. The molecularly optimized double-zeta basis [84] (DZVP-MOLOPT-SR-GTH) together with the corresponding Goedecker-Teter-Hutter pseudopotentials [85-87] were used to form the Kohn-Sham orbitals. Initial coordinates of the ab initio molecular dynamics simulation were obtained by a 1 ns classical molecular dynamics simulation employing the force field of Canongia Lopes and Pádua. The system was equilibrated 19.5 ps before the production run of 32 ps was started. A time step of 0.5 fs was selected in all ab inito molecular dynamics simulations. The abbreviation **AIMD** and black color in graphs will refer to results of the ab initio molecular dynamics simulation.

2.2. Classical Molecular Dynamics Simulations

Following force fields were employed in classical molecular dynamics simulations of 1-ethyl-3-methylimidazolium bis(trifluoromethylsulfonyl)-imide:

- Simulations using all force field parameters of Canongia Lopes and Pádua [55-57] will be abbreviated by **FF-Lopes** and red color is used in graphs in the following.
- Simulations with the force field parameters of Köddermann, Paschek and Ludwig [63] will be abbreviated by **FF-Ludwig** and green color is used in graphs.
- Zhao et al. proposed his refined model of **FF-Lopes** only for 1-butyl-3-methylimidazolium bis(trifluoro-methylsulfonyl)imide [64]. However, parameters of the long nonpolar alkyl chain attached to imidazolium cation are not affected except by the introduced dielectric constant of 1.8. Therefore, his model is easily transferable to [C_2C_1im][NTf_2]. We use the abbreviation **FF-Zhao** or blue color in graphs for this force field model.
- All force field parameters of the bis(trifluoro-methylsulfonyl)imide anion were taken from the work of Liu and Maginn [68]. Similar as Liu and Maginn, the force field parameters for the

1-ethyl-3-methylimidazolium cation were taken from the generalized AMBER force field (GAFF) [69]. The partial charges of the cation were determined by the restrained electrostatic potential method [70] (RESP) and scaled uniformly by 0.8. The Hartree-Fock method in combination with the 6-31G* basis set was employed to obtain the electron density for the RESP calculations. Finally, all force field dihedral potentials of an isolated cation in the gas phase were checked versus energy potential surfaces obtained by the TPSS-D3 functional [88-90] in combination with the 6-31++G** basis set [91,92] and the resolution of identity approximation [93-95]. It was shown that this method produces results which can be hardly improved by post Hartree–Fock methods [13]. In all cases, the difference between the energy potential surface of the force field and the one of density functional theory calculations was less than 2 kJ/mol. Therefore, all dihedral force constants of the cation were taken unchanged from the generalized AMBER force field. The simulations employing this ansatz will use the abbreviation **FF-Maginn** and grey color in graphs in the following.

Each NVT simulation at 350 K included 27 ion pairs and the box length was set to 2295 pm. The temperature was kept constant by a Nosé–Hoover chain thermostat [78-80]. All C-H bonds were constrained by the SHAKE algorithm [96] in the **FF-Lopes**, **FF-Ludwig**, and **FF-Zhao** simulations. The time step was set to 0.5 fs in the production run and data points were collected for the analyzed trajectories every 5 fs. Each system was equilibrated at least 2 ns and the production run time was 0.5 ns. Lennard-Jones and Coulombic interactions were computed up to a cutoff radius of 1100 pm. Coulombic interaction energies beyond the cutoff were computed via the particle-particle particle-mesh solver [97]. LAMMPS [98] was employed for all classical molecular dynamics simulations while TRAVIS [99] was used to analyze the obtained trajectories. Employed atom labels can be found in Fig. 1.

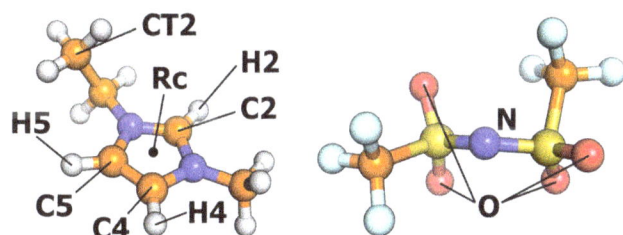

Figure 1. *Ball-and-stick model of ions with used atom labels throughout this work. Rc is the geometric ring center of the imidazolium ring.*

3. Results

Initially, we compared the general structure of the polar and the nonpolar domains of 1-ethyl-3-methylimidazolium bis-(trifluoromethylsulfonyl)imide ($[C_2C_1im][NTf_2]$) in the classical molecular dynamics (CMD) simulations and in the *ab initio* molecular dynamics simulation. As can be seen in Fig. 2a, the shape of the radial pair distribution function (RDF) between the geometric ring center of the imidazolium ring, Rc,

and the mass center of the anion, m_{an}, is very similar in the CMDs and **AIMD**. The most significant difference is the position of the minimum between the first and the second solvation shell which is shifted about 100 pm to larger distances in the CMDs compared to **AIMD**. Since all CMDs are very similar, the Rc-m_{an}-RDF provide no hint which force field should be preferred. The picture slightly changes if one investigates the shape of the RDFs between centers of same charge. As can be seen in the m_{an}-m_{an}-RDF in Fig. 2b, the peak height of the **FF-Zhao** model is too small and also the anions tend to get too close to each other. This might originate from the employed dielectric constant of 1.8 in the **FF-Zhao** model which reduces the Coulombic repulsion of two anions compared to **FF-Lopes**. However, the shape of the m_{an}-m_{an}-RDF of **AIMD** is reproduced in the **FF-Maginn** model very well, which possesses reduced atom charges and, thus, reduced Coulombic repulsion between two anions, as well. Also, the RDFs of the models with integer ion charges, **FF-Lopes** and **FF-Ludwig**, reproduce very well the shape of the **AIMD** reference simulation. Therefore, the m_{an}-m_{an}-RDF is not solely affected by the total ion charge. The Rc-Rc-RDF in Fig. 2c shows that two cations tend to too small distances to each other in the two force field models with reduced Coulomb interactions. Furthermore, the first solvation shell peak is broader than in **AIMD** and the maximum of both force fields is at about 975 pm. The maximum of the **FF-Lopes** and the **FF-Ludwig** model are at about 825 pm while a broad maximum between 750 pm and 1000 pm is visible in **AIMD**. Maybe, a charge reduction of about 0.9 might improve the shape of the Rc-Rc-RDF compared to the investigated force fields. In Fig. 2d can be seen the RDF between the terminal carbon atoms of the ethyl chain attached to the imidazolium ring, CT2. The *ab initio* molecular dynamics simulation uses an empirical dispersion correction and, thus, differences between **AIMD** and the CMDs should not be over-interpreted. However, the forces between the CT2 atoms should be weak compared to the interactions of the polar domains which are very well reproduced by **AIMD**. Thus, this structural motive might be mainly affected by the strong electrostatic interaction which force to arrange the small and weakly interacting nonpolar alkyl chains in a particular fashion. As one can see, the CT2-atoms get closer to each other in **AIMD** than in the CMDs. Additionally, a subpeak of the first solvation shell is visible at about 380 pm in **AIMD** while this subpeak is only visible in **FF-Maginn** and **FF-Zhao** at 415 pm. Nonetheless, all force fields reproduce very well the large shoulder below the maximum of the first solvation shell at about 925 pm.

A common error of ionic liquid all-atom force fields is the hydrogen bond structure between imidazolium cations and strongly coordinating anions such as chloride [14,17]. Therefore, we investigated the structure between C2/C5 of the imidazolium ring and the oxygen/nitrogen atoms of the anion. As can be seen in Fig. 3a, the general structure of the C2-O-RDF is reproduced by all force fields. However, both atoms get about 15 pm too close to each other in **FF-Ludwig** while the distance is about 10 pm too large in **FF-Zhao**. Nonetheless, the broadness and peak hight of the first solvation shell is best reproduced by **FF-Ludwig**. Thus, the Lennard–Jones distance parameter reduction of about 100 pm of H2 seems to bee too large in **FF-Ludwig** compared to **FF-Lopes** which, similar as **FF-Maginn**, reproduces very well the closest C2-O contact. The closest distance between C5 and O of all force field simulations match very well AIMD, see Fig. 3b. Differences are visible in the broadness of the first solvation shell subpeak as well as in the deepness of the two minimas in the first solvation shell where **FF-Ludwig** is superior to all other investigated force fields. Nonetheless, no force field shows a general drastic error compared to the *ab inito* molecular dynamics simulation for the hydrogen bonds between C2/C5 and O.

This is different if the RDFs between C2 of the cation and N of the anion are investigated, see Fig. 3c. **AIMD** shows a large peak, significant above the statistical distribution at 330 pm which is not visible in all force field models. The general shape of the C2-N-RDF is best reproduced by the **FF-Lopes**

model. Especially, the second solvation sphere matches excellent **AIMD**. However, the first solvation sphere peak is at about 360 pm and only slightly above the statistical distribution. In **FF-Zhao**, only a small shoulder indicates the first solvation shell peak found in **AIMD**. Thus, this model can be least recommended. The significant difference between **AIMD** and the force field simulations is also visible in the combined distribution function (CDF) of the C2-N-RDF and the angle distribution function (ADF) of α, see Fig. 4. α is the angle enclosed by the C2–N vector and the C2–H2 bond vector. Nicely visible is the large peak at about 330 pm and 0 ° in **AIMD** which shows that a strong directional hydrogen bond between C2 and N exists in $[C_2C_1\text{im}][NTf_2]$. In **FF-Zhao**, only one peak is visible at 0 ° while the other force field models show at least two separated peaks between 0 ° and 45 ° similar as the *ab initio* molecular dynamics reference simulation.

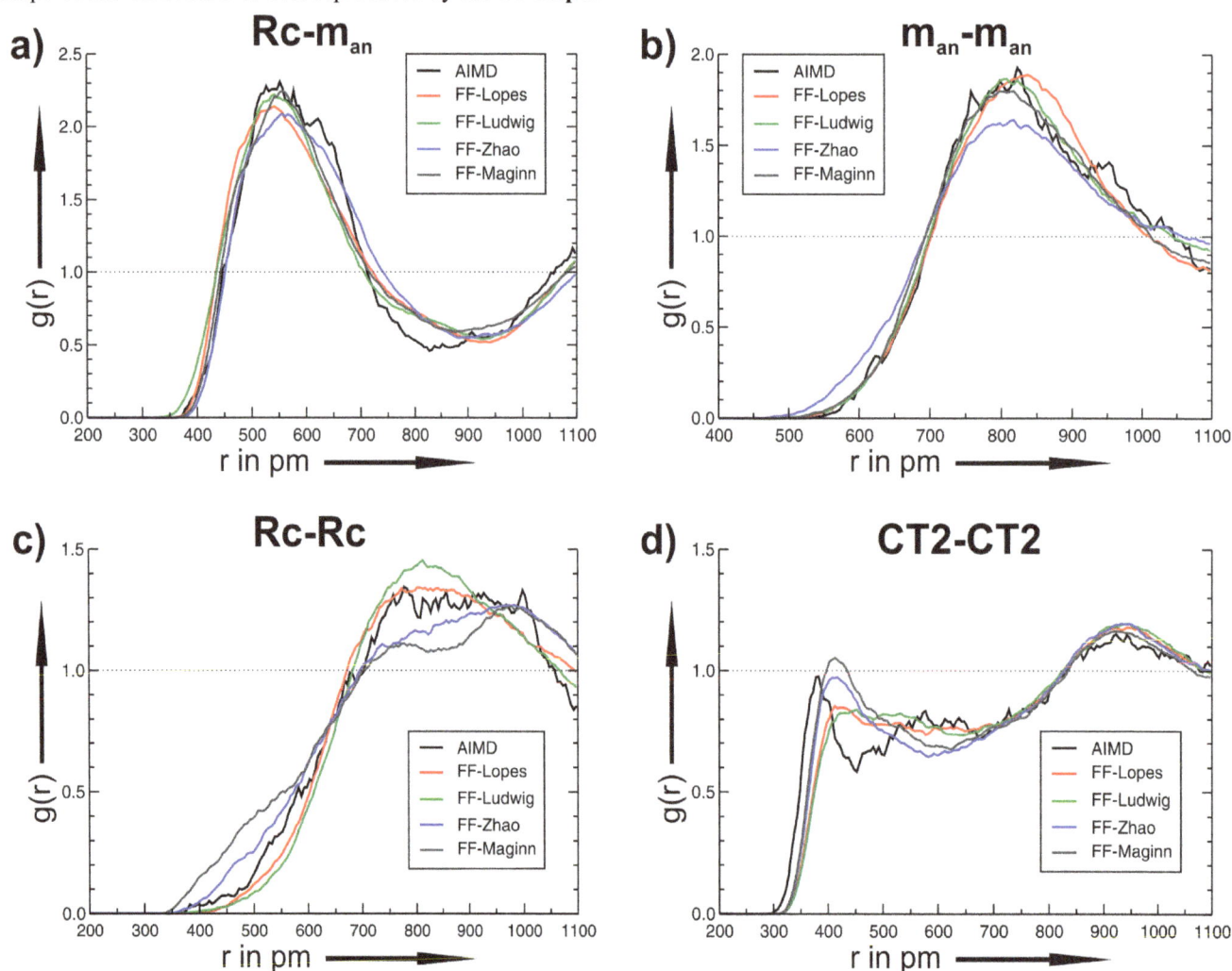

Figure 2. Comparison of RDFs between Rc *and center of mass of the anion,* m_{an}, *(a),* m_{an} *and* m_{an} *(b),* Rc *and* Rc *(c), and* CT2 *and* CT2 *(d). Atom labels can be found in Fig. 1.*

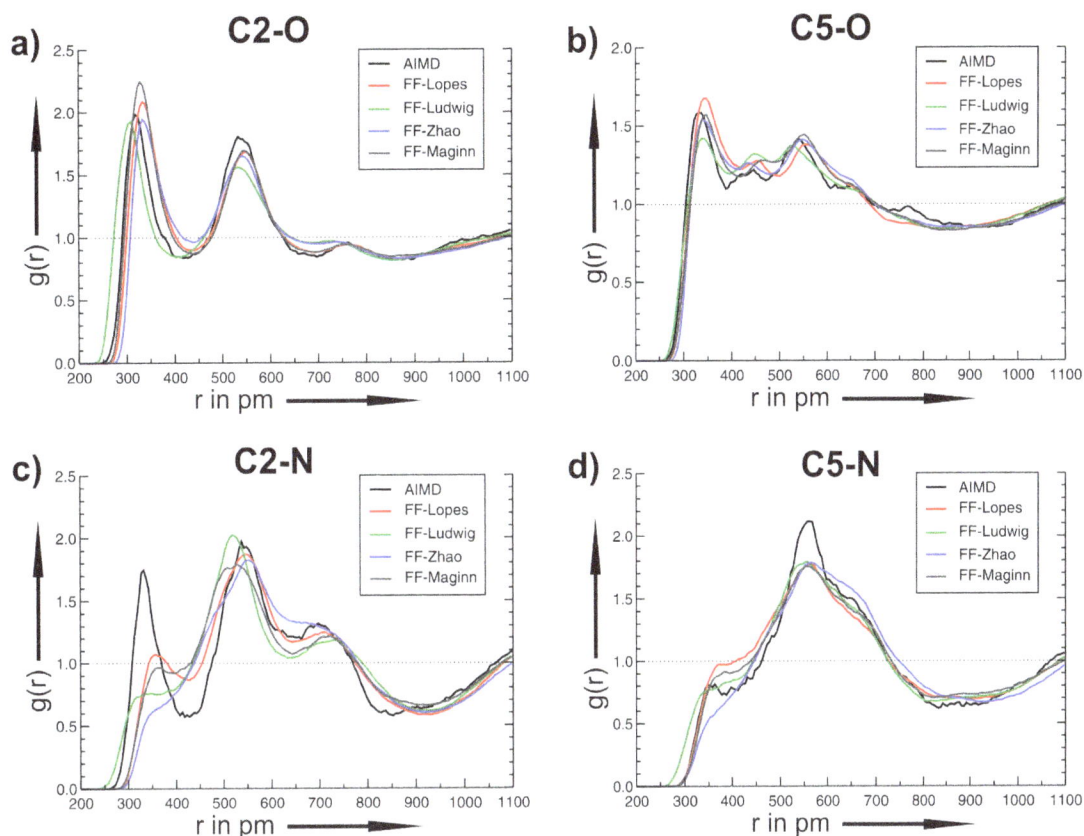

Figure 3. *Comparison of hydrogen bond RDFs. Atom labels can be found in Fig. 1.*

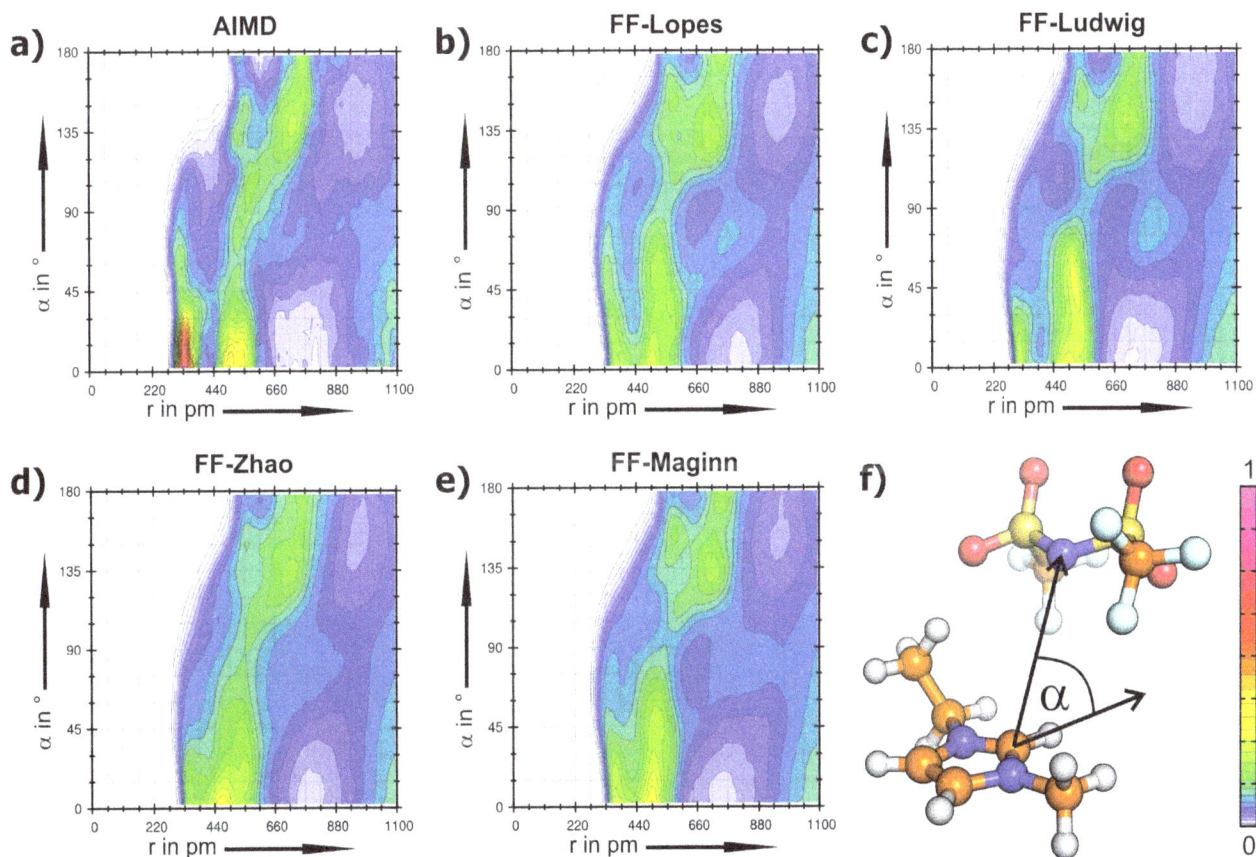

Figure 4. *Comparison of the CDFs of the RDF between C2 of the cation and N of the anion and the ADF of α which is the angle enclosed by the C2–N vector and the C2–H2 bond vector. Atom labels can be found in Fig. 1.*

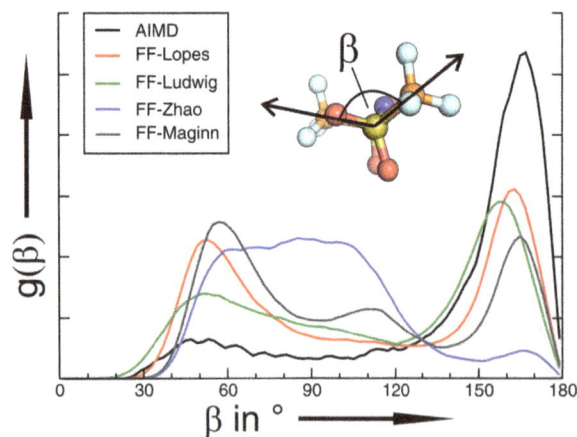

Figure 5. *ADF of β which is the angle enclosed by the two S–C bond vectors of an anion*

Finally, we investigated the angle distribution function of β which is the angle enclosed by the two S–C bond vectors of an anion, see Fig. 5. **FF-Lopes** and **FF-Maginn** adjusted all dihedral potentials to *ab initio* data while **FF-Ludwig** and **FF-Zhao** took only the dihedral angle force constants from **FF-Lopes**. Since **FF-Lopes**, **FF-Ludwig**, and **FF-Zhao** are based on the OPLS-AA force field, Lennard–Jones potentials and partial charge distribution affect dihedral potentials, as well. In **FF-Ludwig**, only Lennard–Jones potentials were fitted to match experimental diffusion coefficients. Their influence on the potential energy surface of the bis(trifluoro-methylsulfonyl)imide anion should be overall small. However, the Coulomb forces were scaled down in **FF-Zhao** by introducing a dielectric constant of 1.8 which should affect significantly dihedral potentials of the anion due to the large absolute atom charges. Indeed, the comparison of β in Fig. 5 reveals that a significant peak at about 165 ° is missing in **FF-Zhao** while this angular value is most preferred in **AIMD**. **FF-Lopes** and **FF-Maginn** are overall very similar and show two peaks at about 50 ° and 160 °. However, the peak at 50 ° is nearly missing in **AIMD**. Thus, it seems that the potential energy surface of the bis(trifluoromethylsulfonyl)imide anion is affected by solvation because dihedral potenial force constants were fitted on *ab initio* data of isolated ions in the gas phase in **FF-Lopes** and **FF-Maginn**. The improved Lennard-Jones parameters in **FF-Ludwig** seems to correct the effect of solvation because the ADF of **FF-Ludwig** and the *ab initio* molecular dynamics reference simulation match best to each other.

4. Summary and Conclusions

Structural properties of four force field models developed for 1-alkyl-3-methylimidazolium bis(trifluoromethylsulfo-nyl) imide ILs were compared to an *ab inito* molecular dynamics study of 1-ethyl-3-methylimidazolium bis(trifluoro-methylsulfonyl)imide. Overall, the structure of polar as well as nonpolar domains are reproduced very well by all force field models. This is also the case for the hydrogen bonds between the acidic hydrogen atoms of the imidazolium ring

and the oxygen atoms of the anion. However, all investigated force fields fail significantly to reproduce the hydrogen bond structure of the most acidic hydrogen atom of the imidazolium ring (H2) and the nitrogen atom N of the anion. A large peak is visible in the C2-N-RDF at 330 pm. This peak is reproduced best by the force field of Canongia Lopes and Pádua (**FF-Lopes**) where a peak is visible at 360 pm slightly above the statistical average. The force field developed by Zhao *et al.* (**FF-Zhao**) can be least recommended because only a small shoulder is visible instead of a peak at 330 pm in the C2-N-RDF. Additionally, no similarities are visible between the force field proposed by Zhao *et al.* and the *ab initio* molecular dynamics reference simulation in the angle distribution of β, which is the angle enclosed by the two S–C bond vectors of an anion.

Acknowledgements

S. Zahn thanks the European Social Fund (100122082) for financial support.

References

[1] Wasserscheid, P.; Keim, W. *Angew. Chem. Int. Ed.* 2000, *39*, 3772–3789.

[2] Plechkova, N. V.; Seddon, K. R. *Chem. Soc. Rev.* 2008, *37*, 123–150.

[3] Walden, P. *Bull. Acad. Sci. St Petersburg* 1914, 405–422.

[4] Wilkes, J. S.; Zaworotko, M. J. *J. Chem. Soc., Chem. Commun.* 1992, 965–967.

[5] Li, C.; Lin, J. *J. Mater. Chem.* 2010, *20*, 6831–6847.

[6] Dupont, J.; Scholten, J. D. *Chem. Soc. Rev.* 2010, *39*, 1780–1804.

[7] Zahn, S.; Uhlig, F.; Thar, J.; Spickermann, C.; Kirchner, B. *Angew. Chem. Int. Ed.* 2008, *47*, 3639–3641.

[8] Zahn, S.; Bruns, G.; Thar, J.; Kirchner, B. *Phys. Chem. Chem. Phys.* 2008, *10*, 6921–6924.

[9] Zahn, S.; Kirchner, B. *J. Phys. Chem. A* 2008, *112*, 8430–8435.

[10] Izgorodina, E. I.; Bernard, U. L.; MacFarlane, D. R. *J. Phys. Chem. A* 2009, *113*, 7064–7072.

[11] Kohanoff, J.; Pinilla, C.; Youngs, T. G. A.; Artacho, E.; Soler, J. M. *J. Chem. Phys.* 2011, *135*, 154505.

[12] Grimme, S.; Hujo, W.; Kirchner, B. *Phys. Chem. Chem. Phys.* 2012, *14*, 4875–4883.

[13] Zahn, S.; MacFarlane, D. R.; Izgorodina, E. I. *Phys. Chem. Chem. Phys.* 2013, *15*, 13664–13675.

[14] Del Pópolo, M.; Lynden-Bell, R.; Kohanoff, J. *J. Phys. Chem. B* 2005, *109*, 5895–5902.

[15] Bühl, M.; Chaumont, A.; Schurhammer, R.; Wipff, G. *J. Phys. Chem. B* 2005, *109*, 18591–18599.

[16] Prado, C. E. R.; Pópolo, M. G. D.; Youngs, T. G. A.; Kohanoff, J.; and, R. M. L.-B. *Mol. Phys.* 2006, *104*, 2477–2483.

[17] Bhargava, B.; Balasubramanian, S. *Chem. Phys. Lett.* 2006, *417*, 486–491.

[18] Ghatee, M. H.; Ansari, Y. *J. Chem. Phys.* 2007, *126*, 154502.

[19] Bhargava, B. L.; Balasubramanian, S. *J. Phys. Chem. B* 2007, *111*, 4477–4487.

[20] Bagno, A.; D'Amico, F.; Saielli, G. *ChemPhysChem* 2007, *8*, 873–881.

[21] Bhargava, B. L.; Balasubramanian, S. *J. Phys. Chem. B* 2008, *112*, 7566–7573.

[22] Bhargava, B. L.; Saharay, M.; Balasubramanian, S. *Bull. Mater. Sci.* 2008, *31*, 327–334.

[23] Spickermann, C.; Thar, J.; Lehmann, S. B. C.; Zahn, S.; Hunger, J.; Buchner, R.; Hunt, P. A.; Welton, T.; Kirchner, B. *J. Chem. Phys.* 2008, *129*, 104505.

[24] Thar, J.; Brehm, M.; Seitsonen, A. P.; Kirchner, B. *J. Phys. Chem. B* 2009, *113*, 15129–15132.

[25] Zahn, S.; Thar, J.; Kirchner, B. *J. Chem. Phys.* 2010, *132*, 124506.

[26] Mallik, B. S.; Siepmann, J. I. *J. Phys. Chem. B* 2010, *114*, 12577–12584.

[27] Krekeler, C.; Dommert, F.; Schmidt, J.; Zhao, Y. Y.; Holm, C.; Berger, R.; Delle Site, L. *Phys. Chem. Chem. Phys.* 2010, *12*, 1817–1821.

[28] Zahn, S.; Wendler, K.; Delle Site, L.; Kirchner, B. *Phys. Chem. Chem. Phys.* 2011, *13*, 15083–15093.

[29] Wendler, K.; Zahn, S.; Dommert, F.; Berger, R.; Holm, C.; Kirchner, B.; Delle Site, L. *J. Chem. Theory Comput.* 2011, *7*, 3040–3044.

[30] Brüssel, M.; Brehm, M.; Voigt, T.; Kirchner, B. *Phys. Chem. Chem. Phys.* 2011, *13*, 13617–13620.

[31] Brüssel, M.; Brehm, M.; Pensado, A. S.; Malberg, F.; Ramzan, M.; Stark, A.; Kirchner, B. *Phys. Chem. Chem. Phys.* 2012, *14*, 13204–13215.

[32] Pensado, A. S.; Brehm, M.; Thar, J.; Seitsonen, A. P.; Kirchner, B. *ChemPhysChem* 2012, *13*, 1845–1853.

[33] Wendler, K.; Brehm, M.; Malberg, F.; Kirchner, B.; Delle Site, L. *J. Chem. Theory Comput.* 2012, *8*, 1570–1579.

[34] Brehm, M.; Weber, H.; Pensado, A. S.; Stark, A.; Kirchner, B. *Phys. Chem. Chem. Phys.* 2012.

[35] Zhang, Y.; Maginn, E. J. *J. Phys. Chem. B* 2012, *116*, 10036–10048.

[36] Bodo, E.; Sferrazza, A.; Caminiti, R.; Mangialardo, S.; Postorino, P. *J. Chem. Phys.* 2013, *139*, 144309.

[37] Hollóczki, O.; Kelemen, Z.; Könczöl, L.; Szieberth, D.; Nyulászi, L.; Stark, A.; Kirchner, B. *Chem.Phys.Chem.* 2013, *14*, 315–320.

[38] Hollóczki, O.; Firaha, D. S.; Friedrich, J.; Brehm, M.; Cybik, R.; Wild, M.; Stark, A.; Kirchner, B. *J. Phys. Chem. B* 2013, *117*, 5898–5907.

[39] Firaha, D. S.; Kirchner, B. *J. Chem. Eng. Data* 2014, DOI: 10.1021/je500166d.

[40] Thomas, M.; Brehm, M.; Hollóczki, O.; Kelemen, Z.; Nyulászi, L.; Pasinszki, T.; Kirchner, B. *J. Chem. Phys.* 2014, *141*, 024510.

[41] Thomas, M.; Brehm, M.; Hollóczki, O.; Kirchner, B. *Chem. Eur. J.* 2014, *20*, 1622–1629.

[42] Payal, R. S.; Balasubramanian, S. *Phys. Chem. Chem. Phys.* 2014, *16*, 17458–17465.

[43] Gabl, S.; Schröder, C.; Steinhauser, O. *J. Chem. Phys.* 2012, *137*, 094501.

[44] Hanke, C. G.; Price, S. L.; Lynden-Bell, R. M. *Mol. Phys.* 2001, *99*, 801–809.

[45] Hunt, P. A. *Mol. Simul.* 2006, *32*, 1–10.

[46] Pádua, A. A. H.; Costa Gomes, M. F.; Canongia Lopes, J. N. A. *Acc. Chem. Res.* 2007, *40*, 1087–1096.

[47] Wang, Y.; Jiang, W.; Yan, T.; Voth, G. A. *Acc. Chem. Res.* 2007, *40*, 1193–1199.

[48] Lynden-Bell, R. M.; Del Pópolo, M. G.; Youngs, T. G. A.; Kohanoff, J.; Hanke, C. G.; Harper, J. B.; Pinilla, C. C. *Acc. Chem. Res.* 2007, *40*, 1138–1145.

[49] Maginn, E. J. *Acc. Chem. Res.* 2007, *40*, 1200–1207.

[50] Bhargava, B. L.; Balasubramanian, S.; Klein, M. L. *Chem. Commun.* 2008, 3339–3351.

[51] Kirchner, B. *Top. Curr. Chem.* 2009, *290*, 213–262.

[52] Borodin, O. *J. Phys. Chem. B* 2009, *113*, 11463–11478.

[53] Maginn, E. J. *J. Phys.: Condens. Matter* 2009, *21*, 373101.

[54] Dommert, F.; Wendler, K.; Berger, R.; Delle Site, L.; Holm, C. *ChemPhysChem* 2012, *13*, 1625–1637.

[55] Canongia Lopes, J. N.; Deschamps, J.; Pádua, A. A. H. *J. Phys. Chem. B* 2004, *108*, 2038–2047.

[56] Canongia Lopes, J. N.; Deschamps, J.; Pádua, A. A. H. *J. Phys. Chem. B* 2004, *108*, 11250–11250.

[57] Canongia Lopes, J. N.; Pádua, A. A. H. *J. Phys. Chem. B* 2004, *108*, 16893–16898.

[58] Canongia Lopes, J. N.; Pádua, A. A. H. *J. Phys. Chem. B* 2006, *110*, 19586–19592.

[59] Canongia Lopes, J. N.; Pádua, A. A. H.; Shimizu, K. *J. Phys. Chem. B* 2008, *112*, 5039–5046.

[60] Shimizu, K.; Almantariotis, D.; Costa Gomes, M. F.; Pádua, A. A. H.; Canongia Lopes, J. N. *J. Phys. Chem. B* 2010, *114*, 3592–3600.

[61] Jorgensen, W. L.; Maxwell, D. S.; Tirado-Rives, J. *J. Am. Chem. Soc.* 1996, *118*, 1225–11236.

[62] Cornell, W. D.; Cieplak, P.; Bayly, C. I.; Gould, I. R.; Merz, K. M.; Ferguson, D. M.; Spellmeyer, D. C.; Fox, T.; Caldwell, J. W.; Kollman, P. A. *J. Am. Chem. Soc.* 1995, *117*, 5179.

[63] Köddermann, T.; Paschek, D.; Ludwig, R. *ChemPhysChem* 2007, *8*, 2464–2470.

[64] Zhao, W.; Eslami, H.; und Florian Müller-Plathe, W. L. C. *Z. Phys. Chem.* 2007, *221*, 1647–1662.

[65] Youngs, T. G. A.; Hardacre, C. *ChemPhysChem* 2008, *9*, 1548–1558.

[66] Bhargava, B. L.; Balasubramanian, S. *J. Chem. Phys.* 2007, *127*, 114510.

[67] Morrow, T. I.; Maginn, E. J. *J. Phys. Chem. B* 2002, *106*, 12807–12813.

[68] Liu, H.; Maginn, E. *J. Chem. Phys.* 2011, *135*, 124507.

[69] Wang, J. M.; Wolf, R. M.; Caldwell, J. W.; Kollman, P. A.; Case, D. A. *J. Comput. Chem.* 2004, *25*, 1157–1174.

[70] Bayly, C. I.; Cieplak, P.; Cornell, W. D.; Kollman, P. A. *J. Phys. Chem.* 1993, *97*, 10269–10280.

[71] Wang, Y.; Voth, G. A. *J. Am. Chem. Soc.* 2005, *127*, 12192–12193.

[72] Canongia Lopes, J.; Pádua, A. *J. Phys. Chem. B* 2006, *110*, 3330–3335.

[73] Triolo, A.; Russina, O.; Bleif, H.-J.; Di Cola, E. *J. Phys. Chem. B* 2007, *111*, 4641–4644.

[74] Xiao, D.; Rajian, J. R.; Cady, A.; Li, S.; Bartsch, R. A.; Quitevis, E. L. *J. Phys. Chem. B* 2007, *111*, 4669–4677.

[75] VandeVondele, J.; Krack, M.; Mohamed, F.; Parrinello, M.; Chassaing, T.; Hutter, J. *J. Comp. Phys. Comm.* 2005, *167*, 103–128.

[76] *CP2K developers group,* http://www.cp2k.org/.

[77] Tokuda, H.; Hayamizu, K.; Ishii, K.; Susan, M. A. B. H.; Watanabe, M. *J. Phys. Chem. B* 2005, *109*, 6103–6110.

[78] Nosé, S. *J. Chem. Phys.* 1984, *81*, 511–519.

[79] Hoover, W. G. *Phys. Rev. A* 1985, *31*, 1695–1697.

[80] Martyna, G. J.; Klein, M. L.; Tuckerman, M. *J. Chem. Phys.* 1992, *97*, 2635–2643.

[81] Becke, A. D. *Phys. Rev. A* 1988, *38*, 3098–3100.

[82] Lee, C.; Yang, W.; Parr, R. G. *Phys. Rev. B* 1988, *37*, 785–789.

[83] Grimme, S. *J. Comput. Chem.* 2006, *27*, 1787–1799.

[84] VandeVondele, J.; Hutter, J. *J. Chem. Phys.* 2007, *127*, 114105.

[85] Goedecker, S.; Teter, M.; Hutter, J. *Phys. Rev. B* 1996, *54*, 1703–1710.

[86] Hartwigsen, C.; Goedecker, S.; Hutter, J. *Phys. Rev. B* 1998, *58*, 3641–3662.

[87] Krack, M. *Theor. Chem. Acc.* 2005, *114*, 145–152.

[88] Tao, J.; Perdew, J. P.; Staroverov, V. N.; Scuseria, G. E. *Phys. Rev. Lett.* 2003, *91*, 146401.

[89] Kossmann, S.; Kirchner, B.; Neese, F. *Mol. Phys.* 2007, *105*, 2049–2071.

[90] Grimme, S.; Antony, J.; Ehrlich, S.; Krieg, H. *J. Chem. Phys.* 2010, *132*, 154104.

[91] Hehre, W. J.; Ditchfield, R.; Pople, J. A. *J. Chem. Phys.* 1972, *56*, 2257–2261.

[92] Francl, M. M.; Pietro, W. J.; Hehre, W. J.; Binkley, J. S.; Gordon, M. S.; DeFrees, D. J.; Pople, J. A. *J. Chem. Phys.* 1982, *77*, 3654–3665.

[93] Baerends, E. J.; Ellis, D. E.; Ros, P. *Chem. Phys.* 1973, *2*, 41–51.

[94] Dunlap, B. I.; Connolly, J. W. D.; Sabin, J. R. *J. Chem. Phys.* 1979, *71*, 3396–3402.

[95] Weigend, F. *Phys. Chem. Chem. Phys.* 2006, *8*, 1057–1065.

[96] Ryckaert, J.-P.; Ciccotti, G.; Berendsen, H. J. C. *J. Comput. Phys.* 1977, *23*, 327–341.

[97] Hockney, R. W.; Eastwood, J. W. *Computer Simulation Using Particles*; McGraw-Hill, 1981.

[98] Plimpton, S. *J. Comp. Phys.* 1995, *117*, 1–19.

[99] Brehm, M.; Kirchner, B. *J. Chem. Inf. Model.* 2011, *51*, 2007–2023.

Hydration Behavior of Composite Cement Containing Fly Ash and Nanosized-SiO$_2$

H. El-Didamony[1], S. Abd El-Aleem[2, *], Abd El-Rahman Ragab[3]

[1]Chemistry Department, Faculty of Science, Zagazig University, Zagazig Egypt

[2]Chemistry Department, Faculty of Science, Fayoum University, Fayoum, Egypt

[3]Quality Department, Lafarge Cement, El Kattamia, El Sokhna, Suez, Egypt

Email address:

saa09@fayoum.edu.eg (S. Abd El-Aleem)

*Corresponding author

Abstract: In recent years, there is a great interest in replacing a long time used materials in concrete structure by nanomaterials (NMs) to produce a concrete with novel functions. NMs are used either to replace a part of cement, producing ecological profile concrete or as admixtures in cement pastes. The great reactivity of NMs is attributed to their high purities and specific surface areas. A number of NMs been explored and among of them nanosilica (NS) has been used most extensively. This work aims to study, the hydration behavior of composite cements containing fly ash (FA) and nanosilica. Different cement blends were made from OPC, FA and NS. OPC was substituted with FA up to 30.0 mass, %, then the FA portion was replaced by equal amounts of NS (2.0, 4.0 and 6.0 mass, %). The hydration behavior was followed by determination of free lime (FL) and combined water (Wn) contents at different curing ages. The required water for standard consistency (W/C), setting times (IST & FST), bulk density (BD) and compressive strength were also estimated. Some selected hydration products were analyzed using XRD and DTA techniques. The results showed that, both of FA and NS improve the hydration behavior and mechanical properties of the investigated cements. But, NS possesses higher improvement level than FA, due to that, both of them behaves not only as filler, but also as activator to promote pozzolanic reaction, which enhances the formation of excessive dense products. The higher beneficial role of NS is mainly due to its higher surface area, seeding effect and pozzolanic activity in comparison with FA. The composite cement containing 70.0% OPC, 26.0% FA and 4.0% NS gave the desirable mechanical properties at all curing ages.

Keywords: Hydration, Mechanical Properties, Composite Cement, Fly Ash, Nanosilica, Curing Time

1. Introduction

Concrete is the second most consumed material after water and it shapes the built environment around the world. Preparation of concrete involves use of natural resources like sand, stone, aggregates, water etc. According to U. S. Geological Survey, mineral commodity summaries January 2015 and the cement production in the world in 2014 is 4.18 billion metric tons [1]. Estimated concrete production in the world in 2009 was more than 25 billion metric tons according to CSI (Cement Sustainability Initiative) report [2]. Since the 2nd half of the 20th century, the addition of supplementary cementitious materials (SCMs) to Portland cement (PC) has received renewed attention. The most commonly used SCMs in cement and concrete are, fly ash (FA), silica fume (SF), blast furnace slag (BFS), metakaolin (MK) and rice husk ash (RHA). SCMs are widely used in concrete either in blended cements or when the concrete is made, due to sustainability action, environmental issues and the technical advantages [3-10].

Thomas et al. [11] concluded that, the superior performance of composite cements over those of PC is mainly due to the following characteristics: i) The pozzolanic reaction of SCMs with Portlandite (CH) produced during cement hydration; ii) The C$_3$A reduction, i.e. dilution effect; iii) The pH reduction therefore, the ettringite becomes less expansive; iv) The formation of additional CSH, which produces a coating layer on the alumina-rich and other reactive phases, thereby hindering the secondary and lastly ettringite formation; and v) The secondary CSH formation also results in pore size refinement, which reduces the permeability as well as the ingress of aggressive ions. Among of the most reactive commonly used SCMs deserve a special place; FA has a beneficial effect.

Fly ash is known as pulverized fly ash (PFA). It is finely divided byproduct that electro-statically precipitated from the combustion of pulverized coal in boilers of power plants. The PFA particles are spherical and have the same fineness as cement so that, the silica is readily available for pozzolanic reaction with lime [12].

The major mineral components of PFA are silicates and aluminates. The silicates are usually present as spherical particles. They are believed to be the melted products of clays, feldspars, quartz, mullite ($3Al_2O_3.2SiO_2$) and the other common minerals in coal [13].

Approximately 33% of FA is used in Europe as a constituent of blended cements and as mineral addition for the concrete production [14]. Previous studies [15–17] have shown that, the replacement of PC by FA improves the workability and durability of concrete, reduces hydration heat and helps in the development of long term compressive strength.

Recently, there is a great interest in replacing a long time used materials in concrete structure by new materials to produce cheaper, harder and durable concrete. There are many applications of nanotechnology in construction engineering field [18-21]. It is being accepted that, by adding a portion of nanoparticles (NPs), even at a very small content, the properties of cement-based material can be enhanced to a great extent in respects of workability, strength gain and durability [22].

The great reactivities of nanomaterials (NMs) are attributed to their high purities and specific surface areas in relation to their volume [23, 24]. Due to their sizes; some researchers have recorded an increased water demand for mixtures containing NMs of the same workability [25-26].

The behavior of NPs in cement can be summarized as follow: i) NPs act as fillers in the empty spaces; ii) NPs act as crystallization and seeding centers of hydrated products, promoting cement hydration; iii) NPs assist the formation of small sized CH crystals as well as homogeneous C-S-H clusters; and iv) NPs improve the ITZ structure [27, 28].

Zhang et al [29], studied the effect of NS on the properties of cement concrete containing high volume of FA. They reported that, NS particles reduce setting times and increase the early age strength of concrete. Both FA and NS are pozzolanic materials, and both adsorb, react with CH that is generated from cement hydration. To get a considerable strength improvement, the NS content shall not be less than 5 mass, % of binder. It is estimated that, the addition of 5g NS can consume almost 50% of the CH produced by 100g of cement when assuming that, NS has been fully hydrated and a total of 20 g of CH can be generated, giving CSH with Ca/Si ratio of 1.7. Since NS and FA compete in consuming CH, but NS is far more reactive than FA.

In a previous work [30], the effect of colloidal nano-SiO_2 (CNS) on FA hydration was studied and it was concluded that, i) the early age strength gain of CNS-FA-cement pastes is mainly due to the NS acceleration effect on both cement and FA hydration; ii) Although CNS can enhance the pozzolanic reaction of FA by increasing the alkalinity of solution in the early age, its later age hydration may be adversely affected and iii) There is a dense coating around FA particles in the CNS modified pastes, which, with a low Ca/Si ratio may result from the reactive CNS hydration at the early age and acts as a barrier that hinders ion penetration and consequently the FA hydration at later ages.

Few researches were done on the use of FA and NS in cement paste, mortar and concrete. This work aims to study the hydration behavior of composite cement containing FA and NS. PC type (I) was partially substituted with FA up to 30.0 mass, %, then the FA portion was replaced by equal amounts of NS (2.0, 4.0 and 6.0 mass, %). The hydration kinetics of cement blends with and without NS was studied and the hydration products were identified using XRD, DTA and SEM techniques.

2. Materials and Experimental Details

The materials used in this investigation were OPC, FA and NS. OPC with Blain surface area of 3000 ± 50.0 cm^2/g was provided from Lafarge Cement Company, Egypt. FA with specific surface area of about 10.672 ± 2 m^2/g, was supplied from Sika Chemical Company, Egypt. NS with average particle size, Blain surface area and purity percentage of about 15.0 nm, 50.0 m^2/g and 99.9%, respectively was supplied from Nanotechnology Lab, Faculty of Science, Beni-Suief University, Beni-Suief, Egypt. The oxide analyses of OPC and FA obtained by X-ray fluorescence (XRF) spectrometry are given in Table 1. The mineralogical composition of OPC is listed in Table 2.

Table 1. Chemical oxide analysis of OPC and FA (mass, %).

Oxides	SiO_2	Al_2O_3	Fe_2O_3	CaO	MgO	SO_3	Na_2O	K_2O	L.O.I	Total
OPC	19.30	3.94	3.80	62.67	1.90	3.22	0.44	0.39	3.04	99.70
FA	63.1	26.54	5.4	2.33	0	0.09	0.85	0.52	0.8	99.63

Table 2. Mineralogical composition of OPC.

Compound	Abbreviation	Chemical formula	Content, %
Tri-calcium silicate	C_3S	$3CaO.SiO_2$	66.08
Di-calcium silicate	C_2S	$2CaO.SiO_2$	5.50
Tri-calcium aluminate	C_3A	$3CaO.Al_2O_3$	4.02
Tetra-calcium aluminoferrite	C_4AF	$4CaO.Al_2O_3.Fe_2O_3$	11.55

The NS used in this study was prepared as described in our previous work [26]. The amorphous glassy nature of FA and NS was verified by different techniques (Figs.1-5). The starting materials were completely dried at 110°C for 2h. Each dry mix was blended in a steel ball mill using some balls for 1h to achieve complete homogeneity.

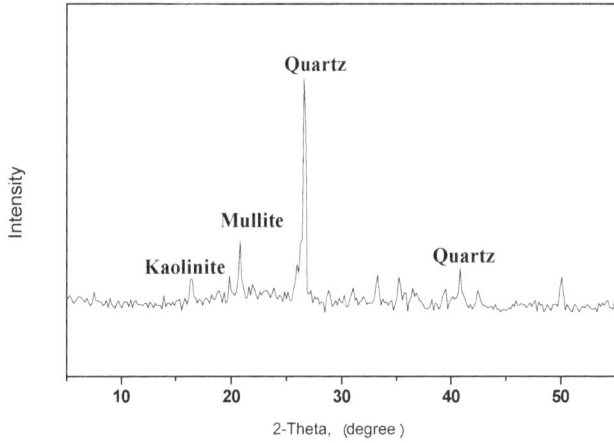

Fig. 1. *XRD pattern of FA.*

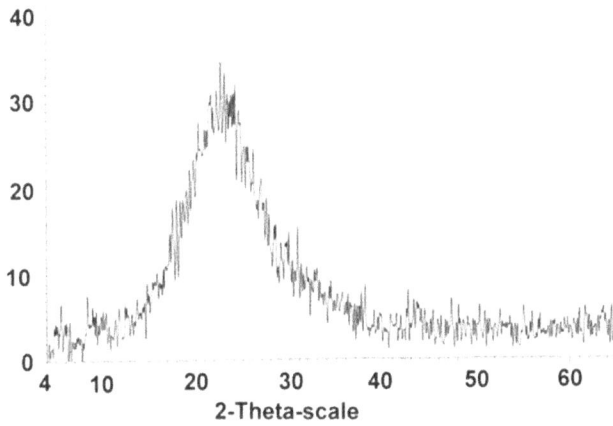

Fig. 2. *XRD pattern of NS.*

Fig. 3. *SEM photograph of NS.*

Fig. 4. *TEM photograph of NS.*

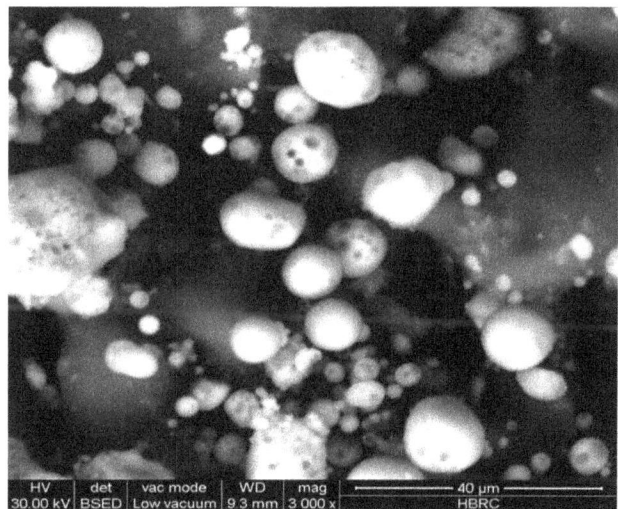

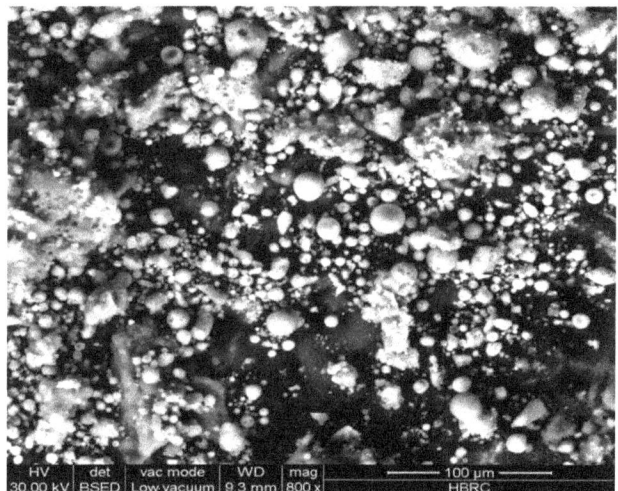

Fig. 5. *SEM photographs of FA.*

The cement blends were mixed in a rotary mixer. NS-particles are not easy to disperse uniformly in water. Accordingly, the NS mixing was performed as follows: (i) NS was stirred with 25% of the required water for standard consistency at speed of 120 rpm for 2min; (ii) The cement containing FA and the residual amount of mixing water were added to the mixture and homogenized at speed of 80rpm for another 2min; (iii) The blend was allowed to rest for 90s, and then mixed for 1 min at speed of 120rpm and (iv)The paste was manually placed, pressed and homogenized in stainless steel molds. After the top layer was compacted, the top surface of the mould was smoothened by the aid of thin edged trowel. For preparation of mortars, the sand was added gradually and mixed at a medium speed for 30s after step (ii). The mortars were prepared according to ASTM (C109-93) by mixing 1 part of cement and 2.75 parts of Lafarge standard sand proportion by weighing with water content that sufficient to obtain a flow of 110±5 with 25 drops of the flowing table [31, 32]. All specimens were cast in stainless steel molds (50×50×50mm cubes), demoulded after 24h, and then cured in fresh tap water at 23.0±2°C until the testing time. The mix compositions of the prepared cements are given in Table 3. The mixing water was measured to get all specimens having the same workability. The water of consistency and setting times for each mix were determined according to ASTM specification [33]. After the predetermined time, the hydration of cement pastes was stopped as described in a previous work [34]. The chemically combined water (Wn), free lime (FL) and bulk density (BD) were determined as mentioned elsewhere [35-37]. The compressive strength was measured according to the ASTM specifications (C-150) [38]. A compressive test was carried out in a hydraulic universal testing machine (3R), Germany, of 150.0 MPa capacity. To verify the mechanism predicted by the chemical and mechanical tests, some selected hydrated samples were examined using XRD, DSC, TG and SEM techniques.

Table 3. *Mix composition of OPC and blended cements, mass%.*

Mix No.	OPC	FA	NS
M1	100	0	0
M2	90	10	0
M3	80	20	0
M4	70	30	0
M5	70	28	2
M6	70	26	4
M7	70	24	6

For XRD, a Philips diffractometer PW 1730 with X-ray source of Cu Ka radiation (k= 1.5418 A°) was used. The scan step size was 2θ, the collection time 1s, and in the range of 2θ from 10° to 55°. The X-ray tube voltage and current were fixed at 40.0 KV and 40.0 mA, respectively. An on-line search of a standard database (JCPDS database) for X-ray powder diffraction pattern enables phase identification for a large variety of crystalline phases in a sample. The DTA was carried out in air using a DT-30 Thermal Analyzer Shimadzu Co., koyoto, Japan. Calcined alumina was used as inert material, about 50 mg (-76μm) of each. The finely ground

hydrated cement paste was housed in a small platinum-rhodium crucible. A uniform heating rate was adopted in all of the experiments at 20°C/min [39]. The microstructure was investigated by SEM, model quanta 250 FEG (Field Emission Gun), with accelerating voltage 30K.V., magnification power 14 x up to 1000000 and resolution for Gun.1n). FEI Company, Netherlands.

3. Results and Discussion

3.1. Characteristics of Composite Cements Containing Fly Ash

The variations of mixing water (W/C, %) as well as initial and final setting times (IST& FST) of the prepared cement pastes are graphically represented in Fig. 6 (A& B). The results show that, the FA-cements require higher water demands and longer setting times (STs) comparing with the OPC. Also, W/C, % and STs increase with FA content. This is mainly due to the higher surface area of FA than OPC. The retardation of setting process may be due to the decrease of cement portion (dilution effect) [40], forming CSH, which has higher setting characteristics [41] in comparison with fly ash pozzolanic cement pastes.

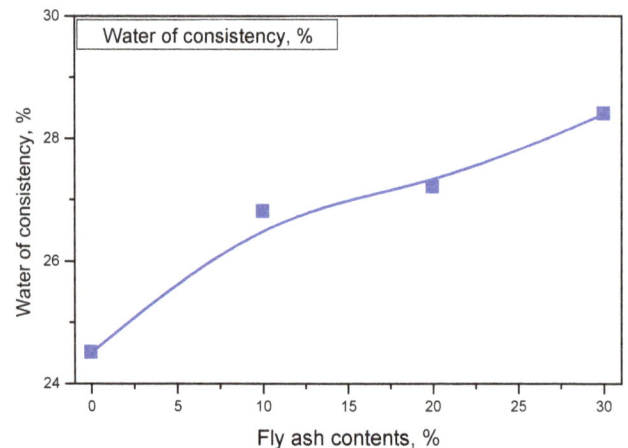

Fig. 6(A). *Water of consistency of OPC-FA pastes.*

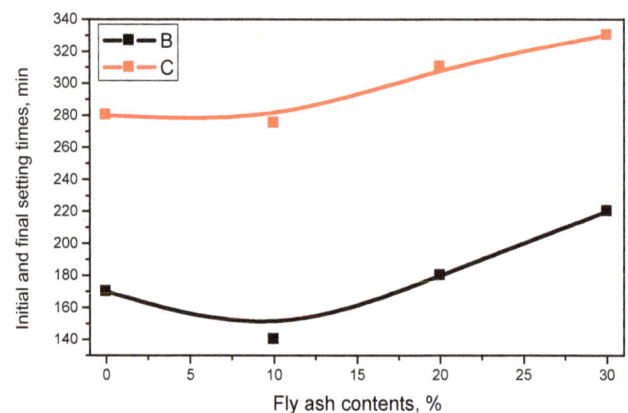

Fig. 6(B). *Initial and final setting times of OPC-FA pastes.*

Also, the slight pozzolanic activity of FA may be adversely affect setting process of FA-blended cement pastes. So cement pastes containing 10 wt. %, of FA tends to shorten the initial setting time than the plain cement paste. This is principally due to the nucleating effect, which accelerates the rate of cement hydration. As the amount of FA increases up to 30 wt. %, the initial setting time is accordingly elongated. This is mainly due to the dilution effect of the cement and the low rate of hydration of FA in comparison with OPC.

The final setting time of FA-blended cement pastes elongates with the amount of FA. This is essentially due to the decrease of the formed CSH gel at early ages and the dilution of the cement content. The hydration of cement is faster than the pozzolanic reaction of FA with Portlandite.

The variation of Wn. % of the hydrated FA-blended cements as well as OPC with curing time up to 90-days and FA content is graphically plotted in Fig. (7). Generally, the values of combined water increase with curing time for all cement pastes, due to the continuous hydration of cement phases and pozzolanic reaction, leading to the formation of hydrated aluminates, silicates and aluminosilicates with high water contents [42]. On the other hand, FA-cement blended cements have lower Wn values comparing to the control (OPC). Also, the amount of Wn decreases with FA content. This is mainly attributed to that; the fly ash sample used in this work is mainly composed of crystalline phases such as quartz and mullite in addition to small amount of amorphous material. Therefore, it has slight pozzolanic activity, which decreases the hydration characteristics. Hence, the combined water content (Wn. %) decreases in accordance with Hanehara et al. [43], but in contrast with other workers [44, 45]. The combined water content of FA cement paste decreases, with FA content, due to that, the CSH formed has low C/S with low water content [46].

The free lime contents of OPC and blended cement pastes cured up to 90d is given graphically illustrated as a function of curing time in Fig. (8): The free lime content of OPC pastes increases with curing time up to 90 days, due to the continuous hydration of alite and belite phases in OPC, liberating protlandite $Ca(OH)_2$ during the hydration period. On the other side, the free lime contents of all pozzolanic cement pastes increase up to 3 days then decrease up to 90 days. This is in a good agreement with earlier work [47, 48]. The initial increase of free lime in pozzolanic cement is mainly due to the fast hydration of the clinker. As fly ash content increases, the liberated portlindite increases, due to the nucleating agent as well as the very low pozzolanicity of FA at one day. The increase of FA content tends to separate the hydrated CSH-gel, which enhances the liberation of portlandite. The decrease after 3d is due to its consumption by the FA. There are two different processes; one tending to increase the portlandite and the other tends to decrease its value due to the pozzolanic reaction [49]. This determination of free lime content is more significant for the demonstration the pozzolanic properties of FA. It provides an indication on the progress of the pozzolanic reaction. These are consistent with whose found that; the fly ash commences reaction with $Ca(OH)_2$ between 3 and 7 days, but considerable amounts of $Ca(OH)_2$ and fly ash still remain un-reacted up to 90 days of

hydration. Therefore, it can be concluded that, this fly ash is poorly pozzolanic material. Both of free lime and combined water contents of FA-blended cement pastes are lower than those of OPC pastes. This may be attributed to the low pozzolanic activity of FA, especially at early hydration ages.

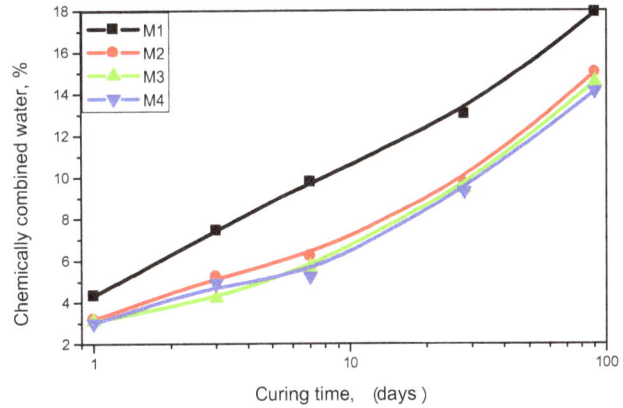

Fig. 7. Combined water contents of OPC-FA pastes with curing time.

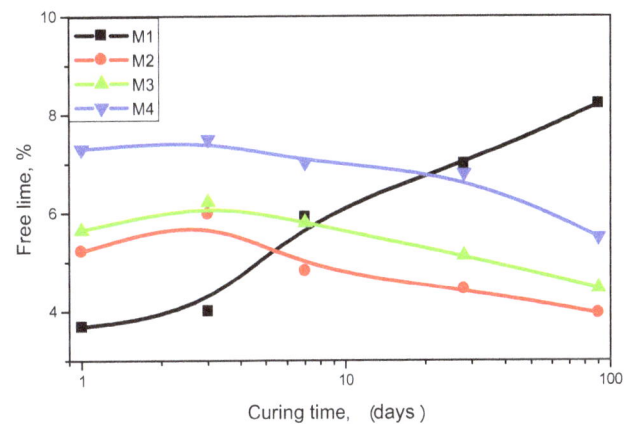

Fig. 8. Free lime contents of OPC-FA pastes with curing time.

The bulk density of OPC as well as Pozzolanic cement pastes cured up to 90 days is given graphically illustrated in Fig. (9). It is clear that, as the curing time proceeds, the bulk density for all hardened pastes increases. This is due to the gradual filling of large pores by the hydration products of cementitious materials. Substitution of OPC with fly ash produces a significant decrease of bulk density [50]. The increase of water of consistency with the fly ash content decreases the bulk density. This is mainly due to the slight pozzolanicity of fly ash. Generally the density of CSH from pozzolanic reaction is lower, than that formed from the hydration of OPC. OPC pastes show high value of portlandite, which gives CSH rich in calcium and water, which increases the bulk density of OPC cement in comparison with pozzolanic cement paste.

The compressive strength data of OPC as well as pozzolanic cement mortars cured up to 90-days is shown the graphically represented as a function of curing time in Fig. (10). The results show that, the compressive strength increases with curing time for all hardened cement mortars. As the hydration proceeds more hydration products and more cementing materials are formed. This leads to an increase of compressive strength of hardened cement mortars. This mainly attributed to that the hydration products possess a large specific volume than un-hydrated cement. Therefore, the accumulation of these hydration products will fill a part of available pore spaces, then giving higher strength. On the other side, as the fly ash content increases, the compressive strength decreases, as observed elsewhere [51]. This is due to the lower pozzolanic activity of the FA-particles. In addition, the decomposing reactions of the glassy phase network of the fly ash would be slowed down. Therefore, these factors affect the structure of the hardened pastes, so that its strength is reduced. The compressive strength was also in harmony with the bulk densely. Also, the compressive strength values are in a good harmony with those of combined water contents. The increase of the hydration products is the main factor of compressive strength.

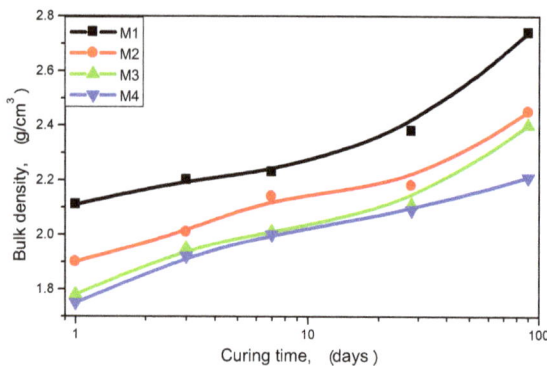

Fig. 9. *Bulk density of hardened FA-cements with time.*

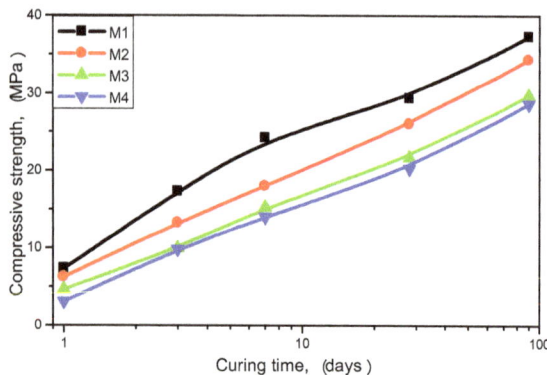

Fig. 10. *Comressive strength of hardened FA-cements with time.*

The effect of curing time up to 90-days on the hydration characteristics of blended Portland cement containing 30 mass, % of FA can be seen from XRD patterns in Fig. (11). It is obvious that, the characteristic peak of CSH increases, whereas the peaks of Portlandite (CH) decrease markedly with curing time. This is mainly due to the reaction of FA portion with the liberated lime, forming additional amounts of CSH. The rate of FA pozzolanic reaction with lime increases with the time, therefore, the rate of lime consumption exceeds the rate of its production. The behavior of CH peaks is in accordance with the results of chemically determined free lime. Also, the XRD patterns show $CaCO_3$

peak that increases with curing age, due to the increase of Portlandite, which is available for carbonation with atmospheric CO_2.

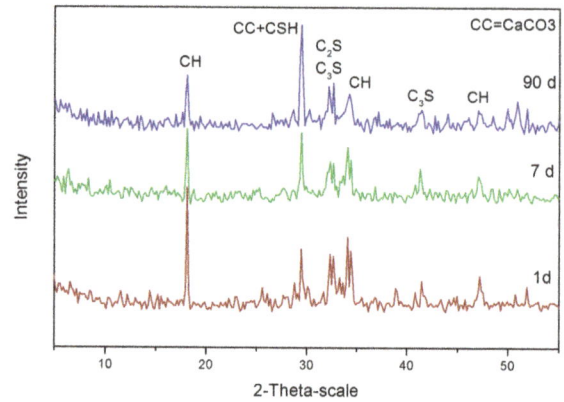

Fig. 11. *XRD patterns of M4 with curing time.*

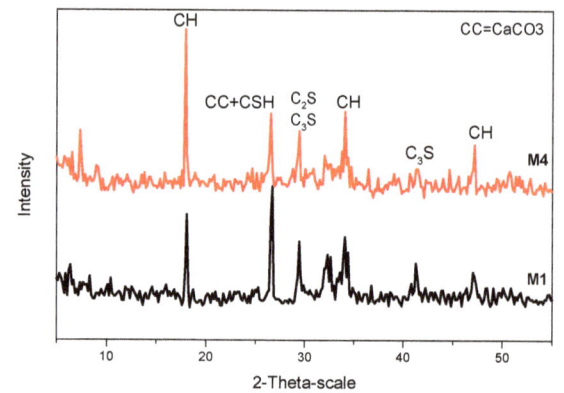

Fig. 12. *XRD patterns of different mixes at 7-days.*

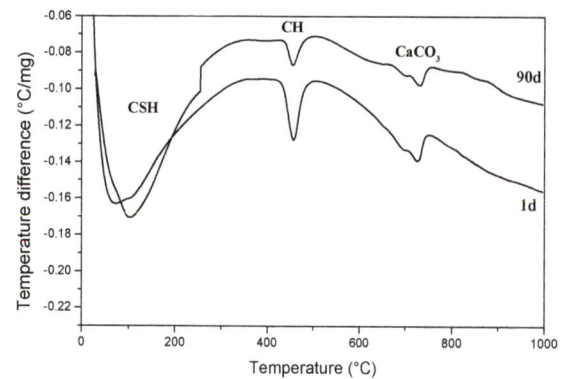

Fig. 13. *DTA thermograms of M4 hydrated as a function of time.*

The XRD patterns of hardened OPC (M1) and 30 mass% FA (M4) cement pastes, at 7 days of hydration are shown in Fig. (12). The results show the presence of un-hydrous silicates (β-C_2S and C_3S), calcium hydroxide (CH), calcite (CC) and calcium silicate hydrate (CSH). The presence of 30% of crystalline FA with Portland cement decreases the amount of OPC. Also, the used FA has low pozzolanic activity, especially at early hydration ages. Therefore, the intensity of CSH and CC peak in OPC pastes is higher than the corresponding peak in A blended cement pastes. Calcium hydroxide behaves in an opposite manner of CSH. The peak of unhydrous silicates decrease with the presence of FA, due to the dilution of clinker phases with 30 mass, % FA. The intensity of Portlandite peak is in accordance with that of chemical analysis.

Figure (13) shows the DTA thermograms of hydrated M4 (30% FA) at 1 and 90-days. There are three main endothermic peaks. The first peak appears at 100°C, which refers to dehydration of CSH. The second peak, in the temperature range 450-500°C, is attributed to $Ca(OH)_2$ decomposition [52]. The intensity of CSH peak increases with curing time. But, the characteristic CH peaks decrease with time, due to the hydration progress and formation of successive amounts of hydrated products (CSH, CAH and CASH). The results of DTA are in a good harmony with each other and with those of XRD, free lime, bulk density and Strength. The endothermic peak of cement paste at one day located round 50 -100°C is due to the removal of moisture of the cement paste. As the hydration precedes the endothermic peak of CSH increases with curing time. The dissociation of Portlandite at 1d occurs at lower temperature than that cured at 90-days. This is mainly due to its amorphous state at one day but, at 90-days, the dehydroxylation of Portlandite occurs at higher temperature due to its crystallinity. This is mainly due to the two process of the formed Portlandite, one is amorphous, which decomposes at lower temperature and crystalline type, which decomposes at relatively higher temperature.

Figure (14) illustrates the TG of FA-cement paste (M4) at 1 and 90-days of hydration. It is clear that, the CSH loss of sample at 90-days is higher than that at 1 day. On the other side, the Portlandite increases with curing time. The loss occurs at 600–700°C is mainly due to the $CaCO_3$ decomposition. These results are in an agreement with those of free lime contents determined by the chemical method and those of XRD analysis.

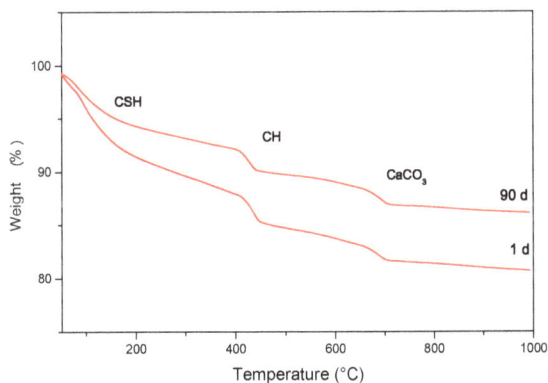

Fig. 14. TG analysis of M4 hydrated at different ages.

3.2. Characteristics of Composite Cements Containing FA&NS

The variations of water of consistency as well as setting times of the investigated cement pastes are graphically plotted in Fig. 15(A&B). The results show that, the water demand and setting times increase with NS, content. The increase in water demand is mainly attributed to the increase

of surface area of NS in comparison with FA [53-55]. Thus, the specimens containing NS require more water to rapid forming of hydrated products [56]. The setting process is elongated due to the increase of water of consistency. As the water of consistency increases, the free water increases, which delays the setting of cement paste [57].

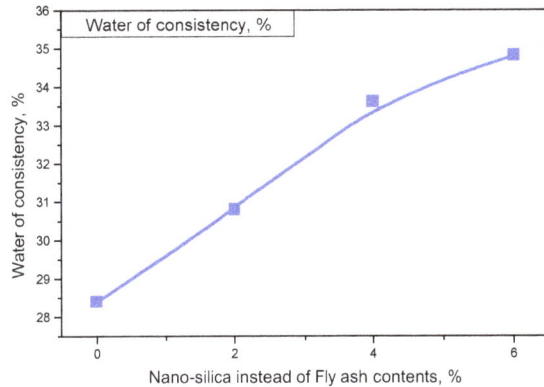

Fig. 15(A). Water of consistency of cement pastes with NS, %.

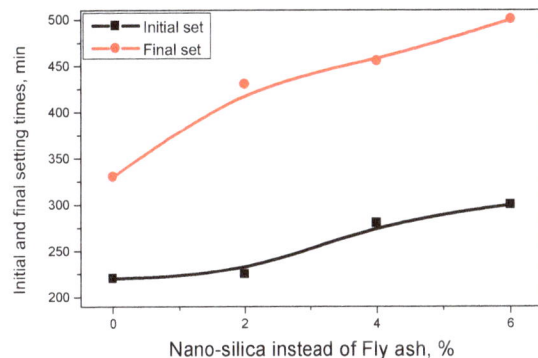

Fig. 15(B). Setting times of cement pastes with NS,%.

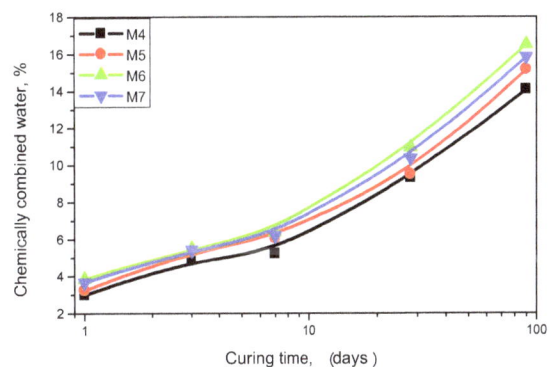

Fig. 16. Combined water of hydrated M4, M5, M6 and M7 with time.

The combined water contents of FA-pozzolanic cement replaced with 2, 4 and 6 wt. %, NS is graphically represented in Fig. (16). It is clear that, the replacement of equal amount of FA with NS is accompanied with gradual increase of combined water content up to 4 wt. % NS, then decrease at 6 wt. %, NS. Generally, the FA-pozzolanic cement pastes give lower values of combined water (Wn, %) than FA-NS cement pastes up to 90-days. This is principally attributed to the high pozzolanic activity of NS portion to react with the liberated CH, leading to the formation of additional amounts of hydrated silicates and alumnino-silicates with high water contents. From the chemical point of view, NS is highly reactive pozzolana which reacts with formed calcium hydroxide (CH) from the hydration of cement clinker phases producing calcium silicate hydrates (CSH) [58]. This is also due to the high surface area and glass content of NS in comparison with FA.

The effect of substituted amounts of FA with NS on the free lime is graphically represented in Fig. (17). It can be seen that, the free lime contents decrease with curing time for all hydrated cement pastes. This mainly attributed to the pozzolanic activity of FA and NS. The pozzolanic reaction of FA and/or NS with the liberated Portlandite through the hydration of cement phases increase with curing time. The substitution of FA with 2 and 4 wt. % NS is accompanied with gradual decrease of free lime, i. e. the cement pastes containing 4 wt. % NS show lower values of F. L, than the plain and FA-cement pastes. At 6 wt. % NS the free lime is increased as previously decreased [56, 57].

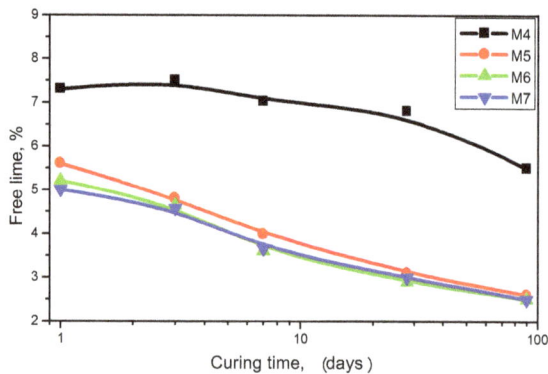

Fig. 17. Free lime of hydrated M4, M5, M6 and M7 with time.

The bulk density (dp) values of FA-NS cement pastes cured up to 90-days is graphically depicted in Fig. (18). The results indicate that, the bulk density (dp) increases with curing time for all hydrated cement pastes, due to the continuous hydration of cement clinker and pozzolanic reaction, leading to the formation and accumulation of excess amounts of hydrated silicates and alumino-silicates which tends to increase the gel/space ratio as well as the bulk density [42]. This leads to the formation of homogeneous and compact microstructure. The bulk density (dp) increases with the nano-silica content up to 4 wt. % and then decreases at 6 wt. %, but still higher than that of the control mix (30 wt. % FA). This can be interpreted as follows [59]. Suppose that, NS particles are uniformly dispersed in cement paste and each particle is contained in a cubic pattern, therefore the distance between nano-particles can be determined. After the hydration begins, hydrated products diffuse and envelop NS particles as kernel [56]. If the NS content and the distance between them are appropriate, the crystallization will be controlled to be a suitable state through restricting the growth of Ca(OH)$_2$ crystals. Moreover, the nano-particles located in cement paste as kernel can further promote cement hydration due to their high activity. This makes the size of Ca(OH)$_2$ crystals smaller, the cement matrix is more homogeneous and compact, then the pore structure is

improved. With increasing the NS content more than 4 wt. %, the improvement of the pore structure of cement paste is weakened. This can be attributed to that, the distance between nano-particles decreases with NS content, and Ca(OH)$_2$ crystals cannot grow up enough, due to limited space, then the crystal quantity is decreased, which leads to the decrease of crystal to gel space ratio [57].

The data of compressive strength of OPC-FA-NS cement pastes cured up to 90-days are shown graphically depicted in Fig. (19). The compressive strengths of the fly ash mortars were increased with the incorporation of the NS in comparison to the corresponding reference mortars containing 30 wt. %, FA at all curing ages. The strength generally increased with the substitution of nano-silica up to 4 wt. %. In particular, the compressive strength was considerably improved for the cement with 26 wt. % fly ash and 4 wt. % NS. This indicates that, the inherently slower rate of strength development of mortars containing fly ash, can be improved by the substitution of small dosages of nano-silica up 4 mass, %. The overall performance of mortar with and without fly ash was significantly improved with the addition of variable dosages of nano-silica. More refinement of the pore structure was achieved with increasing the nano-silica dosage up to 6 mass, % [60].

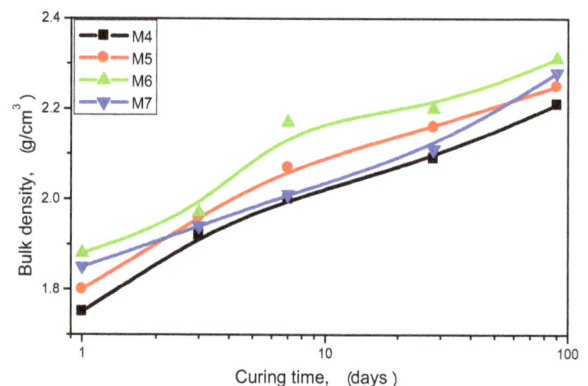

Fig. 18. Bulk density of hardened M4, M5, M6 and M7 with time.

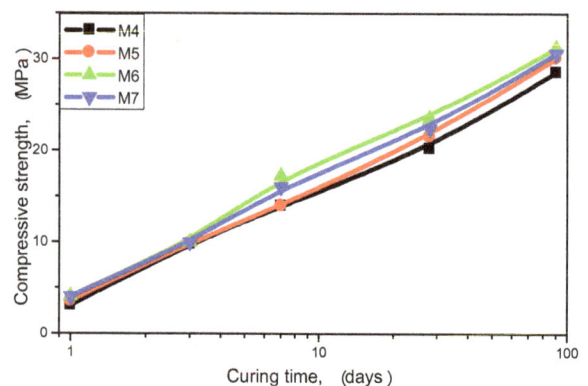

Fig. 19. Strength of hardened M4, M5, M6 and M7 with time.

The effect of 6 mass, % of NS on the hydration of FA-blended cement paste can be seen from the XRD patterns of hydrated M13 (70% OPC + 24 % FA + 6% NS) with curing time in Fig. (20). the results show that, the peaks of CH decrease with curing age whereas with that of CSH and calcite (CC) increases. This mainly due to the continuous hydration of clinker phases as well as the pozzolanic reaction of both FA and NS with the liberated $Ca(OH)_2$. It can be concluded that, NS compensate the poorly pozzolanic action of FA, especially at early ages of hydration. The effect of mix composition on the hydration characteristics of OPC-FA blended cements with and without NS can be shown from XRD patterns of M4 and M7 in Fig. (21). The results indicate that, NS has positively effect on the hydration of FA-cement blends, because nano-sized SiO_2 particles behave not only as nano filler to promote the hydration of clinker silicate phases but also as a good pozzolanic additive. Indeed, NS has very higher pozzolanic reactivity comparing to FA, which has lower amorphous silica content and higher amounts of quartz as well as mullite, which are crystalline phase. These crystalline phases show no or very weak pozzolanic activity. Therefore, the peak of CSH and CC phases increases with the presence of NS.

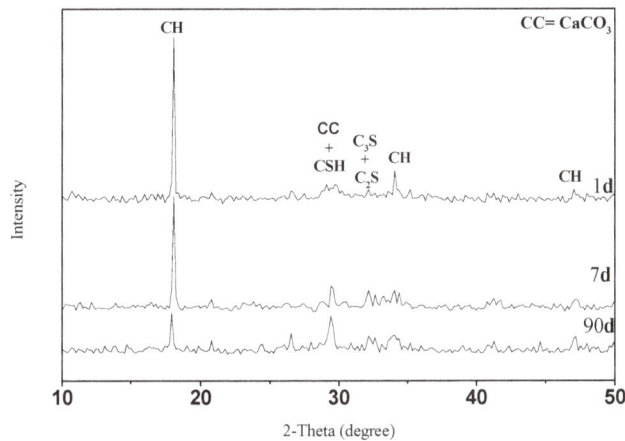

Fig. 20. *XRD patterns of M7 as a function of curing time.*

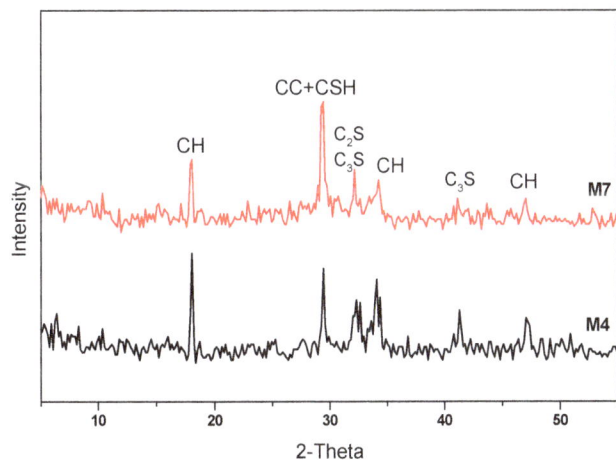

Fig. 21. *XRD patterns of hydrated M4 and M7 at 7-days.*

4. Conclusions

The results obtained in this study indicate that, the substitution of 10 wt. % OPC by FA accelerates the initial

setting time, due to the nucleating effect without any detectable effect on the final setting time. On the other side, the increase of FA up to 30 wt. % elongates the initial and final setting time. The FA increases the water of consistency from 24.5 up to 28.4 % for 0 and 30 wt % FA substitutions. The substitution of OPC with FA up to 30 wt. % has no detectable effect on the values of chemically combined water of cement pastes up to 90-days of hydration. But, these values are lower than those of OPC pastes, due to the low pozzolanic activity of FA. The increase of FA content up to 20 wt. % increases the liberated Portlandite, especially at early ages of hydration up to 3-days. At 30 wt. % FA, the free lime contents of blended-cement pastes are higher than those of only OPC pastes up to 7-days then decreases due to its consumption by FA at later hydration ages.

The replacement of OPC with FA decreases the bulk density as well as the compressive of cement mortars up to 90-days. As the amount of FA increases, the bulk density and compressive strength decrease gradually. It can be said that, 10 wt. % FA is the optimum substitution level.

In OPC-FA blended cement, the replacement of FA with 2, 4 and 6 wt. % NS, increases the amount of water of consistency and elongates the setting times with higher level than FA. This is due to its higher surface area in comparison with FA of micro-particles, whereas the NS has nano-particle size of about 15 nm. The NS is completely amorphous. In the preparation of nano-modified blended cements containing 2, 4 and 6 wt. % NS instead of FA it was found that, 4 wt. % NS enhances the chemical and physico-mechanical properties of blended cement.

References

[1] Jewell, S., and Kimball, S. 2015. Mineral Commodity Summaries 2015. Washington D. C.: U. S. Geological Survey, U. S. Department of Interior.

[2] ACI (American Concrete Institute). 2009. "The Cement Sustainability Initiative." World Business Council for Sustainable Development. Accessed September 8, 2015. http://www.wbcsdcement.org/pdf/CSI%20GNR%20Report %20final%2018%206%2009.pdf.

[3] S. Abd El-Aleem, M. A. Abd-El-Aziz, M. Heikal, H. El-Didamony, "Effect of cement kiln dust substitution on chemical and physical properties and compressive strength of Portland and slag cements", Arabian journal for science and engineering; 30 (2B) (2005), pp. 263-273.

[4] EA. El-Alfi, AM Radwan, S. Abd El-Aleem, "Effect of limestone fillers and silica fume pozzolana on the characteristics of sulfate resistant cement pastes", Ceramics Silikaty; 48 (1) (2004), pp. 29-33.

[5] ASTM International. 2015. "Standard Specification for Coal Fly Ash and Raw or Calcined Natural Pozzolan for Use in Concrete" In ASTM C618-15. West Conshohocken: ASTM International.

[6] Thushara Priyadarshana, Ranjith Dissanayake and Priyan Mendis, "Effects of Nano Silica, Micro Silica, Fly Ash and Bottom Ash on Compressive Strength of Concrete", Journal of Civil Engineering and Architecture 9 (2015) 1146-1152.

[7] Ruben Snellings, Gilles Mertens and Jan Elsen "Supplementary Cementitious Materials" Reviews in Mineralogy & Geochemistry Vol. 74 (2012), pp. 211-278.

[8] S. Abd-El-Aleem, M. A. Abd-El-Aziz, H. El-Didamony, "Calcined carbonaceous shale pozzolanic Portland cement", Egyptian Journal of Chemistry; 45 (3) (2002), pp. 501-517.

[9] S. Abd El-Aleem, "Hydration characteristics of granulated slag with fired by-pass cement dust", Silicates Industriels; 69 (3-4) (2004), pp. 46-52.

[10] E. García-Taengua, M. Sonebi, K. M. A. Hossain, M. Lachemi, J. Khatib, "Effects of the addition of nanosilica on the rheology, hydration and development of the compressive strength of cement mortars" Composites Part B 81 (2015) 120-129.

[11] Thomas, M. D. A., Hooton, R. D., Scott, A. and Zibara, H., "The effect of supplementary cementitious materials on chloride binding in hardened cement paste", Cem. Concr. Res., 42, (2012), pp.1-7.

[12] Hwang, J. Y., "Beneficial Use of Fly Ash", U. S. Department of Energy's Federal Energy Technology Center & Michigan Technological University's Institute Materials Processing's, (1999), pp. 1-23.

[13] Taylor, H. F. W., Cement Chemistry. Academic press Ltd., London, (1990).

[14] Salaheddine Alahrache, Frank Winnefeld, Jean-Baptiste Champenois, Frank Hesselbarth, Barbara Lothenbach, "Chemical activation of hybrid binders based on siliceous fly ash and Portland cement", Cement and Concrete Composites, 66 (2016), pp. 10-23.

[15] V. G. Papadakis, Effect of fly ash on Portland cement systems: Part II. Highcalcium fly ash, Cem. Concr. Res. 30 (10) (2000) 1647–1654.

[16] Afaf Ghais, Duaa Ahmed, Ethar Siddig, Isra Elsadig, S. Albager, Performance of concrete with fly ash and kaolin inclusion, Int. J. Geosci. 05 (12) (2014) 1445–1450.

[17] N. Ghafoori, M. Najimi, H. Diawara, M. S. Islam, Effects of class F fly ash on sulfate resistance of Type V Portland cement concretes under continuous and interrupted sulfate exposures, Constr. Build. Mater. 78 (2015) 85–91.

[18] S. Abd El-Baky, S. Yehia, I. S. Khalil "Influence of nano-silica addition on properties of fresh and hardened cement mortar" NANOCONBrno, Czech Republic, EU, 10 (2013), pp. 16-18.

[19] G. Quercia, P. Spiesz, G. Hüsken, H. J. H. Brouwers, "SCC modification by use of amorphous nano-silica", Cem. Concr. Compos., 45 (2014), pp. 69–81.

[20] M. Stefanidou and I. Papayianni, "Influence of nano-SiO2 on the Portland cement pastes", Composites: Part B; 43 (6) (2012), pp. 2706-2710.

[21] AshwniK. Ranal, "Significance of Nanotechnology in Construction Engineering", International Journal of Recent Trends in Engineering, Vol. 1, No. 4, May (2009).

[22] PengkunHou, JueshiQian, Xin Cheng, Surendra P. Shah, "Effects of the pozzolanic reactivity of nano-SiO_2 on cement-based Materials", Cem. Concr. Compos., 55 (2015), pp. 250–258.

[23] M. Wilson, K. K. G. Smith, M. Simmons, and B. Raguse, "Nanotechnology-Basic Science and Emerging Technologies", Chapman & Hall/CRC; (2000).

[24] M. M. S. Wahsh, A. G. M. Othman, S. Abd El-Aleem, "The influence of nano-silica and zircon additions on the sintering and mechanical properties of in situ formed forsterite", Journal of Industrial and Engineering Chemistry; 20 (2014), pp. 3984-3988.

[25] S. Abd. El. Aleem, Mohamed Heikal, W. M. Morsi "Hydration characteristic, thermal expansion and microstructure of cement containing nano-silica", Constr. Build. Mater.; 59 (2014), pp. 151–160.

[26] SalehAbd El-Aleem and Abd El-Rahman Ragab, "Physico-mechanical properties and microstructure of blended cements incorporating nano-silica" International Journal of Engineering Research & Technology (IJERT), 3 (7); (2014), pp. 339-358.

[27] F. Sanchez, and K. Sobolev, "Nanotechnology in concrete - a review", Constr. Build. Mater.; 24 (2010), pp. 2060-71.

[28] I. Zyganitidis, M. Stefanidou, N. Kalfagiannis, and S. Logothetidis, "Nano-mechanical characterization of cement-based pastes enriched with SiO_2 nano-particles", Mat. Sci. Eng. B 176 (9); (2011), pp. 1580–1584.

[29] Zhang, M. and Islam, J., "Use of nano-silica to reduce setting time and increase early strength of concretes with high volumes of fly ash or slag", Constr. Build. Mater., 29, (2012), pp. 573-580.

[30] Hou, P., Wang, K., Qian, J., Kawashima, S., Kong, D. and Shah, S. P., "Effects of colloidal nano-SiO_2 on fly ash hydration", Cem. Concr. Compos., 34, (2012), pp. 1095–1103.

[31] M. Heikal, S. Abd El-Aleem, and W. M. Morsi, "Characteristics of blended cements containing nano-silica" HBRC Journal (9) (2013), pp. 243–255.

[32] Magdy A. Abdelaziz, Saleh Abd El-Aleem and Wagih M. Menshawy, "Effect of fine materials in local quarry dusts of limestone and basalt on the properties of Portland cement pastes and mortars", International Journal of Engineering Research & Technology (IJERT), 3 (6), (2014), pp.1038-1056.

[33] ASTM Designation: C191, Standard method for normal consistency and setting of hydraulic cement, ASTM Annual Book of ASTM Standards, 04.01, (2008).

[34] Abd El-Aziz, M. A., Abd El-Aleem, S. and Heikal, M., "Physico-chemical and mechanical characteristics of pozzolanic cement pastes and mortars hydrated at different curing temperatures", Constr. Build. Mater., 26, (2012), pp. 310–316.

[35] H. El-Didamony, M. Abd-El. Eziz, and S. Abd. El-Aleem, "Hydration and durability of sulfate resisting and slag cement blends in Qaron's Lake water", Cem. Concr. Res., 35; (2005), pp. 1592-1600.

[36] H. W. Sufee, "Comprehensive studies of different blended cements and steel corrosion performance in presence of admixture", Ph. D. Thesis, Faculty of Science, Fayoum University, Fayoum, Egypt, (2007).

[37] Saleh Abd El-Aleem, Abd El-Rahman Ragab, "Chemical and Physico-mechanical Properties of Composite Cements Containing Micro- and Nano-silica", International Journal of Civil Engineering and Technology; 6 (5) (2015), pp. 45-64.

[38] ASTM C109, "Strength test method for compressive strength of hydraulic cement mortars" (2007).

[39] V. S. Ramachandran, "Thermal Analysis, in; Handbook of analytical techniques in concrete science and technology" Ramachandran V. S. and Beaudoin J. J. Eds., Noyes publications, New Jersey. ISBN: 0-8155; (2001), pp.1473-1479.

[40] Hwang, C.-L. and Shen, D.-H., "The Effects of Blast-Furance Slag and Fly Ash on the Hydration of Portland Cement", Cem. Concr. Res., 21, (1991), pp.410-425.

[41] El-Didamony, H., Heikal, M. and Shoaib, M. M., "Homra Pozzolanic Cement", Silicates Industrials, Ceramic science and Technology,.65, (3-4), (2000), pp.39-43.

[42] Abd El-Aziz, M., Abd El-Aleem, S., Heikal, M. and El-Didamony, H., "Effect of Polycarboxylate on Rice Husk Ash Pozzolanic Cement", Sil. Ind. 69, 9-10, (2004), pp. 73-84.

[43] Henehara, S., Tomosawa, F., kobayakawa, M. and Hwang, K., "Effect of water/power Ratio, Mixing Ratio of fly ash and curing temperature on pozzolanic reaction of fly ash in cement pastes" Cem. Concr. Res., 31(1), (2001), pp. 31-39.

[44] Lam, L., Wong, Y. L. and Poon, C. S., "Degree of hydration and gel/space ratio of high-volume fly ash/cement systems", Cem. Concr. Res., 30, (2000), pp. 747–756.

[45] Singh, N. B., Singh, S. P., Sarvahi, R. and Shukla, A. K., "The Effect of coal dust-fly ash mixture on the hydration of protland cement", Il Cemento.,4, (1993), pp.231-238.

[46] Kar, A., Ray, I., Unnikrishnan, A. and Davalos, F., "Microanalysis and optimization-based estimation of C–S–H contents of cementitious systems containing fly ash and silica fume", Cem. Concr. Compos., 34, (2012), pp. 419–429.

[47] Heikal, M., El-Didamony, H., Sokkary, T. M. and Ahmed, I. A., "Behavior of composite cement pastes containing micro-silica and fly ash at elevated temperature", Constr. Build. Mater., 38, (2013), pp.1180–1190.

[48] Abd-El-Eziz, M. A. and Heikal, M., "Hydration characteristics and durability of cements containing fly ash and limestone subjected to Qaron's Lake Water", Adv. Cem. Res., 21, (3), (2009), pp.91-99.

[49] El-Didamony, H., Salem, T., Gabr, N. and Mohamed, T., "Limestone as a retarder and filler in limestone blended cement" Ceramics-Silikáty39, 15, (1995).

[50] Pandey, S. P. and Sharma, R. L., "The influence of mineral additive on strength and porosity of OPC mortar", Cem. Concr. Res., 30 (1), (2000), pp. 19-23.

[51] Deschner, F. Winnefeld, F. Lothenbach, B. Seufert, S. Schwesig, P. Dittrich, S. Goetz-Neunhoeffer, F. Neubauer, J."Hydration of Portland cement with high replacement by siliceous fly ash", Cem. Concr. Res., 42, (2012).

[52] Trezza, M. A. and Lavat, A. E., "analysis of the system 3CaO. Al2O3–CaSO4.2H2O-CaCO3-H2O by FTIR spectroscopy", Cem. Concr. Rese., 31, (2001), pp. 869-872.

[53] Sobolev, K., Nemecek, J., Smilauer, V. and Zeman, J., "Engineering of SiO2 nanoparticles for optimal performance in nano cement-based materials", Nanotechnology in Constr. Proc. Prague., (2009), pp.139–48

[54] Li, G., "Properties of high volume fly ash concrete incorporating nano-SiO2", Cem. Concr. Res., 34, (2004), pp. 1043–1049.53. Said, A. M., Zeidan, M. S., Bassuoni, M. T. and Tian, Y., "Properties of concrete incorporating nano-silica", Constr. Build. Mater., 36, (2012), pp.838–844.

[55] Bjornstrom, J., Martinelli, A., Matic, A., Borjesson, L. and Panas I., "Accelerating effects of colloidal nano-silica for beneficial calcium–silicate–hydrate formation incement", Chemical Physic Letter, 392, (2004), pp. 242–248.

[56] Legrand, C. and Wirquin, E., "Study of the strength of very young concrete as afunction of the amount of hydrates formed-influence of superplasticizer", Mater.Struct., 166, (1994), pp.106–109.

[57] Li, H., Zhang, M., and Ou, J., "Flexural fatigue performance of concrete containing nanoparticles for pavement", Int. J. Fatigue, 29, (2007), pp. 1292–1301.

[58] Zhang, M-Hand Gjorv, O. E., "Effect of silica fume on cement hydration in low porosity cement pastes", Cem. Concr. Res., 21, (5), (1991), pp. 800–808.

[59] Nazari, A. and Riahi, S., "Splitting tensile strength of concrete using ground granulated blast furnace slag and SiO2 nanoparticles as binder". Composites: Part B 43, (2011), pp. 864–872.

[60] Said, A. M., Zeidan, M. S., Bassuoni, M. T. and Tian, Y., "Properties of concrete incorporating nano-silica", Constr. Build. Mater., 36, (2012), pp. 838–844.

Investigations on the hydrothermal synthesis of pure and Mg-doped nano-CuCrO$_2$

Dirk Friedrich[1, *], **Claudia Wöckel**[1], **Sebastian Küsel**[2], **Robert Konrath**[2], **Harald Krautscheid**[2], **Reinhard Denecke**[1], **Bernd Abel**[1, 3]

[1]Wilhelm-Ostwald-Institute of Theoretical and Physical Chemistry, Department of Chemistry and Mineralogy, University of Leipzig, Leipzig, Germany
[2]Institute for Inorganic Chemistry, Department of Chemistry and Mineralogy, University of Leipzig, Leipzig, Germany
[3]Chemical Department, Leibniz Institute of Surface Modification (IOM), Leipzig, Germany

Email address:
dirkf2283@outlook.de (D. Friedrich), claudia.woeckel@uni-leipzig.de (C. Wöckel), sebastiankuesel@gmx.de (S. Küsel), robert.konrath@gmx.de (R. Konrath), krautscheid@rz.uni-leipzig.de (H. Krautscheid), denecke@uni-leipzig.de (R. Denecke), bernd.abel@uni-leipzig.de (B. Abel)

Abstract: This paper presents some investigations on the hydrothermal synthesis of nano-CuCrO$_2$. Several successively altered synthesis protocols are used to investigate effects of changing the mineralizer amount, lowering reaction temperature and addition of a reducing agent. As a result modified protocols for the hydrothermal synthesis of pure and Mg-doped CuCrO$_2$ are presented. Different washing and annealing steps are used to perform a comparative XRD-study on these materials.

Keywords: Delafossite, CuCrO$_2$, Hydrothermal Synthesis

1. Introduction

The compound CuCrO$_2$ is a promising material for optoelectronic applications. It belongs to a group of ternary oxides that crystallize in the delafossite structure. The structure exists in two polytypes, which are distinguishable by their stacking order along the c-axis. 3R-CuCrO$_2$ exhibits a stacking order AaBbCcAaBbCc, 2H-CuCrO$_2$ a stacking order AaBbAaBb [1]. In both structures there are layers of edge sharing CrO$_6$-octahedra (capital letters in stacking order) which are connected by layers of monovalent copper atoms (small letters in stacking order). Each copper atom exhibits a linear coordination towards oxygen atoms of two different CrO$_6$-layers. Nanosized CuCrO$_2$ and Mg-doped CuCrO$_2$ have been successfully applied in p-type dye sensitized solar cells (DSSCs) [2–5]. Comparison of application of pure and Mg-doped nano-CuCrO$_2$ in p-type DSSCs revealed an optimal performance for CuCr$_{0.9}$Mg$_{0.1}$O$_2$ [3]. The optical transmittance of CuCrO$_2$ based electrodes and the short circuit current density of p-type DSSCs incorporating these electrodes improved upon Mg-doping. This increased current density could be caused by lowering the electrical resistivity upon doping, as it has been reported multiple times [6–10].

However, in this case the decrease in average crystallite size from around 15 nm to 10 nm upon incorporation of Mg has been identified as major improvement. CuCr$_{0.9}$Mg$_{0.1}$O$_2$ based porous electrodes show a higher surface area than undoped ones. This leads to a higher amount of adsorbed dye and ultimately to higher short circuit densities. Therefore, a small reduction in crystallite size resulted in improvement of DSSC performance. There have been several other reports on the hydrothermal synthesis of CuCrO$_2$ [11–17]. These reports all differ in a key component to the aforementioned procedures. They all use Cu$_2$O as starting material, a compound with monovalent copper, whereas the aforementioned procedures use copper(II) nitrate. For the hydrothermal synthesis of the delafossite material CuGaO$_2$ similar procedures based on Cu(NO$_3$)$_2$ are published [18, 19]. However, in these reports the necessary reduction of copper is achieved by the addition of alcohols as reducing agent. As reported, hydrothermal synthesis of CuCrO$_2$ is achieved despite the absence of an apparent reducing agent, a fact that has not been further commented yet. Another aspect that has not been further investigated is the actual dopability of CuCrO$_2$ with magnesium under the applied hydrothermal conditions. Other reports on CuCr$_{1-x}$Mg$_x$O$_2$ involve high temperature annealing steps and the maximum reported amount of Mg that is incorporated into the delafossite host structure varies depending on the applied method [9]. In some case the spinel type material $MgCr_2O_4$ can already be detected as a by-product via XRD for $x > 0.02$ [20], thus limiting Mg-doping. In this publication we report on investigations on the hydrothermal synthesis of pure and Mg-doped nano-$CuCrO_2$. We present a modified hydrothermal synthesis of pure and Mg-doped $CuCrO_2$ with the potential to decrease average crystallite size of these materials.

2. Experimental

2.1. Analytical Measurements

Powder X-ray diffraction data were collected using a STOE STADIP powder diffractometer with Debye-Scherre geometry. Source of radiation was a sealed tube with Cu-anode (λ (Cu-K_α) = 154.06 pm). The radiation was monochromatized using a Ge-single crystal. All measurements were performed at room temperature using a position sensitive detector. Samples were measured in transmission mode with a flat sample geometry. The powdered samples were spread evenly on a circular polymer with diluted glue ("Elmer's white glue"). When the samples had been dried, they were inserted into sample holders and fixed with a mask of 1 cm inner diameter. These samples were measured in a 2θ-ω-coupled mode and continuously rotating around the sample center on an axis perpendicular to the sample surface. Simultaneous thermal analysis was performed using a "STA 449 F1 Jupiter" TG/DTA device (Netzsch) coupled with a "QMS 403 C" electron ionization quadrupole mass spectrometer (Aëolos). The samples were transferred into Al_2O_3 crucibles, the oven was evacuated and flooded with helium carrier gas before measurements were started. The samples were heated from 40-550 °C with a rate of 15 K/min and isothermally treated at this temperature for 20 min. DTA reference was an empty Al_2O_3 crucible. REM pictures have been obtained using a "Ultra 55" (Carl Zeiss SMT) and EDX analysis was performed using a "AXS Quantax" (Bruker AXS Microanalysis).

2.2. Synthetical Procedures

All reagents were of analytical grade (Sigma-Aldrich) and used without further purification. Starting materials: $Cu(NO_3)_2 \cdot 3H_2O$, $Cr(NO_3)_3 \cdot 9H_2O$, $Mg(NO_3)_2 \cdot 6H_2O$, $CuCl_2 \cdot 2H_2O$, $CrCl_3 \cdot 6H_2O$. All hydrothermal reactions have been performed in 20 ml PTFE-lined stainless steel autoclaves. A series of different protocols for the synthesis of $CuCrO_2$ was tested. For each protocol an amount between 1-3 mmol $Cu(NO_3)_2$ is dissolved together with 1-x equivalents of $Cr(NO_3)_3$ and x equivalents of $Mg(NO_3)_2$. We tested x = 0.05 and 0.1. The NaOH mineralizer is added to the homogeneous solution under stirring. For an excess of NaOH it is added in the form of pellets. For the addition of equivalent amounts a solution of known concentration (around 2 mol/l) is used. Autoclaves are filled to 50 -70 % of their volume. The closed

vessels are heated to their reaction temperature within 3 h, this temperature is maintained for 60 h and now the vessels are cooled down to ambient temperature within 3 h. The olive to dark green precipitates are collected by centrifugation. Half of the obtained solid products are washed several times with water and iso-propanol. The other half is once washed with a 0.1M hydrochloric acid and then washed with a 28 % aqueous ammonia solution till the centrifugate remains colorless. After washing several times with water and iso-propanol the final products are dried in air at 100 °C. Differences between the performed protocols are listed in table 1.

Table 1. Protocols for the hydrothermal synthesis of $CuCrO_2$, EG = ethylene glycol

	Temperature	Distinguishing features
P1	240 °C	excess NaOH
P2	240 °C	5 eq NaOH
P3	200-210 °C	5 eq NaOH
P4	240 °C	a: 5 eq NaOH, Cl⁻ instead NO_3^- b: 5 eq NaOH , washed Cl-free
P5	200-210 °C	5 eq NaOH, 0.1-0.5 ml EG
P6	200-210 °C	5-x eq NaOH, x eq Mg, 0.1-0.5 ml EG

3. Results and Discussion

3.1. Hydrothermal Synthesis of $CuGaO_2$ and $CuCrO_2$

As it has been reported, the reduction in the crystallite size of nano-$CuCrO_2$ by Mg-doping is suggested to be a main advantage for application of this material in p-type DSSCs [3]. A comparison of two reports on the hydrothermal synthesis of $CuGaO_2$ and nano-$CuCrO_2$ is depicted in Table 2[2, 18]. Both protocols use a mixture of Cu(II) and Cr(III)-nitrate as starting material. It can be seen, that for the hydrothermal synthesis of $CuGaO_2$ ethylene glycol is used as reducing agent to form monovalent copper. For the formation of $CuCrO_2$ no explicit reducing agent is mentioned and the reaction temperature is 50 °C higher. The pH for the synthesis of $CuGaO_2$ was set to values between 2.6 and 7.5 by addition of mineralizer. With no mineralizer the delafossite phase was only a by-product. Higher crystallinity was achieved by higher pH. According to the authors the optimal amount of mineralizer to be added for lowering the crystallite size is equal to the amount necessary to precipitate $Cu(OH)_2$ and $Ga(OH)_3$. In an attempt to potentially lower the crystallite size of nano-$CuCrO_2$ we have tried to adapt aspects of hydrothermal synthesis of $CuGaO_2$ to the synthesis of $CuCrO_2$. We have tried out several experimental protocols, all using an equimolar solution of Cu(II) and Cr(III) salts.

We added exactly 5 equivalents NaOH to precipitate "CuCr(OH)₅" from a solution of Cu(II) and Cr(III) nitrate in procedure P2. Hydrothermal treatment of this mixture at 240°C for 60 h delivers nanocrystalline 3R-$CuCrO_2$. The literature procedure P1, using excess mineralizer delivers a mixture of 2H and 3R-$CuCrO_2$. The XRD patterns of both products can be seen in Fig. 1. P1 delivers a product, which shows a little more defined crystallinity as compared to P2, which is indicated by slightly broader reflections in the XRD pattern of P2. This observation is similar to findings concerning the hydrothermal synthesis of $CuGaO_2$. Hence, a shift from excess mineralizer to an equimolar amount might result in reduction of crystallite size.

Table 2. Comparison of literature procedures for hydrothermal synthesis of $CuGaO_2$ and $CuCrO_2$

Reference	$CuGaO_2$ [18]	$CuCrO_2$ [2]
Starting materials	$Cu(NO_3)_2$, $Ga(NO_3)_3$,	$Cu(NO_3)_2$, $Cr(NO_3)_3$,
Reducing agent	Ethylene Glycol	-
Temperature	190 °C	240 °C
Duration	56 h	60 h
Mineralizer	KOH or NaOH	NaOH
pH	2.6-7.5	>12
Washing steps	Subsequent: NH_3 (28 %) HNO_3 (diluted) H_2O	Sequence: HCl (diluted) EtOH

3.2. Using an Equimolar Amount of Mineralizer

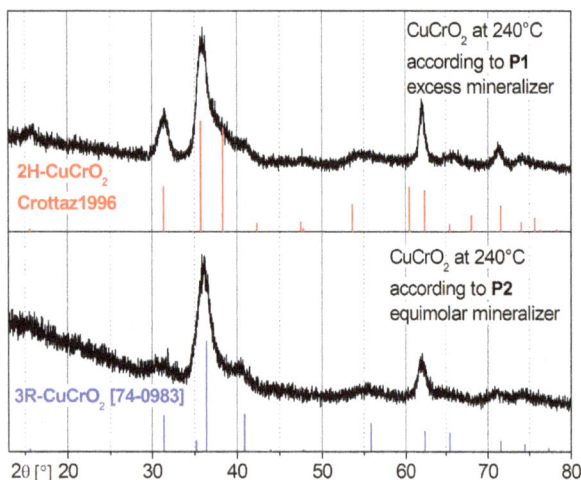

Figure 1. *XRD patterns of hydrothermally synthesized $CuCrO_2$; top: according P1 with reference phase $2H$-$CuCrO_2$ [1]; bottom: according P2 with reference phase $3R$-$CuCrO_2$ [JPCDS 74-0983]*

3.3. Lowering the Temperature

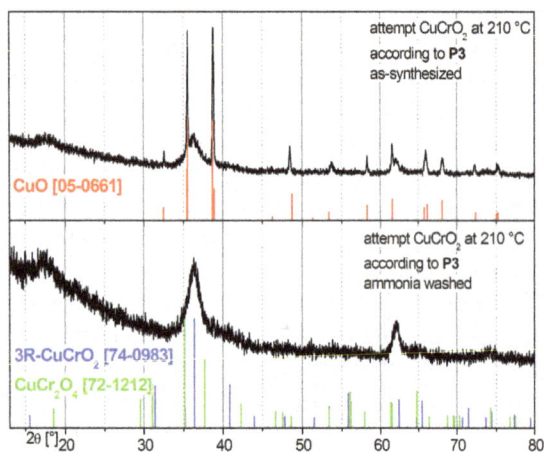

Figure 2. *XRD patterns of hydrothermally synthesized $CuCrO_2$ (according P3); top: water washed sample and reference CuO [JPCDS 05-0661]; bottom: ammonia washed sample and references $3R$-$CuCrO_2$ [JPCDS 74-0983] and $CuCr_2O_4$ [JPCDS 72-1212]*

Several groups using Cu_2O as starting material quote that $CuCrO_2$ can be obtained in a temperature range between 180 and 250 °C [11–14]. With procedure P3 we attempted to lower the reaction temperature to 200-210 °C. However this protocol produced major amounts of CuO, as detected by XRD. When the copper oxide is removed by washing with concentrated ammonia, only a few broad reflections remain in the XRD pattern. XRD data of a products according to P3 is depicted in Fig. 2.

The two most prominent reflections in an ammonia washed sample are very broad and found at 2θ-values of 36.4° and 62.1°. They do fit the major reflections of nano-$CuCrO_2$, the 0 1 2 and 1 1 0 reflection of $3R$-$CuCrO_2$. Hence, the reaction does work at lower temperatures, but it is not completed within the given time of 60 h. The existence of chromium containing by-products cannot be excluded by the XRD data gathered. Other possible by-products can be amorphous or nano-crystalline. Candidates include Cr_2O_3, CrOOH well as spinel-type $CuCr_2O_4$. The most prominent XRD-reflections of the latter compound are close to the most intense one of the delafossite phase. At this level of crystallinity they might just be visible by broadening of the $CuCrO_2$ reflections. Hence, possible incomplete reactions according to (1), (2) and (3) can be proposed. The ratio of CuO to $CuCrO_2$ in the as-synthesized products would deliver a hint on the completeness of reaction. Unfortunately it is not feasible to determine reliable results from as-obtained XRD data. CuO delivers sharp reflections due to a bigger crystallite size. However, the delafossite phase shows very broad reflections and they could be overlapping with reflections due to $CuCr_2O_4$. A Rietveld refinement attempting to determine the product to by-product ratio fails. However, lowering the temperature of hydrothermal synthesis conditions does still deliver $CuCrO_2$ and could be another possibility of crystallite size reduction.

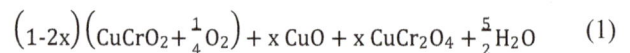

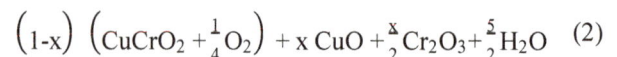

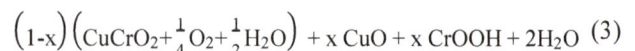

$$Cu(OH)_2 + Cr(OH)_3 \rightarrow$$

$$(1-2x)\left(CuCrO_2 + \tfrac{1}{4}O_2\right) + x\, CuO + x\, CuCr_2O_4 + \tfrac{5}{2}H_2O \quad (1)$$

$$(1-x)\left(CuCrO_2 + \tfrac{1}{4}O_2\right) + x\, CuO + \tfrac{x}{2}Cr_2O_3 + \tfrac{5}{2}H_2O \quad (2)$$

$$(1-x)\left(CuCrO_2 + \tfrac{1}{4}O_2 + \tfrac{1}{2}H_2O\right) + x\, CuO + x\, CrOOH + 2H_2O \quad (3)$$

3.4. The Origin of Copper Reduction

We attempted to gather some more insights on the redox mechanism involved in the hydrothermal formation of $CuCrO_2$ without additional reducing agent. For that purpose we checked if changing the starting material from nitrates to another anion has an effect. We performed experiments using an equimolar solution of $CuCl_2$ and $CrCl_3$. 5 equivalents of NaOH were added to precipitate "$CuCr(OH)_5$". In case P4b the precipitate has been washed chloride free. According to P4a dissolved NaCl remained in the mixture. After hydrothermal treatment at 240 °C nano-$CuCrO_2$ was obtained in both cases as illustrated by XRD patterns in Fig. 3.

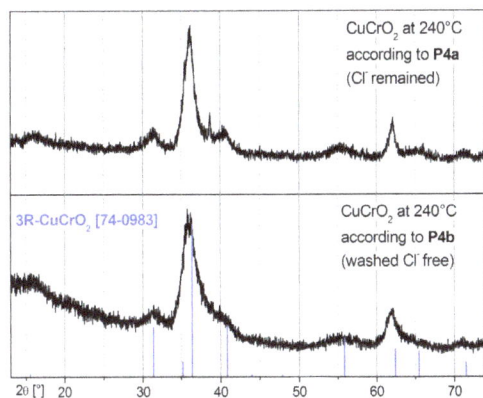

Figure 3. XRD patterns of hydrothermally synthesized CuCrO₂; top: according P4a; bottom: according P4b with reference phase 3R-CuCrO₂ [JPCDS 74-0983]

The success of both reactions indicates that reduction of copper into its monovalent state is not promoted by chloride or nitrate. An explanation can be a redox reaction involving chromium, which is supported by literature about the attempted hydrothermal formation of a related delafossite, AgCrO₂. CrOOH or Cr(OH)₃, unlike other amphoteric transition metal oxide hydroxides and hydroxides, do not react with Ag₂O to form AgCrO₂, but rather these reactions generate a large amount of silver metal. In the presence of Cr(III) which can undergo oxidation into higher valencies, a redox reaction occurs that accelerates the decomposition of Ag₂O to metallic silver and limits the formation of AgCrO₂[11]. Recently it has been claimed that chromium is partially oxidized under hydrothermal reaction conditions for the formation of CuCrO₂ [17]. Higher valent chromium could either be incorporated into the delafossite structure, or it can form mixed-valent oxides [21–23].

3.5. Protocols with Reducing Agent (P5, P6)

Lowering reaction temperature without the addition of a reducing agent resulted in incomplete reaction. For studying the effect of addition of a reducing agent we further adapted CuGaO₂ reaction conditions by adding ethylene glycol (EG) as a reducing agent to the hydrothermal synthesis of CuCrO₂. We developed protocol P5 for the hydrothermal synthesis of CuCrO₂ at temperatures between 200-210 °C. A green product showing a few very broad reflections in XRD analysis is obtained. The most prominent reflections could again be matched to 0 1 2 and 1 1 0 reflections of 3R-CuCrO₂. For CuGaO₂ the ideal EG to water volume ratio is determined to be 0.24. We found that the addition of only 0.1-0.5 ml EG within 10 ml total volume is sufficient. Higher amounts of EG or higher reaction temperature promote the formation of metallic copper. Using this protocol, we attempted the synthesis of doped CuCrO₂ incorporating 5 % of Mg in procedure P6. For that purpose Cr(NO₃)₃ was partially replaced by Mg(NO₃)₂ as a starting material. Comparison to P5 shows an immediate difference. The products incorporating magnesium show more prominent Cu₂O impurities. They can be visually identified after centrifugation as a brick red coloration depicted in Fig. 4.

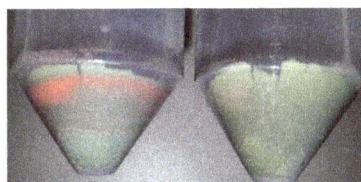

Figure 4. Visual comparison of as-synthesized CuCrO₂ samples; left: Mg-doped according P6; right: undoped according P5; Cu₂O by-products visible as brick red precipitate

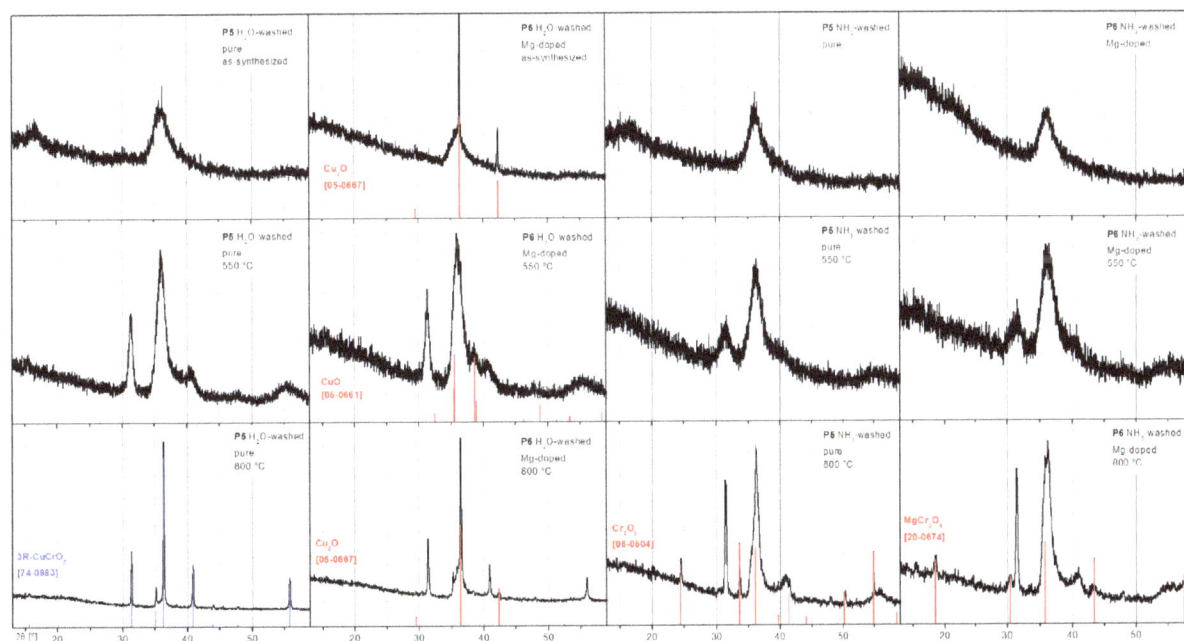

Figure 5 Comparison of XRD patterns of pure CuCrO₂ samples (according to P5) and 5 % Mg-doped CuCrO₂ (according to P6); left part: water washed samples; right part: ammonia washed samples; top row: as-synthesized; second row: post-TG samples (550 °C); bottom row: annealed samples (800 °C); JPCDS entry number of references given in brackets

A comparative XRD study of pure and Mg-doped $CuCrO_2$ is shown in Fig. 5. The native products of $CuCrO_2$ and $CuCr_{1-x}Mg_xO_2$ as-synthesized by hydrothermal reaction at 200 °C were split into two portions. One portion was only washed with water to preserve all insoluble by-products, the other portion was washed with a 28 % aqueous ammonia solution. The resulting XRD patterns are illustrated in the upper row of Fig. 5. As mentioned, it can be seen, that the H_2O-washed $CuCr_{1-x}Mg_xO_2$ contains clearly visible Cu_2O impurities, whereas this side phase is barely traceable in the undoped compound. Copper oxide impurities are removed by ammonia wash. The very broad reflections in the first row of XRD patterns indicate a crystallite size in the range of 5 nm for these samples (estimated using the Scherrer equation). Fig. 6 illustrates SEM picture obtained from as-synthesized water and ammonia washed samples obtained according to P5 and P6. The given resolution of these pictures is limited by the device and does barely allow to make out single crystallites. This confirms that crystallite sizes are well below 100 nm.

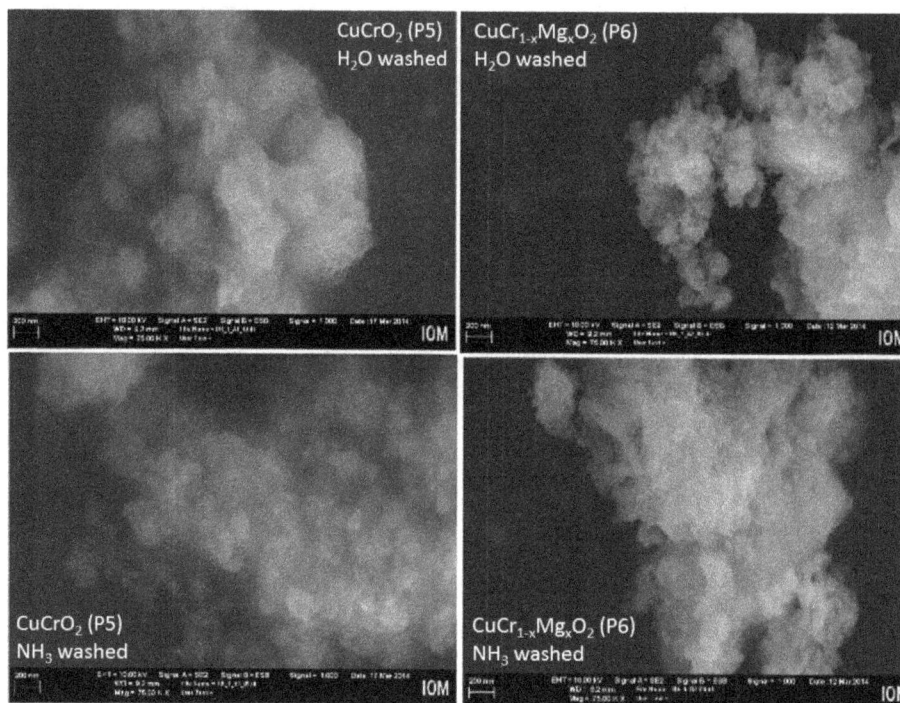

Figure 6 SEM pictures of as-synthesized water and ammonia washed samples according to P5 and P6

3.6. STA Experiments on Nano-CuCrO₂ (P5, P6)

A STA experiment up to 550 °C under He-atmosphere was performed on pure and Mg-doped $CuCrO_2$-samples (according to P5 and P6). (STA = simultaneous thermal analysis = TG coupled with QMS). 550 °C is the final temperature of a sintering process used to produce $CrCrO_2$ based porous electrodes for DSSCs [2]. Fig. 7 illustrates STA results of $CuCrO_2$ samples obtained via P5.

The STA results of the Mg-doped samples are similar. The TG-graphs show mass losses around 15 % for each of the samples. All of the TG-graphs show overlapped two step mass losses. In the water washed samples the first step between 50 to around 400 °C releases water. This is represented by the m/z trace 16 shown in the graphic (m/z 17 and 18 show parallel peaking, but are not shown here). In a second step between 420 and 550 °C an additional release of O2 is observed, as indicated by parallel peaking of m/z 16 and 32. The ammonia washed samples do not show release of molecular oxygen. Instead, in the whole region from 60 to 550 °C the mixed release of water and ammonia is observed, as indicated by m/z 14 for ammonia and 16 for water and ammonia (m/z 15, 17 and 18 show parallel peaking, but are not shown here). The oxygen loss can be caused by a reduction reaction of either divalent to monovalent copper or higher valent to trivalent chromium. Which one it is cannot be distinguished with certainty at this point. The mass loss of around 15 % indicates, that the as-synthesized products have a large surface area that is saturated with chemisorbed water and/or ammonia. The Cu/Cr and Mg/Cr ratio of the post-TG samples were obtained by EDX and are presented in table 3. It can be seen that the water washed products are copper rich, while the ammonia washed ones are copper poor. The Mg/Cr ratio for the doped sample is almost unaltered by the ammonia wash and it fits the ratio as given by the starting materials. XRD patterns of the products of these TG experiments can be seen in the middle row of Fig. 5. The water washed products show increased crystallinity. $3R-CuCrO_2$ can be assigned. The diffractogramms show anisotropic reflection broadening that can be caused by the formation of hexagonal $CuCrO_2$ nano-platelets, as already reported for this material and other delafossites [12, 19]. In the pattern of the water washed $CuCr_{1-x}Mg_xO_2$ traces of Cu-oxide are visible. The ammonia washed samples are seemingly pure after thermal treatment. The reflections can as well be assigned to 3R-CuCrO₂. However, they appear much broader than the ones of the water washed samples. A rough estimation using the Scherrer formula delivers a crystallite size in the range of 10 nm for the water washed samples and in the range of 5 nm for the ammonia washed ones. This means that the average crystallite size of $CuCrO_2$-based electrodes could potentially be lowered by washing as-synthesized nano-$CuCrO_2$ with ammonia solution.

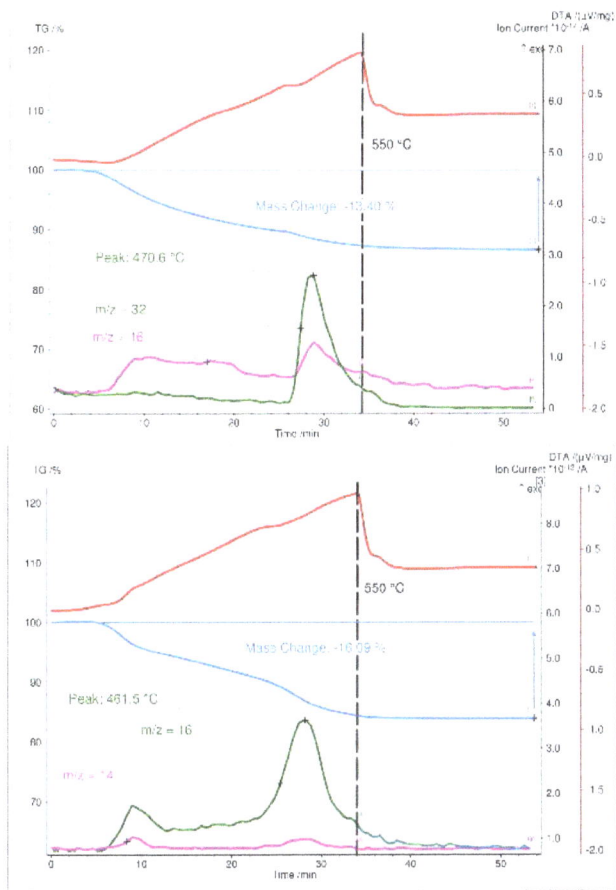

Figure 7. STA results of a CuCrO₂ sample (according to P5); top: water washed sample; bottom: ammonia washed sample

Table 3. *Atomic Cu/Cr and Mg/Cr ratios of water and ammonia washed post-TG samples (P5, P6)*

Protocol: sample	P5: CuCrO₂	P6: CuCr₀.₉₅Mg₀.₀₅O₂
Cu/Cr {Mg/Cr} H₂O-wash	1.09	1.12 {0.049}
Cu/Cr {Mg/Cr} NH₃-wash	0.82	0.77 {0.057}

3.2. Annealing Experiments on Nano-CuCrO₂ (P5, P6)

Annealing experiments performed at 800 °C take place in a temperature range which is applicable for solid state reaction to form $CuCrO_2$. We performed such annealing steps to crystallize any amorphous by-products that might occur. The according XRD-pattern are shown in the bottom row of Fig. 5. Water washed samples form well crystalline $3R$-$CuCrO_2$ upon annealing. No by-products can be detected for undoped

$CuCrO_2$. Mg-doped $CuCrO_2$ shows traces of copper oxide. No magnesium containing by-products can be detected for doping levels up to 10 %. This indicates, that after annealing at 800 °C Mg should be fully incorporated into the delafossite host structure. Charge neutralization upon Mg-doping of $CuCrO_2$ demands that either Cu(I) or Cr(III) have to be formally oxidized into higher oxidation states, or oxygen vacancies have to form. Under the following assumptions: 1. Incorporation of 10 % Mg; 2. All chromium remains in its trivalent oxidation state; 3. One equivalent of Copper in the host structure; 4. No oxygen vacancies; one would obtain the following molecular formula: $Cu^I_{0.9}Cu^{II}_{0.1}Cr_{0.9}Mg_{0.1}O_2$. However, the copper oxide traces demand that either assumption 2, 3 or 4 is not correct. It has been reported for the delafossite $CuRhO_2$, that the incorporation of 10 % of Mg into the host structure causes a Cu(II)/Cu(I) ratio of 0.4 [24]. Transferred to $CuCrO_2$ that would deliver the following molecular formula: $Cu^I_{0.61}Cu^{II}_{0.245}Cr_{0.9}Mg_{0.1}O_2$, which means that 14.5 % of copper used in the starting material would be detected as copper oxide. Another possible explanation for copper oxide impurities would be the incorporation of higher valent chromium into the delafossite host structure, as it has been suggested [17]. However, this should also be the case for undoped $CuCrO_2$. Hence, one would expect trace amounts of Cu oxide in that case too. We could not detect such impurities for the water washed sample of $CuCrO_2$ treated at 800 °C. This does not mean, that no trace amounts of higher valent chromium are present, but it is safe to say, that the incorporation of Mg has a stronger effect. Ammonia washed samples annealed at 800 °C show by-products in their XRD patterns. Cr_2O_3 can be detected for undoped $CuCrO_2$ and the spinel compound $MgCr_2O_4$ for Mg-doped samples. The XRD patterns that belong to the main $3R$-$CuCrO_2$ phase exhibit anisotropic broadening of reflections. This can be explained by the assumption, that the ammonia washing step not only removes copper oxide impurities, but surface copper oxide layers from delafossite crystallites as well.

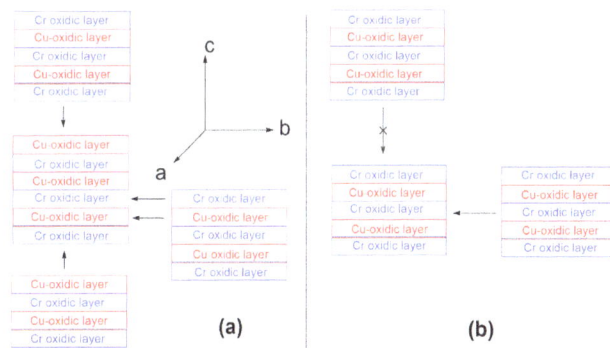

Figure 8. Schematic visualization of proposed growth mechanism of CuCrO₂-nanocrystallites; a: isotropic growth in water washed samples; b: inhibited growth along the c-axis after removal of Cu-oxidic surface layers after ammonia wash

Fig. 8 illustrates how this effect could inhibit the crystallite growth along the c-axis and lead to platelet crystallites, which would cause the observed anisotropic reflection broadening.

The observed by-products can be explained in several ways. If hydrothermal reaction of pure $CuCrO_2$ does not proceed quantitatively, traces of copper oxide and amorphous chromium oxides are left in a product mixture. These by-products react to form $CuCrO_2$ during annealing at 800 °C. However, if copper oxide and surface copper oxides are removed by ammonia wash, the leftover chromium oxides appear as Cr_2O_3 upon annealing. For the Mg-doped $CuCrO_2$ the case is more complicated. In this case, if hydrothermal reaction does not quantitatively produce $CuCr_{1-x}Mg_xO_2$, the question is: Does Mg-doping under these conditions work at all? $MgCr_2O_4$ could quantitatively form under hydrothermal conditions instead of magnesium being incorporated in the delafossite host structure. Equation (4) and (5) deliver a possible explanation why $MgCr_2O_4$ is observed only in ammonia washed annealed samples, but not in water washed ones.

After water wash:

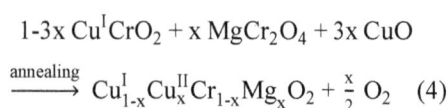

$$1\text{-}3x\ Cu^ICrO_2 + x\ MgCr_2O_4 + 3x\ CuO$$

$$\xrightarrow{\text{annealing}} Cu^I_{1-x}Cu^{II}_xCr_{1-x}Mg_xO_2 + \frac{x}{2}\ O_2 \quad (4)$$

After ammonia wash:

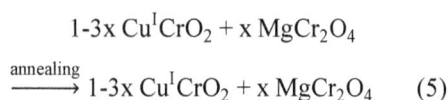

$$1\text{-}3x\ Cu^ICrO_2 + x\ MgCr_2O_4$$

$$\xrightarrow{\text{annealing}} 1\text{-}3x\ Cu^ICrO_2 + x\ MgCr_2O_4 \quad (5)$$

The spinel compound could possibly form much faster than delafossite, since there is no redox reaction involved. Hydrothermally formed $MgCr_2O_4$ might not be detectable as by-product in the as-synthesized product as well as the ones heated up to 550 °C due to broad XRD reflections. When annealed at 800 °C it would react with copper oxide to form $CuCr_{1-x}Mg_xO_2$, according to (4), and show up as $MgCr_2O_4$, according to (5). Another explanation of this observation can be the removal of surface copper-oxide layers during the ammonia wash. Since the surface area of such nanocrystals is very high, a relative high amount of surface copper can be removed from the delafossite crystals. This means that the Cu/Cr ratio is decreased within these washed crystallites, as shown in table 3. When these copper deficient crystallites grow upon annealing, they excrete the surplus of chromium in the form of Cr_2O_3 for pure $CuCrO_2$ and in the form of $MgCr_2O_4$ for Mg-doped $CuCrO_2$. This excretion is therefore necessary to restore a maintainable elemental composition of the delafossite phase. The same effect can cause the appearance of Cr_2O_3 in pure $CuCrO_2$ samples after ammonia wash and annealing. The true cause might be a combination of incomplete hydrothermal reaction and this annealing effect due to the removal of surface copper oxide.

4. Conclusions

We have investigated several protocols for the hydrothermal synthesis of $CuCrO_2$. Our investigations have shown, that using an equimolar amount of NaOH mineralizer

to precipitate "$CuCr(OH)_5$" (P2) instead of using excess NaOH (P1) influences crystallinity of the obtained products. A slight XRD signal broadening, indicating a smaller crystallite size is observed for $CuCrO_2$ prepared according to P2. A lowering of the reaction temperature from 240°C to 200-210 °C causes an incomplete reaction within the same reaction time. This results in major amounts of copper oxide detectable by PXRD. We have also shown that an exchange of anions from nitrates to chlorides (P4a) or only hydroxides (P4b) also produces $CuCrO_2$. All the protocols P1-P4 lead to a reduction from divalent to monovalent copper without the explicit addition of a reducing agent. This leads to the conclusion that a partial oxidation of chromium should occur. The exact nature of this redox mechanism still needs further investigation. We have formulated protocols P5 and P6 for the hydrothermal synthesis of pure and Mg-doped $CuCrO_2$ involving the addition of ethylene glycol as reducing agent. This enabled a lowering of reaction temperature to 200 °C. A comparative study of products of P5 and P6 has shown that an increase in copper oxide by-product is observed upon Mg-doping, which can be removed by washing steps involving an aqueous ammonia solution. This washing step has an influence on the crystallite size when products are administered to a sintering process up to 550 °C, which is applicable for the formation of $CuCrO_2$ based photoelectrodes. The ammonia washing steps can potentially decrease crystallite size in comparison to the annealing of water washed products. Ammonia washed samples annealed at 800 °C exhibit anisotropic reflection broadening and the appearance of Cr_2O_3 or $MgCr_2O_4$ by-products in XRD pattern. Both are explainable by the removal of copper oxide by-products and surface copper oxide from the as-synthesized products by ammonia wash. Our modified protocols P5 and P6 can potentially be used to produce $CuCrO_2$ based photoelectrodes featuring high surface areas. We are currently exploring their application potential for the manufacturing of $CuCrO_2$-based DSSCs.

Acknowledgements

For the preparation of SEM and EDX analysis we are thankful to Andrea Prager from the Leibniz Institute of Surface Modification (IOM). Financial support by ESF is gratefully acknowledged. Funded by the European Union and the Free State of Saxony.

References

[1] O. Crottaz, F. Kubel, "Crystal structure of copper(I) chromium(III) oxide, 2H-$CuCrO_2$" Z. Kristallogr. 211 (1996) 481.

[2] D. Xiong, Z. Xu, X. Zeng, W. Zhang, W. Chen, X. Xu, M. Wang, Y.-B. Cheng, "Hydrothermal synthesis of ultrasmall $CuCrO_2$ nanocrystal alternatives to NiO nanoparticles in efficient p-type dye-sensitized solar cells" J. Mater. Chem. 22 (2012) 24760.

[3] D. Xiong, W. Zhang, X. Zeng, Z. Xu, W. Chen, J. Cui, M. Wang, L. Sun, Y.-B. Cheng, "Enhanced Performance of p-Type Dye-Sensitized Solar Cells Based on Ultrasmall Mg-Doped CuCrO$_2$ Nanocrystals" ChemSusChem 6 (2013) 1432–1437.

[4] X. Xu, B. Zhang, J. Cui, D. Xiong, Y. Shen, W. Chen, L. Sun, Y. Cheng, M. Wang, "Efficient p-type dye-sensitized solar cells based on disulfide/thiolate electrolytes" Nanoscale 5 (2013) 7963–7969.

[5] S. Powar, D. Xiong, T. Daeneke, M.T. Ma, A. Gupta, G. Lee, S. Makuta, Y. Tachibana, W. Chen, L. Spiccia, Y.-B. Cheng, G. Götz, P. Bäuerle, U. Bach, "Improved Photovoltages for p-Type Dye-Sensitized Solar Cells Using CuCrO$_2$ Nanoparticles" J. Phys. Chem. C 118 (2014) 16375–16379.

[6] R. Nagarajan, A.D. Draeseke, A.W. Sleight, J. Tate, "p-type conductivity in CuCr$_{1-x}$Mg$_x$O$_2$ films and powders" J. Appl. Phys. 89 (2001) 8022.

[7] K. Hayashi, K.-i. Sato, T. Nozaki, T. Kajitani, "Effect of Doping on Thermoelectric Properties of Delafossite-Type Oxide CuCrO$_2$" Jpn. J. Appl. Phys. 47 (2008) 59–63.

[8] R. Bywalez, S. Götzendörfer, P. Löbmann, "Structural and physical effects of Mg-doping on p-type CuCrO$_2$ and CuAl$_{0.5}$Cr$_{0.5}$O$_2$ thin films" J. Mater. Chem. 20 (2010) 6562.

[9] Q. Meng, S. Lu, S. Lu, Y. Xiang, "Preparation of p-type CuCr$_{1-x}$Mg$_x$O$_2$ bulk with improved thermoelectric properties by sol–gel method" J. Sol-Gel Sci. Technol. 63 (2012) 1–7.

[10] M.J. Han, Z.H. Duan, J.Z. Zhang, S. Zhang, Y.W. Li, Z.G. Hu, J.H. Chu, "Electronic transition and electrical transport properties of delafossite CuCr$_{1-x}$Mg$_x$O$_2$ ($0 \leq x \leq 12\%$) films prepared by the sol-gel method: A composition dependence study" J. Appl. Phys. 114 (2013) 163526.

[11] W.C. Sheets, E. Mugnier, A. Barnabé, T.J. Marks, K.R. Poeppelmeier, "Hydrothermal Synthesis of Delafossite-Type Oxides" Chem. Mater. 18 (2006) 7–20.

[12] S. Zhou, X. Fang, Z. Deng, D. Li, W. Dong, R. Tao, G. Meng, T. Wang, X. Zhu, "Hydrothermal synthesis and characterization of CuCrO$_2$ laminar nanocrystals" J. Cryst. Growth 310 (2008) 5375–5379.

[13] S. Zhou, X. Fang, Z. Deng, D. Li, W. Dong, R. Tao, G. Meng, T. Wang, "Room temperature ozone sensing properties of p-type CuCrO$_2$ nanocrystals" Sensor Actuat. B-Chem. 143 (2009) 119–123.

[14] M. Miclau, D. Ursu, S. Kumar, I. Grozescu, "Hexagonal polytype of CuCrO$_2$ nanocrystals obtained by hydrothermal method" J. Nanopart. Res. 14 (2012).

[15] D. Ursu, M. Miclau, I. Grozescu, "In situ variable temperature X-ray diffraction studies on size scale of CuCrO$_2$ polytypes with delafossite structure" J. Optoelectron. Adv. M. 15 (2013) 768–773.

[16] D.H. Ursu, R. Miclău, R. Bănică, I. Grozescu, "Hydrothermal synthesis and optical characterization of Ni-doped CuCrO$_2$ nanocrystals" Phys. Scr. T157 (2013) 14053.

[17] D. Ursu, M. Miclau, "Thermal stability of nanocrystalline 3R-CuCrO$_2$" J. Nanopart. Res. 16 (2014).

[18] R. Srinivasan, B. Chavillon, C. Doussier-Brochard, L. Cario, M. Paris, E. Gautron, P. Deniard, F. Odobel, S. Jobic, "Tuning the size and color of the p-type wide band gap delafossite semiconductor CuGaO$_2$ with ethylene glycol assisted hydrothermal synthesis" J. Mater. Chem. 18 (2008) 5647.

[19] M. Yu, T.I. Draskovic, Y. Wu, "Understanding the Crystallization Mechanism of Delafossite CuGaO$_2$ for Controlled Hydrothermal Synthesis of Nanoparticles and Nanoplates" Inorg. Chem. 53 (2014) 5845–5851.

[20] A. Maignan, C. Martin, R. Frésard, V. Eyert, E. Guilmeau, S. Hébert, M. Poienar, D. Pelloquin, "On the strong impact of doping in the triangular antiferromagnet CuCrO$_2$" Solid State Commun. 149 (2009) 962–967.

[21] M.A. Khilla, Z.M. Hanafi, A.K. Mohamed, "Physico-chemical properties of chromium trioxide and its suboxides" Thermochim. Acta 59 (1982) 139–147.

[22] P.G. Harrison, N.C. Lloyd, W. Daniell, "The Nature of the Chromium Species Formed during the Thermal Activation of Chromium-Promoted Tin(IV) Oxide Catalysts: An EPR and XPS Study" J. Phys. Chem. B 102 (1998) 10672–10679.

[23] S. Labus, A. Malecki, R. Gajerski, "Investigation of thermal decomposition of CrO$_x$ ($x \geq 2.4$)" J. Therm. Anal. Calorim. 74 (2003) 13–20.

[24] T.K. Le, D. Flahaut, H. Martinez, N. Andreu, D. Gonbeau, E. Pachoud, D. Pelloquin, A. Maignan, "The electronic structure of the CuRh$_{1-x}$Mg$_x$O$_2$ thermoelectric materials: An X-ray photoelectronspectroscopy study" J. Solid State Chem. 184 (2011) 2387–2392.

Physicochemical and Mineralogical Characterization of Moroccan Clay of Taza and Its Use in Ceramic Technology

A. Er-ramly[*], A. Ider

Laboratory Process of Valorization of the Natural Resources, Materials & Environment, Department of Applied Chemistry & Environment, Faculty of Sciences and Technologies, University Hassan 1st, Settat, Morocco

Email address:

a.erramly@yahoo.fr (A. Er-ramly), a.erramly@gmail.com (A. Er-ramly)

[*]Corresponding author

Abstract: This study concerns the results of Physicochemical and mineralogical characterization of a white clay located in Taza region in Morocco and its use in the ceramics industry. Several techniques were used; in particular X-ray diffraction (XRD), scanning electron microscopy coupled with EDX microanalysis (SEM-EDX), differential thermal and gravimetric analyses (DTA-TGA) and finally infrared Fourier transform (FTIR) and X-ray fluorescence (XRF). The first objective of this work is to put a new line of research that deals with the use of clay in ceramic technology. The second objective was to develop gels of oxides of high purity from these clays. we can say that the white clay of Taza has the same characteristics of clays used in the ceramics industry (medium heat loss, low shrinkage, good flexural strength and good behavior in plasticity), this white clay Taza adding 0.57% sodium carbonate is sufficient to have a good deflocculation and the viscosity is minimum corresponds to the stability of the slip, in his introduction to a formula of slip was successful with a rate of 35 to 45%. The SEM-EDX, X-ray, chemical analysis and Infrared spectroscopy demonstrated and allowed us to identify the different minerals that make up the white clay, compared with the available data, we identified illite and kaolinite as clay minerals, other minerals present as impurities major are quartz, calcite, dolomite and feldspar. These results show the important features to justify its use in the ceramic industry.

Keywords: Clay, Ceramic, DTA, FX, XRD, FTIR

1. Introduction

Today, the use of clays, including those that are rich in SiO_2 and Al_2O_3, is experiencing a boom in new construction, ceramics and crafts, pharmaceuticals foundry and pottery. Aluminosilicate bricks are used in the coating of blast furnaces, refining furnaces and kilns in many laboratory ovens. Also called for the development of ceramic materials that are harder than ceramic, earthenware and traditional as well as the crowns of quartz based on alumina. The clay that is the subject of this work is known as "Taza white Clay" consisting essentially of illite and kaolinite, in its natural state. Most clay deposits in this region of northeastern Morocco are heterogeneous and are composed of some smectites mixed with illite and /or kaolinite and other impurities [1]. In the liquid state, mud clay is defined as a water-clay suspension, the origin of the use of sludge is probably the drilling of oil wells [2]. It allows, due to its rheological properties in order to respond to numerous requests for drilling, such as the stability of the structure (the impregnation of the land and make a cake filter to limit the wall) and spoil disposal [2-3]. In Morocco, earth clays are mainly used for manufacturing traditional and modern construction materials (bricks, tiles, sanitary...) and for pottery. The basic structure of layer silicates and all silicates is the ion SiO_4^{4-}, where the silicon occupies tetrahedral sites. The aluminium ion (Al^{3+}) can substitute for Si^{4+}, but it is generally located in the octahedral sheet.

2. Materials and Methods

2.1. Clay Material

The Liassic aquifer Taza corridor consists of dolomites and limestones Middle Lias Upper Lias. It occupies an area of

1500 km². It is based on an impermeable substratum consisting of Triassic clay-dolerite. It ennoie from south to north below the Miocene marl formations of southern Rif furrow. Its boundaries are formed by the accident southern Rif tight north, Palaeozoic flush with the massive Tazekka and buttonholes in the Middle Atlas Causse. It consists of shales sometimes crossed by quartzite beds and siliceous veins. The physico-chemical analysis shows that this clay is illite.

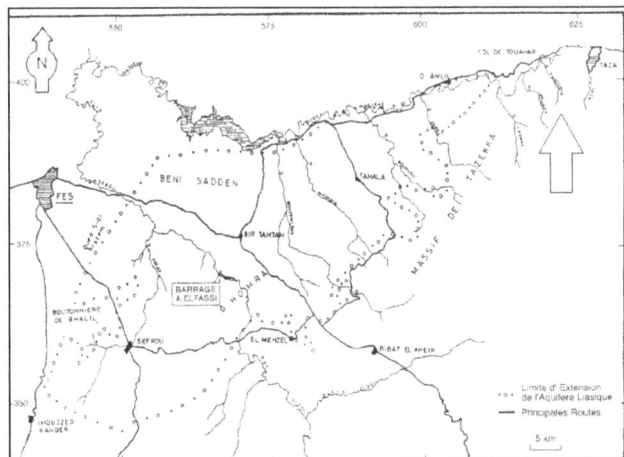

Figure 1. Location map of the city of White Clay of Taza in Morocco.

2.2. Methods

The raw clay and its fine fraction (less than 2 microns in diameter), which is isolated by sedimentation following the experimental procedure [4], were studied using X-ray diffraction (XRD), thermal analysis (DTA and TG), infrared spectroscopy (IR), X-ray fluorescence (XRF), scanning electron microscopy (SEM-EDX). As the mineral composition is variable associated with clay, it seemed necessary to XRD analysis. Spectrometric analysis by SEM, SEM-EDX was performed at the Laboratory of Materials Chemistry IFM, University of Turin (Italy) This is a scanning electron microscope to detect chemical elements, ray analysis X was performed by a diffractometer (45kv, 40mA whose technical characteristics are: Configuration type PW3064, PW3050/60 type goniometer, rotating sample holder (spinner) type PW3064, using either a copper or the anticathode cobalt, the thermograms were carried out by operating a XPERT-PRO inder the following conditions: heating rate = 10°C / min, sample weight = 40 mg, atmosphere: air. The Fourier transform infrared (FTIR) samples were obtained on a spectrometer with a DTGS detector and a KBr beam splitter, the technique of pressed KBr disk (1 mg sample and 200 mg of KBr) was used, the spectra were recorded in the region of 4000 - 400 cm⁻¹.

3. Results and Discussion

3.1. X-ray Analysis (XRD)

Treatise contain minerals
The X-ray analysis diffractometric "Figure 2" of the white

clay of Taza shows that there is a majority phase (illite, kaolinite) and minority phases (feldspar, calcite and dolomite) which identified by the cards ASTM (American Society for Testing materials) which are justified by the characteristic peaks for the phyllosilicates, we note the presence of kaolinite (d = 7.14 Å) and (d = 3.79 Å) and illite (d = 9.98 Å) and (d = 4.48 Å), we note the presence of: quartz (d = 4.25 Å and 3.34 Å =) as the major impurity; dolomite (d = 2, 90 Å), and calcite (d = 3.03 Å).

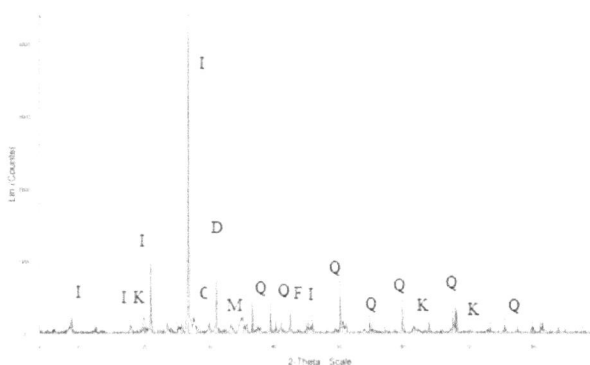

(I): illite, (K): kaolinite, (C): calcite, (D): dolomite, (F): feldspath, (Q): quartz

Figure 2. RX diffractograms of clay white Taza.

3.2. Differential Thermal Analysis and Thermogravimetric Analysis

A substance subject to heat treatment may change its physicochemical properties, such as a phase change, a change in structure, decomposition, a change in volume, etc... [5] the thermal analysis allows observe these changes as a function of temperature. Among the techniques used include differential thermal analysis (DTA), thermogravimetric analysis (TGA).

3.2.1. Differential Thermal Analysis (DTA)

The method involves measuring the temperature difference ΔT between the sample to be studied and a reference sample, inert, both subject to the same warm-up act, used the device can work in a temperature range from 25°C to 1000°C. The heating rate that we have adopted is 10°C / min. The reference sample is alumina. This difference is related to the amount of heat released or absorbed by the material studied. And ΔT is recorded as a function of temperature. This allows the detection of peak endothermic and exothermic changes.

3.2.2. Thermogravimetric Analysis (TGA)

The idea is to continuously monitor the change in mass of a sample as a function of temperature. The sample, placed in an alumina boat suspended from the beam of a balance, is located in a chamber at controlled temperature.

The equilibrium of the balance is provided by an electromagnetic compensation system. The change in mass, given by rebalancing the system, is recorded as a function of the temperature rise.

There are basically three endothermic peaks: the first between 95°C -100°C corresponding to the dehydroxylation of minerals clay and a second at 530°C and the third at 720°C corresponding to the structural reorganization of the clay minerals.

Differential thermal analysis (TGA-DTA) is very useful, especially for groups of clay, the thermal analysis of clay Taza three steps, as shown in Figure 3, the first endothermic peak at 98°C. We initially attributed to the departure of the water which is about 2 wt% clay [6-7] The endothermic reaction that occurs in the range 110°C-630°C due to the gradual exit of water molecules associated with interlayer cations, the structure of water can be removed without destroying the network of clay. Both endo reactions in sequence in the range 500°C-800°C are due to the departure of OH groups of structure (loss of 8% by weight). This suggests that in the range 630°C-830°C, as for the peak located around 710°C, can be attributed to the amount of iron in octahedral sites [8].

The portion of the curve above 900°C, reflecting the phase changes after the destruction of the structure of the clay is quite variable. it appears for the first quartz (α) or (β) and cristobalite, finally, the mullite.

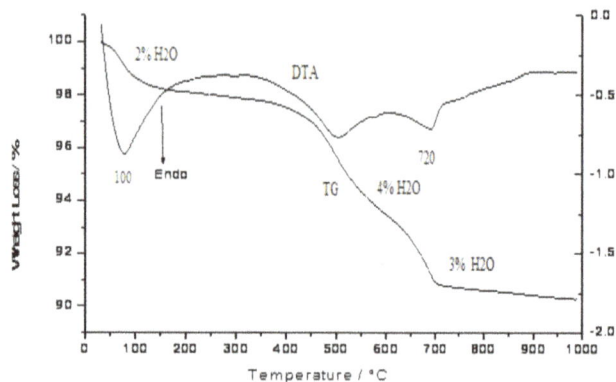

Figure 3. *TG and DTA curves of white clay of Taza.*

3.3. Spectroscopy Fourier Transform Infrared (FTIR)

Results:

From Figure 4, there is an absorption band at 3646 cm^{-1} which is due to stretching vibration of the OH clays, it is a chemical absorption, another band corresponding air also a stretching vibration ν (H_2O) observed around 3383 cm^{-1} but it is a physical absorption of water between the clay layers, a band corresponding to the bending vibration δ (H_2O) of the physical sorption of water observed around 1643 cm^{-1}, there is also a band corresponding to stretching vibration of Si-O band observed around 1032-1210 cm^{-1} group tetrahedron (SiO_4), the band observed around 3430-1430 cm^{-1} is due calcium carbonate, the bands observed around 520 cm^{-1} and 470 cm^{-1} are due to bending vibration of Al-O-Si and Si-O-Si, respectively (TOT) bands observed around 2526, 1817 and 712 cm^{-1} are due to dolomite and observed to 874, 726 cm^{-1} correspond to calcite, vibration bands observed at 920,880 and 841 cm^{-1} correspond to AlAlOH, and AlFeOH AlMgOH respectively [9-10-11].

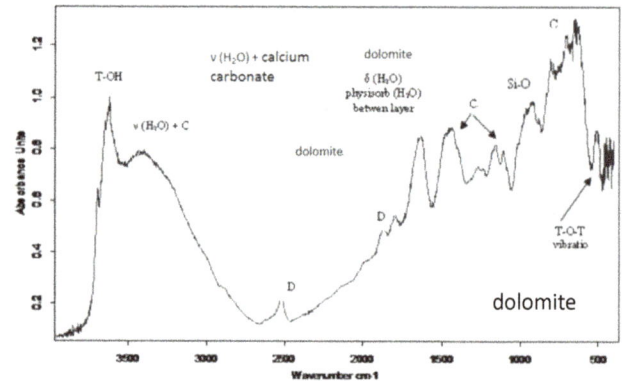

Figure 4. *FTIR white clay of Taza.*

3.4. Elemental Chemical analysis Scanning Electronic Microscope (SEM)

Interpretation:

The elemental chemical analysis of the white clay of Taza (Table 1) shows that there is a significant percentage of SiO_2 (53,19%) of Al_2O_3 (22,49%), FeO (1,29%) and K_2O (3,39%) which proves that clay is rich in illite and a small percentage of TiO_2 (0,18) Na_2O (2,05%). The morphology shows an irregularity of the particles forming the aggregate of clay.

Table 1. *Characterization by Fluorescence X of white clay of Taza.*

compounds	SiO_2	Al_2O_3	FeO	K_2O	CaO	MgO	TiO_2	Na_2O
wt%	53,19	22,49	1,29	3,39	0,92	2,09	0,18	2,05

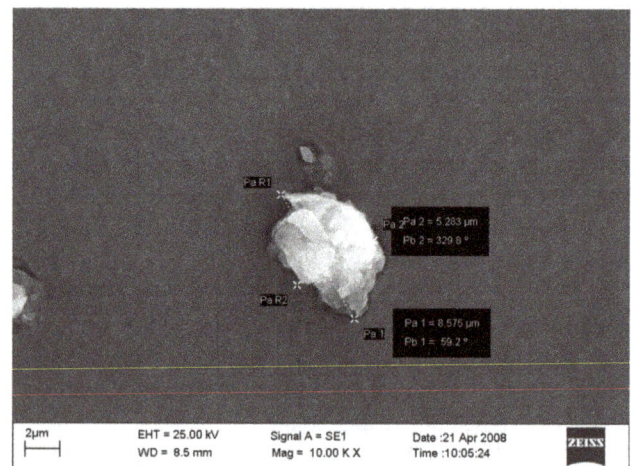

(a)

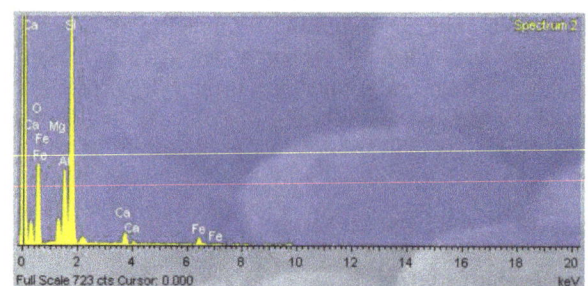

(b)

Figure 5. *SEM ((a) EDS and (b) morphology) of white clay Taza.*

The Figure 5 (a) shows the energy spectra of the element present in the sample corresponding to the chemical analysis of the scanning electron microscopy (SEM), (b) the morphology showed irregularity fine micrometer particle forming the aggregate of the sample.

3.5. Study of the White Clay of Taza

3.5.1. Humidity

The decrease in the weight of the material between the wet and drying at 110°C is determined using the following formula:

$$Humidity \% = 100 \, (mw - m_{110}) / mw$$

with mw the mass of material in the wet state (removed) and m_{110} mass of matter at 110°C.

3.5.2. Weight Loss (LOI)

The decrease in the weight of the material from drying at 110°C and the firing temperature to 1000°C is determined using the following formula:

$$Weight \, loss\% = 100 \, (m_{110} - m_{1000}) / m_{110}.$$

m_{110} with the mass of the material at 110°C and m_{1000} mass of the sample fired at 1000°C temperature. The weight loss can only know the amount of products that may decompose or evaporate during cooking. At 500°C, the product loses its water content. Between 700 and 900°C, the following reactions may occur [12]:

*Oxidation of FeO:

$$FeO \rightarrow Fe_2O_3$$

* Decomposition of carbonates:

$$CaCO_3 \rightarrow CaO + CO_2$$

3.5.3. Linear Shrinkage During Drying

Determining the shrinkage value is by studying the variation of the average lengths of lines recorded on the briquettes between wet and drying at 110 ∘ C. The following formula allows the calculation of drying shrinkage:

Linear shrinkage on drying

$$(\%) = (Lw - L_{110}/Lw) \times 100$$

with the length wet Lw, the length drying at 110°C (L_{110}).

3.5.4. Withdrawal of Cooking

Determining the shrinkage value is by studying the variation of the average lengths of lines recorded on the briquettes from the drying at 110°C.

C and firing at 1000°C. The following formula allows calculation of the withdrawal to cook:

Withdrawal of cooking

$$(\%) = (L_{110} - L_{1000}/L_{110}) \times 100,$$

L_{1000} length with cooking, the length L_{110} drying at 110°C.

3.5.5. Water Absorption Capacity (WAC) (%)

This test is the ratio of the difference between the dry weight after absorption and dry weight cooked, the formula allows the calculation of the absorption capacity of water:

$$\% \, absorption = (W_{dry} - W_{cocked} / W_{cocked}) \times 100$$

with W_{dry} abs dry weight after absorption and W_{cocked} dry weight cooked.

3.5.6. The Plasticity Index (PI)

This test is the ratio of water weight and the weight of dry matter, the formula allows the calculation of the plasticity index (PI):

$$Plasticity \, index\% = (W_w / W_{dm}) \times 100$$

with W_w the weight of water and W_{dm} weight dry matter.

3.5.7. Mechanical Resistance to Flexion (RMF)

The determination of the resistance in kg × f/cm² that can develop the bar against bending under the effect of a load on the bar, is given by the following formula:

$$R \, (kg / cm^2) = P \times 3 \times d / 2 \times w \times e^2$$

R with the mechanical resistance to bending, the force in kg P that causes the breakdown of the bar (60kg), d distance between two supports of the unit (120mm), w the width of the strip (20mm), e the thickness of the bar (10mm).

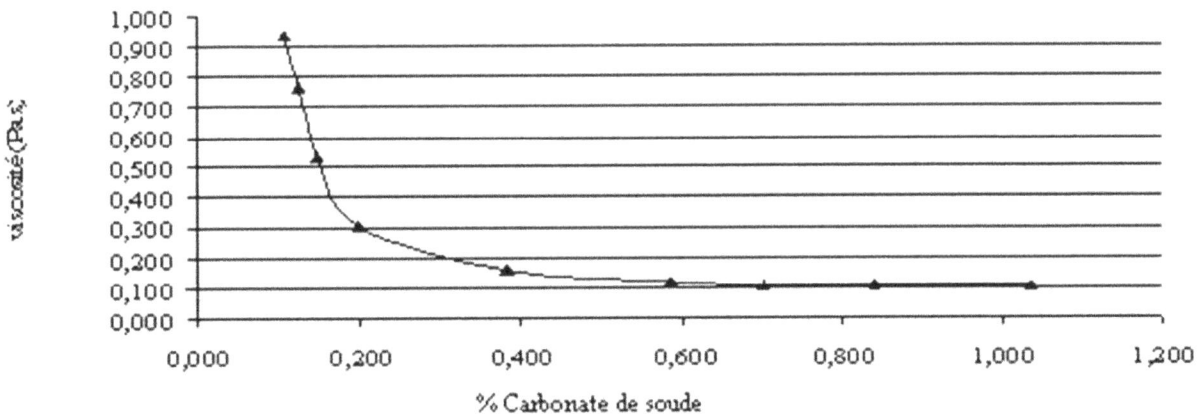

Figure 6. Deflocculation of Taza white Clay with Sodium carbonate (Na₂CO₃).

Interpretation

The white clay of Taza has a medium heat loss, low shrinkage, good flexural strength and good behavior in plasticity (Table 2), In a slurry composed of 320 g clay awhite clay of Tazand 300 ml of water was added gradually increasing amounts of deflocculant (sodium carbonate) to each dosage, was allowed to stir for 20 min and then measured the viscosity of the slip. The result of this study is shown in figure 6. In this study, it was found that from 0.59% of deflocculation, the viscosity is minimum and stable. This suggests that, for the white clay of Taza has a flowability optimal deflocculant of 0.59% sodium carbonate (Na_2CO_3).

Table 2. *Technological characteristics of white clay Taza.*

Characteristic	Result
- Humidity (H) (%)	10,18
- Loss on ignition (PF) (%)	11,05
- Linear shrinkage during drying (RL) (%)	1,32
- Total shrinkage during cooking (RT) (%)	10,42
- Water absorption capacity (WAC) (%)	34,01
- The plasticity index (PI) (%)	20,12
- Mechanical resistance to flexion (RMF) (kg/cm^2)	239

Table 3. *Preparation of slurry with different formulations (25, 35 and 45%).*

Formula (%)	Materiel				Result (% loss of ignition)	prepared slurry
	[1]white clay Taza	[2]chamotte	[3]feldspar	[4]quartz		
Formula 1	45,00	35,00	8,00	12,00	7,41	Good
Formula 2	35,00	45,00	8,00	12,00	7,53	Good
Formula 3	25,00	55,00	8,00	12,00	*	Poor

* not determined
1: white clay Taza. 2: commercial chamotte: commercial feldspar: commercial quartz

4. Conclusion

From the results we can say that the white clay of Taza has the same characteristics of clays used in the ceramics industry (medium heat loss, low shrinkage, good flexural strength and good behavior in plasticity). For this white clay of Taza adding 0,57% sodium carbonate (Figure 6) is sufficient to have a good deflocculation and the viscosity is minimum corresponds to the stability of the slip, in his introduction to a formula of slip was successful with a rate of 35 to 45% (formula 1-2, Table 3). The white clay of Taza has an average loss on ignition is due to the elimination of the water content, the decomposition of certain minerals such as carbonates and associated with the combustion of organic matter in association with minerals such as micas, feldspars or carbonates, the temperature of appearance of a liquid phase during sintering is reduced. The levels of iron oxide and titanium influence the color of ceramic shards. As for organic matter, they affect the rheology of suspensions and behavior of matter at the formatting. The X-ray diffraction patterns allowed us to identify the different minerals that make up the white clay of Taza, compared with the available data, we identified illite and kaolinite as clay minerals, other minerals present as impurities major are quartz, calcite, dolomite and magnetite. We can conclude that these results show the important features to justify its use in the ceramic industry.

Acknowledgements

My sincere thanks go to Pr.Salvatore Goluccia, Dr. Gezley and laboratory IFM staff of the University of Turin (Italy), for helpful advice the study of different samples of Moroccan clay.

References

[1] Laribi, Fleureau, Grossiord and. Kbir-Ariguib, "Comparative yield stress determination for pure and interstratified smectite clays" *Rheol. Acta* 44, 262-269. 2005.

[2] Besq, " Ecoulements laminaires de suspension de bentonite industrielles. Caractérisation rhéométrique - Ecoulements en conduites axisymétriques. Applications aux activités du Génie Civil," Thèse de doctorat de l'Université de Poitiers. 2000.

[3] Jozja,"étude de matériaux argileux Albanais. Caractérisation "multi-échelle" d'une bentonite magnésienne. Impact de l'interaction avec le nitrate de plomb sur la perméabilité." Thèse de doctorat de l'Université d'Orléans 2003.

[4] Alami, Boulmane, Hajjaji, Kacim,. Chemico-mineralogical study of a Moroccan clay. *Ann. Chem. Sci. Master.* 23, 173-176. 1998.

[5] Jouenne A.; Traité de céramiques et matériaux minéraux ; *Ed. Septima*, 1990.

[6] Mackenzie, Caillére, thermal characteristics of soil minerals, In: gieseking, J. E (Ed), *soil components. springer-verlag, new york*, vol. 2, pp. 529-571. 1975.

[7] Todor,. Thermal analysis of minerals. *Abacus pren, kent, UK*, 256 pp. 1976.

[8] Mathieu-Sicaud, and J. Mering, "Etude au microscope de la montmorillonite et de l'hectorite saturées par différents cations" *Bull. Soc. Franc. Miner. Crist.* 74, 439-455 1951.

[9] Farmer, V. C. The layer silicates. In: Farmer, V. C. (Ed), the infraredspectra of minerals. Monograph,. The mineralogical society, vol.4, london, pp.331-363. 1974.

[10] Madejova, Komadel, Cicel,. Infrared study of octaedral site population in smectites. Clay Minerals 29, 319-326. 1994.

[11] Madejova, FTIR technique in clay mineral studies, *Vib. Spectrosc*.31. 1-10. 2003.

[12] Lefort, Analyse élémentaire d'une argile, ENSCI, Limoges, France, pp. 11–12. 1988.

33

Liposome-based nanosensors for biological detection

Changfeng Chen[1, 2, *], **Qiong Wang**[1]

[1]Department of Chemistry, University of Maine, Orono, ME, USA
[2]Kashiv Pharma LLC, Bridgewater, NJ, USA

Email address:

changfeng.chen@umit.maine.edu (Changfeng Chen)

Abstract: Liposomes are self-assembled structures that contain an inner aqueous compartment surrounded by a lipid bilayer. This unique structure inherently provides liposomes with a powerful capability for encapsulating hydrophilic, hydrophobic or amphiphilic molecules or nanoparticles. Combining this property with appropriate signal amplification strategies and transduction techniques results in a variety of in vitro or in vivo biological sensors. In this review article, we discuss the latest trends in engineering and applications of liposome based nanosensors for biological sensing. Particular focus was made on the coupling of liposomes with popular sensor materials (enzymes, quantum dots, metal nanoparticles and other sensor enhancement elements) for highly sensitive and selective detection of chemical and biological species. Such information will be viable in terms of providing a useful platform for designing future ultrasensitive liposome nanosensors.

Keywords: Liposome, Sensor, Nanotechnology, Lipid Bilayer, Ultrasensitive, Biological, Encapsulation

1. Introduction and Background

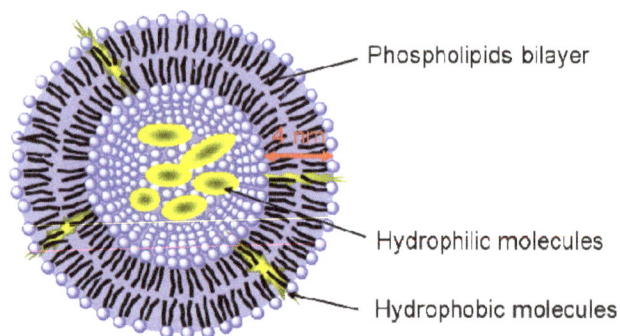

Fig. 1. A schematic drawing of a unilamellar liposome. Hydrophilic molecules can be entrapped inside the liposome, while molecules with hydrophobic portions are oriented within the membrane.

Liposomes are self-assembled structures that contain an inner aqueous compartment surrounded by a lipid bilayer typically composed of phospholipids and sterols. Since it was first discovered by Bangham[1] in 1965, Liposomes are widely used as model systems for cell membranes and as drug carriers in drug delivery systems[2]. Liposomes can be easily prepared by the hydration of lipid thin film followed by downsizing using extrusion or sonication techniques. As described in figure 1, liposomes can simply be modified in a desired manner through the choice of membrane components and it is this property that has made them attractive as model systems for cell membrane. For example, using newly developed IR-based spectroscopy methods in combination with spherical liposomes formulated with different lipids and sterols, Chen et.al. studied the structure-property relationships of liposomes such as membrane fluidity[3], transport [2] and lysis [4]. Furthermore, liposomes possess a unique structure, possessing both a hydrophilic interior and exterior, and a hydrophobic region within the lipid membrane. Such a structure makes it possible to either encapsulate water-soluble molecules in the hydrophilic interior of the liposome or immobilize molecules within the lipid membrane. As a result, liposomes are frequently used as carriers of chemicals, biomolecules and nanoparticles, finding applications in drug delivery, chemical and biological sensors. It is also possible to target specific cells by attaching an appropriate molecule to the liposome surface that binds specifically to a receptor site[5, 6].

Because of this unique structural advance, liposomes also received considerable attention for use as a substrate in sensors for chemical and biological detection. The liposome, combined with effective transduction technology including fluorescence, calorimetric, and optical spectroscopy, could realize enormous signal amplification and achieve ultrasensitive assays. The major functions of the liposomes in these sensors include the following:

1. Working as carrier for active sensor materials or sensor enhancing elements for signal amplification or multi-target detection when carrying multiple sensor materials.
2. Stabilizing the activity of enzymes or proteins for selective detection.
3. Improving the biocompatibility and accessibility of the sensors in biological medium.

This review article is intended to promote the awareness of the function of liposomes in selective and sensitive nanosensor development, and outline the current status and potential applications of liposome-based sensors in environmental sensing and biological screening. It discusses key functions of liposomes when coupling with other sensor materials (enzymes, quantum dots, metal nanoparticles, etc) on current sensor systems for biological sensing. This discussion will encourage the biological sensor community to think about possible applications of liposome-based sensors such as chemical/biological agent detection, early disease diagnosis, and biological species screening.

2. Liposome-Enzyme Based Biosensor

In recent years, because of their green nature, high selectivity, and sensitivity, the use of enzymes in biological sensors received increasing scientific attention. However, current continuous enzyme-based sensing is limited by the instability of most enzymes and their sensitivity to changes in the environment. Enzymes are inherently unstable and tend to lose activity due to folding in free solution. One major challenge in developing enzyme-based biosensors is the stabilization of the enzyme. Many approaches have been developed including the immobilization of enzyme on substrates[7-9] and polymer materials[10]. One promising approach is to protect enzyme from deactivation by encapsulating the enzyme within a liposome membrane or inner aqueous cavity. It has been proven that liposomes can stabilize enzymes in their nano-cavities against unfolding, denaturation and dilution effects due to the hydrophobic force between the enzyme and lipid bilayers[11-13]. This improves the long term stability of enzyme which is critical in biosensor applications.

Vamvakaki and Chaniotakis [14] have developed a liposome-enzyme based nanobiosensor for organophosphorous pesticide detection in drinking water systems. This sensor utilized liposome stabilized enzyme acetylcholinesterase in combination with fluorescence techniques as the optical transduction scheme for trace level detection of dichlorvos and paraoxon. As shown in Figure 2, the inherently unstable enzyme acetylcholinesterase and pH sensitive fluorescence indicator pyranine were encapsulated in eggPC liposomes with an average diameter of 300 nm. Also, Porin was embedded in the liposome membrane to control the selective transport of acetylcholine substrate into the liposomes. The controlled diffusion of the substrate inside the liposome initiates the enzymatic hydrolysis reaction which leads to the production of acetic acid and thus to pH decrease at the local nano-environment of the enzyme. This pH decrease is monitored by a decrease of pyranine fluorescence signal, which can be subsequently correlated to the substrate concentration. Exposing the nano-biosensor to organophosphorous pesticides would block the enzyme activity and decrease the production rate of acetic acid and thus the pH change. As a result, by monitoring the response of the pH sensitive fluorescence indicator signal in the presence of pesticides, the concentration of organophosphorous pesticides can be determined.

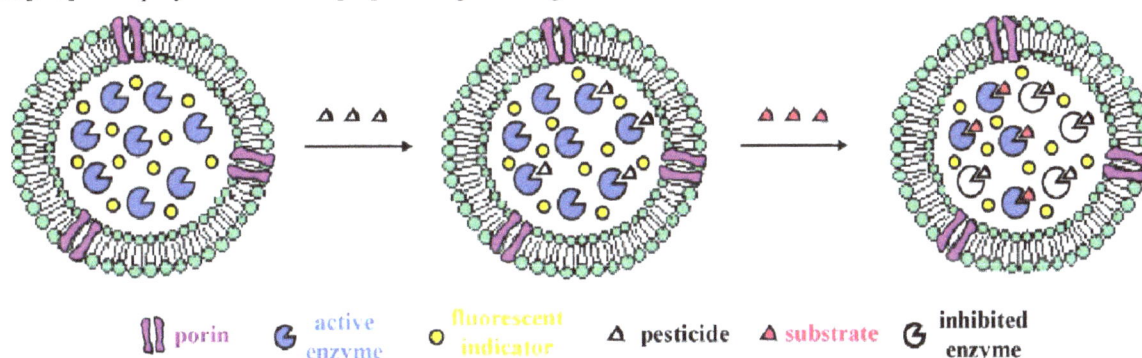

porin   active enzyme   fluorescent indicator   △ pesticide   ▲ substrate   inhibited enzyme

Fig. 2. Schematic diagram of the AChE-based inhibitor liposome biosensor. The scheme was reproduced with the permission of Elsevier.

In this work, the stabilization effect of the liposome on the nano-biosensor was proven by comparing the storage stability of the AChE/liposome nanobiosensor and enzyme in free solution. The liposome encapsulated enzyme retained full activity even after 50 days storage in ambient conditions while the free enzyme lost 66% of its original activity towards acetylcholine substrate. In addition, a detection limit of 10^{-10} M of dichlorvos and paraoxon was achieved by the AchE/Liposome nanobiosensor.

3. Liposome-Quantum Dot Complexes for Single Particle Detection

Semiconductor quantum dots (QDs) since its discovery in 90'[15] immediately draw the attention of researchers in sensor field. Because of their superior optical properties such as broad excitation, narrow emission, high quantum yield, and photochemical stability[16, 17], they become excellent alternatives for organic fluorescent dyes in biomedical research, biological labeling[18, 19], and in vitro/in vivo imaging[20]. A combination of quantum dots with single particle detection techniques has been widely used in DNA sensing[21] and has shown an advantage of a high signal-to-noise ratio, improved sensitivity and low sample consumption. However one major limitation for conventional QD-based nanosensors is that signal enhancement relies on the assembly of multiple target molecules on the surface of a single QD and thus the sensitivity is limited by the availability of both the target molecules and the QDs[22-24].

To overcome this limitation, Zhou et.al[17] reported a new approach using liposome-quantum dots complexes in combination with single particle detection techniques for sensitive detection of attomolar DNA. As shown in figure 3, liposomes were used to encapsulate QDs to form liposome/QDs complexes. The carboxyl-functionalized L/QD complexes and carboxyl-modified magnetic beads were covalently conjugated with the amino-terminated olignucleotides producing the reporter probes and the capture probes, respectively. In the presence of target DNA, a sandwich hybrid structure consisting of a reporter probe, a capture probe and the target DNA was formed. After separating them from the free report probe using magnetic beads, the liposome/QDs were disrupted and the release of QDs was subsequently measured by single particle detection. The advantage of using the lipsome-QDs complex can be summarized as the following: first of all the use of liposomes greatly improves the detection sensitivity by encapsulating hundreds of QDs. Second, the unique nature of liposomes provides accessible functionality and desirable biocompatibility for biomedical applications. In addition, simultaneous detection of multiple DNA targets can be easily realized by using L/QD complexes with different colors.

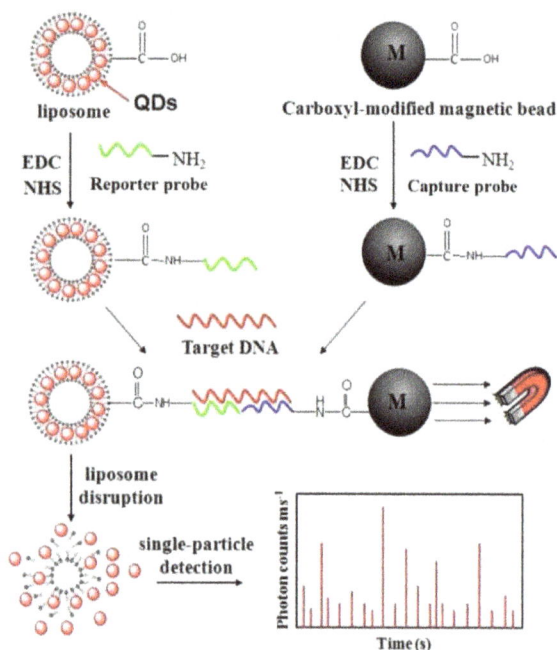

Fig. 3. Schematic illustration of DNA sensing by using liposome-QDs complexes in combination with a single molecular detection technique. This scheme was reproduced with the permission of ACS publications.

4. Liposome-Gold Nanoparticle Nanocomposite

Gold nanoparticles, because of their efficient optical properties and ease of surface functionalization[25], have been widely used in biological detection[26, 27] including DNA hydridization, DNA-ligands, DNA-protein interactions and cell transfection, etc.. Because of the broad application of liposomes in drug encapsulation as well as DNA transfection, a variety of liposome-metal nanoparticle composites were produced and widely used as biological sensing substrates.

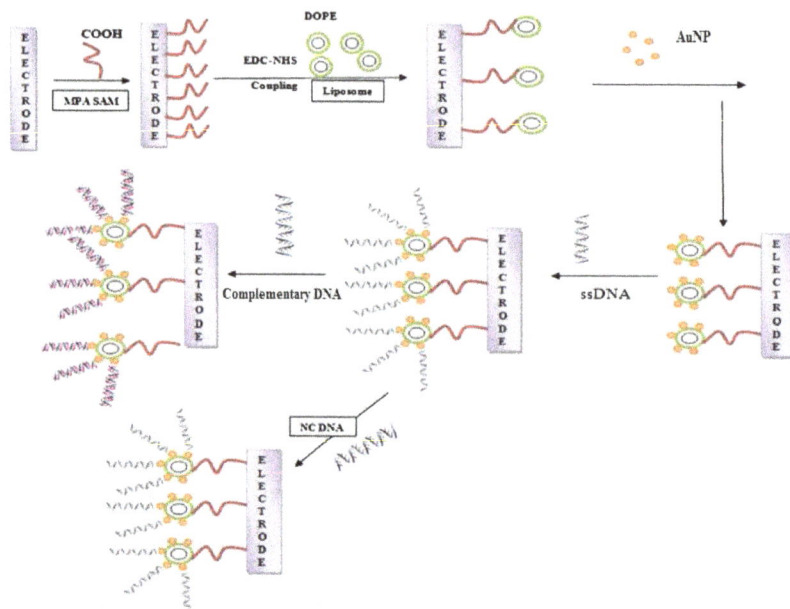

Fig. 4. Schematic illustration of the fabrication of DOPE-AuNPs nanocomposite and DNA detection. This scheme was reproduced with the permission of Elsevier.

Bhuvana et. al.[28] reported liposome-gold NPs nanocomposites that were immobilized on a solid electrode for in situ electrochemical DNA sensing. As described in figure 4. The DOPE liposomes were first immobilized on the gold electrode through covalent bonding with 3-mercaptopropionic acid (MPA) on gold surface. Next the liposome-AuNPs nanocomposites were formed by electroless deposition of AuNPs (4.58 nm) on the amine headgroup of DOPE liposomes. Single stranded probe DNA, were then immobilized on the AuNPs surface through the well-known gold-thiol bond. The resulting sensing substrate in combination with Cyclic voltammetry (CV), electrochemical impedance (EIS), differential pulse voltammetry (DPV) and quartz crystal microbalance (QCM) techniques showed high selectivity and sensitivity for DNA. The limit of detection was determined to be 0.1 femtomolar. The use of liposomes increases the bio-relevance as the liposome-AuNPs nanocomposites are fully exposed to the solution, unlike traditional AuNPs immobilized electrode in which the substrate has limited access to the solution. On the other hand, the abundance of AuNPs on the liposome surface improved the signal-to-noise ratio for DNA sensing.

5. Other Immunoassays Using Signal Enhancer Molecular Encapsulated Liposome

Because of its unique spherical bilayer structure in aqueous systems, liposomes are able to carry hydrophilic, hydrophobic and amphiliphic molecules and release the content upon exposure to surrounding stimulation. Based on this, a variety of biosensors were developed using encapsulation and release of signal enhancer molecules in liposomes. Damhorst, et. al.[29] reported a liposome based ion release impedance sensor for biological detection at the point of care. The core component of the sensor is a micron-sized antibody surface functionalized liposome encapsulating concentrated phosphate buffer saline (PBS). PBS buffer ions were used as the sensor component as they are of significantly low permeability in liposomes in ambient conditions. Chen and Tripp[2] have studied the permeability of different sizes and types of molecules in liposomes and showed that ionic molecules have the lowest release rate. The proof-of-concept experiments were carried out in a microfluidic device for HIV detection. The microfluidic device was pre-functionalized with anti-gp120 antibody and then exposed to HIV virus solution. After sufficient incubation, the unbound HIV virus was removed and IgG-functionalized liposomes in PBS were injected through the device. The unbound liposomes and free PBS ions in the solution were immediately removed by rinsing with DI water. The device was then heated for promoting the release of PBS ions from the liposome, and the impedance was monitored and compared to the control experiment with a virus-free environment. Significantly larger changes in the impedance

after liposome injection in the virus-containing device suggested the capture of liposomes on immobilized HIV virus. The number of HIV viruses can be calculated by the number of bound liposomes which are subsequently calculated using normalized impedance change. Similar detection strategies provide a simple and low cost solution for biological sensing at the point of care.

Using a similar strategy, Mao et.al[30] developed a liposome-based electrogenerated chemiluminescence (ECL) immunoassay for the detection of heart failure biomarker N-terminal pro-brain natriuretic peptide (NT-proBNP). In this detection strategy, cocaine was used as the signal enhancer molecule to enhance the ECL of $Ru(bpy)_3^{2+}$. $Ru(bpy)_3^{2+}$ was encapsulated in the liposome and released for ECL measurement. They specifically designed a sandwich immune sensing platform for the experiments. At the beginning, the glassy carbon electrode (GCE) was surface-modified by the electrodeposition of gold-platinum nanoparticles and then post-modified with capture antibodies (mAb1). The modified electrode was then incubated in the solution of antigen of the NT-proBNP biomarker to load the biomarker. Next the electrode was incubated with a secondary capture antibody (mAb2) functionalized with cocaine encapsulated liposome to allow the liposome to bond to the GCE-mAb1-NT-proBNT hybrid. In the detection procedure, the liposome based sensing sandwich was treated with triton X-100 to release the cocaine which was subsequently detected by the Ru-based ECL aptasensor. With increasing NT-proBNP concentration, the conjugated liposomes increased on the object-electrode, and thus the released cocaine increased as well. This led to the enhancement of the Ru-based ECL intensity and the changes of ECL intensities can be correlated with the concentration changes of NT-proBNP. The core detection strategy in this sensor is that encapsulation in liposomes can increase the amount of the enhancer molecule cocaine, which can multiply the ECL signal response and achieve an ultrasentive assay. The reported NT-proBNP assay exhibited high sensitivity with a linear relationship over 0.01–500 ng/mL range, and a detection limit of 0.77 pg/mL.

Summary

Because of the unique lipid bilayer structure, liposomes showed superior advantages over other biostructures in biosensor fabrication such as excellent biocompatibility, easy preparation and modification. The encapsulation or surface attachment of sensor materials and subsequent release of liposome contents provide a simple and effective approach for signal amplification and transduction. It is ready to be compatible with current sensor technology including semiconductor quantum dots, nanoparticle, immunoassay, electrochemical, fluorescence and optical spectroscopy, etc.. In addition, by incorporating different sensor materials and target molecules, liposome based sensors can easily be modified for multitarget detection and sensing.

References

[1]   A.D. Bangham, M.M. Standish, J.C. Watkins, Diffusion of Univalent Ions across Lamellae of Swollen Phospholipids, Journal of Molecular Biology, 13 (1965) 238-&.

[2]   C. Chen, C.P. Tripp, An infrared spectroscopic based method to measure membrane permeance in liposomes, Biochim. Biophys. Acta Biomembranes, 1778 (2008) 2266-2272.

[3]   C. Chen, C.P. Tripp, A comparison of the behavior of cholesterol, 7-dehydrocholesterol and ergosterol in phospholipid membranes, Biochimica et Biophysica Acta (BBA) - Biomembranes, 1818 (2012) 1673-1681.

[4]   C.F. Chen, C.H. Jiang, C.P. Tripp, Molecular dynamics of the interaction of anionic surfactants with liposomes, Colloids and Surfaces B-Biointerfaces, 105 (2013) 173-179.

[5]   B. Ceh, D.D. Lasic, Kinetics of accumulation of molecules into liposomes, J. Phys. Chem. B, 102 (1998) 3036-3043.

[6]   R. Banerjee, Liposomes: Applications in medicine, J. Biomater. Appl., 16 (2001) 3-21.

[7]   M. Willander, K. Khun, Z.H. Ibupoto, Metal Oxide Nanosensors Using Polymeric Membranes, Enzymes and Antibody Receptors as Ion and Molecular Recognition Elements, Sensors, 14 (2014) 8605-8632.

[8]   Z. Taleat, A. Khoshroo, M. Mazloum-Ardakani, Screen-printed electrodes for biosensing: a review (2008-2013), Microchim. Acta, 181 (2014) 865-891.

[9]   X.H. Shi, W. Gu, B.Y. Li, N.N. Chen, K. Zhao, Y.Z. Xian, Enzymatic biosensors based on the use of metal oxide nanoparticles, Microchim. Acta, 181 (2014) 1-22.

[10] M. Ates, A review study of (bio)sensor systems based on conducting polymers, Materials Science & Engineering C-Materials for Biological Applications, 33 (2013) 1853-1859.

[11] X. Han, G. Li, G. Li, K. Lin, FTIR Study of the Thermal Denaturation of α-Actinin in Its Lipid-Free and Dioleoylphosphatidylglycerol-Bound States and the Central and N-Terminal Domains of α-Actinin in D2O, Biochemistry (Mosc). 37 (1998) 10730-10737.

[12] M. Martí, Zille, A. , Cavaco-Paulo, A. , Parra, J. and Coderch, L., Laccases stabilization with phosphatidylcholine liposomes, Journal of Biophysical Chemistry, 3 (2012) 81-87.

[13] P. Walde, S. Ichikawa, Enzymes inside lipid vesicles: preparation, reactivity and applications, Biomolecular Engineering, 18 (2001) 143-177.

[14] V. Vamvakaki, N.A. Chaniotakis, Pesticide detection with a liposome-based nano-biosensor, Biosens. Bioelectron., 22 (2007) 2848-2853.

[15] W.C.W. Chan, S. Nie, Quantum Dot Bioconjugates for Ultrasensitive Nonisotopic Detection, Science, 281 (1998) 2016-2018.

[16] W.R. Algar, D. Wegner, A.L. Huston, J.B. Blanco-Canosa, M.H. Stewart, A. Armstrong, P.E. Dawson, N. Hildebrandt, I.L. Medintz, Quantum Dots as Simultaneous Acceptors and Donors in Time-Gated Förster Resonance Energy Transfer Relays: Characterization and Biosensing, J. Am. Chem. Soc., 134 (2012) 1876-1891.

[17] J. Zhou, Q.X. Wang, C.Y. Zhang, Liposome-Quantum Dot Complexes Enable Multiplexed Detection of Attomolar DNAs without Target Amplification, J. Am. Chem. Soc., 135 (2013) 2056-2059.

[18] Y.Y. Su, Y.N. Xie, X.D. Hou, Y. Lv, Recent Advances in Analytical Applications of Nanomaterials in Liquid-Phase Chemiluminescence, Applied Spectroscopy Reviews, 49 (2014) 201-232.

[19] X. Gao, W.C.W. Chan, S. Nie, Quantum-dot nanocrystals for ultrasensitive biological labeling and multicolor optical encoding, BIOMEDO, 7 (2002) 532-537.

[20] N. Khemthongcharoen, R. Jolivot, S. Rattanavarin, W. Piyawattanametha, Advances in imaging probes and optical microendoscopic imaging techniques for early in vivo cancer assessment, Adv. Drug Deliv. Rev., 74 (2014) 53-74.

[21] C.L. Wang, Y.X. Zhang, M.D. Xia, X.X. Zhu, S.T. Qi, H.Q. Shen, T.B. Liu, L.M. Tang, The Role of Nanotechnology in Single-Cell Detection: A Review, Journal of Biomedical Nanotechnology, 10 (2014) 2598-2619.

[22] Y. Zhang, C.-y. Zhang, Sensitive Detection of microRNA with Isothermal Amplification and a Single-Quantum-Dot-Based Nanosensor, Anal. Chem., 84 (2011) 224-231.

[23] B. Scholl, H.Y. Liu, B.R. Long, O.J.T. McCarty, T. O'Hare, B.J. Druker, T.Q. Vu, Single Particle Quantum Dot Imaging Achieves Ultrasensitive Detection Capabilities for Western Immunoblot Analysis, ACS Nano, 3 (2009) 1318-1328.

[24] C.-Y. Zhang, H.-C. Yeh, M.T. Kuroki, T.-H. Wang, Single-quantum-dot-based DNA nanosensor, Nat Mater, 4 (2005) 826-831.

[25] S.W. Zeng, D. Baillargeat, H.P. Ho, K.T. Yong, Nanomaterials enhanced surface plasmon resonance for biological and chemical sensing applications, Chem. Soc. Rev., 43 (2014) 3426-3452.

[26] P.D. Howes, R. Chandrawati, M.M. Stevens, Colloidal nanoparticles as advanced biological sensors, Science, 346 (2014) 53-+.

[27] C.J. Feng, S. Dai, L. Wang, Optical aptasensors for quantitative detection of small biomolecules: A review, Biosens. Bioelectron., 59 (2014) 64-74.

[28] M. Bhuvana, J.S. Narayanan, V. Dharuman, W. Teng, J.H. Hahn, K. Jayakumar, Gold surface supported spherical liposome-gold nano-particle nano-composite for label free DNA sensing, Biosens. Bioelectron., 41 (2013) 802-808.

[29] G.L. Damhorst, C.E. Smith, E.M. Salm, M.M. Sobieraj, H.K. Ni, H. Kong, R. Bashir, A liposome-based ion release impedance sensor for biological detection, Biomed. Microdevices, 15 (2013) 895-905.

[30] L. Mao, R. Yuan, Y.Q. Chai, Y. Zhuo, Y. Xiang, Signal-enhancer molecules encapsulated liposome as a valuable sensing and amplification platform combining the aptasensor for ultrasensitive ECL immunoassay, Biosens. Bioelectron., 26 (2011) 4204-4208.

34

Study of thermal behaviour of a fabric coated with nanocomposites

K. Abid, S. Dhouib, F. Sakli

Laboratoire de Génie Textile, Institut Supérieur des Etudes Technologique de Ksar Hellal, Université de Monastir, Avenue Hadj Ali Soua, 5070 Ksar Hellal, Tunisia

Email address:

kaledabid2003@yahoo.fr (K. Abid)

Abstract: In this paper, the thermal insulation of coated fabric by nanocomposites has been studied. In fact, a resin/clay mixture was deposited on a 100 % cotton fabric and tested using a PASOD device for measuring the adiathermic power. The enhancement of fabric thermal insulation was noticed by calculating the difference in temperature between the inside and the outside of fabric. The innovation of this work is that the used clay is a Tunisian natural one which is simply a mixture of several sorts of clays (kaolinite, dolomite, calcite, illite, and quartz) and which has the advantage to be so cheap. Moreover, high clay percentages of 4,17 % to 37,8 % were applied to perform nanocomposites with, which never have been tried before. This clay has been cleaned, purified, dried, and steered with different resins which are actually used in the textile field for several applications such as comfort, elasticity or impermeability. It has been concluded that the increasing quantity of clay enhance significantly the thermal insulation of a 400 g/m^2 sergey fabric 100% cotton. The mathematical equation has proved to be effective in predicting the fabric thermal resistance, simply by knowing the adiathermic power value. In fact, the measure of the thermal resistance demands a long time to be evaluated, but the adiathermic power can be evaluated by a concise operation which lasts only 15 min. This good agreement between these values has been demonstrated by mathematical formulas linking the clay percentage, coating, nanocomposite deposited quantities, and the used resin. The result of theses computations indicates that clay application in nanocomposites proved its importance because the thermal insulation properties of the fabric are really enhanced according to the clay percentage in the coating. The average of this enhancement is about 20 to 30 % and this is upon the used resin, the deposited quantity, and the clay percentage present in the nanocomposite.

Keywords: Nanocomposite, Clay, Coating, Modelling, Thermal Insulation

1. Introduction

Nanotechnology has been heralded as the next major technological leap, as that it is prophesied to yield a variety of substantial advantages in terms of material characteristics: including textile, electronic, and optical and structural characteristics. Nanostructured materials as thin films and coatings possess unique properties due to both size and interface effects (1, 2, 3).

The manufacturing of nanocomposites in this study is considered very delicate because the commercial used resins are frequently mixed with water (problem of mixing with clay which is very hydrophilic). In addition, the analysis of DRX patterns requires a lot of care because the multitude of the spectrum of the different clays is superposed and difficult to analyse. Many researchers tried to give fabrics some new properties by nanoadditives adjunction such as mass spectrometry applications, thermal insulation (4, 5), ignifugation (6), lubricants for space applications (4), flexibility, resistance to organic solvents (8), moisture sensors (9), and surface fonctionnalisation (10) , but their works were carried out using only the montmorillonite clay which has a simple composition (11, 13, 14).

The purpose of this study is to show how the nanocomposite coatings on fabric is carried out and enhance the thermal insulation of this new hybrid fabric in conjunction with the clay percentage and the sort of resin.

2. Experimental

10 g of cleaned and purified Tunisian clay added to 100 mL of methylene chloride (CH$_2$Cl$_2$) have been ultrasonicated

for 2h at 25°C (freq. = 28KHz), in order to have a good dispersion of clay particles. Then the prepared clay solution was added to the resin at different loadings of clay: 4,17%, 14,8%, 25,8% and 37,8%. These clay percentages perhaps seem to be high quantities to perform nanocomposites with, but never have been tried before (8-12). In this study we want to examine the coated fabric thermal respond to large sorts of coated fabric with different clay quantities even the nanocomposites formation did not take place, and only a mixture with a good dispersion clay/resin is well applied as a coating on the fabric surface as a uniform layer.

Five sorts of commercial resins were selected to be mixed with clay: modified Dimethylol dihydroxyethylene urea (DMDHEU), Vinyl-Polyacetate (PVAv), Polyacrylate (PAC), elastic Polyurethane (PU1), and rigid Polyurethane (PU2).

These different mixtures resin/clay were deposited on a cotton fabric Sergey (about 400 g/m²) using a coating apparatus with rake pressure and deposited paste regulations. The nanocomposite formation was confirmed in our previous work (15). The polymerisation of these coatings was carried out at 150°C during 5 min (12) and after a drying operation during 5 min permitting to water and CH_2Cl_2 to evaporate.

Then, the thermal isolation properties of the coated materials were determined by measuring the adiathermic characteristics using a PASOD device for measuring the adiathermic power by measuring the necessary voltage to maintain a temperature difference between the inside and the outside of the fabric equals to 20°C.

3. Results

The deposited quantity of nanocomposite on the fabric (Qc) and the adiathermic power (AP) are shown in figure 1 (a-d).

Table 1. *Measured viscosities of the used resins.*

Resin	Viscosity (cp)
PVAc	30000
PU2	6000
PU1	5000
DMDHEU	1000
PAC	800

It can be noticed that for all resins having high viscosity (cf. table 1), the AP% increases more or less constantly in conjunction with the clay percentage when the last is more then 4,17%. It is probably due to the uniform nanocomposite layer formed on the fabric surface, and there is no significant penetration of the resin into the fabric. That is why for the other resins like PU1 with low viscosity, the variation of their curves is not regular, and it can be remarked that in some points the AP% can drop slightly. It is noticed also that for both resins DMDHEU and PAC, and at low clay filling, the rising of their curves is very slow in compare with the other resins, this is probably due to their low viscosity. Means that when a product with low viscosity is coated on the fabric, its penetration into the fabric is almost total, and the most pores

which were occupied per air are filled now with resin, so that, the AP% will rise but not to the same extent as with high viscosity resin. This fact is more or less general for all resins to explain these phenomenon, but some other effect must be considered to understand well the growing of the AP% like the clay filling, the chemical resin nature etc.

In the case of resins PVAC PU2, the increase of the AP% of the coated fabrics for clay percentages between 0 and 4 % is very significant (20 to 30%). This enhancement becomes less important for the clay percentages from 5% on. It is also noted that for a clay percentages between 5 and 10%, the AP% become greater than the reference's one. The deposited quantity of 32% seems to be the more adequate quatity in order to obtain a better AP%. In fact, the best result is obtained with the PVAc resin (deposited quantity : 38%) even with a 5% as a clay filling.in term of cost and thermal performance.

Finally, and in the case of DMDHEU resin the deposited quantity (32 g/m² to 540 g/m²) does not seem to be not a crucial parameter in increasing the AP% significantly. In fact, the enhancement of thermal insulation expressed in this study by the AP%, is not very expressive (only 1-2 % for all resins). Comprehensive study of this fact is in progress to make sure that the superposition of multilayer structure with several coated fabric (32 g/m² as a deposited quantity of nanocomposite for each fabric) could provide a better thermal insulation than putting the whole quantity of nanocomposite on one fabric layer.

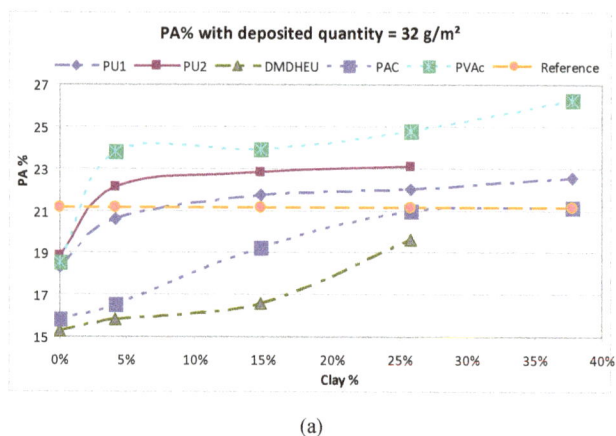

(a)

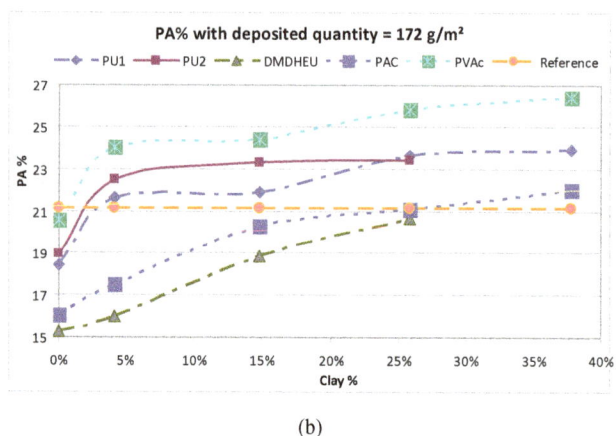

(b)

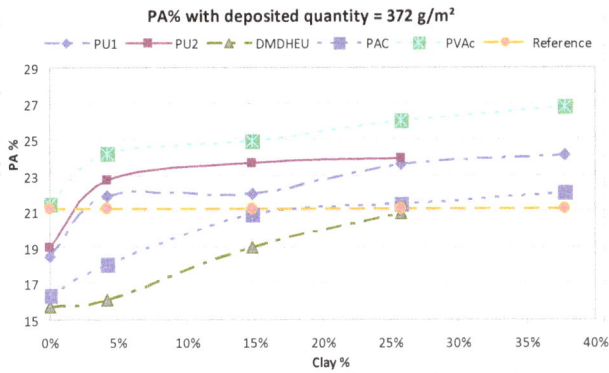

(c)

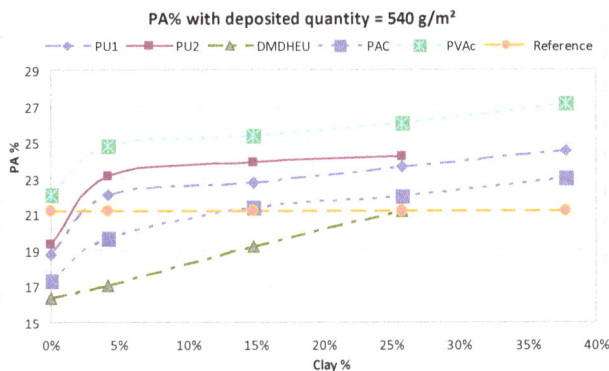

(d)

Fig. 1. *Adiathermic power % in conjunction with the clay quantity, the deposited quantity, and the sort of resin.*

Note: Mixtures PU2/37,8% clay and DMDHEU/70% clay failed because they became too thick.

4. Theoretical Background

In This part, we want to establish a mathematical equation in order to calculate the thermal resistance of coated fabric 100% cotton in conjunction with the heat flow going through, the percentage of clay, and the nanocomposite deposited quantity in order to explain the remarked phenomenon while measuring the adiathermic power. The coating (resin + clay) is performed on one side of the fabric (figure 2):

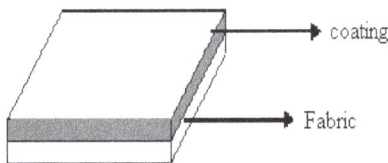

Fig. 2. *Representation of a coated fabric.*

The effective conductivity *Keff* can be determined by considering an analogous circuit model in series, thus:

$$\frac{1}{keff} = \frac{1}{kc} + \frac{1}{kf}$$; where K_c and K_f are respectively the

thermal conductivities of the coating and the fabric.

1 Determination of the fabric conductivity (Kf)

A fabric is a mixture of fibres dispersed randomly in all directions with different sorts of pores (interyarn and intrayarn). It can be represented as shown in figure 3:

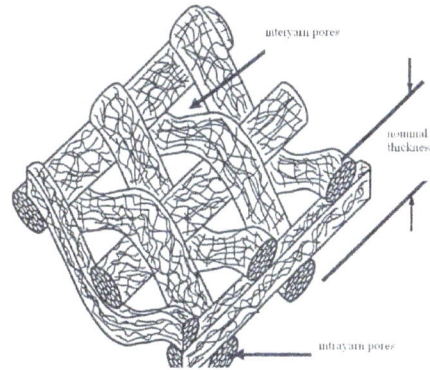

Fig. 3. *Conceptual illustration of a fabric showing interyarn and intrayarn pores(16).*

The thermal conductivity determination will take count of air and fibres conductivity in both radial and longitudinal directions. The heat flow will go through the fabric as shown in figure 4:

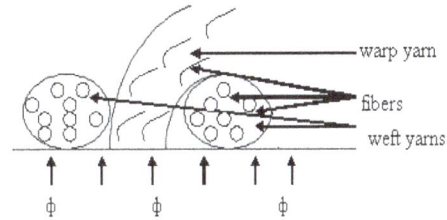

Fig. 4. *Conceptual illustration of heat flow (φ) going through a fabric.*

The heat flow will encounter fibres and air, and its going through the fabric can follow two ways or more specifically two models: series and parallel models. The schema represented in figure 5 shows the function mode of these two models, knowing:

Kf: fabric thermal conductivity,
K1: air thermal conductivity,
K2: fibre thermal conductivity,
Φ: heat flow.

Two temperatures will be recorded T_1 and T_2 respectively below and above the fibrous material in a distance called "dx" which represents the fabric thickness.

The flow equation is represented by the fourrier law equation as below:

$$\Phi = -K_f \frac{dT}{dx}$$

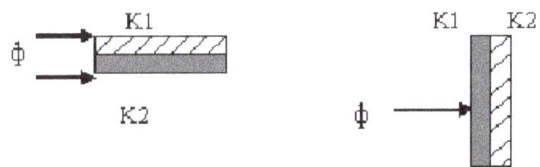

Fig. 5. *Series and parallel models.*

In the case of parallel model $K_{fp} = K_1 + K_2$ and in the case of series model:

$$\frac{1}{K_{fs}} = \frac{1}{K_1} + \frac{1}{K_2}$$

In our study (textile fabric) there is a combined model represented in the same time by a series model and a parallel model is used (figure 6):

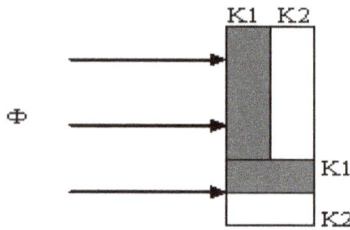

Fig. 6. *Combined model.*

So that, the fabric global and effective conductivity can be expressed as:

$K_f = \alpha K_{fs} + \beta K_{fp}$ where:

α : massic fraction coefficient in relation with the series model,

β : massic fraction coefficient in relation with the parallel model,

K_{fs} : series model thermal conductivity,

K_{fp} : parallel model thermal conductivity.

2 Determination of the coating conductivity (Kc)

The coating is represented by the mixture resin/clay as shown in figure 7.

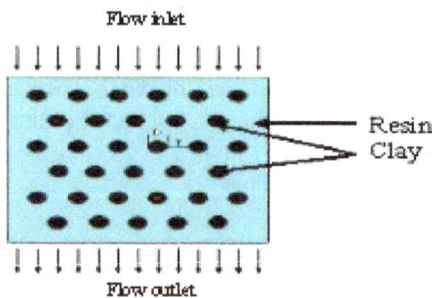

Fig. 7. *Coating resin + clay matrix configuration.*

The coated fabric thermal conductivity Kc follows a combined model:

$Kc = \gamma Kcs + \lambda Kcp$; where:

γ: massic fraction coefficient in relation with the series model,

λ: massic fraction coefficient in relation with the parallel model,

K_{cs} : series model thermal conductivity,

K_{cp} : parallel model thermal conductivity.

3 Determination of the Coated Fabric Theoretical Thermal Resistance (Reff)

We want to calculate the coated fabric thermal resistance R_{eff} in conjunction with the clay percentage put in the resin.

Determination of the adiathermic power in conjunction with the heat flow:

We know that $AP = (1 - (\frac{U_1}{U_0})^2)100$

Where:

U_1 = electric voltage to make a temperature difference of $\Delta T = 20°C$ between the inside and the outside of the fabric.

U_0 = electric voltage to make a temperature difference of $20°C$ in the absence of fabric.

U_1 can correspond to the heat flow Φ ($= \frac{P}{S}$) going through the fabric, and equals to $\sqrt{PR}$, where:

P: calorific power of the wires inside the apparatus,

R: the electric resistance of the wires inside the apparatus.

So, the adiathermic power can be in this form:

$$AP = \frac{100}{U_0^2}(U_0^2 - \phi.S.R)$$

Knowing:

$$\Phi = -\frac{\Delta T}{R_{eff}}$$

We can do the deduction:

$$Reff = \frac{\Delta T.S.P}{U_0^2(1 - AP)}$$

Supposing $\frac{S.R}{U_0^2} = a$

Calculating the arbitrary coefficient "a":

We know by measurement that the thermal conductivity of reference fabric (cellulose), without coating, is 0.04 W/m°K, and in this case, the clay percentage "P" = Qc = 0, so that a = -1025,6.

And

$$Reff = \frac{20}{1025.6(1 - PA)} = \frac{0.0195}{(1 - PA)}$$

From the figure 1(a-d) and the formula above, the coated fabric thermal resistance in conjunction with clay percentage and resin, can be determined (figure 8 (a-e)).

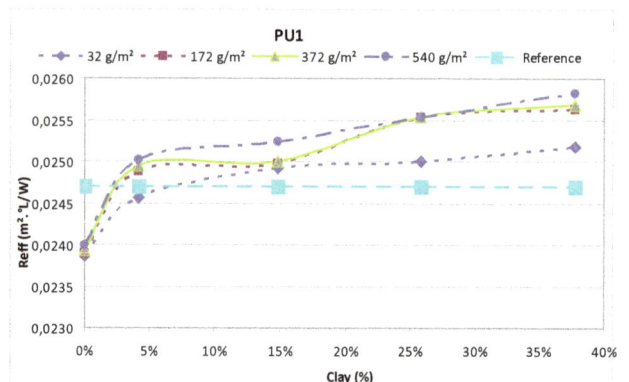

(a)

(b)

(c)

(d)

(e)

Fig. 8(a-e). *Coated fabric thermal resistance ($m^2.°K/w$) in conjunction with clay percentage and resin.*

5. Discussion

According to figure 1(a-d), adiathermic powers of all fabrics increase in conjunction with the clay percentage, but it is specific from one resin to another. For example, PVAc resin present higher AP% than DMDHEU and this is because DMDHEU conductivity is higher than PVAc conductivity. The results of these calculations, given in figure 8 (a-e), indicate that the thermal resistances for all samples are ranging from 0.023 to 0.026 $m^2°K/W$. It is noted that the AP% increases for a constant clay percentage and an increasing deposited quantity. According to figure 8 (a-e), fabric thermal resistances can be classed for each coating resin. The reference fabric thermal resistance is higher than the rest of coated fabric when there is no applied clay (corresponding to an AP = 21.11%) due to the presence of air in the interyarn and intrayarn pores (figure 3) (air conductivity = 0.02 W/m°C). In fact, the resin will replace the air, and the fabric becomes more thermally conductive. The thermal resistance increases in conjunction with the clay percentage for all resins. When a fabric is coated with nanocomposites resin/clay, the thermal resistance becomes higher than the reference when the clay percentage exceeds a specific value (ex. 4,17 % for the PU2 resin). The fabric thermal resistances of the coatings with the nanocomposite DMDHEU/clay are all the time below the reference thermal resistance, even for high clay percentages (37,8 %). This is probably due to the high fluidity of the considered resin, which replaces all the pores which were occupied per air.

In this study, we tried several quantities of nanocomposites deposited on the fabric. Thus, to enlarge this investigation, great quantity of such deposited quantities should be used to derive equations that link the thermal conductivities and resistances to not only the clay percentage but also the quantity of nanocomposites deposited on the fabric, and even more, to the fabric surface weight and its contexture.

6. Conclusion

In this study, we have developed a method of mathematical stimulating the fabric thermal resistance. The clay percentage is a very important parameter since the high clay quantities in the nanocomposite generally present the better thermal resistances. The mathematical equation has proved to be effective in predicting the fabric thermal resistance, simply by knowing the adiathermic power value. In fact, the measure of the thermal resistance demands a long time to be evaluated, but the adiathermic power can be evaluated by a concise operation which lasts only 15 min. This good agreement between these values has been demonstrated by mathematical formulas linking the clay percentage, coating, nanocomposite deposited quantities, and the used resin. The result of theses computations indicates that clay application in nanocomposites proved its importance because the thermal insulation properties of the fabric are really enhanced according to the clay percentage in the coating. The average of this enhancement is about 20 to 30 % and this is upon the

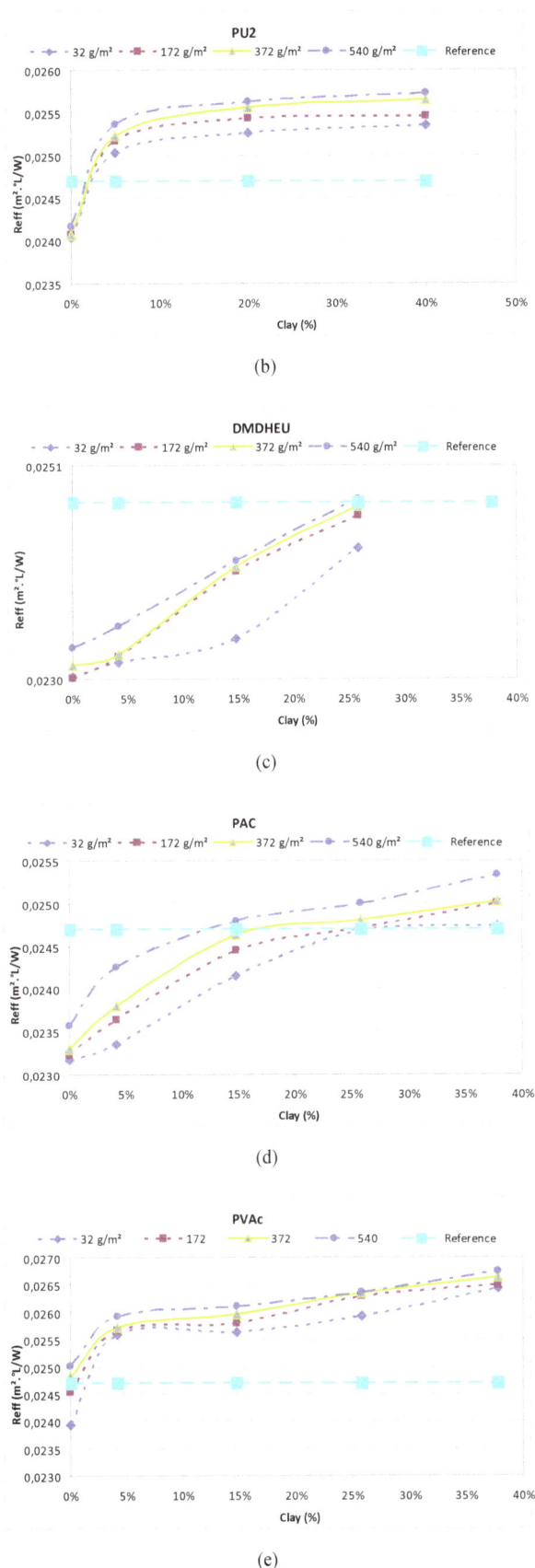

used resin, the deposited quantity, and the clay percentage present in the nanocomposite.

Validation of these results is only possible through direct conductivity measurements. In this case, then, several parameters like α, β, γ, and λ, predicting the percentages of the series and parallel models in the coated fabric, can be determined. Work is in progress to develop some theoretical model to evaluate these findings.

References

[1]   Lomax GR, 1985 „Coated Fabrics: Part 1– Lightweight breathable fabrics", Textiles, Vol 14 No 1 Spring 1985, 2–8 and „Part 2 – Industrial uses", Textiles, Vol 14 No 2, 47–56.

[2]   Lomax GR, 1994 „Coating of Fabrics", Textiles",Vol 2 No 2 1992, 18–23.

[3]   Woodruff FA, 1990 „Environmentally friendly coating and laminating; developments in machinery and processes", Progress in Textile Coating and Laminating, BTTG Conference, Chester 2–3, BTTG Manchester.

[4]   GAO Jing, Weidong Yu, Ning Pan, Structures and Properties of the Goose Down as a Material for Thermal Insulation, Textile College, Donghua University, Shanghai 200051, China, Textile Research Journal, Vol. 77, No. 8, 617-626 (2007).

[5]   Muhammad Juma, Heat Transfer Properties of Cord-reinforced Rubber Composites Journal of Reinforced Plastics and Composites, Vol. 25, No. 18, 1967-1975 (2006).

[6]   Maryline ROCHERY, Optimisation of the Structure of a Polyurethane/clay used as Flame Retartdent Textile Coating, European Coating Conference 22, 23, 24 march 2006, page 3.

[7]   A. Voevodin, J.P. O'Neill, J.S. Zabinski. Nanocomposite tribological coatings for aerospace applications, Surface and Coatings Technology 116–119 (1999), 36–45.OH 45433-7750, USA.

[8]   Dean WEBSTER, Non Isocyanate Polyurethane Coating Via Glycedyl, European Coating Conference 22, 23, 24 march 2006, page 4.

[9]   G. N. Gerasimov, E. I. Grigoriev, A. E. Grigoriev, P. S. Vorontsov, S. A. Zavialov, and L. I. Trakhtenberg, Chem. Phys. Rep. 17, 1247 (1998).

[10] Jurgen VAN HOLEN, Novel Routes to Urethane Acrylate, European Coating Conference 22, 23, 24 March 2006, page 5.

[11] Tae H.Kim, Lee W.jang, Dong C.Lee, Hyoung J.Choi, Myung S.Jhon, Synthesis and Rheology of Intercalated Polystyrene/Na$^+$-Montmorillonite Nanocomposites, Macromolecular Rapid Communications, 2002, 23, 191-195.

[12] HYOUNG J.Choi, Myung S.Jhon, Seong G.Kim, yang H.Hyung. Preparation and Rheological Characteristics of Solvent-Cast (ethylene-oxide)-Montmorillonite Nanocomposites, Macromolecular Rapid Communications, 2001, 22, 320-325.

[13] Sung T.Lim, yang H.Hyung, and Hyoung J.Choi, Synthetic Biodegradable Aliphatic Polyester/Monmorillonite Nanocomposites, American Chemical Society, 2002, 14, 1839-1844.

[14] S. K.Lim, J. W. Kim, I. Chin, W. K. Kwon and Hyoung J.Choi, Preparation and Intercalation Characteristics of with Miscible Polymer Blend of Polyethylene Oxide and PMMA, American Chemical Society, 2002, 14, 1989-1994.

[15] Abid, K ; Dhouib. S, Sakli. F; JTI, Addition Effect of Nanoparticles on the Mechanical Properties of Coated Fabric, in press the journal of textile institute JTI, 2008.

[16] Philips. G, multiphase heat and mass transfer through hygroscopic porous media with application to clothing material ; technical report natic/TR-97/005 ; 1996.

Permissions

List of Contributors

Mohamed Abdul-Aziz Elblbesy and Thamer Abed-Alhaleem Hamdan
Department of Medical Laboratory Technology, Faculty of Applied Medical Science, University of Tabuk, Saudi Arabia, Tabuk, Saudi Arabia

Adel Kamel Madbouly
Department of Biology, Faculty of Science, University of Tabuk, Tabuk, Saudi Arabia

Essam Fadl Abo Zeid
Physics Department, Faculty of Science, Assiut University, Assiut, Egypt
School of Mechanical Engineering, Pusan National University, Pusan, Korea
Physics Department, Faculty of Science &Arts El Mandaq, Al-Baha University, Al Baha, KSA

Yong Tae Kim
School of Mechanical Engineering, Pusan National University, Pusan, Korea

Mohammad Shukri Alsoufi
Mechanical Engineering Department, Collage of Engineering and Islamic Architecture, Umm Al-Qura University, Makkah, Saudi Arabia

Tahani Mohammad Bawazeer
Chemistry Department, Collage of Science, Umm Al-Qura University, Makkah, Saudi Arabia

Mortatha Saadoon Al-Yasiri
Chemical Engineering Department, College of Engineering, University of Baghdad, Baghdad, Iraq

Waleed Tareq Al-Sallami
Department of Air conditioning& Refrigeration, Technical College, Mosul, Iraq

Suresh Sagadevan
Department of Physics, Sree Sastha institute of Engineering and Technology, Chennai, India

Mohamed S. El Naschie
Dept. of Physics, University of Alexandria, Alexandria, Egypt

Taku Saiki and Yukio Iida
Department of Electrical and Electronic Engineering, Faculty of Engineering Science, Kansai University, Osaka, Japan

Amr Atef Elsayed and Omaima Gaber Allam
Textile Research Division, National Research Centre, 33 Bohouth st. Dokki, Giza, Egypt

Sahar Hassan Salah Mohamed and Hussain Murad
Dairy Science, Food industry, Nutrition, National Research Centre, 33 Bohouth st. Dokki, Giza, Egypt

Eldar Mehrali Gojayev
Department of Physics and Research Laboratory of the Department in "Physics and Technology of Nanostructures" Azerbaijan Technical University, Physical and Mathematical Sciences, Honored Scientist of the Republic of Azerbaijan, Baku, Azerbaijan

Khadija Ramiz Ahmadova
Senior Laboratory Laboratory, "Thermophysical Properties of Oil and Petroleum Products", Department of Physics, Azerbaijan Technical University, Baku, Azerbaijan

Sevinc Sarkar Osmanova
Physics, Physical and Mathematical Sciences, Department of Physics, Azerbaijan Technical University, Baku, Azerbaijan

Shujaat Zeynalov Aman
Physical and Mathematical Sciences of Azerbaijan Technical University, Department of Physics, Baku, Azerbaijan

Abdulmajid Abdallah Mirghni
Department of Physics, Faculty of Education, Al Fashir University, Al Fashir, Sudan
Department of Physics, Faculty of Science and Technology, Alneelain University, Khartoum, Sudan

Mohamed Ahmed Siddig
Department of Physics, Faculty of Science and Technology, Alneelain University, Khartoum, Sudan
Department of Medical Physics, Faculty of Medicine, National University, Khartoum, Sudan

Mohamed Ibrahim Omer
Department of Physics, Faculty of Science and Technology, Nile Valley University, Atbara, Sudan

Abdelrahman Ahmed Elbadawi and Abdalrawf Ismail Ahmed
Department of Physics, Faculty of Science and Technology, Alneelain University, Khartoum, Sudan

Natalia Yevlampieva, Mikhail Antipov and Evgeny Ryumtsev
Faculty of Physics, Saint Petersburg State University, Saint Petersburg, Russia

Alexander Bugrov
Faculty of Chemistry, Saint Petersburg State University, Saint Petersburg, Russia
Institute of Macromolecular Compounds, Russian Academy of Sciences, Saint Petersburg, Russia

Tatiana Ana'neva
Institute of Macromolecular Compounds, Russian Academy of Sciences, Saint Petersburg, Russia

Mahendra Kumar Trivedi, Rama Mohan Tallapragada, Alice Branton, Dahryn Trivedi and Gopal Nayak
Trivedi Global Inc., Henderson, USA

Omprakash Latiyal and Snehasis Jana
Trivedi Science Research Laboratory Pvt. Ltd., Bhopal, Madhya Pradesh, India

Mohamed S. El Naschie
Dept. of Physics, University of Alexandria, Alexandria, Egypt

P. D. Nsimama
Dar Es Salaam Institute of Technology, Department of Science and Laboratory Technology, Dar Es Salaam, Tanzania

Aron Varga
Leibniz Insitute of Surface Modification, Permoserstraße 15, D-04318 Leipzig, Germany

Christian Elsner, Andrea Prager, Ulrich Decker, Sergej Naumov and Bernd Abel
Leibniz Institute of Surface Modification, Chemical Department, Permoser Strasse 15, D-04318 Leipzig, Germany

Andreas Neff, Olga Naumov, Bernd Abel, Aron Varga and Katrin R. Siefermann
Leibniz Institute of Surface Modification (IOM), Chemical Department, Permoser Strasse 15, 04318 Leipzig, Germany

Timna-Josua Kühn, Nils Weber and Michael Merkel
FOCUS GmbH, Neukirchner Strasse 2, 65510 Hunstetten, Germany

Mohammed Saad Kamel, Raheem Abed Syeal and Ayad Abdulameer Abdulhussein
Department of Mechanical Techniques, Al-Nasiriyah Technical Institute, Southern Technical University, Thi-Qar, Iraq

Tuhin Subhra Santra
Department of Engineering and Systems Science, National Tsing Hua University, Hsinchu, Taiwan

Fan-Gang Tseng
Department of Engineering and Systems Science, National Tsing Hua University, Hsinchu, Taiwan
Institute of Nanoengineering and Microsystems (NEMS), National Tsing Hua University, Hsinchu, Taiwan
Division of Mechanics, Research Center for Applied Sciences, Academia Sinica, Taipei, Taiwan

Tarun Kumar Barik
Department of Applied Sciences, Haldia Institute of Technology, Haldia, West Bengal, India

Axel Sobottka, Lutz Drößler, Bernd Abel and Ulrike Helmstedt
Leibniz-Institute of Surface Modification, Permoserstrase 15, 04318 Leipzig, Germany

C. Hossbach
Technische Universitat Dresden, Institute of Semiconductors and Microsystems, Nothnitzer Strase 64, 01187 Dresden, Germany

Karina M. M. Carneiro
School of Dentistry, Department of Preventive and Restorative Dental Science, UCSF, San Francisco, USA

Andrea A. Greschner
Institut National de la Recherche Scientifique, Centre d'Énergie, Matériaux et Télécommunications, Varennes, Canada

Hanan Basioni Ahmed and Magdy Kandil Zahran
Chemistry Department, Faculty of Science, Helwan University, Ain-Helwan, Cairo, Egypt

Mohammed Hussein El-Rafie
Textile Research Division, National Research Centre, Dokki, Cairo, Egypt

Xiong Peng, Radoelizo S. A and Liping Liu, Yi Luan
School of Materials Science and Engineering, University of Science and Technology Beijing, 30 Xueyuan Road, Haidian district, Beijing, 100083 (P. R. China)

Stefan Zahn and Richard Cybik
Wilhelm-Ostwald-Institut für Physikalische und Theoretische Chemie, University of Leipzig, Leipzig, Germany

H. El-Didamony
Chemistry Department, Faculty of Science, Zagazig University, Zagazig Egypt

S. Abd El-Aleem
Chemistry Department, Faculty of Science, Fayoum University, Fayoum, Egypt

Abd El-Rahman Ragab
Quality Department, Lafarge Cement, El Kattamia, El Sokhna, Suez, Egypt

Dirk Friedrich, Claudia Wöckel, , Robert and Reinhard Denecke
Wilhelm-Ostwald-Institute of Theoretical and Physical Chemistry, Department of Chemistry and Mineralogy, University of Leipzig, Leipzig, Germany

Bernd Abel
Wilhelm-Ostwald-Institute of Theoretical and Physical Chemistry, Department of Chemistry and Mineralogy, University of Leipzig, Leipzig, Germany
Chemical Department, Leibniz Institute of Surface Modification (IOM), Leipzig, Germany

Konrath, Harald Krautscheid and Sebastian Küsel
Institute for Inorganic Chemistry, Department of Chemistry and Mineralogy, University of Leipzig, Leipzig, Germany

A. Er-ramly and A. Ider
Laboratory Process of Valorization of the Natural Resources, Materials & Environment, Department of Applied Chemistry & Environment,
Faculty of Sciences and Technologies, University Hassan 1st, Settat, Morocco

Changfeng Chen
Department of Chemistry, University of Maine, Orono, ME
Kashiv Pharma LLC, Bridgewater, NJ, USA

Qiong Wang
Department of Chemistry, University of Maine, Orono, ME

K. Abid, S. Dhouib and F. Sakli
Laboratoire de Genie Textile, Institut Superieur des Etudes Technologique de Ksar Hellal, Universite de Monastir, Avenue Hadj Ali Soua, 5070 Ksar Hellal, Tunisia

Index